ADVANCES IN CHEMICAL ENGINEERING
Volume 20

Fast Fluidization

ADVANCES IN
CHEMICAL ENGINEERING

Editor-in-Chief

JAMES WEI

School of Engineering and Applied Science
Princeton University
Princeton, New Jersey

Editors

JOHN L. ANDERSON

Department of Chemical Engineering
Carnegie Mellon University
Pittsburgh, Pennsylvania

KENNETH B. BISCHOFF

Department of Chemical Engineering
University of Delaware
Newark, Delaware

MORTON M. DENN

College of Chemistry
University of California at Berkeley
Berkeley, California

JOHN H. SEINFELD

Department of Chemical Engineering
California Institute of Technology
Pasadena, California

GEORGE STEPHANOPOULOS

Department of Chemical Engineering
Massachusetts Institute of Technology
Cambridge, Massachusetts

ADVANCES IN CHEMICAL ENGINEERING
Volume 20

Fast Fluidization

Edited by

MOOSON KWAUK

*Institute of Chemical Metallurgy
Academia Sinica
Beijing, People's Republic of China*

ACADEMIC PRESS

San Diego   New York   Boston   London   Sydney   Tokyo   Toronto

This book is printed on acid-free paper. ∞

Copyright © 1994 by ACADEMIC PRESS, INC.

All Rights Reserved.
No part of this publication may be reproduced or transmitted in any form or by any means, electronic or mechanical, including photocopy, recording, or any information storage and retrieval system, without permission in writing from the publisher.

Academic Press, Inc.
A Division of Harcourt Brace & Company
525 B Street, Suite 1900, San Diego, California 92101-4495

United Kingdom Edition published by
Academic Press Limited
24-28 Oval Road, London NW1 7DX

International Standard Serial Number: 0065-2377

International Standard Book Number: 0-12-008520-8

PRINTED IN THE UNITED STATES OF AMERICA
94 95 96 97 98 99 BB 9 8 7 6 5 4 3 2 1

CONTENTS

Contributors ix
Preface xi

1. Origins of the Fast Fluid Bed
Arthur M. Squires

I.	Kellogg's Development of Fast-Bed Gasoline Synthesis	4
II.	Commercial Outcomes Respecting Ammonia Synthesis	14
III.	Fast-Bed Reactors at Sasolburg	16
IV.	Pioneering Fast-Bed Research at M.I.T.	19
V.	Fast Beds for Combustion (Affording Low Pressure Drop)	24
VI.	"Bubbleless" Fluidization Studies in China	30
VII.	Contacting in Turbulent and Fast Fluidization	32
	References	35

2. Application Collocation
Yu Zhiqing

I.	Characteristics of Fast Fluidized Bed Reactors	39
II.	Catalytic Processes	41
III.	Non-catalytic Reactions	50
IV.	Other Applications	55
V.	Computer Software for Application Collocation	57
VI.	Conclusions	62
	References	62

3. Hydrodynamics
Youchu Li

I.	Scope of Fast Fluidization	86
II.	Experimental Apparatus, Materials and Instrumentation	94
III.	Experimental Findings	107
	Notation	140
	References	141

4. Modeling
Li Jinghai

I.	Basic Concepts	148
II.	Methodology for Modeling Fast Fluidization	156
III.	Pseudo-fluid Models	157
IV.	Two-Phase Model—Energy-Minimization Multi-scale (EMMS) Model	159
V.	Local Hydrodynamics—Phases	171
VI.	Overall Hydrodynamics—Regions	188
	Notation	197
	References	199

5. Heat and Mass Transfer
Yu Zhiqing and Jin Yong

I.	Introduction	203
II.	Experimental Studies of Heat Transfer	204
III.	Theoretical Analysis and Models for Heat Transfer	223
IV.	Heat Transfer between Gas and Particles	232
V.	Mass Transfer	235
	Notation	235
	References	236

6. Powder Assessment
Mooson Kwauk

I.	Geldart's Classification	241
II.	Powder Characterization by Bed Collapsing	243
III.	Modeling the Three-Stage Bed Collapsing Process	245
IV.	Instrument for Automatic Surface Tracking and Data Processing	247
V.	Qualitative Designation for Bed Collapsing	250
VI.	Quantifying Fluidizing Characteristics of Powders	251
VII.	Improving Fluidization by Particle Size Adjustment	253
VIII.	Measure of Synergism for Binary Particle Mixtures	260
	Notation	264
	References	265

7. Hardware Development
Li Hongzhong

I.	The V-Valve	267
II.	Nonfluidized Diplegs	291
III.	The Pneumatically Actuated Pulse Feeder	314
IV.	Internals	315
V.	Multi-layer Fast Fluidized Beds	320

VI.	Integral CFB	322
	Notation	327
	References	329

8. Circulating Fluidized Bed Combustion
YOUCHU LI AND XUYI ZHANG

I.	Experimental Studies on Coal Combustion	333
II.	Behavior of CFB Combustor	350
III.	Modelling of CFB Combustor	359
IV.	Process Design for CFBC Boilers	371
V.	Improvement of Desulphurization	377
	Notation	384
	References	385

9. Catalyst Regeneration in Fluid Catalytic Cracking
CHEN JUNWU, CAO HANCHANG, AND LIU TAIJI

I.	Basic Requirements and Design Criteria for Catalyst Regeneration in FFB	389
II.	Research and Development on FFB Catalyst Regeneration	395
III.	China's Commercial FFB Regenerators	401
IV.	Commercial Operating Data	406
V.	Analysis of Commercial FFB Regenerator Operation	413
	Notation	418
	References	419

INDEX	421
CONTENTS OF VOLUMES IN THIS SERIAL	433

CONTRIBUTORS

Numbers in parentheses indicate the pages on which the authors' contributions begin.

CAO HANCHANG, *Luoyang Petrochemical Engineering Corporation, SINOPEC, Luoyang, Henan 471003, People's Republic of China* (389)

CHEN JUNWU, *Luoyang Petrochemical Engineering Corporation, SINOPEC, Luoyang, Henan 471003, People's Republic of China* (389)

JIN YONG, *Department of Chemical Engineering, Tsing Hua University, Beijing 100084, People's Republic of China* (203)

MOOSON KWAUK, *Institute of Chemical Metallurgy, Academia Sinica, Zhongguan Village, Beijing 100080, People's Republic of China* (239)

LI HONGZHONG, *Institute of Chemical Metallurgy, Academia Sinica, Zhongguan Village, Beijing 100080, People's Republic of China* (267)

LI JINGHAI, *Institute of Chemical Metallurgy, Academia Sinica, Zhongguan Village, Beijing 100080, People's Republic of China* (147)

YOUCHU LI, *Institute of Chemical Metallurgy, Academia Sinica, Zhongguan Village, Beijing 100080, People's Republic of China* (85, 331)

LIU TAIJI, *Luoyang Petrochemical Engineering Corporation, SINOPEC, Luoyang, Henan 471003, People's Republic of China* (389)

ARTHUR M. SQUIRES, *Virginia Polytechnic Institute and State University, Blacksburg, Virginia 24061* (1)

YU ZHIQING, *Department of Chemical Engineering, Tsing Hua University, Beijing 100084, People's Republic of China* (39, 203)

XUYI ZHANG, *Department of Thermal Energy Engineering, Tsing Hua University, Beijing 100084, People's Republic of China* (331)

PREFACE

Fast fluidization (FF) has become one of the most widely used forms of bubbleless fluidization. It exploits the phenomenon that fine powders do not conform to the correlations for particulate fluidization but can admit far greater gas flow than would be permitted by the terminal velocity of the constituent particles. The particles aggregate into clusters or strands to make way for the rapid fluid flow. These clusters or strands frequently dissolve and reform with fresh particle members, thus leading to high rates of particle–fluid mass and heat transfer that are hardly realizable with bubbling fluidization.

Fast fluidization was first studied in China in the early 1970s, but because of limited contact with the rest of the world, the results have not circulated much outside of its borders. It is the purpose of this book, therefore, to acquaint our international peers with the essential results of our efforts in this research.

The first chapter introduces fast fluidization. In this chapter, A. M. Squires of Virginia Polytechnic Institute and State University recounts the history of fast fluidization, and establishes from records heretofore undisclosed that the Synthol reactor developed by the Kellogg Company in the 1940s actually used the fast fluidized bed long before this regime was recognized and designated as fast fluidization. Nevertheless, the use of FF appears to have a multiplicity of origins. Lurgi Company independently conceived and developed their circulating fluidized bed (CFB) aluminum hydroxide calciner, an FF reactor. In China, the need to process low-grade and complex iron ores in large amounts led to the use of dilute-phase operation in preference to bubbling fluidization. That the concentration of solids increased by recirculation was known from the generalized fluidization chart developed in China in the late 1950s, but the principle was rarely used until the late 1970s when a process for upgrading and comprehensive utilization of the country's vanadium-bearing titanomagnetite was developed.

While many engineers on the international front are turning to FF for better gas–solids contacting (though each is primarily concerned with his or her own process needs), Z. Yu of Tsing Hua University (THU) has studied the common features of different processes. In Chapter 2 she provides a scheme for collocating the multifarious applications of fast fluidization that enables a researcher interested in using FF to have quick access to comparable techniques in different professions. The collocation chart she has prepared is supported by personal computer software.

Chapter 3 relates the hydrodynamics of fast fluidization on the basis of observational techniques and domestically fabricated probes. In this chapter, Y. Li of the Institute of Chemical Metallurgy (ICM) describes his early measurements of the S-shaped axial voidage distribution curves in FF systems, and then introduces his semiempirical equations relating the shape of these curves to the properties of the particles and the operating variables. This correlation is supplemented by another set of semiempirical equations describing radial voidage distribution. Experimental studies on measurement techniques and visualization of cluster formation in FF are also reported.

In Chapter 4, J. Li of ICM gives a brief account of the energy-minimization multi-scale (EMMS) model, which not only is applicable to fast fluidization, but may also be generalized to apply to all particle–fluid systems. This model consists of a nonlinear optimization problem relating eight principal parameters in FF. Analysis of its solution yields quantitative criteria for the transitions of fluidization regimes: bubbling to turbulent, turbulent to fast, and fast to transport. It also places the phenomenologically different species of fluidization—particulate and aggregative—on a common basis, thus pointing out how to "particulatize" a solid–fluid system in order to increase its contacting efficiency.

In Chapter 5, Z. Yu and Y. Jin of THU describe experimental studies of heat transfer between particle suspensions and immersed surfaces, enumerating the effects of variables and of the radial and axial distribution of the heat transfer coefficient. They then present an analysis of the mechanism of heat transfer, particularly in terms of particle convection.

For all the aspects of fast fluidization that have been studied, relatively little attention has been paid to the particles that are being fluidized. In Chapter 6, M. Kwauk of ICM discusses the problem of what kind of powder or powder mixture possesses the qualities needed for fast fluidization, and shows that the fluidization quality of a powdery material can be assessed by the bed collapsing method in a fully automatic instrument, which yields both a three-digit qualitative evaluation index and a quantitative index called the dimensionless time: a zero dimensionless time value signifies bubbling at incipient fluidization, and a large value indicates improved fluidization quality. The effect of particle size distribution is also shown in studies of closely sized binary and ternary mixtures. In the case of a fine component, synergism in improving the quality of fluidization often results.

Chapter 7 gives examples of a number of devices which are of use in handling solids and in reactor design in FF technology. In this chapter, H. Li of ICM describes a number of pneumatically controlled nonmechanical devices for moving and conveying granular materials. In particular, he presents the design and analysis of the trapezoidal V-valve, which is capable of moving solids either from a high-pressure region to a low-pressure region or vice versa, with adequate seal. He elaborates on the dynamics and sealing power of nonfluidized diplegs, and concludes the chapter with descriptions of FF internals and integral CFBs.

With three-quarters of the country's energy needs provided by about a billion tons of coal a year, how to burn coal efficiently and with minimal environmental impact has become a problem of paramount importance in present-day China. In Chapter 8, Y. Li of ICM and X. Zhang of THU describe the efforts of Chinese engineers in experimental studies of coal combustion and of the behavior of the CFB combustor and its modeling.

A significant application of FF on a commercial scale is the FF regenerator in fluid catalytic cracking processes. In Chapter 9, J. Chen, H. Cao, and T. Liu of Luoyang Petrochemical Engineering Corporation (LPEC) describe the development and design of four types of FF regenerators and the results of their commercial operation. More than 30 such units are in operation in China.

In the present volume, the editor attempts to provide an overview of FF technology in China. To be comprehensive, novel features are given as islets in the matrix of the overall picture. In collating manuscripts, the editor has forgone rigid textbook consistency in order to preserve the individual contributors' expertise, style, and literary talent, and has maintained separate tables of symbols and literature references, to which the respective authors are most accustomed.

The editor expresses his appreciation to Y. Hsieh, Director of ICM, for his interest in the present compilation, and especially for his participation in narrowing down the subject matter to the present nine chapters and identifying the appropriate authors; and to Vicki Jennings of Academic Press, in providing an exhaustive review of the writing plan and accepting the efforts of our Oriental scholars in the *Advances in Chemical Engineering* series. Although the number of principal contributors is rather limited, many express their gratitude to their younger collaborators, e.g., to F. Wei of THU for his contribution to Chapters 2 and 5, among many others.

It is the hope of all the authors that this book will constitute a positive contribution toward enriching the international repository of knowledge on fast fluidization.

Mooson Kwauk

1. ORIGINS OF THE FAST FLUID BED

ARTHUR M. SQUIRES

VIRGINIA POLYTECHNIC INSTITUTE AND STATE UNIVERSITY, BLACKSBURG, VIRGINIA, U.S.A.

I. Kellogg's Development of Fast-Bed Gasoline Synthesis	4
II. Commercial Outcomes Respecting Ammonia Synthesis	14
III. Fast-Bed Reactors at Sasolburg	16
IV. Pioneering Fast-Bed Research at M.I.T.	19
V. Fast Beds for Combustion (Affording Low Pressure Drop)	24
VI. "Bubbleless" Fluidization Studies in China	30
VII. Contacting in Turbulent and Fast Fluidization	32
References	35

On September 12, 1946, work began, in anger, that would lead to the first commercial fast fluid bed. Mr. M. W. Kellogg was angry when he convened a strategy meeting of his senior personnel to consider, "What next?" In its answer to this question, the meeting initiated a development that would mature in the 1950s with operation of commercial-scale fast-fluid-bed reactors for gasoline synthesis at Sasolburg, South Africa.

For more than two years, The Standard Oil Company of Indiana and The M. W. Kellogg Company had collaborated in studies of Kellogg's "Synthol" process, an American version of Germany's wartime Fischer–Tropsch synthesis, converting synthesis gas (hydrogen and carbon monoxide) to gasoline. Indiana and Kellogg had exchanged data from pilot plant Synthol reactors, "stationary" fluid beds operating in the bubbling regime: Stanolind, an Indiana subsidiary, had a unit 8 inches in diameter in Tulsa, Oklahoma; and Kellogg had a 2-inch unit at its laboratory in Jersey City, New Jersey. By June 1946, engineers in Kellogg's New York City offices, in the Transportation Building on lower Broadway, were well along in design and cost studies, paid for by Stanolind, for a Synthol installation to produce 6,000 barrels per day of liquid hydrocarbons. The synthesis reactor would be a stationary fluid bed to operate at 1.7 ft/s, and fitted with vertical heat-exchange tubes at metal-to-metal distances of 1.75 inch.

On June, 1946, Kellogg's people experienced a great shock, a bolt from the blue, when George Roberts, Jr., Manager of Stanolind's Research Department, telephoned instructing Kellogg to stop all activites on Indiana's behalf, forthwith. From that day, Indiana would accept invoices from Kellogg only for the packing up and mailing of whatever paperwork already existed from its design effort, without further editing or elaboration.

The summary withdrawal of Indiana's support was all the more galling to Mr. Kellogg because it meant Indiana had selected a rival company, Hydrocarbon Research, Inc., to be its partner in development, design, and construction of a hydrocarbon synthesis facility. Mr. P. C. ("Dobie") Keith had founded Hydrocarbon Research for the specific purpose of developing an American version of Germany's Fischer–Tropsch synthesis. Keith's "Hydrocol" Process would use bubbling fluid bed reactors.

Through the 1930s and into the 1940s, Keith had been The M. W. Kellogg Company's Vice President of Engineering. Early in the 1930s, Keith masterminded the sale, design, and construction of a grass-roots refinery for Continental near Denver—the first refinery ever built from scratch on a greenfield site. Before, refineries had simply grown, adding a unit here, a unit there, without overall logic or master plan. The Continental installation had been such a success that, throughout the 1930s, it gave The M. W. Kellogg Company a head start toward winning contracts for refinery construction, worldwide. In 1940, Kellogg was the premiere builder for the petroleum industry. In early 1942, it was a logical choice as a major player in what would become the Manhattan Project. From early 1943, Keith was President of Kellogg's subsidiary, The Kellex Corporation, responsible for design and construction of the world's first gaseous diffusion plant, enriching U-235, at Oak Ridge, Tennessee (Squires, 1986b).

The New York Times for December 22, 1944, carried the Pullman Company's announcement that it had purchased The M. W. Kellogg Company. Owing Keith much for his role in the latter company's growth, Mr. Kellogg had given Keith an important interest in his company. Keith, however, chose not to remain under Pullman's management, and invested his share of Pullman's tender to Kellogg in his new enterprise, Hydrocarbon Research, Inc. He pulled together a consortium, Carthage Hydrocol, to build a hydrocarbon synthesis plant at Brownsville, Texas. The Texas Company was a major player in Carthage Hydrocol, and its support of Keith's new company may have been a factor in Stanolind's decision to cut its tie with Kellogg and rely upon Hydrocarbon Research for design of a hydrocarbon synthesis plant that Stanolind planned to build at Hugoton, Kansas. I suspect, too, that Keith's friendship with Edward Seubert, Indiana's president, and with George Watts, Indiana's chief engineer and also Keith's chief engineer at Kellex, helped to swing Indiana's favor in Hydrocarbon Research's direction. Although not a member of the Carthage Hydrocol consortium, Stanolind erected a facility at Brownsville to recover oxygenate byproducts from Carthage Hydrocol's synthesis plant.

In the event, Stanolind cancelled Hugoton after all major equipment had been manufactured and site preparation was far along; and the Brownsville synthesis plant was a technological disappointment (Squires, 1982)—a black eye for bubbling fluidization. Standard of Indiana acquired the Brownsville plant in 1952, but stopped operations in 1958.

Internal M. W. Kellogg memoranda (Lobo, 1929–1955) indicated Mr. Kellogg's state of mind on September 12, 1946: he was angry. In 1948, apparently anticipating patent litigation, Kellogg's patent department prepared a list of all of Kellogg's exchanges with Indiana relating to Fischer–Tropsch synthesis from the late 1930s through June 1946.

In 1946, Mr Kellogg was still active in day-to-day operation of Pullman's subsidiary and in decisions affecting its future. It was clear that he wished to "pick up the pieces" of the interrupted Synthol development, and to establish Pullman's position in this new technology.

Kellogg deemed this position to be vital for future competitiveness as a builder for the petroleum industry. At the end of World War II, this industry believed it faced an imminent shortage of oil. Many authorities prophesied that hydrocarbon synthesis would be a centerpiece in the refinery of the future.

Their prophecy would not materialize, for two reasons.

First, in 1948, astonishing discoveries of oil in Saudi Arabia would launch explorations that quickly disclosed vast reserves, postponing for several decades all worries of an oil shortfall.

Second, pipelines carrying natural gas from the southwestern United States to the Northeast had not yet been built, or even visualized by many of the oil industry's decision-makers. In 1946, Carthage Hydrocol was able to write long-term contracts for supplies of natural gas from southern Texas at 3 cents per million Btu's (Spitz, 1988)—a price that would seem absurdly low only a few years later. The going rate for gas quickly climbed far too high for its economic use as a feedstock for synthesis of liquid hydrocarbons. Without Carthage Hydrocol's contract for cheap gas, the plant at Brownsville would be seen to be uneconomic even before it operated. Indeed, I can suspect that Indiana acquired the plant in 1952 primarily to secure Carthage Hydrocol's contract for purchase of natural gas. Between 1952 and 1958, Indiana committed the gas in small amounts to operation of the plant. Later, Indiana resold gas to pipeline companies at the going rate, earning far larger profits than it could recover by continuing gasoline synthesis. Owners of the gas challenged Indiana's contract, asserting that it specifically called for use of the gas for hydrocarbon synthesis—a claim that the U.S. Supreme Court, in time, denied.

I. Kellogg's Development of Fast-Bed Gasoline Synthesis

In 1946, The M. W. Kellogg Company was determined not to permit a Stanolind–Texas Company–Hydrocarbon Research consortium to race ahead with a fluid-bed hydrocarbon synthesis development, unopposed. Mr. Kellogg, in a rare decision for the head of an architect-engineering enterprise, decided to spend some of his company's own money to forge onward with the work interrupted on June 28.

The September 12 meeting over which Mr. Kellogg presided was rare in another sense: it initiated a bout of invention in a "let's start fresh" frame of mind. Engineers seldom have the luxury of tearing up many months of work and starting again with a clean slate. The psychology of "starting again" calls for something new; and, from a sales standpoint, it is always good if an architect–engineering offering—indeed, any sort of offering—is readily distinguishable from the competition. Kellogg's collaboration with Stanolind provided a good picture of how the competition would look. On September 12, Kellogg's people began a search for something different.

Their thoughts, almost inevitably, turned toward catalyst circulation. Management of heat evolution is an important feature in design of a fluid-bed reactor for the highly exothermic Fischer–Tropsch synthesis. If conducted in a fluid bed at elevated pressure, the synthesis produces a heat release per unit bed volume comparable to the heat release in the combustion chamber of a large steam boiler. By 1946, M. W. Kellogg had designed and built a large number of fluid catalytic cracking units, in which catalyst was circulated to carry heat from an exothermic fluid bed for catalyst regeneration to an endothermic fluid bed for oil cracking, each bed operating adiabatically (Squires, 1986a). In early fluid crackers, regeneration had produced more heat than the cracking bed could use, and catalyst undergoing regeneration had been circulated through a catalyst cooler and back to the regeneration bed. In Kellogg's engineering work for Stanolind, terminated on June 28, the reactor design under study had been fitted for circulation of catalyst through an external catalyst cooler, which would remove some 30% or so of reaction heat. On June 14, just before the break in work for Stanolind, Kellogg's Luther R. Hill had disclosed a reactor concept for multiple parallel tubes, with a superficial gas velocity of 5 ft/s in the tubes, and with high catalyst recirculation. (See Fig. 1.) Earlier, on December 7, 1945, a Kellogg–Stanolind meeting in Tulsa had considered three types of reactors: (a) catalyst outside cooling tubes; (b) catalyst inside tubes, these being surrounded by a pool of a circulating cooling oil; (c) a reactor with external cooling by means of catalyst recycle. Hill's June 14 concept combined ideas that had been around for some time.

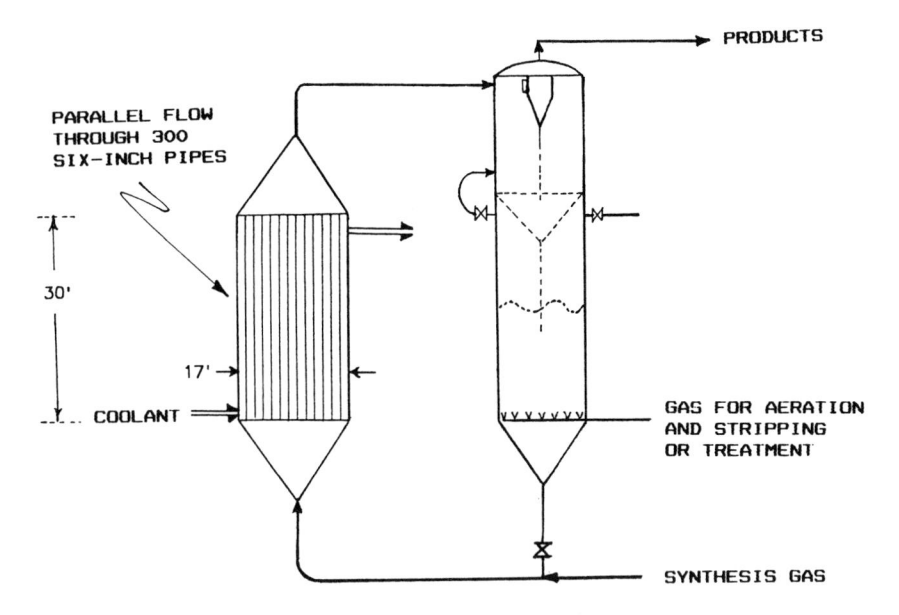

FIG. 1. Luther R. Hill's sketch of June 14, 1946, proposing a circulating-fluid-bed reactor for gasoline synthesis. The arrangement is a poor idea (see Section III for description of troubles that Sasol encountered in operating the heat exchangers depicted in Fig. 4, which also required gas and powder to flow upward via parallel paths).

By September 17, another circulating-bed concept was on the table for discussion: Joseph W. Jewell, Sr., espoused a reactor with high-velocity tubular coolers and with the "bulk of the reaction occurring in the heads and the transfer lines before, between and after the coolers." Jewell and W. Benjamin Johnson had invented a fast fluid bed of a form that would mature in Sasolburg's synthesis reactors. Jewell and Johnson conceived the arrangement on a fine September afternoon, over drinks while sitting on a patio behind Jewell's home (Johnson, personal communication, 1985).

A meeting on September 26 decided to revamp Kellogg's 2-inch stationary fluid bed pilot plant to permit a test of synthesis with catalyst circulation. A slide valve would be installed at the base of the existing reaction zone (12 feet in height; 2 inches in diameter). This would now serve as a standpipe, and synthesis gas would convey catalyst upward, while undergoing reaction, in a 1-inch pipe extending to a height of 20 feet. The pipe would loop downward to a point near the top of the 2-inch standpipe, where reaction products would disengage from catalyst and exit via a pair of ceramic filter elements above the standpipe. The 2-inch section would be jacketed and

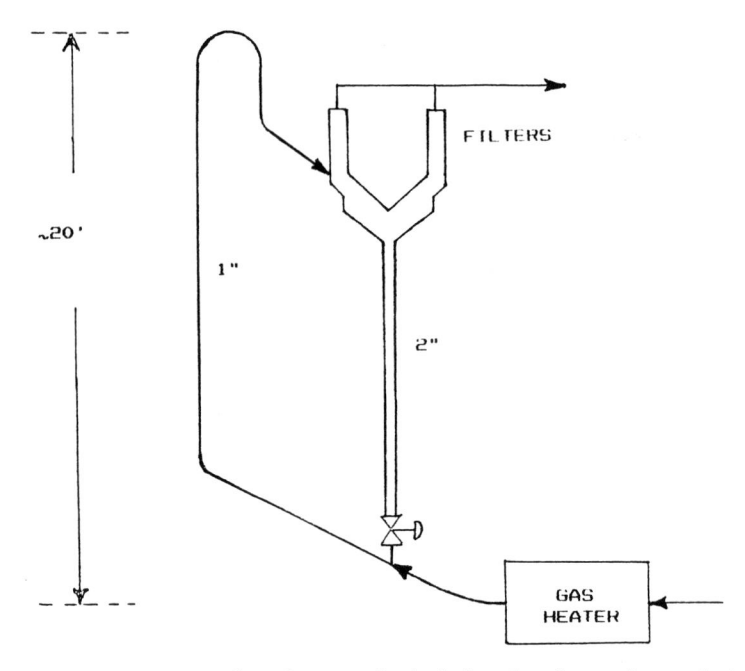

FIG. 2. First M. W. Kellogg Company fast-bed pilot plant for gasoline synthesis.

cooled by Dowtherm, while the insulated 1-inch riser would operate adiabatically. (See Fig. 2.) By October 21, the laboratory was conducting catalyst circulation experiments in the new setup. On October 25, H. G. McGrath wrote, "While maintaining a linear velocity of 15 [ft/s] in the transfer line, catalyst loadings of 3 to 8 [lbs/ft^3] have been observed," the loadings being "based upon the pressure differential across a 15-foot straight section of transfer line."

Synthesis tests began on January 2, 1947, and on February 13, Louis C. Rubin, head of Kellogg's laboratory, reported to Mr. Kellogg that the new reactor's performance "more than fulfills our expectations." Rubin transmitted a report by Martin R. Smith, M. D. Schlesinger, and McGrath, stating, "Most noteworthy among the advantages of the circulating catalyst system are:

"1. It was found possible to operate at low H_2/CO ratio (<2:1) without causing settling or de-aeration of the catalyst.
"2. Hot catalyst at 550–600°F. was mixed with cold synthesis gas at as low as 150°F. without detrimental effects.

"In the [stationary fluid bed] system employed heretofore, catalyst de-aeration troubles were encountered below 4:1–6:1 H_2/CO ratio and when the inlet temperature was reduced below 500°F.... No significant differences in the product distribution and yields from this system and the [stationary fluid bed] system were noted.... Results obtained which closely approximate the desired operating conditions" are shown in the table.

Pressure, psig, lbs/in^2 gauge	150
Temperature, °F (inlet)	600
Transfer line velocity, ft/s	11
Reaction time, s	2.4
Inlet H_2/CO ratio	2.5
Catalyst loading, lbs/ft^3	6
Catalyst density, lbs/ft^3	8
Inlet gas temperature, °F	602
Estimated temperature rise in transfer line, °F[a]	18

[a]Transfer line operates adiabatically.

The reaction time—about one-fifth that of the earlier system—was not sufficient, however, to secure an adequate overall conversion of carbon monoxide.

On November 4, 1946, Mr. Kellogg vetoed a proposal for a Synthol pilot plant for 50 barrels per day of liquid product, to cost $5,000,000. Some of this expense was for an oxygen plant, provided so that synthesis gas could be made by "partial oxidation"—i.e., by reacting a light hydrocarbon with insufficient oxygen for complete combustion, yielding a mixture of carbon monoxide and hydrogen. Rubin offered Mr. Kellogg an alternative: a much smaller pilot plant and a butane–steam reforming furnace as source of synthesis gas.

A memorandum of January 5, 1947 (from J. S. Rearick to Rubin) described three schemes under consideration for use in the pilot plant:

1. A circulating unit "suggested by Messrs. Jewell and Johnson" employing heat removal by means of tubular coolers, with "bulk of reaction occurring in the heads and the transfer lines before, between and after the coolers."
2. A tubular cooler suggested by Hill, to operate at a somewhat lower velocity, and with bulk of reaction in the cooler itself.
3. A laboratory proposal for a high-velocity tubular cooler, but with bulk of reaction in a stationary catalyst fluid bed outside of the cooler.

Rearick notes, "... It does not appear feasible to build a commercial scale [tubular] cooler in which an appreciable reduction in velocity does not occur in the heads."

The first alternative was chosen. The primary reaction zone would be 4 inches in diameter. It would comprise a first, lowermost section 14 ft in length, followed by a cooler section 12 ft long. The cooler consisted of seven 1-inch tubes (outer diameter) extending vertically between two tube sheets, the 1-inch tubes being surrounded by cooling oil. A 7-ft reaction section followed the cooler, and above this there was situated a second cooler, identical with the first. Overall height of the reaction zone (including the coolers) was ~46 ft. Above the second cooler, there was a 2-inch transfer line, in the form of an inverted U, conveying reaction products and catalyst from the top of the primary reaction zone into the top of a knockout chamber. This chamber was ~20 ft in height and was fitted initially with ceramic filter elements ahead of gas outlets. In operation, these elements would display an unfortunate tendency to crack, and would be replaced by metallic filter elements. A standpipe, 2-inches in diameter and ~18 ft in height, delivered catalyst from the knockout chamber via a slide valve to a 2-inch transfer line entering the bottom of the primary reaction zone. (See Fig. 3.)

In early July 1947, Cities Service Corporation joined Kellogg's Synthol effort. Cities Service's refinery in Lake Charles, Louisiana, would operate an experimental steam–methane reforming furnace at 4.4 atmospheres.

In November 1947, Kellogg's laboratory conducted tests of catalyst circulation in its new pilot unit. In these tests, heated air at elevated pressure picked up catalyst below the slide valve. The rate of catalyst circulation was estimated from the temperature and flow rate of air, from the temperature of catalyst in the standpipe, and from the temperature of the catalyst–air mixture in the lower 2-inch transfer line. "Maximum densities in the reactor were approached slowly, increasing the slide valve opening to raise the ΔP over the lower 4" section at the rate of 10" H_2O/Hr. A sharp drop in slide valve ΔP was noted when the limiting reactor density was approached" (Smith and McGrath report of November 21, 1947). A laboratory report of November 29 gave the tabulated data for maximum circulation rates achieved using a magnetite powder at two pressure levels. Operation was maintained for 10 hours at each of the two conditions.

System backpressure, psig	250	150
Reactor, 4-inch section		
Density, lbs/ft^3	27.6	28.0
Velocity, ft/s	8.6	10.1
Pressure, psig	264	174
Temperature at bottom, °F	601	580
Standpipe		
Density, lbs/ft^3	105	115
Velocity, ft/s	0.07	0.09

Temperature at bottom, °F	596	575
Slide valve ΔP, inches water	30	80
Inlet air temperature, °F	648	646
Catalyst flow, lbs/hr	16,500	19,000
Catalyst loading, lbs/ft^3	6.8	6.8

A Roller size analysis of the magnetite charge is shown in another table.

0–10 μm, weight %	1
10–20	13
20–40	14
40–60	20
60+	52

(The Roller analysis is based upon particle suspension velocities.)

Table I summarizes the foregoing magnetite circulation data, together with additional data reported by Kellogg's laboratory for operation at a backpressure of 250 psig. Table I assumes the magnetite to display a specific gravity of 5.2, and the 4-inch reaction zone to comprise a 4-inch nominal Schedule 80 pipe.

The laboratory introduced synthesis gas into the pilot plant in early December 1947. Operations were relatively trouble-free, and the unit was shut down voluntarily on February 11, 1948, in order to replace the upper reactor section and its two coolers with two sections of 4-inch pipe, jacketed with a cooling oil. "Inspection of the reactor sections removed revealed that they were in excellent condition. All cooler tubes were clear and there was practically no catalyst accumulated on the reactor walls. The only case of accumulation worthy of mention was several small nodules of very hard catalyst which appeared in the venturi-shaped section at the bottom of the upper 4-inch section of the reactor. Slight evidence of corrosion was

TABLE I

Gas Velocity	Solid Velocity[a]	Solid Volume Fraction
2.62 m/s	0.055 m/s	0.085
3.08	0.0645	0.086
2.41	0.026	0.111
2.41	0.0245	0.0493
2.16	0.0273	0.1257
1.28	0.0209	0.1763

[a] Solid superficial velocity = solid mass flow rate divided by solid density, where solid mass flow rate = weight of solid per unit time divided by cross-sectional area of transport pipe.

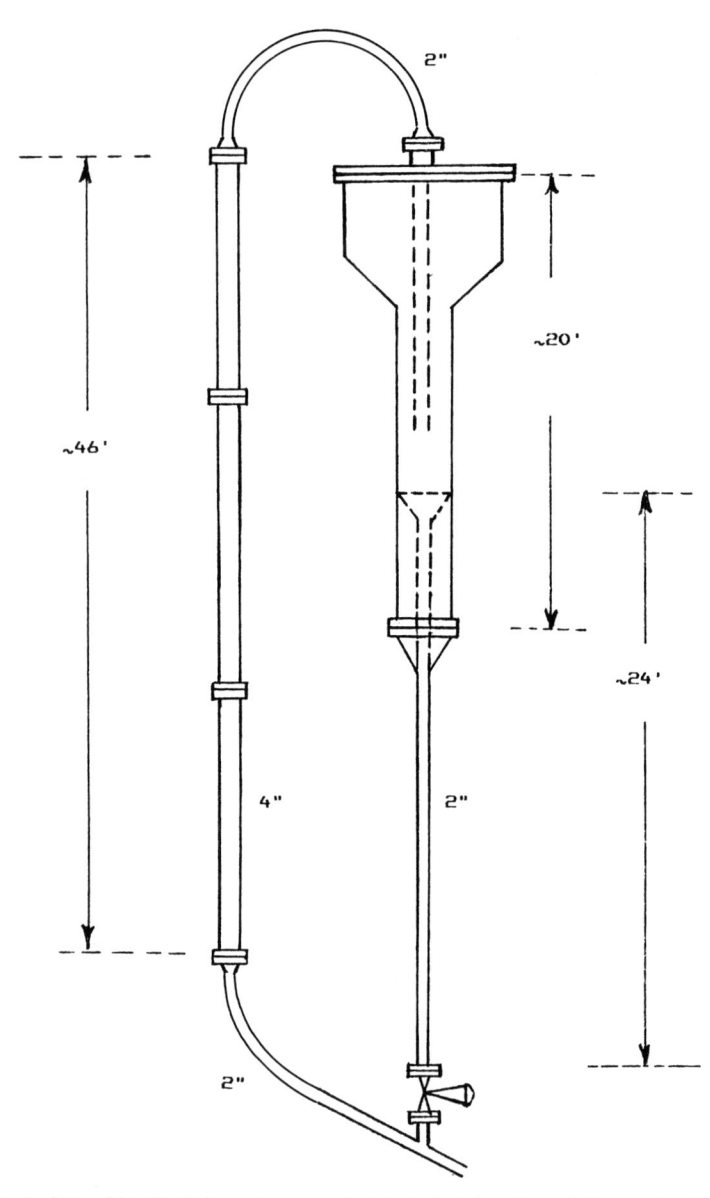

FIG. 3. Second M. W. Kellogg Company fast-bed pilot plant for gasoline synthesis. Initially, the two upper sections of the 4-inch fast bed were fitted with tubular heat exchangers—i.e., vertical tubes, with tube sheets above and below, and with the tubes surrounded by a cooling oil. Later, these exchangers were replaced by oil-jacketed sections of open 4-inch pipe. The dashed lines indicate other changes made after initial shakedown operations, having two effects: lengthening the 2-inch standpipe, and causing catalyst to discharge well below filter elements mounted in the expanded section near the top of the hopper seen on the right.

discernable on the cooler tube sheets, being more marked at the entrance of the lower cooler." The motive for replacing tubular coolers with empty pipe appeared to be to increase the reaction time.

An incident during the operations illustrates the extreme importance of ensuring that the catalyst powder in a fast bed display a sufficiently "Group-A-like" character (Squires et al., 1985). From a practical standpoint, *the significance of this point cannot be over-emphasized* in a tutorial for a development group lacking sophistication in fluid-bed arts and embarking upon a catalytic fluid-bed development (whether the group contemplates using a fast bed or a turbulent bed). A Kellogg laboratory report of January 23, 1948, remarked: "On January 16, 45 pounds of relatively fine particle size catalyst ... was charged to the unit. The condition of intermittent catalyst slugging (which had persisted in both reactor and standpipe since January 15) disappeared shortly after the addition of the fine material, and catalyst flow became much smoother." In the earliest days of fluid catalytic cracking, part of the lore acquired by pioneers was that a slugging ("burping") fluid bed of a nominally Group A powder could be calmed by addition of fines—i.e., by increasing its Group A character. Paradoxically, the addition often *reduced* carryover of powder from the bed (see Braca and Fried, 1956, and Kraft et al., 1956; see also Avidan and Kam, 1985; Squires et al., 1985; Silverman et al., 1986; Pell and Jordan, 1988; Shingles and McDonald, 1988).

A report of February 2, 1948, by Smith and McGrath began: "In view of the great interest within the company concerning the operation of [the new 4-inch circulating synthesis reactor], ... a brief evaluation of results being obtained is in order." Operability was "excellent" (marred only by minor mechanical problems). Catalyst has been circulated at 7 to 10,000 lbs/hr, and at loadings of 5 to 7 lbs/ft^3. "With such loadings, it has been possible to control temperatures easily by introducing process gases at 350–500°F." The report compares results with data from Stanolind's stationary fluid bed at Tulsa (see table).

	Jersey City	Tulsa
Reactor diameter, nominal, inches	4	8
Reaction volume (maximum), ft^3	2.59[a]	5.54
Height, ft	55	20
Catalyst age, hrs	310	530
Temperature (maximum), °F	625	590
Temperature (average), °F	618	588
Pressure, psig	155	165

(*continued*)

	Jersey City	Tulsa
% CO entering reactor	10	10
% CO_2 entering reactor	11.5	7.5
% H_2 entering reactor	66	74
Velocity (minimum), ft/s	5	1.7
Catalyst holdup, lbs	38[a]	247
Catalyst density (maximum), lbs/ft^2	30	60
% Fe in catalyst	71	72
% C on catalyst	7	15
% wax on catalyst	0.3	6
Overall CO conversion, %	(85)	(96)
CO conversion per pass, %	62	94
Standard cubic feet per hour of CO converted per lb Fe	23	5.5
Condensed oil (barrels) per million standard cubic feet of fresh feed	5.6	5.5
Total C_3+ oil	—[b]	20.0
Condensed oil, barrels per day	0.45	0.88
Total C_3+ oil barrels per day	—[b]	3.2

[a] "Including up-flow portion of reactor only. The 4-inch down-flow line represents an additional 1.82 [ft^3], but catalyst concentration in this line is small."
[b] "Results not yet available."

The report commented: At Tulsa, "the usual carbon content curve shows a sharp rise initially and then a continual increase at a slower rate, with very occasional periods where the carbon remains constant." In the new circulating unit, not only were carbon and wax production less, but also carbon level on catalyst actually declined later in a run, something not seen earlier in stationary fluid beds for synthesis.

A report of December 24, 1947, gives catalyst sizes by Roller analysis (see table).

	Charge	At 60 Hours
0–10 μm, weight %	7	4
10–20	11	7
20–40	13	19
40–60	29	35
60+	40	35

	Charge	At 60 Hours
Specific gravity	5.0	4.4
X-ray analysis of catalyst		
Fe, weight %		5
Fe_2C		83
Fe_3O_4		12

(A report of August 20, 1948, gives a similar Roller analysis of catalyst after many hours of operation at 250 psig.)

On September 18, 1948, W. E. Hanford assumed the direction of Kellogg's Synthol activities. Most operations of the 4-inch pilot plant during 1948 appear to have been at 5 ft/s velocity in the reaction zone and at a nominal pressure of 250 psig. In August, the unit operated for a while at 350 psig and at 4 to 4.5 ft/s (density in reaction zone was ~25 lbs/ft^3; in standpipe, 51 lbs/ft^3). Heat transfer coefficients approached 70 Btu/hr-ft^2-°F in the "upper cooler" and 62 in the "lower cooler." Here, "upper" and "lower" refer to the two oil-jacketed sections of 4-inch pipe. The pilot unit was shut down on November 12 (permanently, as near as one can tell from Lobo, 1929–1955).

In August 1948, Kellogg's Engineering Department undertook a design and cost study for a 6,000-barrel-per-day Synthol plant to be erected on the United States Gulf Coast. A memorandum by Johnson provided the design basis and compared this with results obtained from the 4-inch pilot plant (see table).

	Design	Pilot Plant
Average temperature, °F	600	610
Pressure, psig	250	250
Reactor density, lbs/ft^3	30	30
Reactor velocity, ft/s	6.7	4.6
Standpipe density, lbs/ft^3	70	60
Standpipe catalyst velocity, ft/s	3	<1
Slide valve ΔP, lbs/in^2	8	3
Reactor gas contact time, s	15	10
Recycle ratio, recycle/fresh feed	2	2.5
Reactable CO conversion per pass, %	84	73
Reactable CO, overall conversion %	98	98
Average % H_2 in reactor	35	45
Average % CO_2 in reactor	22	19.5

(*Continued*)

	Design	Pilot Plant
Reactor effluent, water–gas-shift equilibrium, °F	600	630
Standard cubic feet per hour of reactable CO converted per lb Fe in reactor	10	12
Gas velocity in disengaging hopper, ft/s	1.5	0.75
Mixed temperature of catalyst and feed gas, °F	550	550

The Engineering Department delivered its design and cost estimate in a report to Hanford dated November 3, 1948. On March 21, 1949, Kellogg delivered an "Engineering Report on Synthol Development 1946–1948" to Cities Service, the authors being Jewell, Johnson, C. E. Slyngstad, and E. J. Syndennis. Johnson and Slyngstad were responsible for process design; Syndennis, for mechanical design; and Jewell, for a cost estimate. For production of 6,000 barrels per day, the design required four synthesis reactors: 6.5 ft in diameter; 103 ft, $11\frac{1}{2}$ inches in height; and $1\frac{1}{16}$ inch wall thickness.

On December 31, 1948, Kellogg and Cities Service stopped pilot plant and engineering activities relating to their joint Synthol development. After about 1950, however, the two companies undertook catalyst development work with the assistance of Sir Hugh S. Taylor, John Turkevich, and Marcel Boudart of Princeton University.

II. Commercial Outcomes Respecting Ammonia Synthesis

Something should be said about sequelae to Kellogg's and Hydrocarbon Research's gasoline synthesis development efforts of the late 1940s.

A disadvantage of steam–methane reforming, as practiced in the U.S. during World War II for ammonia synthesis, was that it furnished synthesis gas at a pressure only a little above atmospheric. In 1944, at the outset of Kellogg's collaboration with Stanolind, Kellogg contemplated increasing the pressure in steam–methane reforming to 4.4 atmospheres absolute, thereby achieving a large decrease in costs for compressing synthesis gas. Kellogg's engineering study for Stanolind, aborted on June 28, 1946, included two alternative routes to synthesis gas: one using steam–methane reforming at 4.4 atmospheres; and a second deriving synthesis gas from partial oxidation of methane at essentially the pressure of the synthesis step. In 1947, Kellogg persuaded Cities Service that steam–methane reforming would be the better choice, and Cities Service built and operated a large pilot reformer at 4.4 atmospheres at its Lake Charles, Louisiana, refinery.

Internal Kellogg memoranda available in Lobo, 1929–1955, give no hint that its preference for steam–methane reforming in its Synthol development reflected a knowledge that increasing reforming pressure would improve Kellogg's competitive position for sale of ammonia synthesis plant. During the early 1950s, availability of reformers operating at higher pressures would give Kellogg a large competitive advantage in sales of ammonia installations.

On November 4, 1946, however, Kellogg's Engineering Department evidently preferred partial oxidation, and included an oxygen plant in its proposal for a 50-barrel-per-day Synthol pilot plant. Throughout Kellogg's collaboration with Cities Service, some Kellogg personnel continued to argue the advantages of partial oxidation.

Hydrocarbon Research Inc., elected partial oxidation for the Carthage Hydrocol plant at Brownsville. After initial experiments that Hydrocarbon Research conducted at Olean, New York, The Texas Company assumed responsibility for further development of partial oxidation at its Montebello, California, laboratory, under duBois ("Dubie") Eastman. For conversion of natural gas to gasoline by Fischer–Tropsch synthesis, partial oxidation's advantage over steam–methane reforming lay in its ability to operate at a pressure approximating that of the synthesis, thereby essentially eliminating need for compression of synthesis gas.

Following his laboratory's development of partial oxidation of natural gas, Eastman turned his attention to partial oxidation of heavy residual oils. Spin-offs from The Texas Company's and Hydrocarbon Research's joint development of partial oxidation would be, first, a few ammonia synthesis installations in the early 1950s employing partial oxidation of natural gas as source of hydrogen and, later, a larger number of ammonia synthesis plants (at sites where natural gas was not available) employing partial oxidation of heavy oils. Eventually, oil partial oxidation was conducted at a pressure higher than the synthesis, eliminating a compressor for synthesis gas.

Over the years, pressures for steam–methane reforming have steadily increased, sustaining this approach's monopoly for ammonia synthesis based upon natural gas. This monopoly may well be the primary outcome, from an economic standpoint, of the joint Kellogg–Cities Service Synthol development effort of 1947–48.

III. Fast-Bed Reactors at Sasolburg

Early in 1951 (probably in late March), Kellogg won a contract to build a gasoline synthesis facility for Sasol employing synthesis gas from coal (Pay, 1980; Dry, 1981). For the plant at Sasolburg, South Africa, the synthesis section would be essentially Kellogg's 1948 design for Cities Service, with minor modifications. Figure 4 is a sketch of a Sasol fast-bed reactor.

A report of May 15, 1953, summarizing a series of discussions of Synthol

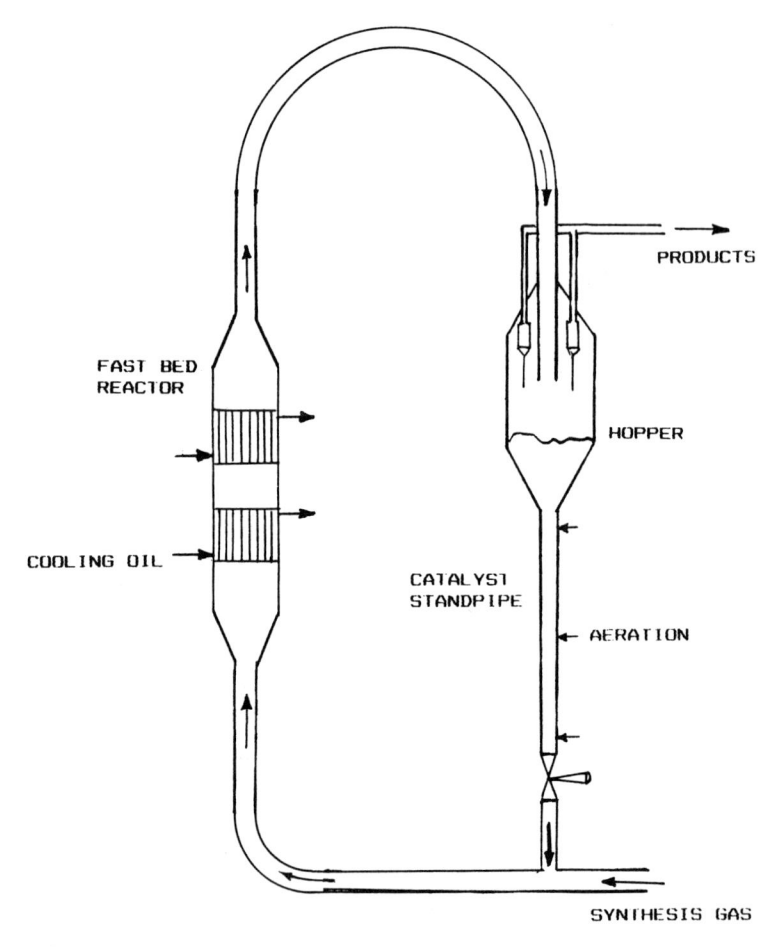

FIG. 4. Sasol fast-bed synthesis reactor. Each of the two cooling sections comprises: two tube sheets; vertical tubes within which gas and catalyst move upward; and a pool of cooling oil, surrounding the tubes.

with Princeton personnel in meetings held in Jersey City, gave the size analysis for powdered iron catalyst (see table).

0–10 μm, weight %	17
10–20	19.5
20–40	24
40–60	32
60+	7.5

This was much finer than the catalyst that seems to have been used in pilot plant operations of 1948. Lobo, 1929–1955, does not appear to contain information permitting one to judge when or why Kellogg decided to use catalyst of a smaller size. The 1953 size analysis, apparently, is roughly in accord with catalyst sizes employed in the synthesis fast-bed reactors at Sasolburg.

Shingles and McDonald (1988) described early operations of the reactors. Considering that they represented a scale-up by a factor of 500 from Kellogg's 4-inch pilot bed of 1947–48, one cannot be surprised that the operations encountered serious troubles.

The worst of these arose from inadequate protection of the reactors from shock; from insufficient means for gas compression; and from features of the process design that were unwise.

Kellogg's operation at a 4-inch bed diameter had given its mechanical engineers little idea of the great impact forces that a fluid bed of a powder as dense as iron catalyst can exert upon its support structure when the bed's inventory shifts suddenly in position. Shock absorber systems initially installed at Sasolburg for both fast bed and catalyst recycle hopper were not adequate and had to be replaced by better designs.

The Sasol plant produced synthesis gas in a chain of process components, each of which had to operate continuously and in step with its neighbors. Some components were novel: Rectisol units employing methanol at subambient temperature for scrubbing acid gases from the raw gas from Lurgi coal gasifiers; and oxygen-fired pressure reformers for partial oxidation of methane in the Lurgi gas. Compression equipment originally provided was not adequate to overcome increases in pressure drop that were encountered during operation, and it became necessary to add booster compressors, both for raising pressure in fresh feed gas and for increasing capability for gas recycle.

In respect to catalyst additions and equipment maintenance, Kellogg contemplated that the practices at Sasol would resemble those developed for fluid catalytic cracking: *viz.*, fresh catalyst was to be added periodically, and operation was to be continuous between reactor maintenance turnarounds at intervals of ~340 days. Not only were transfer means inadequate for adding catalyst during operation, but also the benefit of a catalyst addition was temporary. Accordingly, Sasol elected to operate its reactors on a batch basis. Typically, a batch of catalyst served for 40 days before a voluntary shutdown—although some runs had to be aborted sooner, for a reason to be described later.

Kellogg conceived that the fast-bed reactors were to receive synthesis gas at ambient temperature. In startup, hot catalyst was to be transferred from a lagged storage hopper to a reactor, whereupon synthesis was supposed to ignite "like a fire-cracker." During operation, a significant portion of reaction heat was to be invested in the heating of incoming synthesis gas. In practice, however, unacceptably large temperature gradients developed between catalyst transfer line and bottom of the fast bed. The synthesis reaction is unforgiving of temperature excursions. A temperature that was too high directed the synthesis reaction toward both methane and carbon. The latter developed within a catalyst particle, causing it to decrepitate, yielding a "bug dust" that escaped collection by cyclones. A temperature that was too low directed the reactor toward non-volatile waxes, which caused catalyst particles to become sticky and to form agglomerates or even defluidize. In early operations at Sasol, "cannonballs" formed in the transfer line conducting hot catalyst and cold feed gas from below the standpipe slidevalve into the bottom of the fast-bed reaction zone. Sasol was obliged to add furnaces to preheat synthesis gas. Shingles and McDonald (1988) wrote: "Experience and changes to the reactor internals, still proprietary, led to longer and longer runs but it took 5 years to get the [fast-bed] reactors to work steadily and more or less predictably."

From a reading of Shingles and McDonald (1988), I cannot escape an impression that Sasol's start-up troubles were so great that they might, in ordinary circumstances, have led to a commercial "failure." Would the Sasolburg plant have been abandoned, if it had been privately owned, like Carthage Hydrocol's plant at Brownsville? Can we perhaps regard the Sasol success as a tribute to the determination of the South African government, and to its willingness to spend whatever moneys a success required?

During its operations, Sasol experienced deficiencies in standpipe flow whenever fines were lost—deficiencies that paralleled Kellogg's experience in early 1948, cited earlier.

Sasol's operation encountered a continuing difficulty that arose from an unfortunate feature in the design of the fast bed itself. Operators were obliged to terminate some batch runs, involuntarily, after fewer than the hoped-for 40 days (A. E. McIver, personal communication, 1972).

The Jewell–Johnson design provided parallel flow paths for upward movement of gas and catalyst at a relatively high velocity, each path extending from a common plenum below to a common plenum above, the plenums being fluidized at a lower velocity. (Notice that Hill's concept, too—seen in Fig. 1—incorporated a similar disposition of flow paths.) This arrangement is always a bad idea. Even in the absence of danger from catalyst stickiness, the arrangement is unstable in respect to a transition to a situation wherein a small number of the paths accommodate the entire upward flow in a relatively dilute suspension, while remaining paths act as downcomers, circulating catalyst downward in the dense phase. This imbalanced flow pattern occurs at a lower pressure drop than that required to sustain the desired pattern (wherein upward flows are the same in all paths). Even a slight disturbance can trigger transition to the pattern with the lower pressure drop.

Parallel flow paths of this type are even less wise in a situation where a relatively small downward temperature excursion causes waxes to form and creates danger that deposits of waxy catalyst will block some tubes entirely. Until the mid-1970s, Sasol's operations were plagued by involuntary shutdowns (McIver, personal communication, 1972) for clearing blocked tubes "with jets of high pressure water or more severe mechanical means" (Shingles and McDonald, 1988).

In 1974, Sasol decided to build additional synthesis capacity. In collaboration with The Badger Company—D. H. Jones was a leading player for Badger—Sasol's engineers developed a hairpin heat exchanger to replace Kellogg's tubular coolers. (See Fig. 5.) Sasol tested an experimental hairpin exchanger in a Sasol 1 fast-bed reactor, and the design appears in fast beds at Sasol 2 and 3 (Shingles and Jones, 1986).

Operations commenced at Sasol 2 in 1980, and at Sasol 3 in 1982. Reactors at Sasol 2 and 3 process 3.5 times more gas than the historic reactors at Sasol 1. Pressures and catalyst loadings were higher in the new reactors, and cross-sectional area of a new reactor was only 2.5 times that of the old.

In the late 1980s and early 1990s, Soekor erected yet another gasoline synthesis plant using Sasol fast-bed synthesis technology at Mossel Bay, South Africa. The plant was based upon offshore natural gas.

IV. Pioneering Fast-Bed Research at M.I.T.

In the fall of 1938, eight companies (Standard Oil of New Jersey, Standard of Indiana, Texas Company, Shell, Anglo-Iranian, M. W. Kellogg, Universal Oil Products, and I. G. Farben) organized a consortium, Catalytic Research

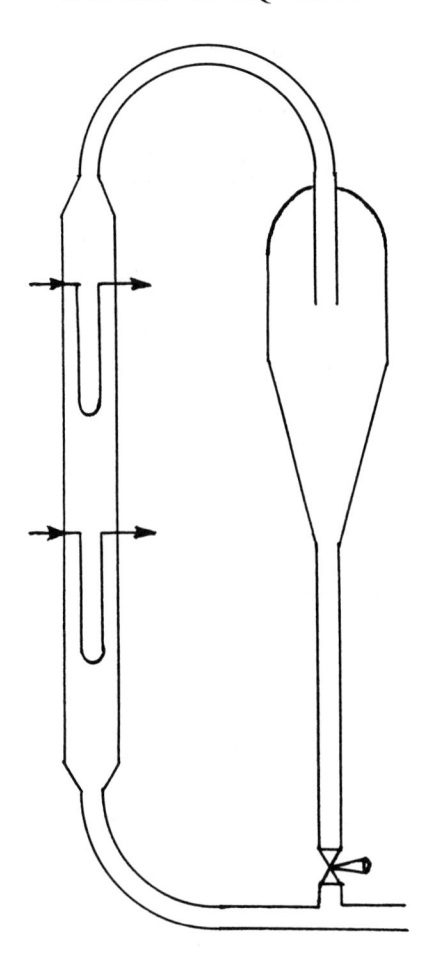

FIG. 5. New hairpin heat exchangers in Sasol reactor.

Associates (CRA), for the purpose of developing a catalytic process for cracking oil that would not infringe upon patents held by Eugene Houdry for oil cracking in beds of relatively large catalyst beads. Squires (1986a) tells the story of CRA's work. Here, I summarize an aspect of the story that relates to the fast fluid bed—*viz.*, work at Massachusetts Institute of Technology (M.I.T.) that would bear fruit in M. W. Kellogg's fast-bed development activities between late 1946 and the end of 1948.

At CRA's first technical meeting, held on November 30, 1938, the eight member companies agreed to develop a process using catalyst in form of a powder.

Warren K. ("Doc") Lewis and Edwin R. Gilliland were present either at this meeting or, a few days later, at a meeting of personnel of Standard Oil Development Company, a Standard of New Jersey subsidiary, under the leadership of Eger Murphree. Murphree's team contemplated oil cracking over powdered catalyst conveyed at high velocity in the dilute phase, but early laboratory experiments in a coil of essentially horizontal tubing encountered troubles with plugging when oil flow was interrupted and restarted. Lewis and Gilliland, professors of chemical engineering at M.I.T., suggested that plugging problems might be less in vertical flow, not yet studied by a member of CRA. The Jersey Standard team was fully engaged, and Murphree offered to provide funds enabling a graduate student to conduct experiments in which air conveyed a fine powder upward in a vertical tube.

This study commenced under some sense of urgency. Lewis phoned a graduate student, John Chambers, during M.I.T.'s 1938 Christmas break, and urged Chambers to return early and begin the study. (Lewis did not tell Chambers the purpose of the study until mid-1940, about a year after Chambers completed his Master of Science degree.)

The sense of urgency, however, soon dissipated. Chambers was left to do the work on his own. In 1986, Chambers recalled that he did not discuss his work with Lewis until April 1939 (although Chambers dwelled in Lewis's home in exchange for tending a coal heating furnace). Chambers had no memory of discussing his work or data with Gilliland on any occasion. Perhaps interest declined because, in early 1939, Jersey Standard's engineers adopted a "snake reactor" design for a pilot cracking installation, to have a capacity of 100 barrels per day and to be built at Standard's Baton Rouge refinery. In Jersey Standard's snake reactor for oil cracking, oil vapor would convey catalyst upward and downward through vertical runs of pipe, adjacent runs being connected at either top or bottom by return blends. Murphree's team may have been less worried about plugging in the snake reactor's vertical pipe runs than in the horizontal tubing of its laboratory coil. In a snake reactor for catalyst regeneration, air would convey catalyst upward and downward in a series of pipe runs.

By early April, Chambers was finishing his work, and beginning to write his master's thesis (Chambers, 1939). He had covered the velocity range from bubbling through turbulent and fast fluidization regimes. Using a simple screw feeder to control rate of solid flow, he had fed a clay powder over a range of flow rates from an aerated hopper into the bottom of a vertical 1-inch glass tube. At his highest air velocity, 7.8 ft/s, he had worked at powder flow rates sufficient to achieve a density of 10 lbs/ft^3.

In late April or early May, Lewis reviewed the data and made his first report to Murphree, who promptly shared it with other members of CRA. The M.I.T. data seemed to attract little attention, but one person, at least, appreciated the potential significance of Chambers's work. Just weeks after receiving the data, on May 20, M.W. Kellogg's Arnold Belchetz filed a catalytic cracking patent application (maturing as Belchetz, 1941) that disclosed use of fast fluid beds in an oil cracking process. In Belchetz's concept, oil vapor or air at velocities as high as 20 ft/s would convey catalyst powder upward at high density. His concept strikingly anticipated the historical trend in fluid catalytic cracking design toward higher velocities: toward use of a "riser" for cracking and a fast bed to complete the conversion of carbon monoxide to carbon dioxide in catalyst regeneration (Saxton and Worley, 1970; Avidan et al., 1990; Avidan and Shinnar, 1990).

Jersey Standard's patent department was slower than Kellogg's in preparing a patent application that read on oil cracking in a fast bed. On January 3, 1940, Jersey Standard filed an application that matured as Lewis and Gilliland (1950).

The Lewis and Gilliland disclosure employed the same means for transferring powder between fluid beds for oil cracking and regeneration that Jersey Standard's engineers had provided in their pilot cracking installation at Baton Rouge. In the pilot plant, Fuller–Kinyon screw pumps elevated the pressure of catalyst powder, overcoming pressure losses in the two snake reactors. In the Lewis and Gilliland teaching, Fuller–Kinyon screws received catalyst from stripping zones fitted with baffles: apparently, Lewis and Gilliland did not contemplate establishing continuous columns of aerated powder in these zones. In January 1940, the aerated standpipe was not yet in view.

Yet Chambers, at the beginning of his work (Chambers, personal communication, 1986), had quickly discovered that he could not feed powder from a hopper to the bottom of a fluid bed with a simple screw unless he aerated the hopper. Scott Walker inherited Chamber's apparatus. (Walker would play a major role in Stanolind's Fischer–Tropsch activities.) While Chambers wrote his master's thesis, he helped Walker replace the screw with a horizontal jet of gas, capable of conveying powder at a controlled rate into the bottom of their 1-inch glass pipe. He also helped Walker revise the equipment for studies of upward conveying of powder by hydrogen or carbon dioxide (Walker, 1940). Recall that neither Lewis nor Gilliland had followed Chambers's work closely. No representative of CRA had visited. I cannot escape the impression that the standpipe might have come into view sooner if Chambers's work had attracted more interest. It remained, however, for Jersey Standard engineers (D. L. Campbell, H. Z. Martin, and C. W. Tyson), in the spring of 1940, to invent the aerated standpipe with a solid-flow control valve at the bottom (Campbell et al., 1948).

This invention came after several months of troubled operation of Jersey Standard's pilot cracking installation at Baton Rouge. Since the unit's Fuller–Kinyon pumps were responsible for many of the troubles, the aerated standpipe was a welcome alternative. Moreover, snake-reactor conversions of oil to gasoline were disappointing. Squires (1986a) told how M. W. Kellogg engineers diagnosed the snake reactor's deficiencies, and how, although installing additional pipe-runs improved snake-reactor performance, sentiment grew for replacing it with fluid beds. For CRA, the M.I.T. data was a godsend, permitting confident design of beds for cracking and catalyst regeneration. They would be bubbling beds, not fast ones, perhaps because heights of fast beds affording requisite reaction times were deemed unacceptable in an installation for 100 barrels per day. (The overall height of the Baton Rouge pilot plant was $88\frac{1}{2}$ ft. After revision of the plant for fluid beds and standpipes, the two reactors would occupy 20 ft of this height. One standpipe would occupy $77\frac{1}{2}$ ft, and a second, $58\frac{1}{2}$ ft. See Fig. 4 in Squires, 1986a.) Snake reactors operated at Baton Rouge until June 19, 1940, when Jersey Standard shut down its pilot plant for work replacing snake reactors and Fuller–Kinyons with fluid beds and standpipes. On July 17, using air as transport medium, catalyst circulation tests commenced in the revised unit. On August 4, oil was introduced, and smooth operations were achieved by August 31.

With invention of the standpipe, the essential elements of Kellogg's fast-bed reactor for Sasolburg were in hand: an aerated standpipe for raising the pressure of a powdered solid (Geldart Type A); a solid-flow control valve at standpipe bottom; and a reaction zone filled with solid at a relatively uniform, relatively high density, conveyed upwards by synthesis gas moving at a speed beyond what Lewis and Gilliland (1950) termed the "blowout" velocity: "a gas velocity sufficient to blow all or substantially all of the solid material out of the reactor in a relatively short time, provided no fresh solid material be introduced during this time."

I have vivid memories from the late 1940s of Stanley Wetterau's demonstration of blowout velocity in glass equipment at Hydrocarbon Research, Inc.'s laboratory, near Trenton, New Jersey. At an air velocity a bit below blowout, he showed me a dense bed of powder, albeit carryover from the bed was evident to the eye. By adjusting the air flow just a minuscule amount upward, Wetterau caused the bed to vanish, almost in an instant. When someone presented Wetterau with a new powder, his first action was to determine blowout. He argued that this velocity was a better indicator of a powder's usefulness for a fluid-bed process than anything else he could measure.

V. Fast Beds for Combustion (Affording Low Pressure Drop)

The M.I.T. fluidization data did not reach the open literature until 1949 (Lewis *et al.*), and it remained for Wilhelm and Kwauk (1948) to provide a report whereby one may infer that fine powders aggregate when buoyed by a rising current of air—clustering to a degree such that velocities to buoy the particles, or transport them upward, were far higher than single-particle terminal velocities (see, especially, Leo Friend's comment on the Wilhelm–Kwauk paper).

Kellogg's fast-bed reactors for Sasol received negligible attention in the European or U.S. engineering literature, and Sasol's engineering achievements in implementing Kellogg's concept attracted little notice.

Accordingly, I could not be surprised when Lothar Reh told me that the M.I.T. data at high gas velocities, as well as the Sasol fast-bed reactor, were far from his thoughts when he invented the Lurgi fast bed for calcining alumina (Reh, 1971; Reh and Rosenthal, 1971). (See Fig. 6.)

This is essentially a device for burning a liquid fuel and using heat of combustion to effect an endothermic reaction, decomposition of aluminum hydroxide to yield alumina feedstock for aluminum production by electrolysis. The alumina powder comprising the fast bed is a Group A solid. Reh's design meets two primary economic requirements: (1) The pressure drop is acceptably low for a combustion process operating at atmospheric pressure. (2) Temperatures are sufficiently uniform (Schmidt, 1972) and residence time of solid is sufficiently long to yield alumina of cell grade.

Primary air flow, supplied at the bottom of the bed, is insufficient for complete combustion of the fuel. The zone below the elevation of secondary-air jets is endothermic, cracking oil to yield carbon and fuel gas species. These burned in the upper, exothermic zone. Alumina product is withdrawn from a standpipe receiving solid from the upper zone, and burn-off of carbon in this zone is sufficient to yield a product that is acceptably white. Fluidizing-gas velocity being lower in the primary combustion zone than in the secondary, density is higher. Provision of the two zones accomplishes two purposes: (1) affording a sufficient solid residence time in the primary zone; and (2) reducing horsepower needed for air compression.

A major difference between the Lurgi and Sasol fast-bed designs is that the former displays a large gradient in solid density (Schmidt, 1972), while the Sasol design is capable of achieving something much closer to uniform catalyst density throughout its height.

In any fluid bed where solid is introduced into the bottom and carried away from the top, there is in principle a separation between a denser region below and a more dilute region above. In a bubbling bed, the separation is

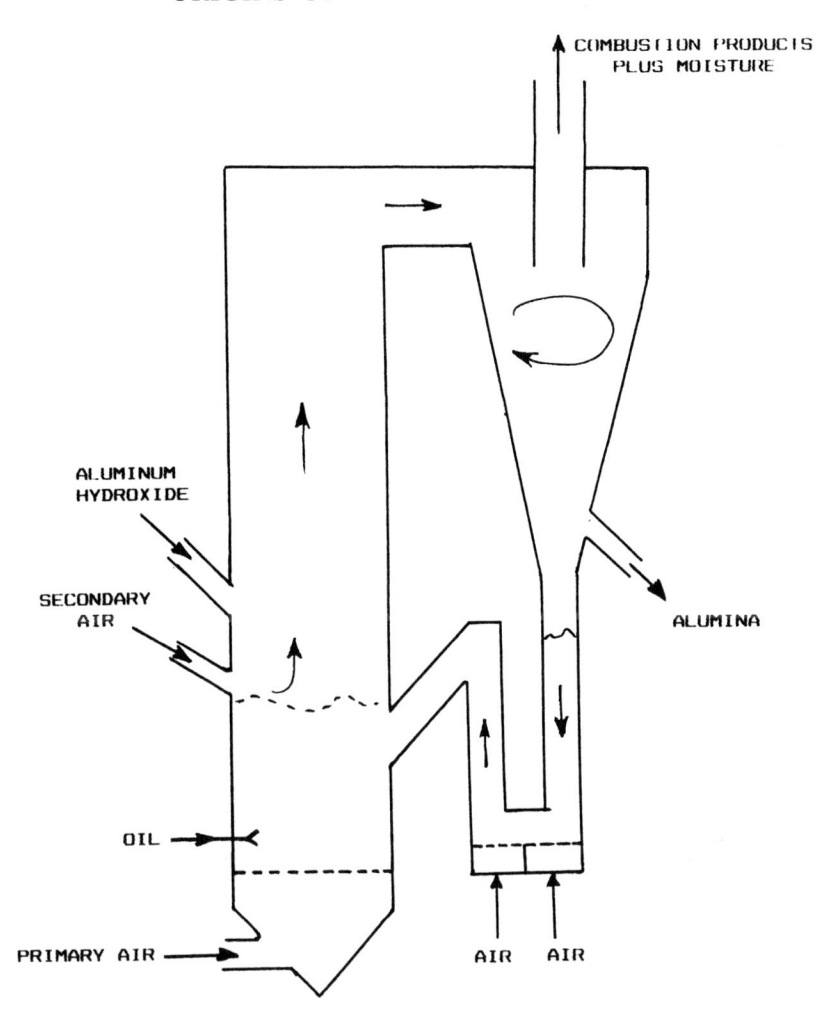

Fig. 6. Lurgi calciner, for converting aluminum hydroxide to cell-grade alumina.

sharp; one can speak of a "bed surface"—one might even best speak of a "meniscus." With increasing velocity, upon transition from bubbling to turbulent and fast regimes (and, as well, to the regime prevailing in a "riser" reactor), the meniscus becomes progressively more diffuse. (See Fig. 7.) In any bed, the level of the meniscus depends simply upon the pressure difference that a standpipe (or other means) imposes upon the bed (Weinstein et al., 1984).

In a reactor for catalysis at an elevated pressure, a designer wishes to

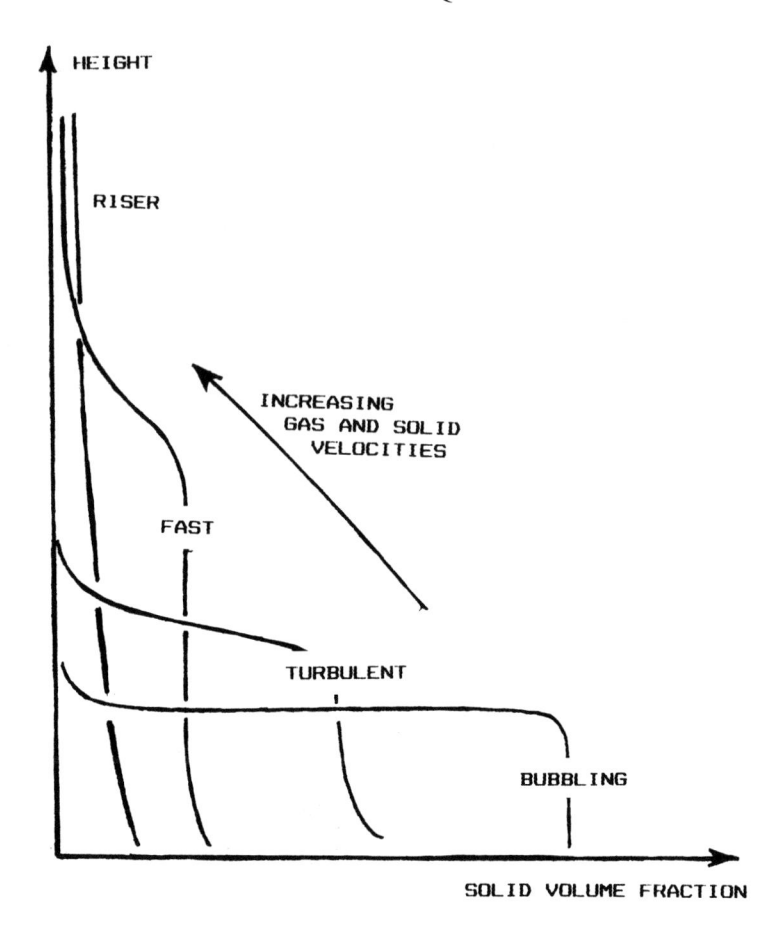

Fig. 7. Schematic diagram indicating, qualitatively, the two regions ordinarily to be seen in a vessel housing a fluid bed. There is a lower region of higher solid density, and an upper region of lower solid density. Separating the two regions is a "meniscus" that becomes progressively more diffuse with increase in gas velocity. Such increase also causes solid density in the lower region to fall, and solid density in the upper region to rise. The increase in velocity also brings about an increase in the minimum solid throughput necessary for maintaining the presence of the lower region of higher solid density. In all instances, external means set the elevation of the meniscus—means that determine the pressure drop across the vessel in question. For example, in a bed with recirculation of solid, the means may be a device for balancing this pressure drop—e.g., an aerated standpipe, or a J-leg. In bubbling and turbulent regimes, gas velocity determines carryover, when the meniscus lies within the vessel in question. If solid is fed to the lower region at a rate higher than the natural carryover, the meniscus is driven upward, and the vessel becomes full of solid at the higher density. In fast and riser regimes, both lower and upper solid densities are functions of solid throughput. In these regimes, too, the meniscus may be driven upward, so that the higher density prevails throughout the vessel.

provide an adequate catalyst inventory, and will seek to impose a pressure difference upon the bed sufficiently large to move the meniscus to the highest possible level in the reactor (or even to move it to a "virtual" level *above* the top of the reaction zone). Accordingly, the Sasol design employed a standpipe with slide valve, achieving a high catalyst inventory.

In the Lurgi fast-bed calciner, in contrast, Reh wished to achieve the lowest practicable horsepower for compression. He wished to maintain a powder inventory at the lowest level necessary to achieve both the desired solid residence time and complete combustion of fuel. He welcomed a decline in density with bed height.

To achieve a pressure balance, Reh did not require a standpipe of considerable height. Providing a slide valve for his operating temperature, $\sim 1,100°C$, might be difficult, and he elected a device with no moving parts, the J-leg, for balancing the pressure drop across his fast bed.

From at least the mid-1940s, the J-leg had been employed in fluid beds for ore roasting, calcinations, and other treatments of "natural" solids where a designer often had little control over particle size. I do not know who invented the J-leg, but it may well have been an engineer at Dorr-Oliver or Lurgi, builders of fluid beds of this general type. In comparison with an aerated standpipe, the J-leg is much more forgiving of excursions in particle size—indeed, it works equally well processing a Group A or Group B solid. Long before 1960, a large number of fluid beds with J-legs processed solids of a wide range of particle sizes and densities. When a J-leg balances the pressure drop in a fluid bed, both pressure drop and inventory can rise and fall, automatically in step, as an operator increases or decreases the inventory of bed solid. (See Fig. 8 for explication of J-leg operation.)

Meeting Reh in the fall of 1967, and almost simultaneously entering academe by joining the Faculty in Chemical Engineering at The City College of New York, I become an enthusiastic spokesman on behalf of fast fluidization and Lurgi's achievement. In my 13 years at Hydrocarbon Research, Inc., and 8 years as consultant to the chemical process industries, I have often witnessed how just a small increase in gas velocity can yield a striking improvement in performance of a fluid-bed process. (I now believe these to be instances where the increase has caused a bed to undergo transition from bubbling to turbulent regime.) These experiences parallel reports of improved process performance of fluid catalytic crackers with increase in throughput several-fold beyond design (Braca and Fried, 1956). A visit to Sasolburg in July 1972—a visit, fortunately, that preceded the renewed interest of late 1973 in gasoline from coal and Sasol's appreciation of the commercial value of its fast-bed-reactor development—reinforced my enthusiasm for fluidization at higher gas velocity.

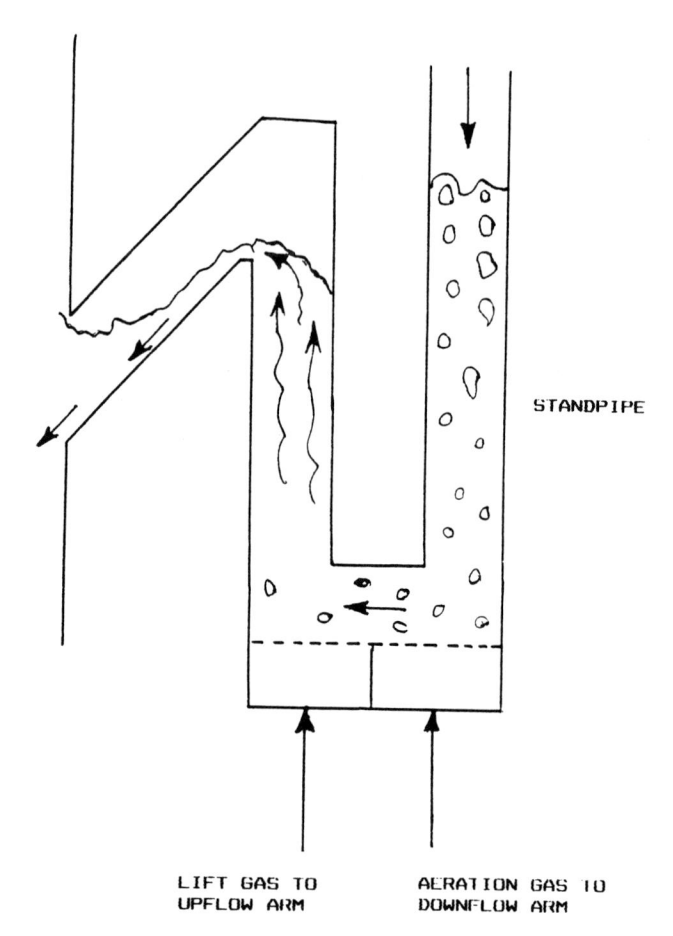

FIG. 8. A J-leg. Flow of lift gas to the J-leg's upflow arm (at the left in the diagram) determines solid flow rate. Aeration gas to the downflow arm (at the right) maintains solid mobility. Height of solid in the standpipe substantially determines the pressure drop that the J-leg imposes upon a fluid bed that receives solid spilling from the upflow arm.

Receiving a grant in 1972 from the RANN Program ("Research Applied to National Needs") of the U.S. National Science Foundation, I organized a team at City College for research on "improved techniques for gasifying coal." One focus of the work was the first academic study of fluidization at high velocities since the work of the early 1940s at M.I.T. (Yerushalmi et al., 1976a, 1976b; Yerushalmi and Squires, 1977; Yerushalmi et al., 1978; Cankurt and Yerushalmi, 1978; Yerushalmi and Cankurt, 1979; Avidan, 1982; Avidan and Yerushalmi, 1982, 1985; Yerushalmi and Avidan, 1985).

In the summer of 1972, the City College team erected a "two-dimensional" fast bed with transparent walls (2 × 24 inches in plan, and 24 ft in height). In the fall of that year, Lurgi's Reh and Schmidt saw this "2-D" bed in operation. Not having modeled the fast bed in transparent equipment, Reh and Schmidt became excited as they watched solid clusters and strands coursing down the wall of the City College bed; they exclaimed that these look just as they expect from their own studies. In the years following 1972, the 2-D bed becomes a teaching tool, introducing many visitors from academe and industry to fast fluidization. In 1973, the City College team shot footage of the bed on motion picture film, with both an ordinary camera and an Edgerton high-speed camera, affording ultraslow motion. In early 1974, Lurgi sent Ludolph Plass for a several-week visit to the City College, to shoot additional footage. A number of commercial firms acquired copies of the edited film that resulted, and its presentation at technical meetings served to introduce the fast bed to many practitioners of fluidization arts.

From the early 1960s, interest grew in devices for burning coal in fluid beds (Howard, 1983; Squires et al., 1985), and it is no wonder that a number of groups commenced work on a fast-bed steam boiler almost simultaneously. First to reach the market was a design, generally resembling Lurgi's calciner of Fig. 6, developed by Finland's A. Ahlstrom Corporation (Engstrom, 1980). At an International Conference on Fluid Bed Combustion held in Atlanta, Georgia, in 1980, Ahlstrom's Fulke Engstrom did me the courtesy of telling me that Ahlstrom had begun work upon its fast-bed boiler in 1973, soon after Engstrom had heard me lecture on fast fluidization for coal gasification, at a meeting of Sweden's Academy of Engineering in Stockholm. At an International Conference on Fluidization held at Asilomar, California, in June 1975, Herman Nack of Battelle Memorial Institute presented data confirming the intuitive expectation that heat transfer in the fast regime varies with solid inventory as well as gas velocity (Kiang et al., 1976). Soon afterward, Lurgi filed a fast-bed-boiler patent application contemplating that an operator would vary solid inventory to maintain an appropriate rate of heat removal as demand for steam varied (this application maturing as Reh et al., 1979). In time, Sweden's Gotaverken Angteknik, Dorr-Oliver, and Foster-Wheeler offered fast-bed boilers generally resembling Lurgi's calciner of Fig. 6. Other groups offered more complex arrangements: Battelle (Nack and Liu, 1978); Sweden's Studsvik Energiteknik (Strömberg, 1979b); and a consortium organized by W. Benjamin Johnson (the same Johnson who was in M. W. Kellogg's employ in the late 1940s (Johnson, 1980; Pell and Johnson, 1981). Workers at Studsvik have contributed valuable data on fast beds of Group B solids and may have been among the earliest to appreciate the advantages of using such solids in a fast-bed coal-fired boiler (Strömberg, 1979b).

Ehrlich and Drenker (1992) reviewed operating data for nine large fast-bed boilers now in service. All of them used a solid displaying a wide particle-size distribution. (I thoroughly deplore the emphasis in so much fluidization research upon particles of uniform size. Almost always, this "academic" approach fails in some important respect to reflect the true situation in commercial practice.) In eight of the units, the fluid beds comprise what are clearly Group B solids, displaying median sizes that vary from about 150 to 200 micrometers (μm). One unit treats a solid borderline between Group A and Group B, having a median size a little over 100 μm. Clearly, the J-leg's flexibility in respect to particle size is a great advantage for fast-bed boiler design.

Today, it appears that designers of large boilers (larger than, say, ~25 megawatts thermal output) prefer the fast-bed design over the older bubbling-bed alternative. Several factors are conductive to the fast bed's advantage:

- The tall structure of the fast bed, with its vertically disposed heat-exchange surface and its relatively small footprint, is more congenial to both purchaser and designer of a large boiler facility.
- In the fast bed, heat-transfer surface usually appears only in bed walls (or in curtain walls that partially divide the bed into segments). Bubbling-bed designs commonly provide horizontal tubes embedded within the bed. Such tubes are far more subject to erosion than the fast bed's vertical surface.
- The bubbling bed requires a far larger number of coal feeds, in order to avoid creation of zones where both carbon and carbon monoxide levels are high. Complex control means must be provided to ensure adequate approach to uniformity of flow of coal through the multiplicity of feeds.
- The smaller number of feeds minimizes exposure of metal surfaces to fluctuation in gas atmosphere between oxidizing and reducing conditions in respect to iron—a circumstance giving rise to rapid erosion and corrosion of the metal.
- The fast-bed design naturally affords the two-stage combustion that is desirable for reducing emissions of nitrogen oxides. Providing for two-stage combustion in a bubbling bed is more difficult.

VI. "Bubbleless" Fluidization Studies in China

In November 1975, a chance meeting in Tokyo brought Mooson Kwauk and me together for the first time in nearly two decades. I found in Kwauk another enthusiast for fluidization at high velocity—indeed, an enthusiast for all forms of "bubbleless" fluidization:

- systems with dilute raining particles
- turbulent and fast fluidization
- shallow fluid beds
- "pulsed" fluidization, with no net fluid flow

As Kwauk pointed out, the Chinese word for a fluid bed, "fu-deng-ceng" is derived from the German "Wirbelschicht" and the Russian "kipiashei sloye"—signifying bubbling. Kwauk thought it to be unfortunate that Chinese, German, and Russian terminology had narrowed the range of vision of fluidization practitioners, while at the same time, fluidization literature in Great Britain, Europe, and the U.S. had focused almost exclusively upon the bubbling regime. Yet failures of bubbling-bed processes as well as growing numbers of studies demonstrating that conversion is better in the "bubbleless" turbulent and fast regimes surely demonstrate that opportunities have been missed through omission of work on options other than the bubbling fluid bed.

Kwauk's Institute of Chemical Metallurgy has conducted research on each of the "bubbleless" fluidization alternatives.

In the 1950s, a major effort had been devoted to systems of dilute raining particles, offering advantages of high gas–particle heat and mass transfer, low pressure drop, and low cost of chimney-like equipment. The systems have been exploited on the pilot or demonstration-plant scale:

- A 15-ton-per-day plant in Dayeh prepares a cupriferous iron ore for extraction of its copper content, prior to smelting in a blast furnace, by subjecting the ore to a sulfatizing roast.
- A 250-ton-per-day plant in Tongguanshan roasts a highly refractory copper ore of low grade. In a mixture of preheated ore and small amounts of anthracite and common salt, copper is volatilized as chloride and transported to the coke surface, where it deposits in the metallic state.
- A 100-ton-per-day plant in Ma'anshan roasts a low-grade iron ore, converting its iron content to Fe_3O_4 for magnetic upgrading.
- A series of industrial roasters in Wanshan recover mercury from extremely low-grade ores ($\sim 0.06\%$ Hg). Later, the roasters are adapted for processing low-grade Guizhou pyrite for sulfuric acid manufacture.

During a visit to the Institute of Chemical Metallurgy in June 1979, I was astonished by the range, sophistication, and aptness of high-velocity fluidization research in progress there, only five years after the Institute had been reinstated with Kwauk as its director. Already, the work had cast important new light upon the behavior summarized in Fig. 7, *viz.*, how a fluid-bed "meniscus" becomes more diffuse as velocity increases.

The Institute had first contemplated use of fast fluidization for the gaseous reduction of Panzhihua vanadium-bearing titanomagnetite, whereby metallic iron could be separated in the molten state, leaving a beneficiated slag, high in both vanadium and titanium, for further processing. The Institute had also employed fast fluidization for calcining magnesite and boron ores, treating vanadium-containing coal from Jianxi, and particularly in the pseudo solid–solid reaction for prereduction of iron ores (for a new iron-and-steel processing using powdered coal and powdered ore).

The Institute's primary interests, however, were the fast-bed boiler and the fluid catalytic cracker regenerator.

In September 1982, I accompanied a group of chemical engineers from the U.S. on its visit to the Institute. During the bus ride back to the group's hotel, I was gratified to hear the spontaneous praise accorded the Institute's work: "world-class" is the adjective members of the group use in characterizing the Institute's activities.

The present volume reports contributions of the Institute of Chemical Metallurgy fluidization research team toward understanding and applying the fast fluid bed.

VII. Contacting in Turbulent and Fast Fluidization

An underlying function of fluidization has often been to afford contact between a gas and a large inventory of solid surface per unit bed volume. In many minds, between about 1950 and the early 1970s, the word "fluidization" became virtually synonymous with operation in the bubbling regime. For processing a fine powder (Geldart Group A), this meant operation, typically, at a superficial gas velocity around 1 ft/s.

Yet Lanneau (1960) provided strong evidence for opportunities to achieve improved contacting at a gas velocity beyond the bubbling regime. Lanneau studied fluidization characteristics of a fine powder in a bed 15 ft deep and 3 inches in diameter, at two pressure levels, 1.7 and 5.1 atmospheres absolute, and at velocities to ~ 5 ft/s. At a given gas velocity, the character of the bed was manifest in recordings from small-point capacitance tubes within the bed. Lanneau noted that when velocity was increased beyond 1 ft/s, the heterogeneous, two-phase character of the bubbling regime gave way to a condition of increasing homogeneity. At velocities from 3 to 5 ft/s, the probe tracings indicated that the bed approached a condition of "almost uniform or 'particulate' fluidization." Lanneau argued that a velocity of ~1 ft/s, where the two-phase, heterogeneous structure of the bubbling bed was most pronounced, was the worst velocity to use from standpoint of efficiency of gas–solid contacting. He provided a rationale for what practitioners had already appreciated (Braca and Fried, 1956), that contacting is better at higher velocity.

In a bubbling bed, gas bypassing is often the main culprit in lowering efficiency of gas–solid contacting, but two other phenomena sometimes play roles:

- In a bubbling bed of a fine powder, gas within the dense phase ("emulsion" gas) is often highly converted, while bubble gas contains unreacted species. In the history of a given particle of the powder, as it moves through the system, it experiences relatively long exposures to highly converted emulsion gas, alternating with relatively brief exposures to relatively unconverted bubble gas.
- In a fluid bed of any particle type (light or heavy, large or small), at any fluidizing-gas velocity, particles travel in concert in large-scale movements that carry them from top of the bed to the bottom and back again. Such movements bring the particles into contact, quasi-cyclically, with both fresh feed and highly converted gas.

Each mechanism exposes the particles to a range of gas atmospheres, the atmosphere varying rapidly in time. Either of the two mechanisms can give rise to reaction kinetic effects, which at times can be hurtful (Squires, 1982), and at other times, helpful (Squires, 1961, 1973).

In a fluid bed of a fine powder, increasing velocity beyond the transition reported by Lanneau (1960) gives rise to "bubbleless" fluidization, thereby greatly reducing both gas bypassing and the gradient in gas composition between emulsion phase and dilute phase.

Because Lanneau had worked in steel equipment, it remained for Kehoe and Davidson (1971) to describe the transition from bubbling to what they dubbed the "turbulent regime" of fluidization.

It remained for Massimilla (1973) to provide laboratory research demonstrating the higher contacting efficiency that a turbulent fluid bed affords. Wainwright and Hoffman (1974) reported excellent contacting for oxidation of orthoxylene in what was probably a fast fluid bed.

What Kehoe and Davidson witnessed, at the transition to turbulence in a tube of relatively small diameter, was a breakdown of a slugging regime into "a state of continuous coalescence—virtually a channelling state with tongues of fluid darting in zig-zag fashion through the bed." The point of breakdown of slugging was not sharp. For several powders with particles sizes below 90 μm, Kehoe and Davidson placed the transition between 1 and 2 ft/s. In Lanneau's experiments, in which solid ranges from 40 to 100 μm, the transition was complete at ~ 3 ft/s.

The work just cited refers to beds of small diameter. Designers and operators of large-scale catalytic fluid beds of Group A powders now appreciate that *all* of these beds function beyond the Lanneau–Kehoe–Davidson transition (Avidan, 1982; Squires *et al.*, 1985). Most are turbulent beds; Sasol reactors and some fluid catalytic cracking regenerators are fast beds. Sasol engineers reported successful development of a turbulent bed for hydrocarbon synthesis (Steynberg *et al.*, 1991).

Engineers at Mobil Oil Corporation are satisfied that a one-dimensional analysis is suitable for treating reaction kinetics in these beds, simply using an appropriate Peclet number to represent the effective axial gas diffusivity (Avidan, 1982; Krambeck *et al.*, 1987). Inputs for Mobil's analysis are two: (1) the Peclet number expected for a commercial fluid bed in question—they estimate this to be ~ 7 for beds they contemplate for carrying out Mobil's methanol-to-gasoline or methanol-to-olefin reactions—and (2) kinetic data from a pilot fluid bed, which can be expected to reflect, reasonably well, whatever top-to-bottom mixing of powder will occur in the commercial bed.

Notice that an engineer can never, a priori, expect to produce a confident design from steady-state differential rate data obtained in a laboratory fixed bed. For a design from first principles, an engineer would require knowledge of the history of a typical particle in a bed in question—how its position and gas environment vary in time; also, the engineer would need unsteady-state differential rate data reflecting whatever quasi-cyclic variations in gas atmosphere a particle may encounter. Studies of solid mixing in fluid beds are few (Avidan, 1980; Avidan and Yerushalmi, 1985); to my knowledge, no one has attempted to obtain, even for one catalytic reaction, the unsteady-state differential rate data that the engineer would require for a design from first principles.

I caution, indeed, that even Mobil's analysis cannot faithfully reflect reaction-kinetic effects, if any, arising from top-to-bottom mixing of catalyst in a turbulent- or fast-bed design in question. The analysis may mislead if a given reaction is particularly sensitive to quasi-cyclic extremes in gas atmosphere.

At Virginia Polytechnic Institute & State University, my colleagues and I (Thomas and Squires, 1989; Benge and Squires, 1994) are attempting to develop a vibrated-bed microreactor providing capability for varying two parameters (each, independently, over a wide range) that may be significant for kinetic performance of a large fluid bed:

- axial Peclet number
- front-to-back mixing of powder

If the attempt is successful, an engineer whose assignment is to select a reactor for a new catalytic process could quickly determine whether choice of a fluid bed might bring into play either of two dangers: that the reaction in question is highly sensitive to Peclet number, thereby creating danger of failure if an incorrect Peclet number is used in design; or that the reaction is highly vulnerable to top-to-bottom circulation of catalyst.

References

Avidan, A. A. "Bed expansion and solid mixing in high-velocity fluidized beds," Ph.D. dissertation, The City University of New York (1980).

Avidan A. A. "Turbulent fluid bed reactors using fine powder catalysts," *Proc. J. Meet. Chem. Eng., Chem. Ind. Eng. Soc. China and Am. Inst. Chem. Engrs. 1982*, 411–423 (1982).

Avidan, A. A., and Kam, A. Y. "Process for the conversion of alcohols and oxygenates to hydrocarbons in a turbulent bed reactor," U.S. patent 4,513,160 (April 23), filed March 28, 1983 (1985).

Avidan, A. A., and Shinnar, R. "Development of catalytic cracking technology. A lesson in chemical reactor design," *I&EC. Research* **29**, 931–942 (1990).

Avidan, A. A., and Yerushalmi, J. "Bed expansion in high velocity fluidization," *Powder Technol.* **32**, 223–242 (1982).

Avidan, A. A., and Yerushalmi, J. "Solids mixing in an expanded top fluid bed," *AIChE J.* **31**, 835–841 (1985).

Avidan, A. A., Edwards, A., and Owen, H. "Innovative improvements highlight FCC's past and future," *Oil & Gas Journal* **88**(2), 33–36+ (January 8, 1990).

Belchetz, A. "Catalytic conversion of hydrocarbons," U.S. patent 2,253,486 (August 19), filed May 20, 1939 (1941).

Benge, G. G., and Squires, A. M. "Microreactor simulating reaction scene in turbulent fluid bed of Group A powder," paper presented at AIChE Annual Meeting, San Francisco, CA, November 13–18, 1984.

Braca, R. M., and Fried, A. A. "Operation of fluidization processes," *in* "Fluidization" (D. F. Othmer, ed.), pp. 117–138. Reinhold, New York, 1956.

Campbell, D. L., Martin, H. Z., and Tyson, C. W. "Method of contacting gases and solids," U.S. patent 2,451,803 (October 19), filed October 5, 1940 (1948).

Cankurt, N. T., and Yerushalmi, J. "Gas backmixing in high velocity fluidized beds," *in* "Fluidization" (J. F. Davidson and D. L. Keairns, eds.), pp. 387–393. Cambridge University Press, Cambridge, U.K., 1978.

Chambers, J. W. "Flow characteristics of air–fine particle mixtures in vertical tubes," Master of Science thesis, Massachusetts Institute of Technology, Cambridge (received M.I.T. Library, November 1, 1939).

Dry, M. E. "The Fischer–Tropsch synthesis," *in* "Catalysis: Science and Technology" (J. R. Anderson and M. Boudart, eds.) Vol. 1, pp. 160–255. Springer Verlag, New York, 1981.

Ehrlich, S., and Drenker, S. "Performance results and fluidization parameters of large FBC boilers," presentation at International Conference on Fluidization VII, Australia, 1992.

Engstrom, F. "Development of commercial operation of a circulating fluidized bed combustion system," *in* "Proceedings International Conference on Fluid Bed Combustion," 6th, Vol. 2. U.S. Department of Energy (CONF-800428), Washington, D.C., 1980.

Howard, J. R., ed. "Fluidized Beds: Combustion and Applications." Applied Science Publishers, London, 1983.

Jewell, J. W., and Johnson, W. B. "Method for synthesis of organic compounds," U.S. patent 2,543,974 (March 6), filed July 21, 1947 (1951).
Johnson, W. B. "Fluidized-bed compact boiler and method of operation," U.S. patent 4,240,377 (December 23), filed January 31, 1977 (1980).
Kehoe, P. W. K., and Davidson, J. F. "Continuously slugging beds," *Instn. Chem. Engrs. (London) Symp. Ser.* **33**, 97–116 (1971).
Kiang, K. D., Liu, K. T., Nack, H., and Oxley, J. H. "Heat transfer in fast bed," *in* "Fluidization Technology," Vol. 2 (D. Keairns, ed.), pp. 471–483. Hemisphere Press, Washington, D.C., 1976.
Kraft, W. W., Ullrich, W., O'Connor, W. "The significance of details in fluid catalytic cracking units: Engineering design, instrumentation, and operation," *in* "Fluidization" (D. F. Othmer, ed.), pp. 184–211. Reinhold, New York, 1956.
Krambeck, F. J., Avidan, A. A., Lee, C. K., and Lo, M. N. "Predicting fluid-bed reactor efficiency using adsorbing gas tracers," *AIChE J.* **33**, 1727–1734 (1987).
Lanneau, K. P. "Gas–solids contacting in fluidized beds," *Trans. Inst. Chem. Engrs.* **38**, 125–143 (1960).
Lewis, W. K., and Gilliland, E. R. "Conversion of hydrocarbons with suspended catalyst," U.S. patent 2,498,088 (February 21), filed January 3, 1940 (1950).
Lewis, W. K., Gilliland, E. R., and Bauer, W. D. "Characteristics of fluidized particles," *Ind. Eng. Chem.* **41**, 1104–1117 (1949).
Lobo, W. (1929–1955). Microfilm reels of files accumulated during Walter Lobo's employment at The M. W. Kellogg Company. In 1955, Lobo's position was Director of Kellogg's Chemical Engineering Division. Deposited in Newman Library's Special Collections, Virginia Polytechnic Institute & State University, Blacksburg, the reels are available for consultation by scholars. Reels 54–56 and 59 contain matter on Kellogg's Synthol development; reels 6–7, 42–44, 47, and 50–51 have matter on development of fluid catalytic cracking.
Massimilla, L. "Behavior of catalytic beds of fine particles at high velocities," *AIChE Symp. Ser.* **69**(128), 11–13 (1973).
Nack, H., and Liu, K. T. "Operating method," U.S. patent 4,084,545 (April 18), filed October 21, 1975 (1978).
Pay, T. D. "Foreign Coal Liquefaction Technology Survey and Assessment: Sasol—Commercial Experience." Oak Ridge National Laboratory (ORNL/Sub-79/13837/4, November), Oak Ridge, Tennessee, 1980.
Pell, M., and Johnson, W. B. "Stone & Webster solids circulating fluidized bed combustion," *in* "Proceedings Coal Technology 81," Vol. 5, pp. 16–28 (1981).
Pell, M., and Jordan, S. P. "Effects of fines and velocity on fluid bed reactor performance," *AIChE Symp. Ser.* **84**(262), 68–73 (1988).
Reh, L. "Fluidized bed processing," *Chem. Eng. Progr.* **67**(2), 58–63 (1971).
Reh, L., and Rosenthal, K. "Production of alumina from aluminum hydroxide," U.S. patent 3,565,408 (February 23), filed June 3, 1968 (1971).
Reh, L., Hirsch, M., Collin, P. K., and Flink, S. N. "Process for burning carbonaceous materials," U.S. patent 4,165,717 (August 23), filed September 5, 1975 (1979).
Saxton, A. L., and Worley, A. C. "Modern catalytic cracking unit design technology," *Oil & Gas J.* **68**(20), 82+ (18 May, 1970).
Schmidt, H. W. "Combustion in a circulating fluid bed," *in* "Proceedings of Third International Conference on Fluidized Bed Combustion." Environmental Protection Technology Series, EPA-650/2-73-053, pp. II-1-1-14, 1973.
Shingles, T., Jones, D. H. "The development of Synthol circulating fluidized bed reactors," *ChemSA* **12**, 179–182 (1986).
Shingles, T., and McDonald, A. F. "Commercial experience with Synthol CFB reactors," *in* "Circulating Fluidized Bed Technology II," pp. 43–50. Pergamon Press, Oxford, 1988.
Silverman, R. W., Thompson, A. H., Steynberg, A., Yukawa, Y., and Shingles, T. *in*

"Fluidization V" (K. Ostergaard and A. Sorensen, eds.) Engineering Foundation, New York, 1986.

Spitz, P. W. "Petrochemicals: The Rise of an Industry," p. 305. John Wiley, New York, 1988.

Squires, A. M. "Steam–oxygen gasification of fine sizes of coal in a fluidised bed at elevated pressures, parts I, II, and III," *Trans. Instn. Chem. Engrs. (London)* **39**, 3–27 (see especially pp. 14–15) (1961).

Squires, A. M. "Role of solid mixing in fluidized-bed reaction kinetics," *AIChE Symp. Ser.* **69**(127), 8–10 (1973).

Squires, A. M. "Contributions toward a history of fluidization," *Proc. J. Meet. Chem. Eng., Chem. Ind. Eng. Soc. China and Am. Inst. Chem. Engrs. 1982*, 322–353 (1982). In its account of the development of fluid catalytic cracking, this citation has mistakes that Squires (1986a) corrects.

Squires, A. M. "The story of fluid catalytic cracking: The first 'circulating fluid bed,'" *in* "Circulating Fluidized Bed Technology," Proceedings of First International Conference on Circulating Fluidized Beds, Halifax, Nova Scotia, November 18–20, 1985 (Prabir Basu, ed.), pp. 1–19. Pergamon Press, New York (1986a).

Squires, A. M. "The Tender Ship: Governmental Management of Technological Change," pp. 13–46. Birkhäuser Boston, Boston, 1986b.

Squires, A. M., Kwauk, M., and Avidan, A. A. "Fluid beds: At last, challenging two entrenched practices," *Science* **230**, 1329–1337 (1985).

Steynberg, A. P., Shingles, T., Dry, M. E., Jager, B., and Yukawa, Y. "Sasol commercial scale experience with Synthol [fixed fluid bed] and [circulating fluid bed] Fischer–Tropsch reactors," *in* "Circulating Fluidized Bed Technology III" (P. Basu, M. Horio, and M. Hasatani, eds.), pp. 527–532. Pergamon, New York, 1991.

Strömberg, L. "Experiences of coal combustion in a fast fluidized bed," Studsvik report E4-79/83 (1979a).

Strömberg, L. "Operational modes for fluidized beds," Studsvik report E4-79/85 (1979b).

Thomas, B., and Squires, A. M. "Vibrated-bed microreactors simulating catalytic fluid beds," *in* "Fluidization VI" (J. R. Grace, L. W. Shemilt, and M. A. Bergougnou, eds.), pp. 375–382. Engineering Foundation, New York, 1989.

Wainwright, M. S., and Hoffman, T. W. "The oxidation of *o*-xylene in a transported bed reactor," *Adv. Chem. Ser.* **133**, 669–685 (1974).

Walker, S. W. "Flow characteristics of fluid–fine particle mixtures," Master of Science thesis, Massachusetts Institute of Technology, Cambridge (received M.I.T. Library, July 24, 1940).

Weinstein, H., Graff, R. A., Meller, M., and Shao, M. J. "The influence of the imposed pressure drop across a fast fluidized bed," *in* "Fluidization" (D. Kunii and R. Toei, eds.), pp. 299–306. Engineering Foundation, New York, 1984.

Wilhelm, R. H., and Kwauk, M. "Fluidization of solid particles," *Chem. Eng. Progr.* **44**, 201–218 (for L. Friend's discussion, see p. 218) (1948).

Yerushalmi, J., and Avidan, A. A. "High-velocity fluidization," *in* "Fluidization," 2nd Ed. (J. F. Davidson, R. Clift, and D. Harrison, eds.), pp. 225–291. Academic Press, New York, 1985.

Yerushalmi, J., and Cankurt, N. T. "Further studies of regimes of fluidization," *Powder Technology* **24**, 187–205 (1979).

Yerushalmi, J., and Squires, A. M. "The phenomenon of fast fluidization," *AIChE Symp. Ser.* **73**(161), 44–50 (1977).

Yerushalmi, J., Gluckman, M. J., Graff, R. A., Dobner, S., and Squires, A. M. "Production of gaseous fuels from coal in a fast fluidized bed," *in* "Fluidization Technology" (D. Keairns, ed.), Vol. 2, pp. 437–469. Hemisphere Publishing, Washington, D.C., 1976a.

Yerushalmi, J., Turner, D. H., and Squires, A. M. "The fast fluidized bed," *I&EC Process Design & Development* **15**, 47–53 (1976b).

Yerushalmi, J., Cankurt, N. T., Geldart, D., and Liss, B. "Flow regimes in vertical gas–solid contact systems," *AIChE Symp. Ser.* **74**(176), 1–13 (1978).

2. APPLICATION COLLOCATION

YU ZHIQING

DEPARTMENT OF CHEMICAL ENGINEERING
TSING HUA UNIVERSITY
BEIJING, PEOPLE'S REPUBLIC OF CHINA

I. Characteristics of Fast Fluidized Bed Reactors	39
II. Catalytic Processes	41
A. Catalytic Cracking Processes	41
B. Synthesis Reactions	47
III. Non-catalytic Reactions	50
A. Circulating Fluidized Combustion of Coal	50
B. Circulating Fluidized Regenerator	52
C. Gasification	53
D. Calcination	53
E. Prereduction of Iron Ores	55
IV. Other Applications	55
A. Drying	55
B. Gas Cleaning	56
V. Computer Software for Application Collocation	57
VI. Conclusions	62
References	62

Fast fluidized bed reactors have been widely used in the petrochemical industry, coal combustion and calcining processes. In order to formulate and explore potential uses of FFB, features of present-day applications will be critically examined and then compared in a collocation chart for both commercial-scale processes and laboratory/pilot-scale studies on new processes.

I. Characteristics of Fast Fluidized Bed Reactors

Important features of an FFB reactor and its advantages over alternative fixed bed or conventional fluidized bed reactors are summarized as follows:

- High throughput and essentially plug flow of gases

- Excellent interparticle and interphase heat and mass transfer
- Temperature uniformity and/or controllability
- Independently adjustable solids and gas retention times
- Ease of catalyst circulation through separate zones for continuous regeneration, periodic operation or other applications
- Reaction control along riser axis for periodic operation or temperature profiling.

Large gas throughput, essentially plug flow of gas giving higher selectivity for intermediate products, and excellent interparticle and interphase heat and mass transfer are derived from the fineness of particles and a high mean relative gas–solids velocity. When solids from the riser top are returned directly to the riser bottom via a standpipe, more uniform temperature may be ensured. Chemical reactions may be carried out almost adiabatically, with the high solids flux providing either a heat source or a heat sink to control the riser temperature. Control of circulation of particles is likewise relatively easy. It is thus an ideal reactor for reactions in which a catalyst is prone to getting deactivated or the reaction is strongly endothermic or exothermic, because the continuous recirculation of solids offers the possibility of regeneration in separate vessels. These vessels may be used to carry out different functions as well: reaction, absorption, heat transfer or stripping of the solids. FFB also provides a practical approach to implementing periodic operation. The high heat transfer coefficient between the suspension and the heat exchange surface may also lead to a uniform temperature profile throughout the reactor.

Probably one of the least stressed advantages of fast fluidized bed reactors is the ability to independently adjust the solids and gas retention times. This feature can help to increase the turndown ratio of the reactor or to manage those processes with varying feed quality or product requirements.

In 1942 the so called "upflow" fluid catalytic cracking unit was put into commercial operation. Few realized it was the first application of FFB and was destined to have great influence on the world's petroleum refining processes. Today, many applications of FFB are put into use, such as riser catalytic cracking, coal combustion, aluminium hydroxide calcining, etc. A collocation chart is now proposed to outline the range of these applications, listing processes that generally have the following characteristics:

1. Reactions with high or very high reaction rate
2. Reactions with reversible catalyst deactivation calling for continuous catalyst regeneration
3. Reactions where the intermediate products are desired, e.g., (1) partial oxidation of hydrocarbons, (2) hydrocarbon cracking
4. Highly exothermic or endothermic reactions

5. Processes with varying feed and product requirements
6. Processes where heat removal or supply can be achieved by solid circulation
7. Reactions with staged reactants inlets
8. Reactions that require independent solid and gas retention-time adjustment
9. Reactions that require temperature uniformity
10. Reactions that could benefit from periodic operation, e.g., oxidative reactions in stoichiometric redox mode

These features will be further discussed with some examples of catalytic and non-catalytic reactions as well as other FFB applications.

II. Catalytic Processes

A. CATALYTIC CRACKING PROCESSES

Catalytic cracking is the process of upgrading gas oil or even residual oil (heavy oil) to produce gasoline, distillates, light olefines, etc. Commercialized processes include: fluid catalytic cracking (FCC), residual oil catalytic cracking (RFCC), and catalytic pyrolysis, etc.

These processes are dominated by two features:

1. The reactions are highly endothermic; in order to supply the heat needed for the reactions, the catalyst also acts as a heat carrier, and the reactions are often carried out adiabatically;
2. The reactions are always accompanied by coke deposition on the catalyst, which causes catalyst deactivation, and the catalyst must be regenerated continuously.

Generally, the cracking process has two sections: One is for the heat sink, where cracking reactions and coke deposition take place; the other is to burn off the coke deposited on the catalyst, which acts as a heat source and catalyst regenerator. The circulating solids are the means for catalysis and heat transport. The only way to carry out the process efficiently is by use of a solids circulation system involving one or more fluidized beds. Usually, the cracking reaction takes place in an adiabatic riser reactor.

1. Fluid Catalytic Cracking (FCC)

Fluid catalytic cracking has been one of the key processes in petroleum refining for the last 50 years. The first commercial fluid catalytic cracking process employed an FFB reactor. Because of some mechanical problem, it was shut down afterwards. Instead, the conventional fluidized bed reactor took its place. In the 1960s, the newly developed molecular sieve totally changed the property of the FCC catalyst—the activity of the new zeolite catalyst was raised by about two orders in magnitude, with selectivity and gasoline yield also greatly improved. These features allowed the cracking reaction to complete itself within several seconds but also caused rapid catalyst deactivation. By taking advantage of this remarkable catalyst, the FFB riser has become an ideal reactor for the FCC process, resulting in a higher gasoline yield, higher C_3–C_5 olefine content, and less carbon formation.

Today more than 600 commercial FFB reactors are in operation worldwide, over 350 of which are riser FCC plants. These riser FCC units produce over one-half of the world commercial gasoline. In China, over 90% of the FCC plants are riser reactors. The total number is more than 83 units, and the processing capacity ranks second, following that of the United States.

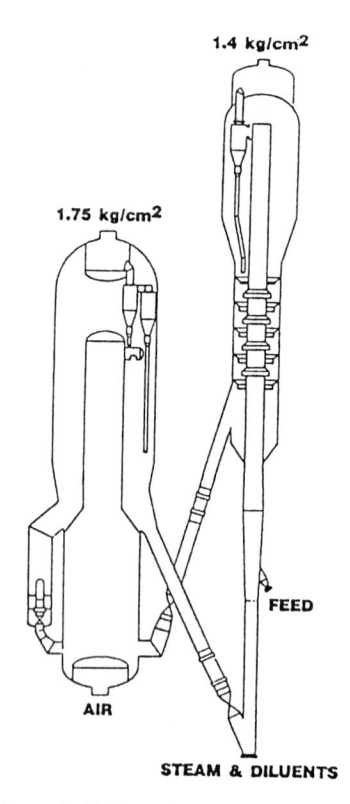

FIG. 1. Riser FCC unit. From the UOP meeting report, SINOPEC, Beijing, China, 1990.

According to the statistics of SINOPEC in 1991, about 70% of gasoline and 31% of distillates on the Chinese commercial market were provided by the FCC process.

In the FCC process, the heavy hydrocarbons come in contact with the hot cracking catalyst and crack into lower-molecular-weight compounds. The key to a successful cracking process is a method for supplying the large amount of heat needed for the endothermic cracking and an effective way of rapidly regenerating tens of tons of deactivated catalyst per minute. The riser FCC does this efficiently and simply by carrying out the catalyst regeneration step in the downer, which in turn supplies the heat for the reaction in the riser. See Fig. 1.

In the riser cracker, the feed oil is sprayed into the fast upflowing stream of regenerated catalyst. The cracking reaction occurs entirely in this nearly plug flow riser, and the selectivity of desired hydrocarbon fractions is thereby markedly improved. Catalyst circulates smoothly between the regenerator and the riser reactor. The regenerator can be an ordinary fluidized bed or a combination of risers with ordinary fluidized beds.

The advantages claimed for the riser-cracker are as follows:

- High conversion in very short contact time.
- Closeness to plug flow for both solids and gas, improved reaction selectivity and least overcracking, and resulting higher yields of gasoline.
- High activity of the zeolite catalyst can be effectively utilized.
- Formation of liquid products is enhanced and formation of coke is reduced.

2. Heavy Oil Cracking

Upgrading heavy oil (atmospheric residue) can make better use of limited petroleum resources. Because the feedstock has a high Conradson carbon index (4–10%) and contains much sulfur and heavy metals (S: 0.2–3.5%, V and Ni: 6–170 ppm), its upgrading in FFB encounters the following problems:

- Coke deposition on the catalyst is increased. There needs to be efficient ways to regenerate the catalyst and to remove the excess heat produced in the regenerator.
- The catalyst is rapidly poisoned by vanadium and nickel, which deactivate the catalyst permanently and lower reaction selectivity.
- Additional flue gas cleaning is required to remove the SO_2 formed.
- An effective way is required to recover thermal energy released in the burning of coke.

Because of the increase in coke deposition, the regenerator is different

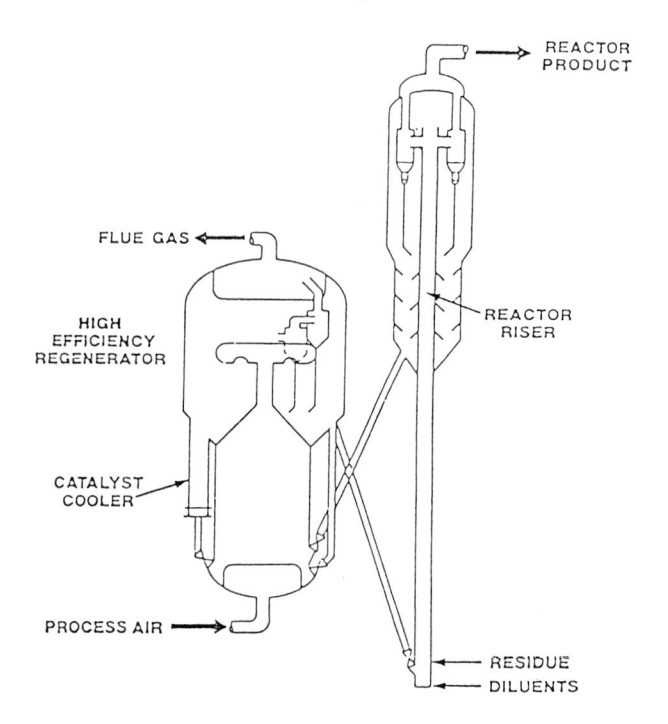

Fig. 2. Heavy oil cracker. From the Ph. D. dissertation of Muo Weijian, University of Petroleum, Beijing, China, 1988.

from the ordinary FCC unit in two respects. Firstly, the load for the regenerator is greatly increased, so that ordinarily one can not fulfil the task of effective catalyst regeneration, and an FFB regenerator, called a high-efficiency regenerator, is often employed to ease the catalyst regeneration. Secondly, the regenerator must have an effective way to recover the thermal energy, such as a catalyst cooler or a flue gas turbine. Because of the giant scale and the worldwide importance of heavy oil cracking, many efforts have been made to improve the regenerator, and various kinds of regenerators have been developed, among which the FFB regenerator has achieved great success. See Fig. 2.

3. Fluid Catalytic Pyrolysis

In contact with catalyst at temperatures higher than those for FCC, heavy petroleum fractions crack to produce propylene and butene, which are useful starting materials for organic synthesis and polymerization. The catalytic cracking has many advantages over the thermal cracking of naphtha to

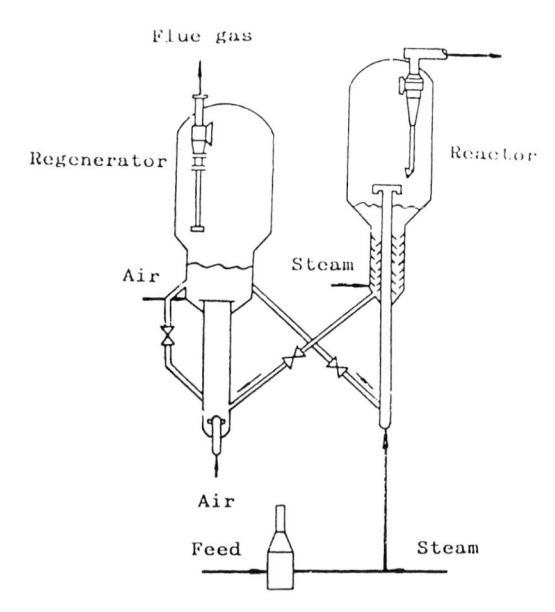

FIG. 3. Fluid catalytic pyrolysis. From the Research Institute of Petroleum Processing, Beijing, China, 1991.

produce olefines, among which high reaction selectivity and less coke formation are very attractive. Recently a new riser catalytic pyrolysis process to produce olefins has been developed by SINOPEC in China. See Fig. 3. The process has high selectivity to propylene and butene by using a special fine and high-activity zeolite as catalyst. A commercial unit producing 15,000 tons per year of propylene was built and operated in Jinan Refinery, China, in 1991. Other such units have since been constructed in China.

4. FCC Using Quick Contact Reactor

In the just-mentioned catalytic cracking processes, the reaction takes place in the upflow riser. The relatively long retention time and large gas and solids backmixing in the riser cannot meet the demand for high reaction selectivity and least overcracking. One latest development is the new type of cracking reactor called the "quick contact" (QC) system to overcome the preceding shortcomings in a riser. See Fig. 4. This system can allow feed and catalyst to contact, react, separate and quench in altogether less than half a second. The downflow reactor substantially eliminates the problems of backmixing and non-uniform distribution of catalyst, which are significant in a conventional upflow riser. The downward direction of flow of the catalyst

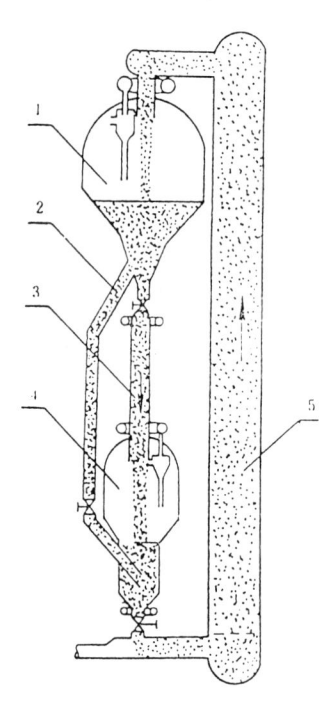

FIG. 4. Quick contact reactor, concurrent downflow cracker. Modified from P. K. Niccum and D. P. Bunn, U. S. Pat. 4,514,284 (1985). 1, catalyst storage; 2, recirculating pipe; 3, quick contact reactor; 4, steam stripper; 5, riser regenerator.

and the feed facilitates uniform distribution of the catalyst throughout the feed in a relatively short time, thereby decreasing coke formation and enabling rapid separation of the catalyst from the converted feedstock at the bottom portion of the riser. The net result of providing a downflow reactor in the FCC plant is decreased amount of coke deposition, increased gasoline selectivity, the production of higher octane value gasoline at substantially the same gasoline efficiency, and increased catalyst efficiency.

In order to promote the commercial adoption of the downflow reactor in FCC processes, much effort has been given to fundamental research at Tsing Hua University, China. Such research includes hydrodynamics, gas and solids mixing, modelling, etc. A novel ballistic rapid gas–solid separator with an efficiency of more than 97% has also been developed. In cooperation with the Research Institute of Petroleum Processes (RIPP), a pilot test was carried out, though no commercial plants are yet in operation. Understanding and modelling highly non-isothermal mixing systems in which residence time is very short are still a challenge to Chinese chemical engineers. They are confident that innovation of FCC technology and high economic profit would result if the conventional upflow riser were replaced by the downflow reactor. This is particularly true in heavy oil processing and light oil cracking to produce olefines.

B. SYNTHESIS REACTIONS

The main reasons for choosing the fast fluidized bed rather than the fixed bed or the traditional bubbling fluidized bed for solid-catalyzed gas-phase synthesis reactions are the demand for high selectivity, the need to reoxidize catalyst in a separate unit, and the necessity of strict temperature control of the reaction zone. These requirements are based on the following possibilities: The reaction may be explosive outside a narrow temperature range; the yield of desired products may be sensitive to the temperature of operation; hot spots in the catalyst bed may lead to rapid deterioration and deactivation of the otherwise stable catalyst that normally does not require regeneration. These reactions are generally highly exothermic, and thus make temperature control difficult.

1. Synthol Reactor

The synthesis of hydrocarbons from H_2 and CO is strongly exothermic and proceeds in a narrow temperature range, around 340°C. Kellogg and Sasol jointly developed a synthetic gasoline process based on a fast fluidized bed system in 1955. This process was very successful and has been expanded on a giant scale to meet most of the liquid fuel needs of South Africa. In the Synthol FFB reactor as shown in Fig. 5, the reactant gases, H_2 and CO, carry suspended catalyst in a vertical riser upward at 3–12 m/s, the average voidage being between 0.85 and 0.95. The large amount of solids circulation removes the reaction heat, and therefore controls the temperature and the reaction time. The conversion of synthesis gas in the reactor can reach 85% or more.

2. Laboratory/Pilot Scale Studies

With the improvement of catalyst, the activity of many synthesis reactions has been increased. This leads to a change in reactors from traditional fluidized bed or packed bed to FFB for better reaction selectivity and high throughput. A comparison of selectivity for various reactors is shown in the table.

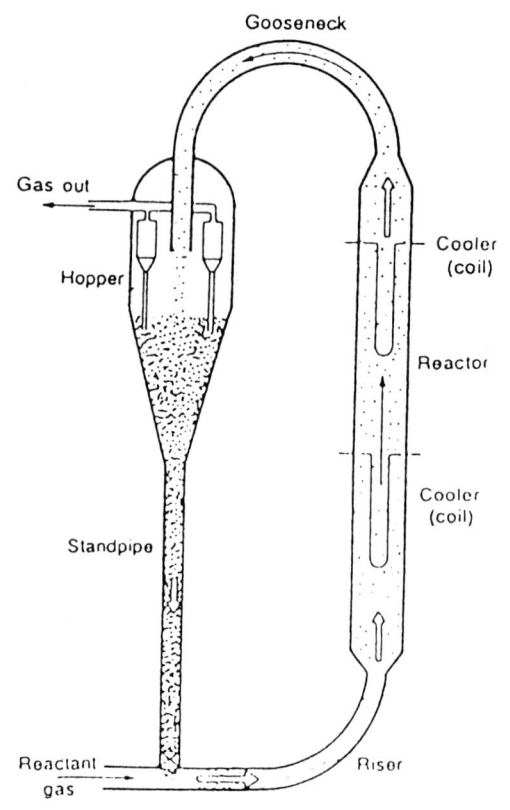

FIG. 5. Synthol circulating solids reactor (modified from Shingles and Jones, 1986).

Processes	Fixed Bed	Conventional Fluidized Bed	Fast Fluidized Bed
o-Oxylene oxidation to phthalic anhydride	−71	—	75
n-Butane oxidation to maleic anhydride	—	66–75	85–94
Propylene ammoxidation to acrylonitrile	—	−76	84
Oxydehydrogenation of butene to butadiene		87–90	>90
Ethylene epoxidation[a]	77		−84
Deacyloxylation of vicinal hydroxyesters to oxirane compounds	83	55	86

[a]The limiting selectivity based on the stoichiometry of this reaction is 85.7%.

a. Oxydehydrogenation of Butene to Butadiene. The Institute of Coal Chemistry, Academic Sinica, has recently developed a process for the oxydehydrogenation of butene to butadiene in a laboratory fast fluidized bed, with a higher yield of butadiene than the conventional bubbling/turbulent fluidized bed. This process promises to be a greatly improved means for utilizing the C_4 fraction produced in ethylene manufacture. In the reaction, an intermediate and a high selectivity are desired. Therefore, the fast fluidized bed has been considered the most ideal reactor. Experimental studies have given the following optimal operation conditions:

Temperature: 355 to 365°C
Superficial gas velocity: 3.5 to 4.5 m/s
$C_4:O_2:H_2O$ (mol): 1:(0.85–0.97):8

Results obtained in the fast fluidized bed reactor show that both the conversion of butene and the yield of butadiene are increased by about 15 to 25% as compared to the turbulent fluidized bed reactor. It is reported that this process is being commercialized in Liaoning Province, in northeast China.

b. Vapor Phase Catalytic Oxidation of Butane to Maleic Anhydride. Contractor and co-workers (1988, 1987a, b, c) reported on a new catalytic process to produce maleic anhydride by selective oxidation of *n*-butane in an FFB. In this process, *n*-butane is oxidized in a riser reactor using an oxidized catalyst, and the reduced catalyst is reoxidized in a separate fluidized bed regenerator. All process requirements have been well met by using a fast fluidized bed reactor with a plurality of air injections.

Good selectivity and conversion could be achieved at temperatures lower than those of conventional known processes. Du Pont recently was reported (Alperowicz *et al.*, 1990) to have constructed a multimillion dollar pilot plant to demonstrate the technology and announced that they planned to commercialize the process by 1994 at a site in Europe.

c. Propylene Ammoxidation to Acrylonitrile. Acrylonitrile, an important silk-like fiber, can be produced by the strongly exothermic catalytic oxidation of propylene and ammonia at a temperature between 375°C and 525°C. Presently, nearly 90% of the acrylonitrile in the world is being produced by the conventional fluidized bed reactor called the Sohio process.

Beuther *et al.* (1978) claimed that excellent yields of acrylonitrile could be obtained by using an FFB reactor instead. The reactant gases can be added into the reactor in stages, while the lower part of the reactor serves as a regeneration zone. For example, an oxygen-containing gas, such as air, can be introduced stepwise to improve selectivity and to maintain a low partial pressure of oxygen in the reactor.

Typical results obtained in a laboratory reactor, 4.8 mm i.d. and 18.3 m in height, show that more than 84% yield of acrylonitrile is obtained. In comparison with that of the conventional fluidized bed reactor, the yield is improved by about 9%.

It appears that in most cases the FFB is the reactor of choice whenever exothermic heat is great, or when selectivity is of importance, or when high capacity is the goal. Other processes, such as phthalic anhydride by oxidizing *o*-xylene with air, and unsaturated aldehydes from an unsaturated olefine, are also reported to use FFB reactors.

III. Non-catalytic Reactions

A. CIRCULATING FLUIDIZED COMBUSTION OF COAL (For additional details, see Chapter 8.)

Fast or circulating fluidized bed combustion systems have become popular since the late 1970s and now represent the current level of activity in the area. It is clearly recognized that this technology has a stable future in the boiler market. In competition with conventional bubbling beds, the FAB/CFB boiler demonstrates the following advantages:

- Fuel flexibility:
  coal, lignite, brown coal, peat, coal washing rejects, wood waste, bark, petroleum coke, oil and gas, industrial and sewage sludges
- High combustion efficiency:
  usually 98–99% for various fuels, excess air lower than 20%.
- Excellent contribution to pollution control:
  efficient sulfur removal (around 90% capture of sulfur at Ca/S ratio of 1.5–2).
  low NO_x emissions around 100–300 vppm
- Good turndown and load following (up to 5:1)
- Simpler fueld handling and feed system:
  CFB boilers receive solid fuels in fairly coarse sizes, and only one or a few feed points are needed.
- High availability (more than 90%)

A typical CFB boiler, shown in Fig. 6, operates with a two-stage air supply. Primary air supplied through a distributor is substoichiometric, and combustion is completed with secondary air in the upper portions of the boiler. Staging of combustion brings several benefits. Combustion is

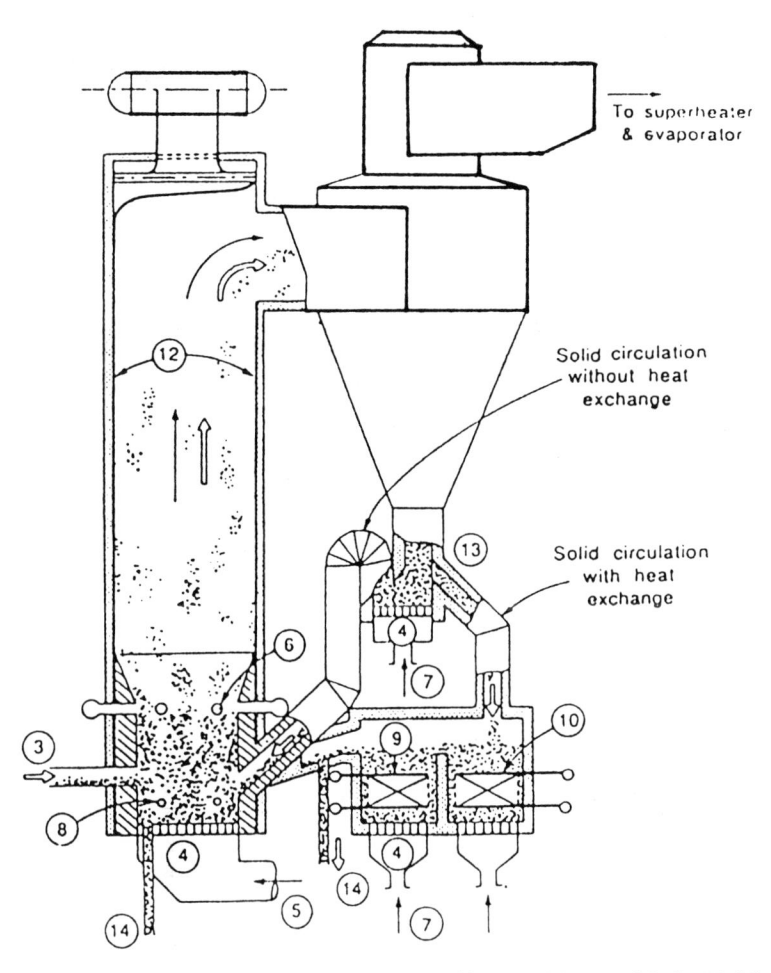

FIG. 6. Circulating fluidized bed boiler. From Kunii, D., and Levenspiel, O. "Fluidization Engineering" 2nd Ed., Butterworth-Heinemann Series in Chemical Engineering, 1991.

essentially complete with small excess of air, and NO_x emissions is relatively low. Splitting the air makes the bottom of the bed denser and highly turbulent. This serves well for the introduction and dispersion of the coal feed and provides higher solid residence time and greater stability of operation. In order to improve the performance of FFB boilers, great efforts have been made to develop the pressurised FFB boilers.

In coal combustion in China, both fundamental research and industrial experience have resulted in improved understanding of CFB combustion. It is reported that more than 100 commercial CFB coal combustors have been in operation in China, though the capacities of these units are relatively small, and further work, especially the control of particle size and solids circulation rate, is needed.

B. CIRCULATING FLUIDIZED REGENERATOR (For additional details, see Chapter 9.)

The circulating fluidized regenerator in the past decade has become a troubleshooter for the regeneration of residue fluid catalytic cracking (see Fig. 7). Operating the regenerator in the fast fluidized bed regime improves the efficiency of coke burning rate to more than three times higher that of the ordinary fluidized regenerator. More than 30 units of this type of regenerator are in operation in China. The circulating solids serve as a heat carrier to increase the bottom temperature of the regenerator as well as solids concentration. The high gas velocity greatly enhances gas solids contact, and hence improves coke combustion. Consequently, this regenerator is called a high-efficiency regenerator.

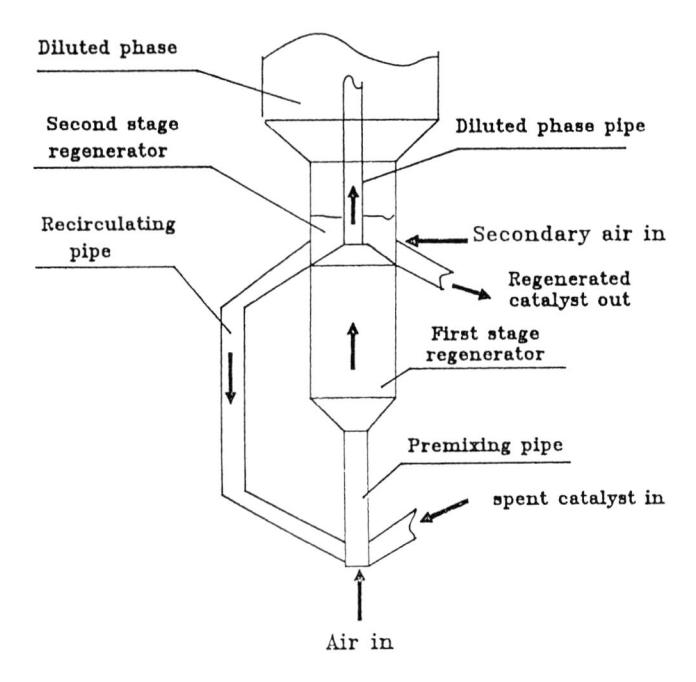

FIG. 7. High-efficiency FCC regenerator (Wei et al., 1993a).

A pilot demonstration FCC riser regenerator (inside diameter 300 mm) with a capacity of 6,000 tons per year was developed and put into operation in 1992 through the cooperation of the Beijing Design Institute of SINOPEC, Tsing Hua University and Changling Refinery. The regenerator employs a gas velocity higher than that of the so-called high-efficiency regenerator operating between the turbulent and fast fluidized regime, and staging of air inlets provides high coke burning efficiency. Collateral fundamental research, especially simulation, has been carried out at Tsing Hua University since 1988.

C. Gasification

Based on experience with endothermic and exothermic reactions in FAB/CFB calciners and combustors, the fast fluidized bed has been used in the gasification of biomass (wood chips, bark, etc.) and coals with low and high ash contents. The major advantages of FAB gasification are:

- High specific gas production because of excellent heat and mass transfer
- Flexible partial load behavior
- Low pollution
- Minimal by-products (tar, oil, etc.) because of rapid heating up of the fresh fuel
- Uniform temperature in the whole FAB-gasifier through internal and external recirculation of solids
- No limitation on particle size distribution, allowing for fines and low grade fuels

Figure 8 shows a flow scheme of an FAB gasification plant for biomass. The high moisture content of these materials, like some lignites, calls for a dryer in front of the gasifier. The solid fuel is charged into the gasifier via a screw feeder. Gasification agents (air, air/steam, oxygen/steam) are introduced from below to fluidize the solids. Air and steam are preheated by heat exchange with the hot product gas. A gasifier of 3.6 m inside diameter has a design capacity of 4,000 tons of coal per day, almost 10 times that of other types of gasifiers (see Fig. 8).

D. Calcination

It is well known that the first calcination unit was developed by Verinigte Aluminium Werke AG, and Lurgi GMBH, Germany, in the 1960s for alumina (Al_2O_3). After its success in the aluminum industry, fast fluidized bed calcination also found application in other calcining operations, such as

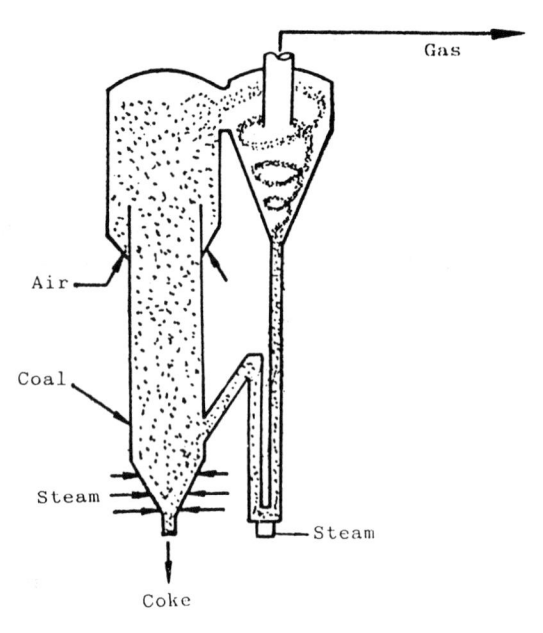

FIG. 8. Fast fluidized bed gasifier. From A. M. Squires, U. S. Pat. 3,840,353 (1973).

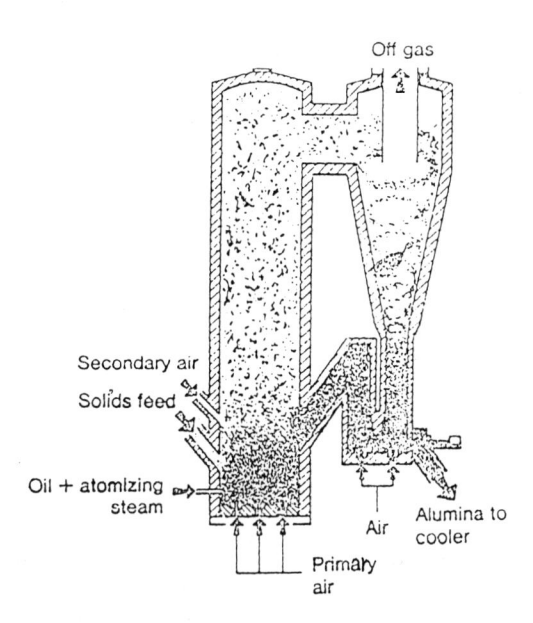

FIG. 9. Lurgi's circulating fluid bed calciner for aluminium hydroxide (Yerushalmi and Avidan, 1985).

cement raw meal, phosphate rock, clay, limestone, etc., where higher conversion yield and combustion efficiency were needed.

Figure 9 shows the flow scheme of a fast fluidized bed alumina calciner. The FAB calciner itself is operated stagewise by burning oil or gas for supplying heat to the endothermic calcining reaction. In the case of alumina, calcination is carried out in the FAB furnace at a temperature of about 1,000 to 1,100°C by direct combustion of an ash-free fuel (oil or gas) with the fluidizing air. Gas velocity near the furnace exit is about 3–4 m/s, and the average solids concentration in the furnace ranges from 100 to 200 kg/m^3. The hot gas from the furnace is used to preheat the fresh solids by heat exchange, and the precipitated solids are recirculated via a siphon-seal to the lower section of the FAB reactor. Heat is thus efficiently recovered and the lowest possible fuel consumption is achieved. The FAB calciner is also notable for the high-quality product and the low nitrogen oxide emission. Today, about 27 FAB caliners with capacities up to 1,600 tons of alumina per day and thermal loads up to 55 MW are in operation all over the world. Figure 10 shows the pyritic calciner reported by Wang *et al.* It combined a slow reaction and a fast one together to produce concentrated SO_2 up to 94%. FeS_2 reacts slowly with Fe_2O_3 in the downcomer to form Fe_3O_4 and SO_2. The reduced Fe_3O_4 is then quickly oxidized with air in the riser.

E. PREREDUCTION OF IRON ORES

Examples of metallurgical processes using the circulating fluidized bed are provided by the prereduction of iron ores and lateritic nickel ores.

In the circulating fluidized bed, prereduction of finegrained iron ore concentrate together with powdered coal proceeds without sticking at temperatures above 950°C. Sticking is prevented by an excess of carbon (char) in the bed material and by the turbulent mixing behavior of the CFB. As raw material, fine-grained concentrate with an iron content preferably above 65% and coal of different grades (anthracite, bituminous coals or even lignites) can be used. The coal is crushed and dried before being injected into the furnace. The final reduction then takes place in an electric furnace, where liquid iron is produced. The process incorporates a power plant as well to generate electricity by using the fuel gases of the two reduction stages.

IV. Other Applications

A. DRYING

The fast fluidized bed dryer is used extensively in a wide variety of industries. The fast fluidized bed dryer works well for cohesive and sticky

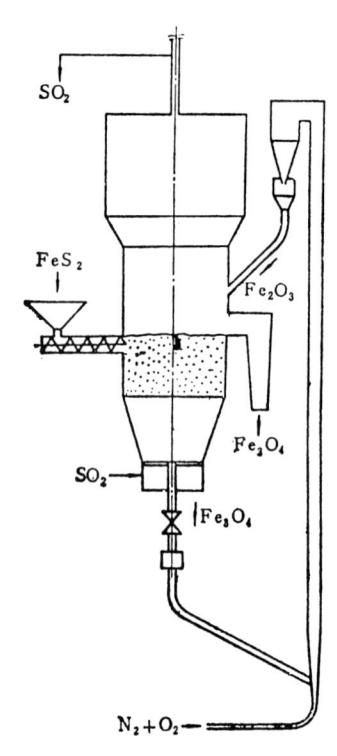

FIG. 10. CFB pyritic calcination reactor (Wang et al., 1989).

solids that aggregate or stick to metal surfaces. Granular minerals or salts that are only surface-wetted can be effectively dried in fast fluidized dryers, called flash driers. Recently, flash driers have been employed in the initial stage drying of PVC and MBS to remove the large amount of water in the material in a very short time. It is reported that this process can reduce energy consumption significantly.

B. Gas Cleaning

In recent years the fast fluidized bed has also found applications in other processes, such as low-temperature adsorption of pollutants (e.g, HCl, HF, SO_2) from different process waste gas streams (Reh, 1986), noise reduction (Yang, 1992), etc. for example, in the aluminum smelter of VAW/Lurgi today, 3×10^6 m^3/hr of fluorine-laden potline offgas are cleaned to less than 1 mg F/m^3 by adsorption on alumina at a temperature of 70°C.

All the processes have proven the fast fluidized bed to be an efficient tool in environmental control.

V. Computer Software for Application Collocation

For all the examples so far enumerated for FFB applications, there are more that have not been quoted in the above brief review, and there will be even more as time goes on. It would be highly desirable to identify quickly literature sources of specific applications, such as internals for improving flow, common experiences derived both in petroleum refining and in calcination on particle-to-gas heat transfer, or pressure fluctuation measurements in commercial units of any of the known industries now using FFB. Also, engineers engaged in developing a new process often wish to verify their conceptualizations against prior experiences on specific aspects of FFB application.

A system, shown in Table I, has been devised consisting of nine categories, each with a number of keywords for a computerized search for multiple identical features from a data bank of international FFB literature. The so-called categories are represented by digit numbers:

1. scale
2. industry
3. flow
4. heat transfer
5. mass transfer
6. mathematical modeling
7. equipment/devices
8. measurement/instrument
9. author

Each category is assigned a number of keywords denoting certain specific contents (all different for the nine different categories). For instance, for category 2, *industry*, the following keywords have been assigned:

combustion
gasification
petroleum refining
petrochemicals
gas cleaning
absorption
quick contacting

TABLE I
REFERENCE DESIGNATION BY COLLOCATORS

Collocation	Digit Number							
	1	2	3	4	5	6	7	8
	Scale	Industry	Flow	Heat trans	Mass trans	Mathematic modeling	Equipment devices	Measure instrum
: any \ none								
a	laboratory	combustion	dynamics	fluid–part	fluid–part	empirical	reactor	press drop
b	bench	gasificatn	structure	syst–wall	syst–wall	analytic	internals	press fluc
c	pilot	petrol ref	pres bal			numerical	solid feed	fluid vel
d	commercial	petrochem	void distr					part vel
e		dehydratn						voidage
f		calcinatn						visualizn
g		roasting						
h		ion exchan						
i		cat reactn						
j								
k								
l								
m								
n								
o								
p								
q								
r								
s								
t								
u								
v								
w								
x								
y								
z								

Example of reference designation:
bd\ \\a b\ = bench-scale, petrochemical, empirical modeling, internals
  ↑ ↑
  digit #4 = heat transfer
  collocation \ = none

digit #2 = industry
collocation d = petrochemical

Exit   Start searching

non-catalytic reaction
catalytic reaction
calcination
roasting

This designation system facilitates the choice of a keyword or a string of keywords for a computerized search of any required literature reference. The following example will illustrate the operation.

Suppose one needs information on empirical modeling on petroleum refining based on pilot-scale studies. First you click on *Scale* on the menu bar with your mouse. Then, on the pulldown *Scale* menu you choose *Pilot* and click again. A dialog box appears with *Pilot*. Click *O.K.* Then click on the menu bar *Industry*, and on the pulldown menu click *Petroleum Refining*. The dialog box appears for the second time with *Pilot, Petroleum Refining*. Click *O.K.* Pull down the *Mathematical Modeling* menu, and click *Empirical* on the menu list. Then click *O.K.* on the dialog box which now appears for the third time. Then click the *Start Searching* button on the lower right-hand corner of the screen to start the computer searching process. The return of this particular search soon appears on the screen: *reference number: 0*, showing that the literature data bank contains no investigation on empirical modeling on petroleum refining based on pilot-scale studies.

The scope of literature search can be narrowed by repeated clicking in of keywords, including the author's last name. Indexing of the author's name also permits multiple authors. For instance, one may write the author's last name *Bai* in the blank space of the *Author* pulldown menu to retrieve all publications by Bai on FFB. You may also click *Bai*, enter, and then click *Yu* to retrieve all publications on FFB that were coauthored by Bai and Yu.

The Windows-based computer program called RFFA (Retrieval Fast Fluidization Application) and described above, requires the support of Microsoft Windows 3.1 or higher on any IBM-compatible PC. RFFA is menu-driven and possesses a friendly interface with the user (see Table II). Users can see directly the subtitles they are looking for from pulldown menus on the screen, and can gain access to the desired group of literature references by just clicking on the subtitles. At the moment, the FFB application data bank contains, as an initial phase, 406 literature references. The complete contents of keywords in the pulldown menus and the two operating buttons which appear on the screen are given below:

TABLE II
FFB Application Collocation Chart

No.	Process & Reaction	T °C	Ug m/s	Gs kg/m² s	t_c s	Conversion & Yield (%)	1	2	3	4	5	6	7	8	9	Scale	Year of First Unit
A. Catalytic																	
A-1	fluid catalytic cracking	450–550	4–30	400–1,000	1–5	70–85 / 65–80	*	*	*	*	*	*	*			Com.	1942
A-2	heavy oil cracking	480–550	4–30	400–1,000	1–5	70–85 / 65–8	*	*	*	*	*	*	*			Com.	1975
A-3	fluid catalytic pyrolysis	540–610	4–30	400–1,000	1–5	70–90 / 18–30	*	*	*	*	*	*	*			Com.	1990
A-4	Fischer–Tropsch synthesis	330–500	3–12	—		85 / 85	*	*	*	*	*	*				Com.	1955
A-5	oxydehydrogenation of butene	355–365	3–4.5			~90 / 76–81		*	*	*	*	*		*		Pil.	1983
A-6	maleic anhydride	360–420			1–7	−47.7 / 83–90.1	*	*	*	*		*	*		*	Pil.	1987
A-7	phthalic anhydride	300–500	12,	—	2		*		*	*		*	*		*	Com.	1981
A-8	oxidative coupling methane						*										
A-9	propylene ammoxidation	450–490	2–4.5	15.6		98.5 / 84.5	*	*	*	*	*		*		*	Lab.	1978
A-10	ethylene epoxidation	200	0.7–3				*		*	*	*		*		*	Lab.	1986
A-11	refining raw gas from gasification of carbonaceous																
A-12	propylene oxide from propylene glycol monoacetate	350–450															
B. Non-Catalytic																	
B-1	calcination of Al(OH)₃	1,000–1,100	3–4	100–200						*	*			*	*	Com.	
B-2	calcination of phosphate	650–850								*	*			*	*		
B-3	calcination of clay	650								*	*			*	*		

B-4	pre-calcination of cement raw meal	850							
B-5	synthesis of AlF_3	530							
B-6	synthesis of SiC_4	400							
B-7	sulfate decomposition	950–1,050			*	*	*		
B-8	chloride decomposition	850			*	*	*		
B-9	carbonate decomposition	850			*	*	*		
B-10	combustion of shale	700		*	*	*	*		
B-11	coal combustion	850–1,150			*	*	*	Com.	
B-12	coal gasification	850–1,150	5–6	10–40	98–99	*	*	*	Com. 1970
B-13	biomass gasification	800–900				*	*	*	Com.
B-14	wood gasification	800–900				*	*	*	Com.
B-15	pyrohydrolysis of spent potlining	1,200				*	*	*	
B-16	regeneration of FCC catalyst	600–800	0.7–10	10–1,000	90	*	*		Com. 1976
B-17	$NaHCO_3$ decomposition					*	*	*	
B-18	pyrolysis of coals	650–700				*	*	*	Pil.
B-19	pyrolysis organic materials					*	*	*	
B-20	reduction of fine ores	1,050–1,150				*			

C. Others

C-1	coal gas	400–950							
C-2	waste gases from aluminium electrolysis (HF)	70							
C-3	incinerating off gas (HCl, HF, SO_2)	150–250							
C-4	pulverised-fuel boiler off gas SO_2	100				*	*	*	Com. 1986
C-5	drying of PVC and MBS	50–120	10–20	20–200		*	*	*	Com.

VI. Conclusions

The fast fluidized bed reactor can offer several considerable advantages over alternative reactors for many catalytic and non-catalytic reactions, especially for very fast exothermic/endothermic reactions. With the mushrooming of high activity catalysts and the ever increasing pressure for energy conservation, environmental controls, etc., FFB can play more and more important roles in these areas. More potential commercial applications of FFB in the near future include hydrocarbon oxidations, ammoxidation, gasoline and olefines production by concurrent downflow FFB and basic operation for organic chemical productions.

References

Anders, R., Beisswenger, H., and Plass, L. "Clean and Low Cost Energy From Atmospheric and Pressurized Lurgi CFB System," *in* "Circulating Fluidized Bed Technology III" (Basu, P., Horio, M., and Hasatani, M., eds.), pp. 431–438. Pergamon Press, Oxford (1991).

Andersson, B. A., and Leckner, B. "Local Lateral Distribution of Heat Transfer on Tube Surface of Membrane Walls in CFB Boilers," *in* "Circulating Fluidized Bed Technology VI" (Amos, A., and Avidan, ed.), pp. 368–373. Somerset, Pennsylvania (1993).

Anthony, E. J., Couturier, M. F., and Briggs, D. W. Gas sampling at the Point Tupper AFBC facility, Report 1986 CANMET-86-70(TR), 66 pp., Energy Res. Abstr. 1988, 13(11), Abstr. No. 25743, CANMET, Ottawa, Ont., Can. (1986).

Arena, U., Cammarota, A., Marzochella, A., and Massimilla, L. "Solids flow structure in a two-dimensional riser of a circulating fluidized bed," *J. of Chem. Eng. of Japan*, Vol. 22, pp. 236–241 (1989).

Arena, U., Marcozhella, A., Bruzzi, V., and Massimilla, L. "Mixing Between A Gas–Solids Suspension Flowing in a Riser and a Lateral Gas Stream," *in* "Circulating Fluidized Bed Technology IV" (Amos A. Avidan, ed.), pp. 660–664. Somerset, Pennsylvania (1993).

Arpalahti, Olli, "Recirculating fluidized-bed reactor for the removal of particulate material from flue gases," Finland; FI 77165 B, 881031, FI 86244 (860120), 11 pp. (1988).

Asai, M., Aoki, K., Shimoda, H., Makino, K., Watanabe, S., and Omata, K. "Optimization of Circulating Fluidized Bed Combustion," *in* "Circulating Fluidized Bed Technology III" (Basu, p., Horio, M., and Hasatani, M., eds.), pp. 379–384. Pergamon Press, Oxford (1991).

Aubert, E., Barreteau, D., Gauthier, T., and Pontier, R. "Pressure Profiles and Velocities in a Co-current Downflow Fluidized Reactor," *in* "Circulating Fluidized Bed Technology IV" (Amos A. Avidan, ed.), pp. 490–495. Somerset, Pennsulvania (1993).

Avidan, A. A., Edwards, M., and Owen, H. "Innovative Improvements Highlight FCC's Past and Future," *Oil & Gas Journal*, Jan. 8 (1990).

Avidan, A. A., Edwards, M., and Owen, H. *Oil and Gas J.*, Vol. 88, pp. 33–58 (1990); Avidan, A. A., Gould, R. M., and Kam, A. Y., "Operation of a Circulating Fluidized Bed Technology" (Basu, P., ed.), p. 287. Pergamon Press, Canada (1986).

Avidan, A. A., and Shinnar, R. "Development of Catalytic Technology. A Lesson in Chemical Reactor Design," *Ind. Eng. Chem. Res.*, vol. 29, No. 6 (1990).

APPLICATION COLLOCATION

Baerns, Mleczko, L., Tjitjopoulos, G. J., and Vasalos, I. A. "Comparative Study of the Oxidative Coupling of Methane in Circulating and Bubbling Fluidized Bed Reactors," *in* "Circulating Fluidized Bed Technology IV" (Amos A. Avidan, ed.), pp. 509–524. Somerset, Pennsylvania (1993).

Bai, D., Jin, Y., and Yu, Z. "The two-channel model for fast fluidization," *in* "Fluidization '88: Science and Technology" (Kwauk, M., and Kunii, D., eds.), pp. 155–164. Science Press, Beijing (1988).

Bai, D., Jin, Y., and Yu, Z. "Observation of Cluster in Two-Dimensional Gas–Solid Fast Fluidized Bed," (in Chinese), The Proceeding of 5th National Conference on Fluidization, pp. 139–142 (1990).

Bai, D., Jin, Y., Yu, Z., and Cao, C. "An experimental study on heat transfer in circulating fluidized bed," *in* China–Japan Proc. of Chem. Eng., Tianjin, p. 701 (1992).

Bai, D., Jin, Y., Yu, Z., Gan, N., and Yang, Q. "The Influence of Bed Inlet Structures on the Behaviour in the Fast Fluidized Bed" (in Chinese), The Proceeding of 5th National Conference on Fluidization, pp. 94–98 (1990).

Bai, D., Jin, Y., Yu, Z., and Gan, N. "Circulating Fluidized Bed Technology III" (Basu, P., Horio, M., and Hasatani, M., eds.), pp. 157–162. Pergamon Press, Oxford (1991).

Bankel, J., and Olsson, E. "Measurements of Particle Velocity in a Cyclone Separator Using Laser Doppler Anemometer," *in* "Circulating Fluidized Bed Technology IV" (Amos A. Avidan, ed.), pp. 630–635. Somerset, Pennsylvania (1993).

Bao, Y., Li, H., and Wang, W. "Application of Gas Suspension Calciner for Alumina" (in Chinese), The Proceeding of 5th National Conference on Fluidization, Beijing, pp. 362–365.

Basak, A. K., Koskinen, J., Isaksson, J., and Sellakumar, K. "Sulphur Capture in Ahlstrom Pyroflow PCFB," *in* "Circulating Fluidized Bed Technology IV" (Amos A. Avidan, ed.), pp. 266–271. Somerset, Pennsylvania (1993).

Baskakov, A. P., Maskaev, V. K., Usoltsev, A. G., Ivanov, I. V., and Zubkov, V. A. "An Influence of a Secondary Air on Convective Heat Transfer Between a Riser's Wall and Circulating Fluidized Bed (CFB)," *in* "Circulating Fluidized Bed Technology IV" (Amos A. Avidan, ed.), pp. 380–383. Somerset, Pennsylvania (1993).

Bassi, A. S., Briens, C. L., and Bergougnou, M. A., "Short Contact Time Fluidized Reactors (SCTFRs)," *in* "Circulating Fluidized Bed Technology IV" (Amos A. Avidan, ed.), pp. 17–22. Somerset, Pennsylvania (1993).

Basu, P. "Design Consideration for Circulating Fluidized Bed Combustors," *J. Inst. Energy.* vol. 59, p. 179 (1986).

Basu, P. Heat transfer in high temperature fast fluidized beds, *Chem. Eng. Sci.*, vol. 45, pp. 3123–3136 (1990).

Basu, P., and Ritchie, C. D. "Burning rate of Suncor Coke in a Circulating Fluidized Bed," The 8th Int. Conf. on Fluidized Bed Combustion House, Texas (1985).

Basu, P., Ritchie, C., and Halder, P. K. "Burning Rates of Carbon Particles in a Circulating Fluidized Bed," *in* "Circulating Fluidized Bed Technology" (Basu, P., ed.), p. 229. Pergamon Press, Canada (1986).

Beisswenger, H. *et al.* "Circulating Fluidized Bed as a High-Temperature Reactor," *Chem.-Ing.-Tech.*, vol. 53, p. 289 (1981).

Beisswenger, H., Barner, H. E., and Herbertz, H. "Applications of CFB Technology for Utilization of Refinery-Generated Fuels," *Proc.-Refin. Dep., Am. Pet. Inst.*, vol. 65, p. 561 (1987).

Belin, F. "CFB Boiler with an Impact Particle Separator-Design and Operating Experience," *in* "Circulating Fluidized Bed Technology IV" (Amos A. Avidan, ed.), pp. 233–238. Somerset, Pennsylvania (1993).

Bellon, M., Jacob, T., Joos, E., and Laurent, B. "In-depth Tests Carrier out by EDF on the 125 MWe CFB Boiler of the Emile Huchent Power Plant," *in* "Circulating Fluidized Bed Technology IV" (Amos A. Avidan, ed.), pp. 226–232. Somerset, Pennsylvania (1993).

Bender, D. J., Saize, F. A., and Warfe, W. A. "The Chartham Station Circulating Fluidized Bed Boiler Projects," Proc. Am. Power Conf., vol. 48, p. 99 (1986).

Bengtsson, L. *et al.* "Commercial Experience in a Circulating Fluidized Bed System for Regeneration," Proc. Am. Power Conf., vol. 43, p. 953 (1981).

Berggen, J. C. *et al.* "Gasification of Uranium-Bearing Block Shale in a Circulating Fluidized Bed Reactor," Can. J. Chem. Eng., vol. 59, p. 614 (1981).

Berggren, J. C. *et al.* "Application of Chemical and Physical Operations in a Circulating Fluidized Bed System," Chem. Eng. Sci., vol. 35, pp. 446 (1980).

Beuther, H., Innes, R. A., and Swaft, H. E. Process for Preparing Acrylonitrile, U.S.P., 4,102,914.

Bi, H., Jin, Y., Yu, Z. Q., and Bai, D. R. "An investigation on heat transfer in circulating fluidized bed," *in* "Circulating Fluidized Bed Technology III" (Basu, P., Horio, M., and Hasatani, M., eds.), pp. 233–238. Pergamon Press, Oxford (1991).

Bi, X., Yu, Z., and Jin, Y. "The Influence of Probe Direction on Measuring Heat Transfer Coefficients in Fast Fluidized Bed" (in Chinese), The Proceeding of 5th National Conference on Fluidization, pp. 168–171 (1990).

Bielfeldt, K. "Use of a Circulating Fluidized Bed in the Aluminum Industry," Erzmetall, vol. 37, p. 421 (1984).

Blackadder, W., Morris, M., Rensfelt, E., and Waldheim, L. "Development of an Integrated Gasification and Hot Gas Cleaning Process Using Circulating Fluidized Bed Technology," *in* "Circulating Fluidized Bed Technology III" (Basu, P., Horio, M., and Hasatani, M., eds.), pp. 511–518. Pergamon Press, Oxford (1991).

Blasetti, A., Ng, S., and de Lasa, H. I., "Catalytic Cracking of Gas Oil in a Novel FCC Pilot Plant Unit with Heat Exchange: Reactor Performance," *in* "Circulating Fluidized Bed Technology IV" (Amos A. Avidan, ed.), pp. 553–558. Somerset, Pennsylvania (1993).

Bodelin, P., Molodtsof, Y., and Delebarre, A. "Flow Structure Investigations in a CFB," *in* "Circulating Fluidized Bed Technology IV" (Amos A. Avidan, ed.), pp. 151–156. Somerset, Pennsylvania (1993).

Boersma, W. H., and Jansen, X. X. "Modelling of Dispersed Two-Phase Solid-Fluid Riser Flow," *in* "Circulating Fluidized Bed Technology IV" (Amos A. Avidan, ed.), pp. 454–459. Somerset, Pennsylvania (1993).

Bouillard, J. X., and Miller, A. "Experimental Investigations of Chaotic Hydrodynamic Attractors in Circulating Fluidized Beds," *in* "Circulating Fluidized Bed Technology IV" (Amos A. Avidan, ed.), pp. 80–85. Somerset, Pennsylvania (1993).

Bouillard, J. X., and Lyczkowski, R. W. "Hydrodynamics/Heat-Transfer/Erosion Computer Predictions for a Cold Small-Scale CFBC," *in* "Circulating Fluidized Bed Technology IV" (Amos A. Avidan, ed.), pp. 442–447. Somerset, Pennsylvania (1993).

Boyd, T. J., and Friedman, M. A. "Operation and Test Program Summary at the 110 MWe Nucla CFB," *in* "Circulating Fluidized Bed Technology III" (Basu, P., Horio, M., and Hasatani, M., eds.), pp. 297–312. Pergamon Press, Oxford (1991).

Brereton, C. M. H., Grace, J. R., and Yu, J. "Axial gas mixing in a circulating fluidized bed," *in* "Circulating Fluidized Bed Technology II" (Basu, P., and Large, J. F., eds.), pp. 307–314. Pergamon Press (1988).

Bruce, J., Milne, Franco Berruti, and Leo A. Behie. "The Hydrodynamics of the Internally Circulating Fluidized Bed at High Temperature," *in* "Circulating Fluidized Bed Technology IV" (Amos A. Avidan, ed.), pp. 29–34. Somerset, Pennsylvania (1993).

Brue, E., Moore, J., and Brown, R. C. "Process Model Identification of Circulating Fluid Bed Hydrodynamics," *in* "Circulating Fluidized Bed Technology IV" (Amos A. Avidan, ed.), pp. 535–540. Somerset, Pennsylvania (1993).

Buchanan, John S., and Schoennagel, Hans J. "Laboratory fluid catalytic cracking riser and particulate delivery system," PATENT: United States US 4978441 A, Date: 901218 (1990).

Buyan, Frank M., and Ross, Mark S. "Fluid catalytic cracking reactor multi-feed nozzle system," USP 4650566 (1987); Campbell, H. W., and Hurn, E. J. "Circulating Fluidized Bed Combustion on Stream at California Portland Cement Company," Coal Technology (Houston), 8th (3–4), p. 19 (1985).

Cao, B. et al. "Experimental Study on the Internal Separator of a Circulating Fluidized Bed Combustor," *Huagong Xuebao* (Chinese), vol. 37, p. 478 (1987).

Cao, B. "Internal Separator for Circulating Fluidized Bed Combustion Boiler," *in* "Circulating Fluidized Bed Technology" (Basu, P., ed.), p. 385. Pergamon Press, Canada (1985).

Cao, B., Song, Z., Li, W., Zhang, J., Geo, W., and Pan, G. "Circulating Fluidized Bed Boiler with an Internal Separator of Three Vortex Chambers," *in* "Circulating Fluidized Bed Technology III" (Basu, P., Horio, M., and Hasatani, M., eds.), pp. 471–478. Pergamon Press, Oxford (1991).

Cao, C., Jin, Y., Yu, Z., and Wang, Z. "The Gas–Solid Velocity Profiles and Slip Phenomenon in a Concurrent Downflow Circulating Fluidized Bed," *in* "Circulating Fluidized Bed Technology IV" (Amos A. Avidan, ed.), pp. 496–501. Somerset, Pennsylvania (1993).

Chen, B., Fang, G., Shen, T., and Zhou, R. "Study of the Wearing Features on the Equipment of a FCC Unit," *Petroleum Processing* (Chinese), No. 10, pp. 50–56 (1988).

Chen, L., and Weinstein, H. "The Radial Pressure Gradient in a Circulating Fluidized Bed Riser," *in* "Circulating Fluidized Bed Technology IV" (Amos A. Avidan, ed.), pp. 187–192. Somerset, Pennsylvania (1993).

Cheng, D. "A Study on the Fluidized Bed Reactor for Manufacture of Butadiene from Butene" (in Chinese), The Proceeding of 5th National Conference on Fluidization, pp. 353–356 (1990).

Clive, M. H., Brereton, X. X., and John R. Grace. "End Effects in Circulating Fluidized Bed Hydrodynamics," *in* "Circulating Fluidized Bed Technology IV" (Amos A. Avidan, ed.), pp. 169–174. Somerset, Pennsylvania (1993).

Coating apparatus with circulating powder bed, *European Pat.*, EP 273573 (1988).

Collins, Michael E., and Mann, Michael D. "Impact of Fuel Properties on CFBC N2O Emissions," *in* "Circulating Fluidized Bed Technology IV" (Amos A. Avidan, ed.), pp. 748–753. Somerset, Pennsylvania (1993).

Contractor, R. M., Patience, G. S., Garnett, D. I., Horowitz, H. S., Sisler, G. M., and Bergna, H. E. "A New Process for *n*-Butane Oxidation to Maleic Anhydride Using a Circulating Fluidized Bed Reactor," *in* "Circulating Fluidized Bed Technology IV" (Amos A. Avidan, ed.), pp. 466–471. Somerset, Pennsylvania (1993).

Contractor, R. M. "Improved Vapor Phase Catalytic Oxidation of Butane to Maleic Anhydrige," U.S.P., 4,668,802.

Contractor, R. M., and Chaouki, J. "Circulating Fluidized Bed as a Catalytic Reactor," *in* "Circulating Fluidized Bed Technology III" (Basu, P., Horio, M., and Hasatani, M., eds.), pp. 39–48. Pergamon Press, Oxford (1991).

Courutier, M. F., Karidio, I., and Steward, F. R. "A Study on the Rate of Breakage of Various Canadian Limestones in a Circulating Transport Reactor," *in* "Circulating Fluidized Bed Technology IV" (Amos A. Avidan, ed.), pp. 788–793. Somerset, Pennsylvania (1993).

Couturier, M. F., and Stevens, D. "Capture Efficiency of an Industrial CFB Cyclone," *in* "Circulating Fluidized Bed Technology IV" (Amos A. Avidan, ed.), pp. 654–659. Somerset, Pennsylvania (1993).

Crouse, F. W., Nazemi, A. H., and Lewis, J. P. "Performance Analysis of a Pressurized Circulating Fluidized Bed Boiler Combined Cycle Power Plant," Proc. Int. Conf. Fluid. Bed Combust., Volume Data 1982, 7th(1), p. 232 (1983).

Cui, P., and Ying, S. "Development and Application of an Optimal Control Model for a FCC Unit," Petroleum processing (Chinese), No. 4, pp. 14–18 (1989).

Ciu, X., Jiang, D., Jin, Y., Bu, Z., and Liu, H. "Performance of Trilocular-Type Catalytic-Cracking Gas–Solid Fast Separator in Laboratory Model and Commercial FCC Unit," Petroleum processing (Chinese), No. 2, pp. 32–36 (1991).

Daradimos, G., and Hirschfelder, H. "Longterm Experience with Commercially Operating Large Stream CFB Generators," in "Circulating Fluidized Bed Technology IV" (Amos A. Avidan, ed.), pp. 211–219. Somerset, Pennsylvania (1993).

Darling, S. L., Beisswenger, H., and Wechsler, A. "The Lurgi/Combustion Engineering Circulating Fluidized Bed Boiler Design and Operation," in "Circulating Fluidized Bed Technology" (Basu, P., ed.), p. 297. Pergamon Press, Canada (1986).

Dasggpta, S., Jackson, R., and Sundaresan, S. "Turbulent Gas-Particle Flow in CFB Risers," in "Circulating Fluidized Bed Technology IV" (Amos A. Avidan, ed.), pp. 448–453. Somerset, Pennsylvania (1993).

Dassori, C. G., and Shah, Y. T. "Effects of particle characteristics on the performance of a fast fluidized bed reactor," *Chem. Eng. Commun.*, vol. 100, pp. 29–45 (1991).

Davison, J. E., Cross, P. J., and Topper, J. M. "Application of CFBC to Power Generation in the UK," in "Circulating Fluidized Bed Technology III" (Basu, P., Horio, M., and Hasatani, M., eds.), pp. 445–450. Pergamon Press, Oxford (1991).

Deng, X., Kong, J., and Pan, H. "Application of Fluidized Bed Preheating Calciner" (in Chinese), The Proceeding of 5th National Conference on Fluidization, Beijing, pp. 357–361 (1990).

Deng, X. J. "CFPC Process," in "Circulating Fluidized Bed Technology IV" (Amos A. Avidan, ed.), pp. 472–477. Somerset, Pennsylvania (1993).

Divilio, R. J., and Thomas J. Boyd. "Particle Implications of the Effect of Solid Suspension Density on Heat Transfer in Large-Scale CFB Boilers," in "Circulating Fluidized Bed Technology IV" (Amos A. Avidan, ed.), pp. 402–411. Somerset, Pennsylvania (1993).

Donsi, G., and L. Sesti Osseo. "Gas Solid Flow Pattern in a Circulating Fluid Bed Operated at High Gas Velocity," in "Circulating Fluidized Bed Technology IV" (Amos A. Avidan, ed.), pp. 696–701. Somerset, Pennsylvania (1993).

Dou, S. M., Harandi, X. X., Schipper, P. H., and Owen, H. "Multi-Solid Fluid Bed Cold-Flow Study," in "Circulating Fluidized Bed Technology IV" (Amos A. Avidan, ed.), pp. 484–489. Somerset, Pennsylvania (1993).

Dry, R. J., and Lanauze, R. D. *Chem. Eng. Prom.*, vol. 86, pp. 31–48 (1990).

Dry, R. J., and Nause, R. D. "Combustion in Fluidized Beds," Chemical Engineering Process, July (1990).

Dry, R. J., White, R. B., and Joyce, T. "Correlation of Solids Circulation Rate in Circulating Fluidized Bed Systems," in "Circulating Fluidized Bed Technology IV" (Amos A. Avidan, ed.), pp. 732–737. Somerset, Pennsylvania (1993).

Duerrfels, H., and Failing, K. H. "Circulating Fluidized-Bed Combustion in Boiler Firing with Low-Grade Fuels and Coal," Comm. Eur. Communities, [Rep.] EUR, EUR 9404, 77 (1985).

Duerrfeld, H., and Failing, K. H. "Further Experimental Development of Circulating Fluidized-Bed Combustion for Boiler Firing," Comm. Eur. Communities, [Rep.] EUR, EUR 9236, Energy Conserv. Ind., vol. 1, p. 302 (1985).

Dutta, S. "A Novel CFB Application: A Chemical Compressor," in "Circulating Fluidized Bed Technology III" (Basu, P., Horio, M., and Hasatani, M., eds.), pp. 465–470. Pergamon Press, Oxford (1991).

Dutta, S., Charistian, B. R., and Raterman, M. F. "Circulating Fluid Bed Reactor Model Developed for FCC/ART Regenerator," in "Fluidization VI" (Grace, J. R., Schemilt, L. W., and Bergougnou, M. A., eds.), Engineering Foundation, New York (1989).

Engstrom, F. "Apparatus for Separating Solids from Wastes Gases in a Reactors with Circulating Fluidized Bed," Ger. Offen. DE 3,517,986.

Engstrom, F. "Operating Experience of Commercial Scale Pyroflow Circulating Fluidized Bed

Combustion Boilers," Proc. Int. Conf. Fluid. Bed Combust., Volume Data 1982, 7th(2), p. 1136.

Engstrom, Folke. "Apparatus for separating solid material from flue gases in a circulating fluidized bed reactor," CA 1261122.

Engstrom, F., Ahlstrom, O. A., and Sahagian, J. "Operating Experiences with Circulating Fluidized Bed Boiler," *in* "Circulating Fluidized Bed Technology" (Basu, P., ed.), p. 309. Pergamon Press, Canada (1986).

Engstrom, F., Hakulin, B., and Ruttu, S. "Fluidized Combustion of Fuel," Brit. UK Pat. Apl. GB 2,159,432 (1985).

Engstrom, Folke, Henricson, Kaj, and Lundqvist, Ragnar. "Solids removal from waste gases using a recirculating fluidized bed reactor," Germany (East), DD 253945.

Engstrom, Folke, and Lundqvist, Ragnar. "Circulating fluidized-bed reactor," CA 1261121.

Engstrom, Folke. "Method and apparatus for separating solid particles from flue gases in a circulating fluidized bed reactor," PATENT: PCT International; WO 8603986 A1 DATE: 860717, APPLICATION: WO 86FI2 (860109) *FI 85134 (850111), 11 pp. (1986).

Evardsson, C. M., and Alliston, M. G. "Control of NO_x and N_2O Emission from Circulating Fluidized Bed Boilers," *in* "Circulating Fluidized Bed Technology IV" (Amos A. Avidan, ed.), pp. 754–759. Somerset, Pennsylvania (1993).

Fan, L.-S. *Chem. Eng. J.*, vol. 21, p. 179 (1981).

Feugier, A., Gaulier, C., and Martin, G. "Hydrodynamics, heat transfer, and combustion of natural gas in a circulating fluidized bed," *Rev. Inst. Fr. Pet.*, vol. 43, pp. 439–47 (1988).

Fitting, Arno, and Hirsch, Martin. "Sulfur recovery from sulfidic materials in a recirculating fluidized bed," PATENT: Germany Offen.; DE 3926723 A1 DATE: 910214, APPLICATION: DE 3926723 (890812), 6 pp. (1991).

Freel, B. A. *et al.* "The Kinetics of the Fast Pyrolysis Cellulose in a Fast Fluidized Bed Reactor," *AIChE Symp. Ser.* vol. 83, p. 105 (1987).

Fifiwara, N., Inokuchi, K., Matsuoka, S., and Yamada, T. "A Study on Fragmentation of Coal Particles in a Circulating Fluidized Bed Combustor," *in* "Circulating Fluidized Bed Technology III" (Basu, P., Horio, M., and Hasatani, M., eds.), pp. 373–378. Pergamon Press, Oxford (1991).

Fujima, Y., Fujioka, Y., Hino, H., and Takamoka, H. "Circulating Fluidized Bed Technology" (Basu, P., ed.), pp. 239–246. Pergamon Press, Canada (1986).

Fusey, I., Lim, C. J., and Grace, J. R. "Fast Fluidized in a Concentric Circulating Bed," *in* "Circulating Fluidized Bed Technology" (Basu, P., ed.), p. 409. Pergamon Press, Canada (1986).

Gamblin, B., Newton, D., and Grant, C. "X-Ray Characterization of the Gas Flow Patterns from FCC Regenerator Air Rising Nozzles," *in* "Circulating Fluidized Bed Technology IV" (Amos A. Avidan, ed.), pp. 595–600. Somerset, Pennsylvania (1993).

Garside, R. J. "Fluidization VI" (Grace, J. R., Schemilt, L. W., and Bergougnou, M. A., ed.), Engineering Foundation, New York, Gelbeln, A. P., Phthalic Anhydride Reaction System, U.S.P., 4,261,899 (1989).

Gendreau, R. J., and Raymond, D. L. "An Assessment of Burning Waste Fuel in Circulating Fluidized Bed Boilers," *Plant/Oper. Prog.*, vol. 6, p. 46 (1987).

Gianetto, A., Pagliolico, S., Rover, G., and Ruggeri, B. "Theoretical and Practical Aspects of Circulating Fluidized Bed Reactor (CFERs) for Complex Chemical Systems," *Chem. Eng. Sci.*, Vol. 45, pp. 2219–2225 (1990).

Gidaspow, D., and Therdthianwong, A. "Hydrodynamics & SO_2 Sorption in a CFB Loop," *in* "Circulating Fluidized Bed Technology IV" (Amos A. Avidan, ed.), pp. 436–441. Somerset, Pennsylvania (1993).

Giese, M. "Some Aspects of the Performance of Three Different Types of Industrial Circulating

Fluidized Bed Boiler," *in* "Circulating Fluidized Bed Technology III" (Basu, P., Horio, M., and Hasatani, M., eds.), pp. 347–352. Pergamon Press, Oxford (1991).

Glicksman, L. R., Westphalen, D., Brereton, C., and Grace, J. R. "Circulating Fluidized Bed Technology III" (Basu, P., Horio, M., and Hasatani, M., eds.), pp. 119–24. Pergamon Press, Oxford (1991).

Goedicke, F., Nicolai, R., Tanner, H., and Reh, L. "Particle Induced Heat Transfer Between Walls and Gas–Solid Fluidized Beds," *in* "Circulating Fluidized Bed Technology IV" (Amos A. Avidan, ed.), pp. 356–361. Somerset, Pennsylvania (1993).

Gounderin, X. X. "Ahlstrom Pyropower's Reheat Experience," *in* "Circulating Fluidized Bed Technology IV" (Amos A. Avidan, ed.), pp. 205–210. Somerset, Pennsylvania (1993).

Graf, R. "First operating experience with a dry flue gas desulphurization (FGD) process using a circulating fluid bed (FGD-CFB)," *in* "Circ. Fluid. Bed Technol." (Basu, P., and Prabir, X., eds.), pp. 317–27. Pergamon, Toronto (1986).

Graf, Rolf, Riley, John D. "Dry/semi-dry flue gas desulphurization using the Lurgi circulating fluid bed absorption process," Brown, Lloyd Chauncy, Besenbruch, Gottfried Ernest Alfred, Adams, Charles Cornelious, Electr. Power Res. Inst. (Rep.) EPRI CS, EPRI CS-5167, Proc. Symp. Flud Gas Desulphurization, 10th, 2, 9/147–9/162 (1986).

Graf, Rolf, Riley, John D., and Erdoel Kohle. "Dry/semi-dry flue gas desulphurization Lurgi circulating fluid bed absorption process," Erdgas, Petrochem., Vol. 40, pp. 263–8 (1987).

Graham, R. G., Freel, B. A., Overend, R. P., Mok, L. K., and Bergougnou, M. A. "The ultra-rapid fluidized-bed reactor: application to determine the kinetics of the fast pyrolysis (ultrapyrolysis) of cellulose," *in*: Oestergaard, Knud, Soerensen, Ans. (1986).

Gulyuurtlu, I., and Cabrita, I. "Combustion of Woodwastes in a Circulating Fluidized Bed," *in* "Circulating Fluidized Bed Technology" (Basu, P., ed.), p. 247. Pergamon Press, Canada (1986).

Halder, P. K., Chattopadhyay, R., Datta, A., and Dutta Gupta, S. "Axial and Radial Variation of Bed to Wall Heat Transfer in a Circulating Fluidized Bed," *in* "Circulating Fluidized Bed Technology IV" (Amos A. Avidan, ed.), pp. 374–379. Somerset, Pennsylvania (1993).

Halder, P. K., and Datta, A. "Modelling of Combustion of a Char in a Circulating Fluidized Bed," *in* "Circulating Fluidized Bed Technology IV" (Amos A. Avidan, ed.), pp. 92–97. Somerset, Pennsylvania (1993).

Hallstrom, C., and Karlsson, R. "Waste Incineration in Circulating Fluidized Bed Boiler Test Results and Operating Experience," *in* "Circulating Fluidized Bed Technology III" (Basu, P., Horio, M., and Hasatani, M., eds.), pp. 417–422. Pergamon Press, Oxford (1991).

Hamiltilon, C. J., and Mokenzie, A. B. "The Circulating Fluidize Bed: An Efficient and Flexible Process for Combustion of Low-Grade Fuels," Chemeca 85; 13th Aust. Chem. Eng. Conf., p. 477 (1987).

Hannes, J., Svoboda, K., Renz, U., and Cor, M. van den Bleek. "Modelling of Size Distributions and Pressure Profiles in CFBC," *in* "Circulating Fluidized Bed Technology IV" (Amos A. Avidan, ed.), pp. 199–204. Somerset, Pennsylvania (1993).

Hansen, P. F., Dam-Johansen, K., Bank, L. H., and Ostergaard, K. "Sulphur Capture on Limestone Under Periodically Changing Oxidizing and Reduce Conditions," *in* "Circulating Fluidized Bed Technology III" (Basu, P., Horio, M., and Hasatani, M., eds.), pp. 411–416. Pergamon Press, Oxford (1991).

Hao, F. *et al.* The Proceeding of 4th National Conference on Fluidization, Beijing, p. 285 (1987).

Harris, B. J., and Davidson, J. F. "Modelling Options for Circulating Fluidized Beds: A Core/Annulus Deposition Model," *in* "Circulating Fluidized Bed Technology IV" (Amos A. Avidan, ed.), pp. 35–40. Somerset, Pennsylvania (1993).

Harris, B. J., Davidson, J. F., Xue, Y. "Axial and Radial Variation of Flow in Circulating Fluidized Bed Risers," *in* "Circulating Fluidized Bed Technology IV" (Amos A. Avidan, ed.), pp. 137–142. Somerset, Pennsylvania (1993).

Hasatani, M. et al. "Combined Fast Fluidized-bed/Moving-bed Combustor for Low-Grade Solid Fuels," in "Circulating Fluidized Bed Technology" (Basu, P., ed.), p. 47. Pergamon Press, Canada (1986).

He, Y., and Rudolph, V. "Control Mechanism for Gas Crossflow in a Compartmented Dense Phase CFB," in "Circulating Fluidized Bed Technology IV" (Amos A. Avidan, ed.), pp. 684–689. Somerset, Pennsylvania (1993).

Helmrich, H. "Reaction Technical Studies on the Sodium Hydrogen Carbonate Decomposition in a Circulating Fluidized Bed," Chem.-Ing.-Tech., Vol. 53, p. 137 (1981).

Helmrich, H., Schuegeri, K., and Hanssen, K. "Decomposition of Sodium Bed Biacarbonate in Laboratory and Bench Scale Circulating Fluidized Reactors," in "Circulating Fluidized Bed Technology" (Basu, P., ed.), p. 161. Pergamon Press, Canada (1986).

Henricson, Kaj, Engstrom, Folke, and Faming Zhuanli Shenqing Gon. "Circulating fluidized bed reactor and method of separating solid material from flue gases," CN 86108105 (1987).

Herbert, P. K., Baguss, I., and Loeffler, J. C. "Low-Btu gas from bark and waste via CFB (circulating fluidized bed) gasification," Energy Biomass Wastes, Vol. 12, pp. 731–51 (1989).

Highley, J., and Kaye, W. G. "Fluidized Bed Industrial Boiler and Furnaces," in "Fluidized Beds: Combustion and Applications" (Howard, J. R., ed.), p. 127. Applied Science, London–New York (1983).

Hildebrandt, R. "System Study on a Horizontally Circulating Fluidized Bed for Preliminary Degasification and Combustion of Hard Coal," Forschuhgsber. -Bunderminist. Forsch. Technol., Technol. Forsch. Entwickl.. BMFT-FB-T 81-137, 61 (1981).

Hirota, T., Oshita, T., Kosugi, S., Nagato, S., Makiyama, Y., and Hirose, T. "Characteristics of the Internally Circulating Fluidized Bed Boiler," in "Circulating Fluidized Bed Technology III" (Basu, P., Horio, M., and Hasatani, M., eds.), pp. 491–496. Pergamon Press, Oxford (1991).

Hirsch, M., Janssen, K., and Serbent, H. "Circulating Fluidized Bed Technology" (Basu, P., ed.), p. 329. Pergamon Press, Canada (1986).

Hofbuer, H. "Studies on a Circulating Fluidized Bed with a Centrally Positioned Tube." Chem.-Ing.-Tech., Vol. 54, p. 529 (1982).

Horio, M. "Hydrodynamics of Circulating Fluidization-Present Status and Research Needs," in "Circulating Fluidized Bed Technology III" (Basu, P., Horio, M., and Hasatani, M., eds.), pp. 1–2. Pergamon Press, Oxford (1991).

Horio, M. et al. "Coal Combustion in a Transparent Circulating Fluidized Bed," in "Circulating Fluidized Bed Technology" (Basu, P., ed.), p. 229. Pergamon Press, Canada (1986).

Horio, M., Ishii, H., Kobukai, Y., and Yamanishi, N. J. of Chem. Eng. of Japan, Vol. 22, pp. 587–592 (1989).

Horio, M., Kuroki, H., and Ogasawara, M. "The Flow Structure of a Three-Dimensional Circulating Fluidized Bed Observed by the Laser Sheet Technique," in "Circulating Fluidized Bed Technology IV" (Amos A. Avidan, ed.), pp. 86–91. Somerset, Pennsylvania (1993).

Horio, M., Morishita, K., Tachibana, O., and Hurata, H. "Solid distribution and movement in circulating fluidized beds," in "Circulating Fluidized Bed Technology II" (Basu, P., and Large, J. F., eds.), pp. 147–154. Pergamon Press (1988).

Horio, M., and Takei, Y. "Macroscopic structure of recirculating flow of gas and solid in circulating fluidized beds," in Preprints for 3rd Int. Conf. on CFB, Nagoya, Japan (Oct. 15–18, 1990).

Htano, H., Suzuki, Y., and Kido, N. "Flow Structure in Circulating Fluidized Bed Combustors," in "Circulating Fluidized Bed Technology IV" (Amos A. Avidan, ed.), pp. 157–162. Somerset, Pennsylvania (1993).

Huang, Z., and Wang, Z. Shiyiu Huagong (Chinese), Vol. 17, pp. 730–735 (1988).

Huang, Z. et al. "Study of Fast Fluidized Bed for Converting Butene to Butadiene by Oxidative

Dehydrogenation" (in Chinese), The Proceeding of 4th National Conference on Fluidization, Beijing, p. 154 (1987).
Hyppanen, Timo, Palonen, Juha, and Rainio, Aku. "Pyroflow Compact—Ahlstrom Pyropower's Second Generation CFB," in "Circulating Fluidized Bed Technology IV" (Amos A. Avidan, ed.), pp. 131–136. Somerset, Pennsylvania (1993).
Institute of Mechanics, Academic Sinica, Refinery of Beijing Yanshan Petrochemical Company. "Study and Commercial Application of Novel Feed Ejectors in FCCU," Petroleum processing (Chinese), No. 1, pp. 30–37 (1990).
Iordache, O., Bloise, Y., Chaouki, J., and Legros, R. "Kinetic Theory for Circulating Fluidized Bed Reactors," in "Circulating Fluidized Bed Technology IV" (Amos A. Avidan, ed.), pp. 418–423. Somerset, Pennsylvania (1993).
Isaksson, J., Sellakumar, K. M., and Provol, S. J. "Development of PYROFLOW Pressurized Circulating Fluidized Bed (PCFB) Boilers for Utility Application," in "Circulating Fluidized Bed Technology III" (Basu, P., Horio, M., and Hasatani, M., eds.), pp. 439–444. Pergamon Press, Oxford (1991).
Ishida, K., Hasegawa, Y., Nishioka, K., Kono, M., Kiuchi, H., Shintani, K., Yoshinaka, K., and Uemoto, M. "Experience with 300 T/h Mitsui Circulating Fluidized Bed Boilers," in "Circulating Fluidized Bed Technology III" (Basu, P., Horio, M., and Hasatani, M., eds.), pp. 313–320. Pergamon Press, Oxford (1991).
Ishii, H., and Murakami, I. "Circulating Fluidized Bed Technology III" (Basu, P., Horio, M., and Hasatani, M., eds.), pp. 125–130. Pergamon Press, Oxford (1991).
Ishii, H., Nakajima, T., and Horio, M. "The structure of flow fields in circulating fluidized beds," in "Fluidized-Bed and Three-Phase Reactor" (Yoshida, K., and Morooka, S., eds.), pp. 139–146. Tokyo, Japan (1988).
Ishii, H., Nakajima, T., and Horio, M. "The clustering annular flow model of circulating fluidized beds," J. of Chem. Eng. of Japan, vol. 22, pp. 484–490 (1989).
Itoh, S., Okatada, Y., Higuchi, G., and Furubayashi, K. "Combustion Properties of Wide Range Fuels on CIRCUFLUID," in "Circulating Fluidized Bed Technology III" (Basu, P., Horio, M., and Hasatani, M., eds.), pp. 485–490. Pergamon Press, Oxford (1991).
Jiang, P., Cai, P., and Fan, L. "Transient Flow Behaviour in Fast Fluidization," in "Circulating Fluidized Bed Technology IV" (Amos A. Avidan, ed.), pp. 143–150. Somerset, Pennsylvania (1993).
Jiang, P. J., Inokuchi, K., Jean, R. H., Bi, X. T., and Fan, L. S. "Ozone Decomposition in a Catalytic Circulating Fluidized Bed Reactor," in "Circulating Fluidized Bed Technology III" (Basu, P., Horio, M., and Hasatani, M., eds.), Pergamon Press, Oxford, pp. 557–562 (1991).
Jiang, Peijun, Bi, Hsiaotao, Jean, Rong Her, and Fan, Liang Shih. "Baffle effects on performance of catalytic circulating fluidized bed reactor," AIChE J., vol. 37, No. 9, pp. 1392–1400 (1991).
Jiang, X. L., Keener, T. C., Khang, S. J., and Hao, J. M. "Experimental study and modelling of sulphur dioxide removal in a circulating fluidized bed absorber by sodium bicarbonate sorbet," Proc.-A&WMA Annu. Meet., 83rd vol. 7, 90/110.2 (1990).
Jin, B., Lan, J., Fan, X., and Zhang, M. "The Experimental Investigation of Coal Combustion Characteristics in a FBC with Fly Ash Recycle," J. of Southeast University (Chinese), vol. 20, No. 2, pp. 49–55 (1990).
Jin, H., Park, Jang, J., Park, Jong, H., Choi, In, S. Chang, Yong Kang, and Sang D. Kim. "Hydrodynamic Characteristics of Fine Particle in Riser and Standpipe of a Circulating Fluidized Bed," in "Circulating Fluidized Bed Technology IV" (Amos A. Avidan, ed.), pp. 181–186. Somerset, Pennsylvania (1993).
Jin, Y., Yu, Z. Q., Bai, D. R., and Yao, W. H. CHN., 15143 (1987).
Jin, Y., Yu, Z. Q., Bai, D. R., Qi, C., and Zhong, X. "Theoretical and Experimental Studies on Gas–Solid Cocurrent Downward Flow Fast Fluidization," (Chinese), The Proceeding of

5th National Conference on Fluidization, Beijing, pp. 134–138 (1990).
Jin, Y., Yu, Z.-Q., Zi, C.-M., and Bai, D.-R. "The influence of exit structures on the axial distribution of voidage in fast fluidized bed," *in* "Fluidization '88: Science and Technology" (Kwauk, M., and Kunii, D., eds.), pp. 165–173. Science Press, Beijing (1988).
Jing, P., Bi, H., Jean, R.-H., and Fan, L.-S. AIChE J., Vol. 37, No. 9, pp. 1392–1400.
Joachim, Werther. "Fluid Mechanics of Large-Scale CFB Units," *in* "Circulating Fluidized Bed Technology IV" (Amos A. Avidan, ed.), pp. 1–16. Somerset, Pennsylvania (1993).
John, S., and Shoennagel, Hans J. "Laboratory fluid catalytic cracking riser and particulate delivery system," USP, 4978441 (1988).
Johnsson, J. E., Amand, L. E., and Leckner, B. "Modelling of NO_x Formation in a Circulating Fluidized Bed Boiler," *in* "Circulating Fluidized Bed Technology III" (Basu, P., Horio, M., and Hasatani, M., eds.), pp. 405–410. Pergamon Press, Oxford (1991).
Jones, O., and Seber, E. "Initial Operating Experience at Conoco's South Texas Multi-Solids FBC Steam Generator," Proc. Int. Conf. Fluid. Bed Combust., Volume Data 1982, 7th(2), 1136 (1983).
Lambert, Joseph M. Jr., and Townsend, Eric J. "Application of Various Design Criteria for Circulating Fluid Bed Reactors in Pilot Plants," *in* "Circulating Fluidized Bed Technology IV" (Amos A. Avidan, ed.), pp. 502–508. Somerset, Pennsylvania (1993).
Joshi, P. A., Chidambaram, M., and Shankar, H. S. "Circulating Fluidized Bed Technology III," Pergamon Press, pp. 431–436 (1986).
Judd, M. R. "The Circulating Fluidized Bed," Ger. Ofen. DE3,536,871.
Kagwa, H., Mineo, H., Yamazaki, R., and Yosjhida, K. "Circulating Fluidized Bed Technology III" (Basu, P., Horio, M., and Hasatani, M., eds.), pp. 551–556. Pergamon Press, Oxford (1991).
Kaczmarzyk, Grzegorz and Sciazko, Marek. "Coal pyrolysis in a circulating fluidized-bed reactor for hybrid gas–steam power plants," *Koks, Smola, Gas*, vol. 35, No. 8, pp. 175–84 (1990).
Kajiyama, R., Toyama, T., Inaba, R., Takatsuka, T., and Shibagaki, T. "CFB-PTSA Process for Recovering CO_2 from Power Plant Flud Gas," *in* "Circulating Fluidized Bed Technology IV" (Amos A. Avidan, ed.), pp. 478–483. Somerset, Pennsylvania (1993).
Kataoka, A., Nowak, W., Ihara, T., Matsuda, H., Win, K., and Hasatani, M. "A Study of the Flow Structure in a CFB with Application to Gas Combustion," *in* "Circulating Fluidized Bed Technology IV" (Amos A. Avidan, ed.), pp. 678–683. Somerset, Pennsylvania (1993).
Katoh, Y., Kaneko, S., and Miyamoto, M. "Flow Patterns and Velocity Distributions on Fluidized Particles in a Riser of Circulating Fluidized Beds," *in* "Circulating Fluidized Bed Technology IV" (Amos A. Avidan, ed.), pp. 332–337. Somerset, Pennsylvania (1993).
Keairns, D. L. *et al.* "Advanced Atmospheric Fluidized-Bed Combustion Design: Internally Circulating AFBC," Report, DOE/MC/19329-1405: Order No. DE8830110973, 379 pp. (1983).
Keener, Tim C., and Jiang, Xiaolin. "Low-temperature application for control of sulphur dioxide with a circulating fluidized bed absorber," Proc.-Annu. Int. Pittsburgh Coal Conf. 6th(1), pp. 441–450 (1989).
Keener, T. C., and Jiang, X. L. "Application of a circulating fluidized bed absorber for retrofit control of sulphur dioxide," Proc.-APCA Annu. Meet., 81st(5), Paper 88/86.2 (1988).
Kelly, A., and Engstrom, F. "Startup and Operation of the Kauttua 200,000 lbs of Stream per Hour Cogeneration Ahlstrom Circulating Fluidized Bed Combustion," *Coal Technol.* (Houston), 4th (vol. 5), p. 7 (1981).
Kelly, A. J., and Okakes, E. "Circulating Fluidized Bed Combustion Systems and Commercial Operating Experience," Proc. Tech. Assoc. Pulp Pap. Ind., Eng., Conf., Bk. (3), p. 523 (1982).
Kelly, A. J., and Okakes, E. "Circulating Fluidized Bed Combustion Systems and Commercial Operating Experience," Annu. Meet.-Tech. Assoc. Pulp Pap. Ind., p. 301 (1982).
King, D. "Model Predicting Average Densities in Commercial FCC CFB Regenerators," *in*

"Circulating Fluidized Bed Technology IV" (Amos A. Avidan, ed.), pp. 601–606. Somerset, Pennsylvania (1993).

Kohno, T., and Shimano, S. "Application of Fluidization. Application of Optimal Control to Circulating Fluidized-Bed Calciner for Alumina," *Kagaku Kogaku*, vol. 49, No. 5, p. 352 (1985).

Korbro, H., and Brereton, C. "Control and Flexibility of Circulating Fluidized Bed," in "Circulating Fluidized Bed Technology" (Basu, P., ed.), p. 263. Pergamon Press, Canada (1986).

Koreberg, J., and Scaroni, A. W. "Low Temperature Combustion of Alternate Fuels in Circulating Fluidized Beds," in "Circulating Fluidized Bed Technology" (Basu, P., ed.), p. 273. Pergamon Press, Canada (1986).

Krishna, Ashok S. "Gasoline octane enhancement in fluid catalytic cracking with split feed injection to a riser reactor," PCT International, WO 8702695 (1987).

Kuehle, K. "Calcining Cement Raw Meal in the Circulating Fluidized Bed at the Works Port La Nouvelle, France," Zem.-Kalk-Gips, Ed. B, vol. 37, No. 5, p. 219 (1984).

Kuehle, K., Eschenburg, J., and Beck, O. "Commissioning a dry flue-gas desulfurizing plant operating on the principle of a circulating fluidized bed (ZWS)," *ZKG Int.*, Ed. B, vol. 42, No. 9, pp. 441–3.

Kullendroff, A., Herstad, S., and Olofason, J. "Control of NO and SO Emissions by Combustion in Circulating Fluidized Beds—Methods and Operating Experiences," VGB Conf: Power Station Environment 1985, Essen, West Germany Kullendroff, A. (1985).

Herstad, S., and Olofsson, J. "Operating Experience of Circulating Fluidized Bed Boiler," The 8th Int. Conf. on Fluidized Bed Combustion, Houston, Texas (1985).

Kunii, D. "A Study of Particle-Circulating Fluidized Bed System for the Complete Utilization of Coal," *Asahi Garasu Kogyo Gijustsu Shoreikai Kenkyu Hokoku*, vol. 29, p. 235 (1976).

Kwauk, M., Wang, N.-D., Li, Y., Chen, B.-Y., and Shen, Z.-Y. "Fast fluidization at ICM," in "Circulating Fluidized Bed Technology" (Basu, P., ed.), p. 33. Pergamon Press, New York (1986).

Lee, G. S., Han, G. Y., and Kim, S. D. "Coal Combustion Characteristics in a Circulating Fluidized Bed," *Korean J. Chem. Eng.* vol. 1, p. 71 (1984).

Lewnard, J. J., Herb, B. E., and Tsao, T. R. "Effect of Design and Operating Parameters on Cyclone Performance for Circulating Fluidized Bed Boilers," in "Circulating Fluidized Bed Technology IV" (Amos A. Avidan, ed.), pp. 636–641. Somerset, Pennsylvania (1993).

Li, H., Xia, Y., Tung, Y., and Kwauk, M. "Micro-visualization of two-phase structure in a fast fluidized bed," in Preprints for 3rd Int. Conf. on CFB, Nagoya, Japan (Oct. 15-18, 1990).

Li, J., Chen, A., Yan, Z., Xu, G., and Zhang, X. "Particle-Fluid Contacting in Circulating Fluidized Beds," in "Circulating Fluidized Bed Technology IV" (Amos A. Avidan, ed.), pp. 49–54. Somerset, Pennsylvania (1993).

Li, J., Tung, Y., and Kwauk, M. "Method of energy minimization in multi-scale modelling of particle-fluid two-phase flow," in "Circulating Fluidized Bed Technology II" (Basu, P., and Large, J. F., eds.), p. 75. Pergamon Press (1988).

Li, S., and Xu, Y. "Fluidization in FCC Process," Petroleum Processing (Chinese), No. 9, pp. 17–21 (1982).

Li, Y., and Kwauk, M. "Fast Fluidized Bed Combustion of Coal," Huagong Yejin (Chinese), No. 4, pp. 87–94 (1981).

Li, Y., and Kwauk, M. "The Dynamics of Fast Fluidization," in "Fluidization" (Grace, J. R., and Matsen, J. M., eds.), pp. 537–544. Plenum Press, New York (1980).

Li, Y., Wang, F., and Kwauk, M. "An Improved Circulation Fluidized Bed Combustor," in "Circulating Fluidized Bed Technology III" (Basu, P., Horio, M., and Hasatani, M., eds.), pp. 359–364. Pergamon Press, Oxford (1991).

Li, Y., Wang, F., and Zeng, Q. "A New Process of Preparing Anhydrous Boric Oxide by Dehydration of Boric Acid in a Fast Fluidized Bed," Chem. Eng. Reac. Eng. & Tech. (Chinese), Vol. 6, No. 2, pp. 43–49 (1990).
Li, Y. C., Wang, F. M., and Zeng, Q. X. "The New Process for Calcination of Boric Acid in a Fast Fluidized Bed (Chinese)," The Proceeding of 5th National Conference on Fluidization, pp. 371–374 (1990).
Lin, X., and Li, Y. "A Two-Phase Model for Fast Fluidized Bed Combustion," *in* "Circulating Fluidized Bed Technology IV" (Amos A. Avidan, ed.), pp. 547–552. Somerset, Pennsylvania (1993).
Lin, W., and Van Den Bleck, C. M. "Circulating Fluidized Bed Technology III" (Basu, P., Horio, M., and Hasatani, M., eds.), pp. 545–550. Pergamon Press, Oxford (1991).
Linder, C. "Gasification of Self-Renewing Materials, Particularly Wood," Comm. Eur. Communities, [Rep.] EUR, EUR 8245, Energy Biomass, p. 799 (1983).
Lints, M. C., and Glicksman, L. R. "Parameters Governing Particle-to-Wall Heat Transfer in a Circulating Fluidized Bed," *in* "Circulating Fluidized Bed Technology IV" (Amos A. Avidan, ed.), pp. 350–355. Somerset, Pennsylvania (1993).
Liu, C., Yu, L., and Xie, B. "Demonstration of China's First Two-Stage Separation CFB Boiler," *in* "Circulating Fluidized Bed Technology IV" (Amos A. Avidan, ed.), pp. 220–225. Somerset, Pennsylvania (1993).
Liu, D., Kwauk, M., Li, H., and Zheng, C. "Preliminary Investigation on an Integral Circulating Fluidized Bed," *in* "Circulating Fluidized Bed Technology IV" (Amos A. Avidan, ed.), pp. 642–647. Somerset, Pennsylvania (1993).
Liu, J. S., Zhang, J. R., Zhao, X., Sun, D. K., Zhao, S. K., Wang, J. F., and Yang, G. L. Shiyiu Huagong (Chinese), Vol. 4, No. 1, pp. 37–43 (1988).
Liu, Y. Z., Li, C. N., and Hao, X. R. Catalytic Cracking Twenty Years in China (Chinese), Beijing, pp. 181–197 (1985).
Lotze, J., and Wargalla, G. "Characteristic Data and Utilization Possibilities of Ash from a Circulating Fluidized Bed Furnace. Part 1: Process Engineering and Characteristics," *Zem.-Kalk-Gips*, Ed. B, 38(5), 239 (1985).
Lotze, J., and Wargalla, G. "Characteristic Data and Utilization Possibilities of Ash from a Circulating Fluidized Bed Furnace. Part 2: Suitability and Application in the Construction Industry," *Zem.-Kalk-Gips*, Ed. B, 38(7), p. 374 (1985).
Louge, Michel Y. "Boundary Conditions at the Wall of Circulating Fluidized Bed Risers," *in* "Circulating Fluidized Bed Technology IV" (Amos A. Avidan, ed.), pp. 424–429. Somerset, Pennsylvania (1993).
Lu, C., and Wang, Z. "Experimental Studies on Commercial Fast Fluidized Bed at High Temperature (in Chinese)," The Proceeding of 5th National Conference on Fluidization, pp. 122–125 (1990).
Lu, H., Yang, L. D., Bao, Y. L., and Chen, L. Z. "The Radial Distribution of Heat Transfer Coefficient Between Suspensions and Immersed Vertical Tube in Circulating Fluidized Bed (Chinese)," The Proceeding of 5th National Conference on Fluidization, pp. 164–167 (1990).
Lucat, P., Morin, J. X., and Guilleux, E. "Development of Very Large CFB Boilers for Power Stations," *in* "Circulating Fluidized Bed Technology IV" (Amos A. Avidan, ed.), pp. 794–802. Somerset, Pennsylvania (1993).
Lucat, P., Semedard, J. C., Rollin, J. P., and Beisswenger, H. "Stauts of the Industries 125 MWe Emile Huchet CFB Boiler," *in* "Circulating Fluidized Bed Technology III" (Basu, P., Horio, M., and Hasatani, M., eds.), pp. 329–334. Pergamon Press, Oxford (1991).
Lui, J., Zhang, R., Luo, G., and Yang, G. "The Macrokinetic Study on the Oxidative Dehydrogenation of Butene to Produce Butadiene by Fast Fluidized Bed," *Chem. Rec. Eng. & Tech.* (Chinese), vol. 5, No. 1, p. 1–9 (1989).

Lui, Y. "Adoption of Two-Stage Regeneration Technique on a FCC Unit," *Petroleum Processing* (Chinese), No. 11, pp. 20–24 (1988).

Lund, T. "Lurgi Circulating Fluidized Bed Boiler: Its Design and Operation," The 7th Int. Conf. on Fluidized Bed Combustion, Philadelphia, Pennsylvania (1982).

Lundqvist, R., Puhakka, M., and Stahl, K. "Pressurized CFB Biomass Gasification: The Bioflow Energy System," *in* "Circulating Fluidized Bed Technology IV" (Amos A. Avidan, ed.), pp. 254–259. Somerset, Pennsylvania (1993).

Mahalingam, M., and Ajit, Kumar Kolar. "Experimental Correlation for Average Heat Transfer Coefficient at the Wall of a Circulating Fluidized Bed," *in* "Circulating Fluidized Bed Technology IV" (Amos A. Avidan, ed.), pp. 390–395. Somerset, Pennsylvania (1993).

Martin, G., Dolignier, J. C., Flament, P., and Nougier, L. "The AUDE Boiler-Experience on A 10 MW Semi-industrial Unit and Modelling Results," *in* "Circulating Fluidized Bed Technology IV" (Amos A. Avidan, ed.), pp. 124–130. Somerset, Pennsylvania (1993).

Matthews, F. T. *et al.* "Conceptual Design and Assessment of a Pressurized Circulating Fluidized Bed Boiler. Final Report," Report, ERRI-CS-3206, 124 pp. (1984).

Matts, D. *et al.* "Circulating Fluidized Bed Combustion for Industrial Combined Heat and Power Applications," IEE Conf. Publ., 272 (Int. Conf. Ind. Power Eng., 1986), p. 195 (1986).

Mehrling, P., and Reimert, R. "Fuel and Synthesis Gas from Biomass via Gasification in the Circulating Fluidized Bed," Comm. Eur. Communities, [Rep.] EUR, EUR 10024, Energy Biomass, p. 905 (1985).

Merchiuk, Yosef. "Method and device for fluidizing solid particles particularly in loop reactors," PATENT: Israel; IL 79655 A1, DATE: 890630 APPLICATION: IL 79655 (860807), 16 pp. (1989).

Minet, Ronald G., and Tsotsis, Theodore T. "Recirculating fluidized-bed process and apparatus for the recovery of chlorine from hydrogen chloride," ES 2010473 (1989).

Moe, Thomas, A., Mann, Michael D., Henderson, Ann K., and Hajicek, Douglas R. "Pilot-Scale CFBC Systems: A Valuable Tool for Design and Permitting," *in* "Circulating Fluidized Bed Technology IV" (Amos A. Avidan, ed.), pp. 116–123. Somerset, Pennsylvania (1993).

Molerus, O. "Fluid Dynamics and Its Relevance for Basic Features of Heat Transfer in Circulating Fluidized Beds," *in* "Circulating Fluidized Bed Technology IV" (Amos A. Avidan, ed.), pp. 338–343. Somerset, Pennsylvania (1993).

Mori, S., Kashima, N., Wakazono, Y., and Tsuchiya, M. "Particles Behaviour and Flow Pattern in a Circulating Fluidized Bed," *in* "Circulating Fluidized Bed Technology IV" (Amos A. Avidan, ed.), pp. 308–313. Somerset, Pennsylvania (1993).

Moritomi, H., Suzuki, Y., Kido, N., and Ogisu, Y. "NO_x Emission and Reduction from a Circulating Fluidized Bed Combustor," *in* "Circulating Fluidized Bed Technology III" (Basu, P., Horio, M., and Hasatani, M., eds.), pp. 399–404. Pergamon Press, Oxford (1991).

Morooka, S., Kawamura, H., Yan, S., Okubo, T., and Kusakabe, K. "Surface Modification of Milled Carbon Fibre by Low-Temperature Oxygen Plasma in a Fast Fluidized Bed," *in* "Circulating Fluidized Bed Technology III" (Basu, P., Horio, M., and Hasatani, M., eds.), pp. 456–464. Pergamon Press, Oxford (1991).

Mueller, P., and Reh, L. "Particle Drag and Pressure Drop in Accelerated Gas–Solid Flow," *in* "Circulating Fluidized Bed Technology IV" (Amos A. Avidan, eds.), pp. 193–198. Somerset, Pennsylvania (1993).

Mu, T. *Sulphuric Acid Industry* (Chinese), No. 4, p. 21 (1991).

Munts, V. A., Baskakov, A. P., Fendorenko, Yu. N., and Lekomtseva, Uu. G. "The Optimum Size of Particles in Circulating Fluidized Bed Combustor," *in* "Circulating Fluidized Bed Technology III" (Basu, P., Horio, M., and Hasatani, M., eds.), pp. 365–372. Pergamon Press, Oxford (1991).

Muschelknautz, X. X. "Fundamental and Practical Aspects of Cyclones," *in* "Circulating

Fluidized Bed Technology IV" (Amos A. Avidan, ed.), pp. 23–28. Somerset, Pennsylvania (1993).

Muschelknautz, M., and Greif, V. "A New Impeller Probe for Measurement of Velocity at High Temperatures and Dust Loadings," *in* "Circulating Fluidized Bed Technology IV" (Amos A. Avidan, ed.), pp. 648–653. Somerset, Pennsylvania (1993).

Na, Y., Lu, Q., and Wanf, T. "The Model Test of Staged Circulating Fluidized Bed Combustion Boiler with Louver Type Separators," *in* "Circulating Fluidized Bed Technology IV" (Amos A. Avidan, ed.), pp. 613–617. Somerset, Pennsylvania (1993).

Namiki, T. *et al*. "Development of in-furnaces De-NO_x Combustion in the Circulating Fluidized Bed Combustion System," *Mitsubishi Juko Giho*, vol. 22, No. 1, p. 52 (1985).

Nandjee, F., and Jacquet, L. "The Retrofitting of EDF 250 MWe Oil-Fired Units with Circulating Fluidized Bed Boiler," *in* "Circulating Fluidized Bed Technology III" (Basu, P., Horio, M., and Hasatani, M., eds.), pp. 321–328. Pergamon Press, Oxford (1991).

Nag, P. K., and Mukhtar, Sallam Ali. "An Experimental Study of the Effect of Fins on Heat Transfer in a High Temperature Circulating Fluidized Bed," *in* "Circulating Fluidized Bed Technology IV" (Amos A. Avidan, ed.), pp. 362–367. Somerset, Pennsylvania (1993).

Naruse, I., Imanari, M., Koizumi, K., Kuramoto, K., and Ohtake, K. "Separate Measurement of Bubble and Emulsion Phase by Conditional Gas Sampling in Fluidized Bed Combustion," *in* "Circulating Fluidized Bed Technology IV" (Amos A. Avidan, ed.), pp. 326–331. Somerset, Pennsylvania (1993).

Nowak, W., Matsuda, H., Win, K. K., and Hasatani, M. "Diagnosis of Multi-Solid Fluidized Beds by Power Spectrum Analysis of Pressure Fluctuations," *in* "Circulating Fluidized Bed Technology IV" (Amos A. Avidan, ed.), pp. 163–168. Somerset, Pennsylvania (1993).

O'Brien, Thomas J., and Syamlal, Madhava. "Particle Cluster Effects in the Numerical Simulation of a Circulating Fluidized Bed," *in* "Circulating Fluidized Bed Technology IV" (Amos A. Avidan, ed.), pp. 430–435. Somerset, Pennsylvania (1993).

Okes, E. J., Javen, J., and Engstrom, F. "Startup and Operating Experience at the Kattua 22-MWe Cogeneration Circulating Fluidized Bed Combustion Plant," Proc. Am. Power Conf., Vol. 44, p. 64 (1982).

Orjala, M., and Haukka, P. "Effect of CFB Conditions on Alkali Metals Removal from Hot Gas," *in* "Circulating Fluidized Bed Technology III" (Basu, P., Horio, M., and Hasatani, M., eds.), pp. 423–428. Pergamon Press, Oxford (1991).

Ormston, O., Robinson, E., and Buckle, D. "Stream Raising Circulating Fluidized Bed Plant," Inst. Energy Symp. ser. (London), 4 (Fluid. Combust.: Syst. Appl.), IIA/1-IIA/1/10 (1980).

Ouyang, S., and Potter, O. E. "Modelling Chemical Reaction in a 0.254 m I.D. Circulating Fluidized Bed," *in* "Circulating Fluidized Bed Technology IV" (Amos A. Avidan, ed.), pp. 515–520. Somerset, Pennsylvania (1993).

Owen, Hartley. "Heavy oil catalytic cracking process and apparatus" PATENT: United States; US 4917790 A, DATE: 900417 (1990).

Pan, Z., Ren, A., and Wang, T. "Upper Space Combustion Share and Heat Transfer in a Coal-Fired CFB Combustor," *J. of Engineering Thermophysics* (Chinese), Vol. 9, No. 1, pp. 81–85 (1988).

Pappal, D. A., and Schipper, P. H. "ZSM-5 in catalytic cracking: riser pilot plant gasoline composition analyses," Prepr.-Am. Chem. Soc., Div. Pet. Chem. vol. 35, No. 4, pp. 678–834 (1990).

Pappal, D. A., and Schipper, P. H. "Increasing motor octanes by using ZSM-5 in catalytic cracking," Riser pilot plant gasoline composition analyses. ACS Symp. Ser., 452, Fluid Catal. Cracking, pp. 45–55 (1991).

Park, D. W., and Gau, G. "Simulation of Ethylene Epoxidation in a Multitubular Transport Reactor," *Chem. Eng. Sci.*, vol. 41, No. 1, pp. 143–150 (1986).

Park, S. S., Choi, Y. T., Lee, G. S., and Kim, S. D. "Coal Combustion Characteristics in an Internal Circulating Fluidized Bed Combustor," *in* "Circulating Fluidized Bed Technology III" (Basu, P., Horio, M., and Hasatani, M., eds.), pp. 497–504. Pergamon Press, Oxford (1991).

Pell, M., and John, B. "Conoco-Stone & Webster Solids Circulating Fluidized Bed Combustion," *Coal Technology* (Houston), 4th (vol. 5), p. 15 (1981).

Pintsch, Stephan, and Gudenau, H. W. "Waste oil gasification in fluidized bed and circulating fluidized bed," *Erdoel Kohle, Erdgas, Petrochem.*, vol. 43, No. 12, pp. 494–502 (1990).

Plass, L. *et al.* "Environmentally Safe Coal Combustion in the Circulating Fluidized Beds," *Tech. Mitt.*, vol. 77, No. 6–7, p. 313 (1984).

Plass, L. *et al.* "Environmentally Safe Coal Combustion in the Circulating Fluidized Bed," VDI-Berr., 495, 181 (1984).

Plass, L. *et al.* "Concepts and Experiences in Operation Power Plant with Circulating Fluidized Bed Combustion," *VGB Kraftwerkstech.*, vol. 66, No. 9, p. 801 (1986).

Plass, L., and Anders, R. "Industrial Fluidized Beds Gain Significance," *Chem. Ind.* (Duesseldorf), vol. 36, No. 10, p. 601 (1985).

Plass, L., Daradimos, G., and Lienhard, H. "Power Plants with Circulating Fluidized-Bed Combustion: Plant Design and Operating Experience," VDI-Ber., 574, 125 (1985).

Plasynki, S. I., and Mathur, M. P. "Pressure Fluctuations Investigation for High Pressure Vertical Pneumatic Transport," *in* "Circulating Fluidized Bed Technology IV" (Amos A. Avidan, ed.), pp. 260–265. Somerset, Pennsylvania (1993).

Pugsley, T. S., Berruti, F., Godfroy, L., Chaouki, J., and Patience, G. S. "A Predictive Model for The Gas–Solid Flow Structure in Circulating Fluidized Bed Risers," *in* "Circulating Fluidized Bed Technology IV" (Amos A. Avidan, ed.), pp. 41–48. Somerset, Pennsylvania (1993).

Qi, C., and Ihab, H. Farag. "Heat Transfer Mechanism Due to Particle Convection in Circulating Fluidized Bed," *in* "Circulating Fluidized Bed Technology IV" (Amos A. Avidan, ed.), pp. 396–401. Somerset, Pennsylvania (1993).

Qi, C. M., Jin, Y., Yu, Z. Q., Bai, D. R., and Zhong, X. X. *Huagong Xuebao* (Chinese), vol. 41, No. 3, pp. 281–290 (1990).

Qi, C. M., Yu, Z. Q., Jin, Y., Bai, D. R., and Yao, W. H. *Huagong Xuebao* (Chinese), vol. 41, No. 3, pp. 273–280 (1990).

Qi, C. M., Yu, Z. Q., Jin, Y., Cui, X. L., and Zhong, X. X. Shiyiu, Lianshi (Chinese), No. 12, pp. 51–56 (1989).

Qian, G., and Li, J. "Particle-Velocity Measurement in CFB with an Integrated Probe," *in* "Circulating Fluidized Bed Technology IV" (Amos A. Avidan, ed.), pp. 320–325. Somerset, Pennsylvania (1993).

Qiu, K., Jin, B., Lan, J., and Zhang, M. "Desulphurization Experiments and Mathematical Model for FBC with Fly Ash Recycle," J. of Southeast University (Chinese), Vol. 20, No. 2, pp. 56–62 (1990).

Rassudov, N. S., and Gardenina, G. N. "Fuel Combustion in a Recirculating Fluidized Bed," Teploenergetika (Moscow), No. 12, p. 69 (1986).

Reddy, Karri S. B., and Knowlton, T. M. "The Effect of Fines Content on the Flow of FCC Catalyst in Standpipes," *in* "Circulating Fluidized Bed Technology IV" (Amos A. Avidan, ed.), pp. 589–594. Somerset, Pennsylvania (1993).

Reddy, Karri S. B., and Knowlton, Ted M. "Comparison of Group A Solids Flow in Hybrid Abgled and Vertical Standpipes," *in* "Circulating Fluidized Bed Technology IV" (Amos A. Avidan, ed.), pp. 296–301. Somerset, Pennsylvania (1993).

Redick, H. "Circulating Atmospheric Fluidized-Bed Combustion," VDI-Ber. 574, 71 (1985); Reh, L. "The Circulating Fluidized Bed Reactors—Its Main Features and Applications," *Chem. Eng. Process.*, vol. 20, No. 3, p. 117 (1986).

Reh, L. "Circulating in the Circulating Fluidized Bed," *Zem.-Kalk-Gips*, Ed. B, vol. 36, No. 9, p. 512 (1983).
Reh, L. "Fluidized Bed Processing," *Chem. Eng. Prog.*, vol. 67, No. 2, pp. 58–63 (1971).
Reh, L. et al. "Circulating Fluidized Bed Combustion, an Efficient Technology for Energy Supply and Environmental Protection," Inst. Energy Symp. (London), 4 (Fluid. Combust.: Syst. Appl.), VI/2/1–VI/2/11 (1980).
Reh, Lothar. "The circulating fluid bed reactor—a key to efficient gas/solid processing," *in* "Circ. Fluid Bed. Technol." (Basu, Prabir, ed.), pp. 105–118 (1986).
Reidick, H. "EVT-Achlstroem Circulating Fluidized-Bed Firing System in Steam Generators," VDI-BER., 601, 203 (1986).
Reimert, R., and Mehrling, P. "Wood Gasification in Lurgi Circulating Fluidized Bed Boiler Combustor," *Huagong Xuebao* Vol. 38, No. 4, p. 478 (1987).
Reimert, Rainer, and Mehrling, Peter. "Wood gasification in Lurgi circulating fluidized bed," Symp. For. Prod. Res. Int.: Achiev. Future, vol. 5, CSIR, Pretoria, S. Afr., pp. 1–18 (1985).
Rensfelt, Erik, and Ekstroem, Clas. "Fuel gas from municipal waste in an integrated circulating fluid-bed gasification/gas-cleaning process," *Energy Biomass Wastes*, No. 12, pp. 891–906 (1989).
Rhodes, M. J., Mineo, H., and Hirama, T. "Particle motion at the wall of the 305 mm diameter riser of a cold model circulating fluidized bed," *in* Preprints for 3rd Int. Conf. on CFB, Nagoya, Japan (Oct. 15–18, 1990).
Rhodes, M. J., Hirama, T., Cerutti, G., and Geldart, D. "Non-uniformities of solids flow in the risers of circulating fluidized beds," *in* "Fluidization VI" (Grace, J. R., Shemilt, L. W., and Bergougnou, M. A., eds.), pp. 73–80. Engineering Foundation, New York (1989).
Rhodes, M., and Cheng, H. "Operation of an L-Valve in a Circulating Fluidized Bed of Fine Solids," *in* "Circulating Fluidized Bed Technology IV" (Amos A. Avidan, ed.), pp. 284–289. Somerset, Pennsylvania (1993).
Roques, Y., Gauthier, T., Pontier, R., Briens, C. L., and Bergougnou, M. A. "Residence Time Distributions of Solids in a Gas–Solids Downflow Transported Reactor," *in* "Circulating Fluidized Bed Technology IV" (Amos A. Avidan, ed.), pp. 672–677. Somerset, Pennsylvania (1993).
Sahagian, J., and Engstrom, F. "Combustion of Petroleum Coke in an Ahistrom Pyroflow Circulating Fluidized Bed Boiler," Proc.-Refin. Dep. Am. Pet. Inst., 65, p. 585 (1987).
Saraf, A. V., Silerman, M. A., and Ross, J. L. "FCC Modeling Based on Advanced Feed Characterization Techniques," *in* "Circulating Fluidized Bed Technology IV" (Amos A. Avidan, ed.), pp. 559–564. Somerset, Pennsylvania (1993).
Saraiva, P. C., Azevedo, J. L. T., and Carvalho, M. G. "Modelling the Flow, Combustion and Pollutants Emission in a Semi-industrial CAFBC," *in* "Circulating Fluidized Bed Technology IV" (Amos A. Avidan, ed.), pp. 72–79. Somerset, Pennsylvania (1993).
Sato, Shuzo, Suzuki, Kozo, Morimoto, Yasuyuki Hashimoto, Hideo, Takahashi, Tsutomu, kodama, Seiji, and Takatsuka, Tooru. "Apparatus for fluidized cracking of residual oils," Japan Kokai Tokkyo Koho: JP 86196164 (1986).
Schaub, G., Remimert, R., and Hafke, C. "Utilization of Petroleum Residues for Combustion in the Circulating Fluid Bed and for Gasification," Erdoel Koohle, Erdgas, Petrochem., Vol. 40, No. 3, p. 123 (1987).
Shell, G. J., and Chen, C. L. "Fast Fluidized Bed Coal Gasification in a Process Development Unit," Intersoc. Energy Convers. Eng. Conf., 15th(1),642(1980)87. Fraely, L. D., Do, L. N., and Hasiao, K. H. "Circulating Fluidized Bed Boiler," Proc. Intersoc. Energy Covers. Eng. Conf., 15th(1), p. 337 (1980).
Shen, T. "Influence of Apparent Velocity upon the Regenerator Performance of a FCC Unit," *Petroleum Processing* (Chinese), No. 9, pp. 1–8 (1988).

Shen, X., and Xu, Y. "Study on Operation Performances of Pressurized Circulating Fluidized Bed," *J. of Southeast University* (Chinese), vol. 20, No. 2, pp. 23–28 (1990).

Shen, X., Zhou, N., and Xu, Y. "Experimental Investigation in to Heat Transfer in a Pressurized Circulating Fluidized Bed," *J. of Southeast University* (Chinese), vol. 20, No. 2, pp. 29–34 (1990).

Shen, Z., and Kwauk, M. "Mass transfer between gas and solid in fast fluidization," Institute of Chemical Metallurgy, Academic Sinica. (1985).

Sherwin, M. B., and Frank, M. E. "Process for Preparing Oxirane Compounds," U.S.P., 4,399,295.

Shimizu, T., Inagaki, M., and Furusawa, T. "Effects of Sulphur Removal and Ammonia Injection on NO_x Emission from a Circulating Fluidized Bed Combustor," *in* "Circulating Fluidized Bed Technology III" (Basu, P., Horio, M., and Hasatani, M., eds.), pp. 393–398. Pergamon Press, Oxford (1991).

Shingles, T., and Jones, D. H. "The Development of Synthol Circulating Fluidized Bed Reactors," *ChemSA*, vol. 12, No. 8, p. 179 (1986).

Shingles, T., and Silverman, R. W. "Determination of Standpipe Pressure Profiles and Slide Valve Office Discharge Coefficients on Synthol Circulating Fluidized-Bed Reactors," *Powder Technology*, vol. 47, No. 2, p. 129 (1986).

Shingles, T., and McDonald, A. F. "Commercial Experience with Synthol CFB Reactors," Preprings for the 2nd Int. Conf. on CFB, Compiegne, France, March 14–18 (1988).

Silverman, R. W., Thompson, A. H., Steynberg, A., Ukawa, Y., and Shing, J. "Fluidization V" (Ostergaard, K., and Sorenson, A., eds.), Engineering Foundation, New York (1986).

Silverman, M. A. "New Stone & Webster FCC Riser Terminator Commercial Performance," *in* "Circulating Fluidized Bed Technology IV" (Amos A. Avidan, ed.), pp. 577–582. Somerset, Pennsylvania (1993).

Soong, C. H., Tuzla, K., and Chen, J. C. "Identification of Particle Cluster in Circulating Fluidized Bed," *in* "Circulating Fluidized Bed Technology IV" (Amos A. Avidan, ed.), pp. 726–731. Somerset, Pennsylvania (1993).

Spooner, H., Peace, K., Martin, B., Chisty, T., and Brown, T. "Case History of Piney Creek Circulating Fluid Bed Project," *in* "Circulating Fluidized Bed Technology IV" (Amos A. Avidan, ed.), pp. 239–247. Somerset, Pennsylvania (1993).

Squires, A. M. "Gasification of Coal in High-Velocity Fluidized Beds," *in* "Feature Energy Prod. Syet., Lect. Pap. Int. Semin., Meeting Data 1975, Volume 2,509" (Denton, J. C., and Afgan, N. H., eds.), Academic: New York. (1986).

Squires, A. M. "The City College Clean Fuels Institute: Programs for (I) Gasification of Coal in High-Velocity Fluidized Beds and (II) Hot Gas Cleaning," Pap.-Clean Fuels Coal Symp., 2nd, 681. Inst. Gas Technol.: Chicago, III (1975).

Squires, A. M., U.S.P., 3840353.

Squires, A. M., U.S.P., 3957457.

Squires, A. M., U.S.P., 3957458.

Stappen, Michel L. M. van der, Schouten, Jaap C., and Bleek, Cor M. van den. "Application of Deterministic Chaos Analysis to Pressure Fluctuation Measurement in A 0.96 m^2 CFB Riser," *in* "Circulating Fluidized Bed Technology IV" (Amos A. Avidan, ed.), pp. 55–60. Somerset, Pennsylvania (1993).

Steynberg, A. P., Shingles, T., Dry, M. E., and Yukawa, Y., and Sasol, X. X. "Commercial Scale Experience with Synthol FFB and CFB Catalytic Fischer–Tropsch Reactors," *in* "Circulating Fluidized Bed Technology III" (Basu, P., Horio, M., and Hasatani, M., eds.), pp. 527–532. Pergamon Press, Oxford (1991).

Stomberg, L. "Fast Fluidized Bed Combustion of Coal," The 7th Int. Conf. on Fluidized Bed Combustion, Philadelphia, PA (1982).

Stringfellow, T. E., Sage, W. L., and Atabay, K. "Modifying Existing Furnace Designs to CFB," Proc. Am. Power Conf., pp. 47, 88 (1985).

Stroemberg, L. "Experience with Coal Combustion in a Fast Fluidized Bed," *Arch. Combust.*, vol. 1, No. 1–2, p. 95 (1981).

Suzuki, T., Hirouse, R., Takemura, M., Morita, A., Yano, K., and Hyvarinen, K. "Comparison of NO_x Emission Between Laboratory Modelling and Full Scale Pyroflow Boilers," *in* "Circulating Fluidized Bed Technology III" (Basu, P., Horio, M., and Hasatani, M., eds.), pp. 387–392. Pergamon Press, Oxford (1991).

Tadrist, L., and Cattieuw, P. "Analysis of Two-Phase Flow in a Circulating Fluidized Bed—Local Measurements by Using Phase Doppler Analyzer," *in* "Circulating Fluidized Bed Technology IV" (Amos A. Avidan, ed.), pp. 702–707. Somerset, Pennsylvania (1993).

Takasuka, H., Yabune, S., Miyamoto, S., Ichimura, S., Kanehira, S., and Togashira, K. "Mitsubishi–Lurgi Circulating Fluidized Bed Boiler," *in* "Circulating Fluidized Bed Technology III" (Basu, P., Horio, M., and Hasatani, M., eds.), pp. 335–340. Pergamon Press, Oxford (1991).

Takeuchi, H., and Hirama, T. "Flow visualization in the riser of a circulating fluidized bed," *in* Preprints for 3rd Int. Conf. on CFB, Nagoya, Japan (Oct. 15–18, 1990).

Talukdar, J., and Basu, P. "Sensitivity Analysis of a Performance Predictive Model of Circulating Fluidized Bed Boiler Furnace," *in* "Circulating Fluidized Bed Technology IV" (Amos A. Avidan, ed.), pp. 541–546. Somerset, Pennsylvania (1993).

Tan, R., Montat, Mareux, H., and Couturier, S. "About the Effect of Experimental Set-up Characteristics on the Validity of Heat Transfer Measurements in Circulating Fluidized Bed Mock-ups," *in* "Circulating Fluidized Bed Technology IV" (Amos A. Avidan, ed.), pp. 384–389. Somerset, Pennsylvania (1993).

Tang, J. T., Curran, R. A., and Taylar, T. E., "Challenges and Strategies for CFB Boilers to Meet Stringent Emissions," *in* "Circulating Fluidized Bed Technology IV" (Amos A. Avidan, ed.), pp. 766–773. Somerset, Pennsylvania (1993).

Tawara, Y., Furuta, M., Hiura, F., Ishida, Y., Uetani, J., and Harajiri, H. "NSC CFB Boiler: Pilot Plant Test and Industrial Application in Japan," *in* "Circulating Fluidized Bed Technology III" (Basu, P., Horio, M., and Hasatani, M., eds.), pp. 353–358. Pergamon Press, Oxford (1991).

Taylor, E. S. "Development, Design, and Operational Aspects of a 150 MW(e) Circulating Fluidized Bed Boiler Plant for the Nova Scotia Power Corporation," *in* "Circulating Fluidized Bed Technology" (Basu, P., ed.), p. 363. Pergamon Press, Canada (1986).

Tesch, M., Meili, R. T., and Reh, L. "Short Time Constant Dosing and Jet Mixing for the Preparation of Homogeneous Gas/Solid Suspensions," *in* "Circulating Fluidized Bed Technology IV" (Amos A. Avidan, ed.), pp. 666–671. Somerset, Pennsylvania (1993).

Tsukada, M., Nakanishi, D., and Horio, M. "Effect of Pressure on 'Transport Velocity' in a Circulating Fluidized Bed," *in* "Circulating Fluidized Bed Technology IV" (Amos A. Avidan, ed.), pp. 248–253. Somerset, Pennsylvania (1993).

Tung, Y., and Kwauk, M. "Fast Fluidization—A Growing Technology," *in* "Fluidization—Science and Technology" (Kwauk, M., and Kunii, D., eds.), pp. 106–129. Science Press, Beijing (1988).

Ushiki, K., Mizuno, M., Sakakibara, H., Utsumi, R., Hata, T., and Horiuchi, T. "Experimental Study on the Performance of Very Small Cyclones for the Recycle of Ultra-Fine Powder," *in* "Circulating Fluidized Bed Technology IV" (Amos A. Avidan, ed.), p. 607–612. Somerset, Pennsylvania (1993).

Ushiki, Kenichi, Nakamura, Takashi, Tanaka, Yasutaka, and Yuu, Shinichi. "A rise in solid holdup in an ascending flow type gas–solid contactor with the use of inertial gas–solid separator," *Chem. Eng. Commun.* vol. 108, pp. 185–199 (1991).

Van Der Ham, A. G. J., Prins, X. X., and Van Swaaij, W. P. M. "Regenerative, High Temperature Desulphurization of Coal Gas in a Circulating Fluidized Bed," *in* "Circulating Fluidized Bed Technology IV" (Amos A. Avidan, ed.), pp. 774–779. Somerset, Pennsylvania (1993).

Van Hook, J., and Robertson, A. "Circulating Pressurized Fluidized Bed Pilot Plant," *in* "Circulating Fluidized Bed Technology IV" (Amos A. Avidan, ed.), pp. 278–283. Somerset, Pennsylvania (1993).
Van Swaaij, W. P. M. "Chemical Reaction Engineering Review" (Luss, D., and Weekman, V. W., Jr., eds,), p. 193. ACS, Washington, D.C. (1978).
Wang, D. *et al.* "2.8 MWt Circulating Fluidized-Bed Combustor," *Gongcheng Rewuli Xuebao* (Chinese), vol. 7, No. 3, p. 266 (1986).
Wang, D., Tang, W., Zhang, Y., and Pan, Z. "The 10 T/h Circulating Fluidized Bed Boiler," *J. of Engineering Thermophysics* (Chinese), vol. 10, No. 4, pp. 459–461 (1989).
Wang, F., Li, Y., and Zeng, Q. "A Experimental Study on the New Type Fast Fluidized Bed for Clean Combustion (Chinese)," The Proceeding of 5th National Conference on Fluidization, pp. 375–378 (1990).
Wang, T., Yang, G., Zhang, Y., Pan, Z., Jiang, Z., Jiang, H., and Tang, W. "The Circulating Fluidized Bed Combustion With Staged Soiled Circulating," *in* "Circulating Fluidized Bed Technology III" (Basu, P., Horio, M., and Hasatani, M., eds.), pp. 479–484. Pergamon Press, Oxford (1991).
Wang, W. "Development of Fluidized Bed Calciner for Alumina (Chinese)," The Proceeding of 5th National Conference on Fluidization, Beijing, pp. 366–370 (1990).
Wang, D. X., Jin, Y., Yu, Z. Q., and Zhang, J. P. "A Study on Gas Liquid and Solid Holdups in a Three-Phase Fluidized Bed with Baffle (Chinese)," The Proceeding of 5th National Conference on Fluidization, pp. 168–171 (1990).
Wang, X. S., and Gibbs, B. M. "Combustion Behaviour of British Coals in a Circulating Fluidized Bed," *in* "Circulating Fluidized Bed Technology IV" (Amos A. Avidan, ed.), pp. 98–103. Somerset, Pennsylvania (1993).
Wang, Z. "Investigations on String Operation of Fast and Turbulent Fluidized Bed (Chinese)," The Proceeding of 4th National Conference on Fluidization, p. 201 (1987).
Wang, Z., Bai, D., and Jin, Y. *Powder Technology*, vol. 70, No. 3, pp. 271–275 (1992).
Wargalla, G., and Lotze, J. "Environmentally Safe Combustion of Low-Grade Coal in a Circulating Fluidized Bed," *Erzmetall*, vol. 38, No. 6 (1985).
Wedel, G. von, and Hirschfelder, H. "LLB Approach Pressurised Circulating Fluidized Bed Combustion Current Status and Future Development," *in* "Circulating Fluidized Bed Technology IV" (Amos A. Avidan, ed.), pp. 272–277. Somerset, Pennsylvania (1993).
Wei, F., Jin, Y., and Yu, Z. "The Visualization of Macro Structure of the Gas–Solids Suspension in CFB," *in* "Circulating Fluidized Bed Technology IV" (Amos A. Avidan, ed.), pp. 708–713. Somerset, Pennsylvania (1993).
Wei, F. "Comparison of the calculation methods of solids flux in commercial FCC units in: Refinery Design" (in Chinese), No. 5, pp. 40–43 (1990).
Wei, F., Liu, J., Jin, Y., and Yu, Z. "Gas Mixing in Cocurrent Downflow Circulating Fluidized Bed," *in* Proceeding of 6th National Meeting on Fluidization (in Chinese), Wu Han (1993).
Wei, F., Lou, X., Jin, Y., and Yu, Z. "One dimensional optical fibre cluster image system," *in* Proceeding of 6th National Meeting on Fluidization (in Chinese), Wu Han (1993).
Wei, F., Chen, W., Jin, Y., and Yu, Z. "Axial solids mixing in cocurrent downflow circulating fluidized bed," *in* Proceeding of 5th National Meeting on Chemical Reaction Engineering (in Chinese), Tianjing (1993).
Wei, F., Lin, S., and Yang, G. "Solids and Gas Mixing in Commercial FCC Regenerator," *Chemical Engineering & Technology*, vol. 16, 3, pp. 55–61 (1993a).
Weimer, R. F., Bixer, A. D., Pettit, R. D., and Wang, S. I. "Operation of a 49 MW Circulating Fluidized Bed Combustor," *in* "Circulating Fluidized Bed Technology III" (Basu, P., Horio, M., and Hasatani, M., eds.), pp. 341–346. Pergamon Press, Oxford (1991).
Wein, W. *et al.* "Stream Generator with Circulating Atmospheric Fluidized-Bed Combustion,"

Forschongsber.-Bundesminist. Forch. Technol., Technol. Forsch. Entwickle., BMFT-FB-T 82-134, 58 pp. (1982).
Weinell, C. E., Johansen, K. D., and Johnsson, J. E. "Single Particle Velocities and Residence Times in Circulating Fluidized Bed," in "Circulating Fluidized Bed Technology IV" (Amos A. Avidan, ed.), pp. 690–695. Somerset, Pennsylvania (1993).
Weiss, V., and Fett, F. N. "Modelling the decomposition of solidum bicarbonate in a circulating fluidized bed reactor," in "Circulating Fluidized Bed Technology" (Basu, P., ed.), pp. 167–172. Pergamon Press, Canada (1986).
Weiss, V., Fett, F. N., Helmrich, H., and Janssen, K. "Mathematical modelling of circulating fluidized bed reactors by reference to a solids decomposition reaction and coal combustion," *Chem. Eng. Process.* vol. 22, No. 2, pp. 79–90 (1987).
Weisweiler, W. "Environmentally friendly denitrification catalysts based on iron/manganese oxide/sulfate for use in simultaneous denitration and desulphurization of waste gases in a circulating fluidized bed," DECHEMA:Monogr., 118 (Katalyse), pp. 81–103 (1989).
Werdermann, Cord C. and Werther, J. "Heat Transfer in Large-Scale Circulating Fluidized Bed Combustors of Different Sizes," in "Circulating Fluidized Bed Technology IV" (Amos A. Avidan, ed.), pp. 521–528. Somerset, Pennsylvania (1993).
Werner, Kurt Friedrich Joachim. "Steam gasification of coal using a pressurized circulating fluidized bed," Ber. Kern-forschungsanlage Juelich, Juel-2312 (1989).
Westphalen, D., and Glicksman, L. "Experimental Verification of Scaling for a Commercial-Size CFB Combustor," in "Circulating Fluidized Bed Technology IV" (Amos A. Avidan, ed.), pp. 529–534. Somerset, Pennsylvania (1993).
White, C. C., and Dry, R. J. "The Effect of Particle Size on Gas–Solid Contact Efficiency in a Circulating Fluidized Bed," in "Circulating Fluidized Bed Technology III" (Basu, P., Horio, M., and Hasatani, M., eds.), pp. 569–574. Pergamon Press, Oxford (1991).
Wirth, K. E. "Prediction of Heat Transfer in Circulating Fluidized Beds," in "Circulating Fluidized Bed Technology IV" (Amos A. Avidan, ed), pp. 344–349. Somerset, Pennsylvania (1993).
Wittmann, E. "Preliminary Results with an Expanded Circulating Fluidized Bed for Cleaning the Flue Gases from the Schwanderf Trash-Burning Plant," *Chem.-Ing.-Tech.*, vol. 56, No. 9, p. 692 (1984).
Wu, B. "A Preliminary Dynamic Analysis on Phase Distribution Phenomena in FFB," in "Circulating Fluidized Bed Technology IV" (Amos A. Avidan, ed.), pp. 460–465. Somerset, Pennsylvania (1993).
Wu, R. L., Grace, J. R., Lim, C. J., and Brereton, C. M. "Suspension to Surface Heat Transfer in a Circulating Fluidized Bed Combustor," *AIChE J.*, vol. 35, pp. 1685–1691 (1989).
Wu, S., Alliston, M., Evardsson, C., and Probst, S. "Size Reduction, Residence Time and Utilization of Sorbet Particles in a Circulating Fluidized Bed Combustor," in "Circulating Fluidized Bed Technology IV" (Amos A. Avidan, ed.), pp. 780–787. Somerset, Pennsylvania (1993).
Xu, X., and Mao, J. "Mathematical Model and Simulation of CFBC Boilers," in "Circulating Fluidized Bed Technology IV" (Amos A. Avidan, ed.), pp. 104–109. Somerset, Pennsylvania (1993).
Yamazaki, R., Asai, M., Nakajima, M., and Jimbo, G. "Characteristics of Transition Regime to a Turbulent Fluidized Bed," in "Circulating Fluidized Bed Technology IV" (Amos A. Avidan, ed.), pp. 720–725. Somerset, Pennsylvania (1993).
Yamaoka, Hideyuki. "Flow characteristics of gas and powders at circulating fluidized bed reactor," *Tetsu to Hagane*, vol. 75, No. 3, pp. 424–431 (1989).
Yan, G. et al. "Experiments in 0.8 m^2 × 0.8 m^2 Circulating Fluidized Bed Cold Model (Chinese)," The Proceeding of 5th National Conference on Fluidization, Beijing, pp. 292–295 (1990).

Yang, G., Huang, Z., Chen, D., and Zhao, S. "Study of Fast Fluidized Bed for Converting Butene to Butadiene by Oxidative Dehydrogenation," *Petrochemical Technology* (Chinese), vol. 16, No. 10, pp. 680–685.
Yang, G. L., Huang, Z., Cheng, D. B., Zhao, S. K., and Liu, J. S. *Shiyiu Huagong* (Chinese), vol. 17, No. 1, pp. 38–42 (1988).
Yang, W. "Incipient Fast Fluidization Boundary and Operational Maps for Circulating Fluidized Bed Systems," *in* "Circulating Fluidized Bed Technology IV" (Amos A. Avidan, ed.), pp. 61–71. Somerset, Pennsylvania (1993).
Yang, Y. L., Jin, Y., Yu, Z. Q., and Wang, Z. W. *Shiyiu Huagong* (Chinese), vol. 20, No. 11, pp. 765–770 (1991).
Yang, Z. "Application of Heat Exchange by Gas–Solid Dilute Fluidization in Cement Industry (Chinese)," The Proceeding of 2nd National Conference on Fluidization, pp. 349–350 (1980).
Yaslik, A. D. "Circulating Difficulties in Long Angled Standpipes," *in* "Circulating Fluidized Bed Technology IV" (Amos A. Avidan, ed.), pp. 583–588. Somerset, Pennsylvania (1993).
Yasuna, J., Elliott, S., Westersund, H., Gamwo, I., and Sinclair, J. "Assessing the Predictive Capability of a Gas–Solid Flow Model which Includes Particle–Particle Interactions," *in* "Circulating Fluidized Bed Technology IV" (Amos A. Avidan, ed.), pp. 412–417. Somerset, Pennsylvania (1993).
Yerushalmi, J. "Circulating Fluidized Bed Boilers," Fuel Process. Technol., Vol. 5, No. 1–2, p. 25 (1981).
Yerushalmi, J. *et al.* "Production of Gaseous Fuels from Coal in the Fast Fluidized Bed," *Fluid. Technol.*, Proc. Int. Fluid. Conf., Meeting Data 1975, vol. 2,437. Keairns, D. L. (ed.), McGraw-Hill: New York (1976).
Yerushalmi, J., and Avidan, A. "High-Velocity Fluidization," *in* "Fluidization" (Davidson, J. F., Harrison, D., and Clift, R, eds.), pp. 225–291. Academic Press, London (1985).
Yerushalmi, J., Tuner, D. H., and Squires, A. M. "The Fast Fluidized Bed," *Ind. Eng. Chem., Process Des. Dev.*, vol. 15, No. 1, p. 47–53 (1976).
York, Y. P., Tsuo, Arun K. Basak, and Yam, Y. Lee. "Simulation of NO_x Emissions from Circulating Fluidized Bed Boilers," *in* "Circulating Fluidized Bed Technology IV" (Amos A. Avidan, ed.), pp. 760–765. Somerset, Pennsylvania (1993).
Yue, G., Li, Y., and Zhang, X. "The Design of 50 MW CFB Boiler with Planar Flow Separator," *in* "Circulating Fluidized Bed Technology IV" (Amos A. Avidan, ed.), pp. 625–629. Somerset, Pennsylvania (1993).
Yue, G., Wang, L., Li, Y., and Zhang, X. "Ash Size Formation Characteristics in CFB Coal Combustion," *in* "Circulating Fluidized Bed Technology IV" (Amos A. Avidan, ed.), pp. 110–115. Somerset, Pennsylvania (1993).
Zethraeus, B., and Ljungdahl, B. "A First Aproach to the Development and Behaviour of Particle Aggregates in the Central Part of a CFB Riser," *in* "Circulating Fluidized Bed Technology IV" (Amos A. Avidan, ed.), pp. 738–747. Somerset, Pennsylvania (1993).
Zhang, R. "Adoption of Two-Stage Regeneration Technique on a FCC Unit," Petroleum Processing (Chinese), No. 9, pp. 8–13 (1988).
Zhang, J. P., Jin, Y., and Yu, Z. Q. "Estimation of General Wake Model Parameters for a Three-Phase Fluidized Bed by Bed Collapse Technique (Chinese)," *in* The Proceeding of 5th National Conference on Fluidization, pp. 327–331 (1990).
Zhang, J.-X., and Pang, J.-Q. *Yiuqi Jiagong* (Chinese), vol. 2, No. 1, pp. 29–40 (1992).
Zhang, J., and Rudolph, V. "Flow Cessation of Standpipes in Circulating Fluidized Systems," *in* "Circulating Fluidized Bed Technology IV" (Amos A. Avidan, ed.), pp. 302–307. Somerset, Pennsylvania (1993).
Zhang, J. S., and Shi, W. Y. Report of Research Institute of Petroleum Processing (1987).
Zhang, J., and Wang, F. "Application and Development of Circulating Fluidization in Petroleum

Processing in China (Chinese)," *in* The Proceeding of 5th National Conference on Fluidization, Beijing, pp. 379–383 (19990).

Zhang, J., and Wang, F. "The Application and Development of Circulating Fluidization in China's Oil Refinery Industry," *in* "Circulating Fluidized Bed Technology III" (Basu, P., Horio, M., and Hasatani, M., eds.), pp. 533–536. Pergamon Press, Oxford (1991).

Zhang, W., Johnsson, F., and Leckner, B. "Characteristics of the Lateral Particle Distribution in Circulating Fluidized Bed Boilers," *in* "Circulating Fluidized Bed Technology IV" (Amos A. Avidan, ed.), pp. 314–319. Somerset, Pennsylvania (1993).

Zhang, Y., Arastoopour, H., Wegerer, D. A., Lomas, D. A., and Charles L. Hemler. "Experimental and Theoretical Analysis of Gas and Particles Dispersion in a FCC Unit," *in* "Circulating Fluidized Bed Technology IV" (Amos A. Avidan, ed.), pp. 571–576. Somerset, Pennsylvania (1993).

Zhang, Y., Li, Y., Zhang, X., Jin, D., and Duan, J. "Investigation on an Inertia Collector Using Slotting Tubes," *in* "Circulating Fluidized Bed Technology IV" (Amos A. Avidan, ed.), pp. 619–624. Somerset, Pennsylvania (1993).

Zhao, C., Chen, X., Jin, B., and Yang, M. "Experimental Investigation on Anthracite Combustion Characteristics in Fluidized Bed with Flyash Recycle," *J. of Southeast University* (Chinese), vol. 22, No. 5, pp. 83–89 (1992).

Zhao, C., Lan, J., Jin, B., Li, F., and Lu, Y. "Investigation of the Fluidized Bed Combustion with Flyash Recycle and Its Industrial Applications," *J. of Southeast University* (Chinese), vol. 20, No. 2, pp. 42–48 (1990).

Zhao, S., Liu, J., Wang, J., Sun, D., Zhao, X., and Yang, G. "Optimum Distribution of Gaseous Products of Oxydehydrogenation of Butene to Butadiene in a Fast Fluidized Bed," *Petroleum Technology* (Chinese), vol. 17, No. 1, pp. 38–42 (1988).

Zhao, S., Liu, J., Wang, J., Sun, D., Zhao, X., and Yang, G. *Shiyiu Huagong* (Chinese), vol. 17, No. 1, pp. 38–42 (1988).

Zhao, W. "Investigation on the Problem of Coke Deposition in FCCU," *Petroleum Processing* (Chinese), No. 9, pp. 26–30 (1992).

Zhao, W., and Xu, X. "The Adoption of Complete Combustion Two-Stage Regeneration Technology on a Resid FCC Unit," *Petroleum Processing* (Chinese), No. 4, pp. 1–14 (1989).

Zheng, Q., Ma, Z., and Wang, A. "Experimental Study on the Flow Pattern and Flow Behaviour of Gas–Solid Two Phase Flow in L-Valve," *in* "Circulating Fluidized Bed Technology IV" (Amos A. Avidan, ed.), pp. 290–295. Somerset, Pennsylvania (1993).

Zheng, Q., and Zhang, H. "Experimental Study of the Effect of Bed Exits with Different Geometric Structure on Internal Recycling of Bed Material in CFB Boilers," *in* "Circulating Fluidized Bed Technology IV" (Amos A. Avidan, ed.), pp. 175–180. Somerset, Pennsylvania (1993).

Zhong, J., Wang, S., and Jiang, W. *Huagong Xuebao* (Chinese), vol. 39, No. 4, pp. 461–468 (1988).

Zhou, Jainhua, Wang, Shieyu, and Jiang, Weisun. "A steady-state mathematical model for a catalytic cracking riser," *Chemical Engineering & Industry* (in Chinese), vol. 39, No. 4, pp. 461–468 (1988).

Zhu, J. X., and Bi, X. T. "Development of CFB and High Density Circulating Fluidized Beds," *in* "Circulating Fluidized Bed Technology IV" (Amos A. Avidan, ed.), pp. 565–570. Somerset, Pennsylvania (1993).

Zou, B., Li, H., Xia, Y., and Ma, X. "Cluster Structure in a Circulating Fluidized Bed," *in* "Circulating Fluidized Bed Technology IV" (Amos A. Avidan, ed.), pp. 714–719. Somerset, Pennsylvania (1993).

3. HYDRODYNAMICS

YOUCHU LI

INSTITUTE OF CHEMICAL METALLURGY
ACADEMIA SINICA
ZHONGGUAN VILLAGE, BEIJING, PEOPLE'S REPUBLIC OF CHINA

I. Scope of Fast Fluidization	86
A. Phenomena of Fast Fluidization	86
B. Onset of Fast Fluidization	89
C. Transition to Pneumatic Transport	91
D. Mapping Fast Fluidization	91
E. Conditions for Existence of Fast Fluidization	93
F. Essential Characteristics	94
II. Experimental Apparatus, Materials and Instrumentation	94
A. Types of Apparatus	95
B. Test Materials	97
C. Measurement Methods and Instrumentation	97
III. Experimental Findings	107
A. Voidage Profiles	107
B. Bed Structure	116
C. Particle Velocity	122
D. Gas Velocity	123
E. Mixing	124
F. Effect of Boundaries	135
Notation	140
References	141

The calcination of aluminium hydroxide in circulating fluidized beds, successfully developed by Lurgi Company, Germany, in the late 1960s (Reh, 1971), led to the exploitation of the special fluidization behaviors of fine particles, though its ability to admit high gas velocities had been recognized in even earlier studies on fluidization (Wilhelm and Kwauk, 1948; Lewis *et al.*, 1949). The longitudinal solids distribution in fast fluidized beds—dilute at the top and dense at the bottom with a possible point of inflection in the middle—could not be accounted for by one-dimensional analysis on accelerative motion for particulate fluidization (Kwauk, 1964). Yet, up to the mid-1970s, little had been done to elucidate the dynamics of fast fluidization.

Yerushalmi et al. (1976) reported on experimental observations on this subject for FCC catalyst systems. Experiments of fast fluidization on five kinds of solid particles were conducted at the Institute of Chemical Metallurgy (ICM), Academia Sinica (Li et al., 1979, 1981), and a fresh approach was proposed in the form of dynamic equilibrium between solid particles diffusion and segregation fluxes in the vertical direction (Li and Kwauk, 1980a), which led to an analytical expression of longitudinal solids concentration distribution for convenient application in engineering problems. A wave of studies on fast fluidization on the international front has since rapidly developed, their emphases being mostly focused on hydrodynamics and flow structure, i.e., axial and radial voidage profiles (Yerushalmi et al., 1976, 1978; Li et al., 1979, 1981; *Avidan and Yernshalmi*, 1982; Weinstein et al., 1984, 1986a, 1986b; Monceaux et al., 1986a, 1986b; Dry, 1987; Tung et al., 1988; Bai et al., 1987, 1990a; Choi et al., 1991), local and overall structure of the bed (Li et al., 1980b; Hartge et al., 1986, 1988; Monceaux et al., 1986a; Arena et al., 1991; Li et al., 1990; Azzi et al., 1991), velocity profiles of gas and solids (Monceaux et al., 1986b; Bader et al., 1988; Horio et al., 1988; Hartge et al., 1988; Zhang, 1990; Yang et al., 1991), effect of boundaries (Li et al., 1978; Jin et al., 1988; Xia and Tung, 1989; Mori et al., 1991), as well as mixing of gas and solids (Cankurt and Yerushalmi, 1978; Yang et al., 1984; Bader et al., 1988; Brereton et al., 1988; Adams, 1988; Wu, 1988; Li and Wu, 1991; Werther et al., 1991; Chesonis et al., 1991). At the same time, fast fluidization (or circulating fluidized bed) has been applied to coal combustion and many other industrial processes (see Chapter 2).

The hydrodynamics of G/S two phases flow, which often dominates mass transfer, heat transfer and overall chemical reaction, has in general played a decisive role in practical design. Based on experimental results, the related parameters of the model mentioned earlier were then correlated with physical properties of the solid particles and the flowing gas (Li et al., 1982). This flow model has since been used in the profession, though somewhat and often modified by other investigators in various alternative forms.

I. Scope of Fast Fluidization

A. PHENOMENA OF FAST FLUIDIZATION

The definition of fast fluidization by different investigators (Yerushalmi et al., 1978; Li and Kwauk, 1980a; Takeuchi et al., 1986; Reddy and Knowlton, 1991) has been far from consistent. However, the consensus has been that fast fluidization denotes a new type of bubbleless G/S contacting, mainly characterized by the aggregation of particles into clusters. Phenomenologically, fast fluidization possesses the following distinguishing features.

Axial Structure—Three types of axial voidage profile have been observed in fast fluidized beds (Li and Kwauk, 1980a), as shown in Fig. 1a. Curve 2 may be described as a complete profile with the voidage approaching a maximum ε^* at the top and an asymptotic minimum ε_a at the bottom, with a point of inflection z_i somewhere in the middle. Curve 1 may be understood to possess a point of inflection z_i located beyond the top of the fast column, and curve 3 has its z_i below the bottom of the apparatus. Physically, in the case of curve 1, the whole column is filled with solids in the dense phase; the flow behavior for curve 2, typical of fast fluidization, is characterized by the co-existence of a dilute phase region in the upper section and a dense phase region in the lower section; curve 3 belongs to the essentially uniform distribution of discrete solid particles in the dilute phase along the bed height.

Local Structure—Curve 1 on the left indicates that when the gas velocity is less than U_{tf}, bubbling (or slugging, in the case of small-diameter columns), or even turbulent fluidization would occur, as shown in Fig. 1b. In this case, the recordings of the capacitance or optical fiber probe installed in the bed would display its heterogeneity as low frequency signals, as shown in Fig. 1c (Li et al., 1982; Li et al., 1990). As gas velocity exceeds U_{tf}, fast fluidization ensues, while local heterogeneity monitored by the probe would diminish remarkably, demonstrating signals of increased frequencies, as shown in Fig. 1c. Inasmuch as this incipient fast fluidization velocity U_{tf} depends not only on the properties of the solids but is also affected by solids circulation rate (Li and Kwauk, 1980a; Li and Wang, 1985), solids need to be replenished at the bottom as fast as they are carried out at the top of the column, that is, equal to the saturation entrainment at U_{tf}. Unfortunately, this necessary prerequisite for fast fluidization was not recognized by some investigators (Yerushalmi et al., 1978; Perales et al., 1991).

As gas velocity increases to U_{pt}, the dense phase region at the bottom of the column would disappear, and the entire bed would be transformed into dilute-phase pneumatic transport, while the average voidage climbs to greater than 0.95, as shown by curve 3 in Fig. 1a. This transformation usually occurs with a sudden jump in voidage (Li and Kwauk, 1980a). However, the difference in capacitance or optical fiber signals would not warrant a clear distinction between the two, as can be recognized in Fig. 1c.

Radial Structure—However, experiments showed that the radial voidage profile for fast fluidization is quite different from that for pneumatic transport, as shown in Fig. 1d: a steep profile with an average voidage of less than 0.95 for the former, and a flat profile for the latter with an average voidage greater than 0.95 (Li et al., 1980b; Li and Wang, 1985; Tung et al., 1988). Therefore,

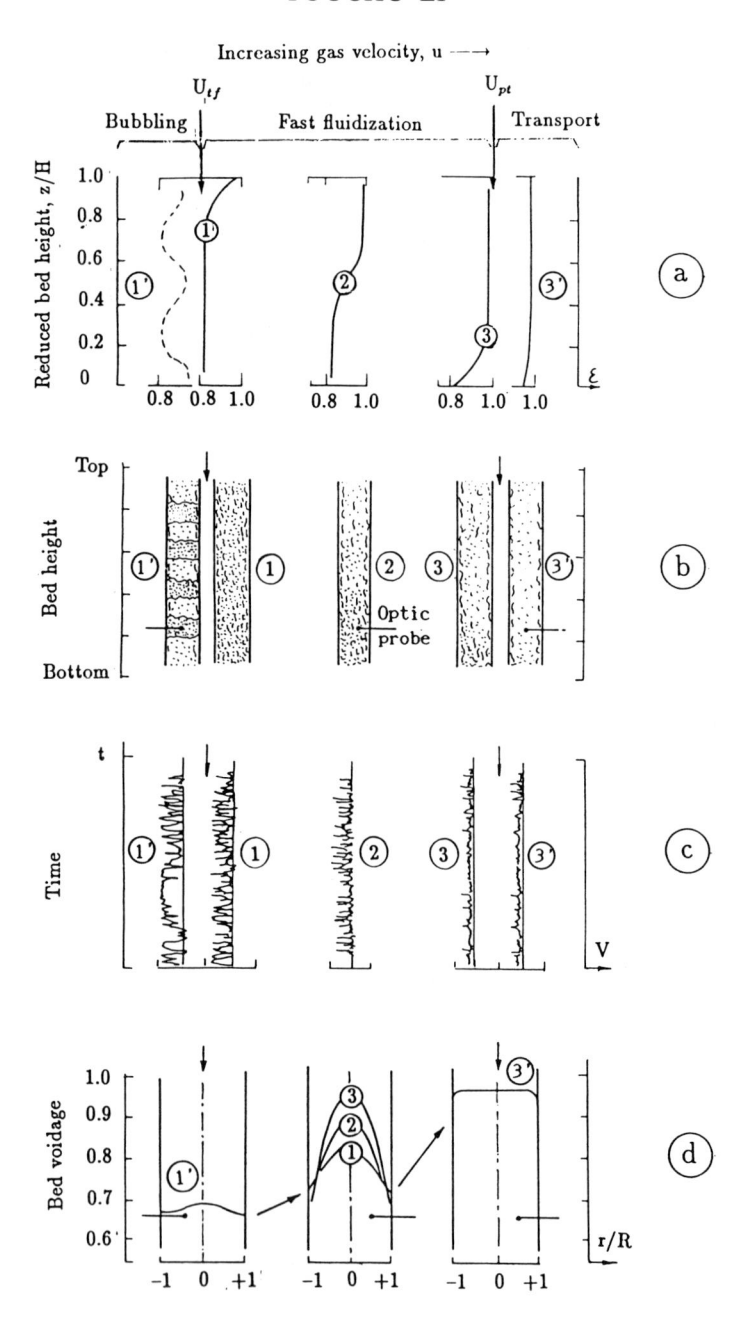

FIG. 1. Main phenomena of fast fluidization (after Li et al., 1982).

radial heterogeneity can serve as a better criterion for distinguishing pneumatic transport from fast fluidization.

B. Onset of Fast Fluidization

While the average voidage for classical fluidization is only related to gas velocity, the variation of bed voidage for fast fluidization is determined not only by gas velocity, but also by solids circulation rate, as mentioned earlier. The experimental data for FCC catalyst in Fig. 2 indicate that fast fluidization requires a gas velocity greater than 1.25 m/s and a solids circulation rate beyond 12 kg/m² s. At gas velocities less than 1.25 m/s, bed voidage is not much affected by solids circulation rate, as for classifical fluidization, while as gas velocity approaches 1.25 m/s with solids circulation rate below 12 kg/m² s, pneumatic transport with an average voidage greater than 0.95 would take place. As solids circulation rate gradually approaches 12 kg/m² s, a dense phase region would start to form at the bottom of the column, and this dense phase region would expand upward with further increase in the solids circulation rate. With any given solids circulation rate, as gas velocity further increases, this dense phase region would diminish, or even disappear throughout the column.

Based on these findings, it can well be appreciated that conditions for the onset of fast fluidization would change with gas and solids properties. Table I gives experimentally measured parameters at incipient fast fluidization, indicating that the U_{tf} is roughly 3.5 to 4 times that of the fall-free velocity

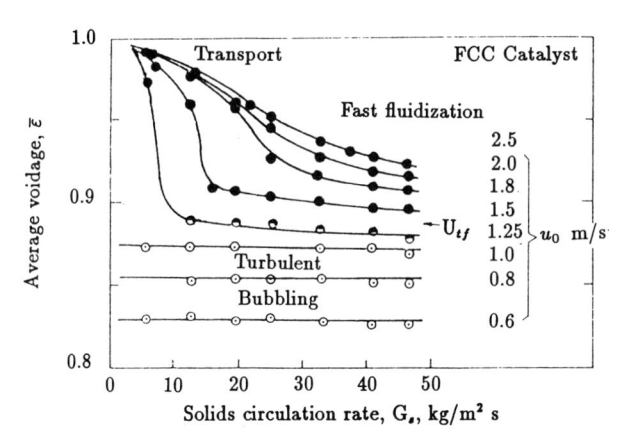

Fig. 2. Average voidage as a function of solids circulation rate and gas velocity (after Li and Wang, 1985).

TABLE I
PARAMETERS FOR INCIPIENT FAST POINTS

Materials	d_p μm	Ar	U_t m/s	U_{tf} m/s	G_{sm} kg/m².s	Re_c	Re_m
FCC catalyst	58	12.39	0.368	1.25	12	4.756	38.67
Fine alumina	54	17.75	0.485	1.70	18	6.022	54.0
Pyrite cinder	56	19.11	0.509	1.80	24	6.612	74.67
Coarse alum.	81	58.11	0.712	2.90	32	5.41	144.0
Iron ore	105	186.3	1.240	4.90	60	33.75	350.0

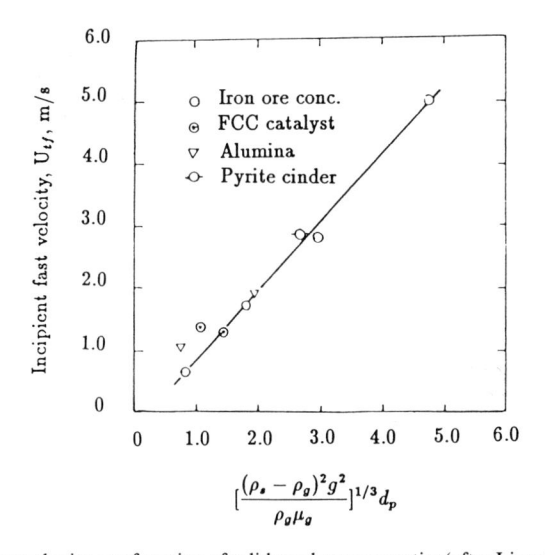

FIG. 3. Incipient velocity as a function of solids and gas properties (after Li and Wang, 1985).

of the particles. The incipient fast fluidization velocity U_{tf} has been correlated by the following equation (Li and Wang, 1985) to the physical properties of the gas and the solids, such as d_p, ρ_s, ρ_g, and μ_g, and shown in Fig. 3:

$$U_{tf} = 1.10\left[\frac{(\rho_s - \rho_g)^2 g^2}{\rho_g \mu_g}\right]^{1/3} d_p - 0.328, \tag{1}$$

or

$$Re_c = 1.10 Ar^{2/3} - 0.328\frac{d_p \rho_g}{\mu_g} \tag{2}$$

$$\approx 0.7584 Ar^{0.73}, \tag{2a}$$

where

$$\text{Re}_c = \frac{d_p \rho_s U_{tf}}{\mu_g}. \tag{3}$$

Similar relations have been presented by other authors (Horio, 1991; Perales et al., 1991). The minimum solids circulation rate can be estimated from the following empirical formula:

$$\text{Re}_m = 6.08 \text{Ar}^{0.78}, \tag{4}$$

where

$$\text{Re}_m = \frac{d_p \rho_s u_d}{\mu_g} = \frac{d_p G_{sm}}{\mu_g}. \tag{5}$$

C. Transition to Pneumatic Transport

For any given solids circulation rate, as gas velocity increases to U_{pt}, the dense-phase region at the bottom of the bed will disappear. Point U_{pt} shifts to the right as solids circulation increases, thus expanding the regime of fast fluidization, as can be seen in Fig. 4. Transition from fast fluidization to pneumatic transport is not subject to accurate observation and measurement. From experiments it has been found that the transition voidage at U_{pt} corresponds to an average value between ε_a and ε^*, that is, $\bar{\varepsilon} = (\varepsilon_a + \varepsilon^*)/2$, which generally ranges from 0.945 to 0.960. When both gas velocity and solids circulation rate are high, the two regions may not necessarily coexist, and when the average voidage of the bed lies below 0.95, fast fluidization rather than pneumatic transport is favored. Thus, by substituting the voidage value of 0.945 into Eq. (35), the transition velocity U_{pt} could be estimated:

$$18 \text{Re}_{pt} + 2.7 \text{Re}_{pt}^{1.687} - 20.23 \text{Ar} = 0, \tag{6}$$

in which

$$\text{Re}_{pt} = \frac{d_p \rho_g U_{pt}}{\mu_g} - 17 \left(\frac{\rho_g}{\rho_s} \right) \text{Re}_m. \tag{7}$$

D. Mapping Fast Fluidization

The influence of gas velocity and solids circulation rate on flow regime transition was measured in a fast fluidization, 90-mm i.d. and 8-m high (Li and Kwauk, 1980a). Figure 4 shows the change of the average voidage, $\bar{\varepsilon}$,

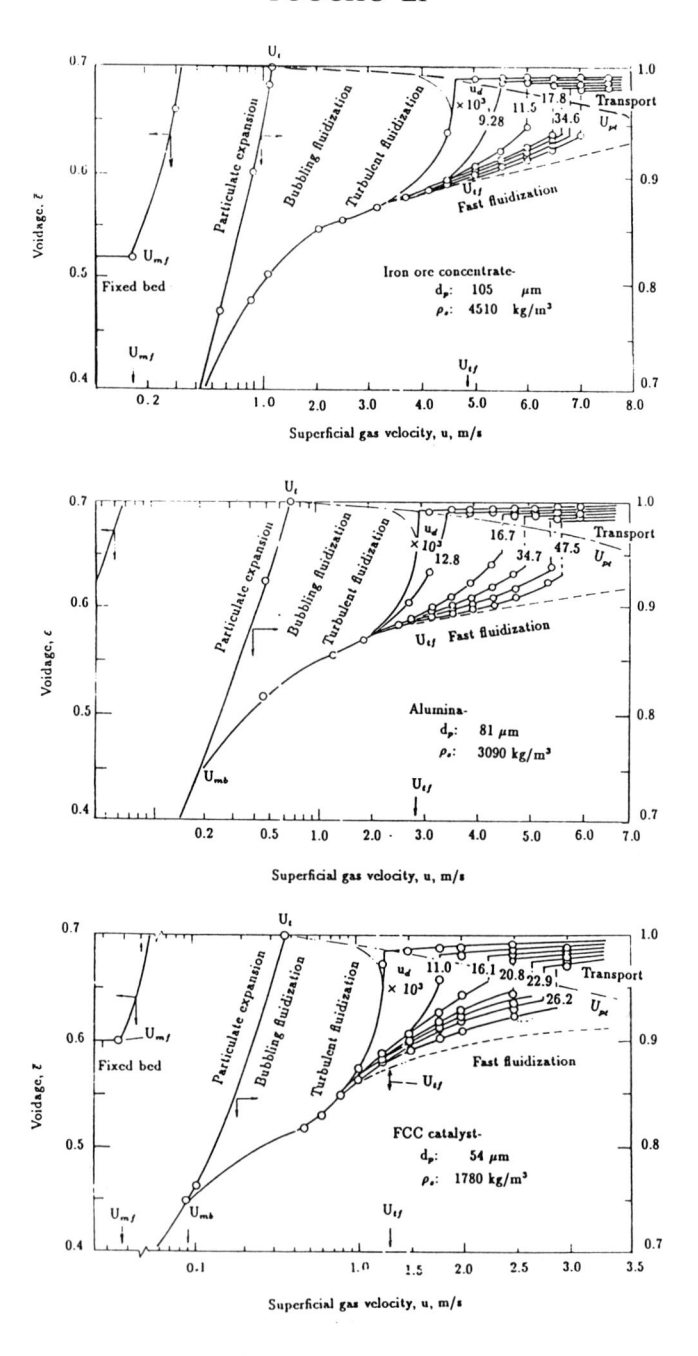

FIG. 4. Flow regimes for different materials (after Li et al., 1980a, b).

the voidage for the lower dense region $\bar{\varepsilon}_A$, and that for the upper dilute region, $\bar{\varepsilon}_B$. These maps for different solids materials demonstrate that the bed starts to expand from the fixed bed, followed by particulate expansion, bubbling fluidization, turbulent fluidization, fast fluidization and pneumatic transport in succession. While expanding, fine particles aggregate into clusters. Owing to the reduction of drag force on the particles when clusters form, it is possible to maintain dense phase fluidization even at gas velocities considerably in excess of the free fall, as shown in the bulged bed expansion curve of Fig. 4.

E. Conditions for Existence of Fast Fluidization

The necessary prerequisites to ensure fast fluidization can therefore be summarized as follows (Li and Kwauk, 1980a; Li and Wang, 1985):
- Sufficiently large solids circulation rate: $G_s > G_{sm}$
- Corresponding range of gas velocity: $U_{tf} < U < U_{pt}$
- Appropriate properties of solids

Comparison of the flow regime maps of Fig. 4 shows that with increasing particle size and/or particle density, the range for fast fluidization contracts, that is, the difference between ε_{tf} and ε_{pt} becomes smaller. The fluidization of coarse particles needs not consider interparticle forces which are small compared to gravity and drag. For fine particles, according to Mutsers and Rietem (1977), the measured cohesive force between 80 μm glass spheres, which belong to Geldart's Class A (1973), could be 10 to 100 times the particle weight. Van Deemter (1980) concluded from his experiments that the cavity structure for a fluidized bed of fine particles could easily break down into particle aggregates, or clusters, consisting of 10 to 100 particles under somewhat more severe flow conditions. With the larger and/or heavier Geldart's Class B particles, cohesion has much less influence, and clusters would consist of perhaps 2 to 10 particles. Class C represents very cohesive powders, the sizes of clusters of which could probably be beyond 1,000 or even more particles. Class D would move as individual particles because of negligible interparticle cohesive forces. Mechanistically, it seems therefore reasonable to expect that cohesion may result in clusters with their inherent ability of admitting more gas flow, thus leading to a wider regime of fast fluidization. Fujima *et al.* (1991) have conceptually studied the process of fast fluidization formation.

The preceeding observation and analysis on its flow structure indicate that fast fluidization should be referred to as a new type of bubbless gas–solids contacting with high operating parameters—high gas velocity, high solids throughput, high solids concentration, all resulting from the aggregative nature of fine powders.

F. Essential Characteristics

Fast fluidization is a regime intermediate between bubbling fluidization and pneumatic transport, possessing many of the advantages of both, but divorced from many of their disadvantages. It has the following unique characteristics:

- As a phase-reversed counterpart of bubbling fluidization, bubbles are replaced by clusters, which are subject to rapid dissolution and reformation, thus leading to improved G/S contacting.
- The operating gas velocity is severalfold larger than the terminal velocity of the solid particles, thus resulting in increased throughput and processing capacity of reactors.
- The high slip velocity between the gas and the particles favors both G/S mass and heat transfer and treatment of cohesive solids.
- High solids circulation rate results in high solids concentration, thus ensuring high volumetric utilization of equipment and increased bed-to-immersed-surface heat transfer.
- Low gas backmixing and axial dispersion favor improved conversion and selectivity of chemical reactions, especially for fast and complex reactions.
- Rapid solids mixing ensures uniform bed temperature, and therefore optimal conversion.
- Less pronounced effect of reactor diameter and gas distribution design on operation makes scale-up easier.

These attractive features bespeak the high efficienty of fast fluidized beds for both physical and chemical processing.

II. Experimental Apparatus, Materials and Instrumentation

The hydrodynamics of fast fluidization depends greatly on the boundaries of the retaining vessel, e.g., the wall, the inlet and the outlet. Inasmuch as different geometric structures were employed by different investogators in their experimental apparatus, inconsistencies in the resulting data are unavoidable. This will be the main topic for the present section.

In measurement and visualization of gas–solid flow in fast fluidization, various techniques have been developed and used, such as pressure gradient, pressure fluctuation, capacitance probe, optical fiber probe, momentum probe, laser Doppler velocimeter, night-television, and video camera. The resulting data on local gas and solid velocity, solids concentration and its

HYDRODYNAMICS 95

distribution, as well as the size and structure of clusters, provide a physical basis for modeling of the two-phase flow.

A. Types of Apparatus

Figure 5 shows typical test facilities used by different investigators. In configuration, they all consist of a fast column or riser, a gas–solids separator, a downcomer for solids recycle, and a loop seal valve and/or an additional controlling device for adjusting solids circulation rate, mounted in positions appropriate with the inlet and outlet geometry. These facilities can be grouped into three types.

1. Type A, Independent Free System

The main features of this type of apparatus, shown in Fig. 5a (Li et al., 1978), are:

- Firstly, the solids rate controlling device is placed at the top of the downcomer, and hence solids rate will not be affected by gas–solid flow in the system;

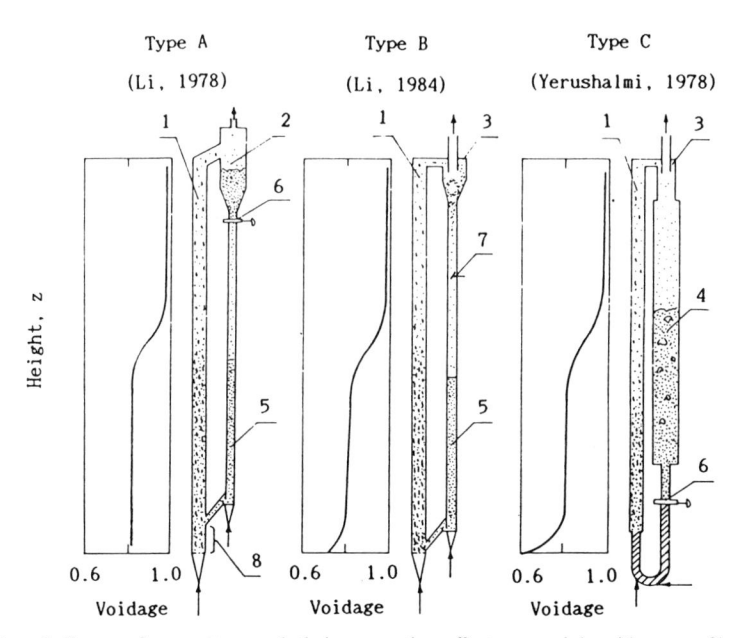

FIG. 5. Types of apparatus and their respective effects on axial voidage profile. 1, fast fluidized bed; 2, intermediate hopper; 3, cyclone; 4, slow fluidized bed; 5, downcomer; 6, solids rate controlling device; 7, solids rate measurement device; 8, suspension section.

- Secondly, an intermediate hopper is mounted above the solids rate controlling device to accommodate the change of solids inventory for different operation conditions;
- Thirdly, the pneumatically actuated valve, such as the V-valve installed at the lower end of the downcomer, only serves as a loop seal rather than a control, thus resulting in a "free system," that is, it does not influence gas–solids flow;
- Fourthly, the port for solids return into the fast column is located at a far enough distance from the gas distributor to minimize the effect of the inlet on flow.

As a result, gas velocity and solids circulation rate can be adjusted separately and independently. For any given gas velocity and solids circulation rate, there is only a unique equilibrium solids inventory in the fast fluidized column, which determines the height of the dense-phase region at the bottom.

2. Type B, Dependent Free System

As Fig. 5b (Li et al., 1984) shows, there is no intermediate solids hopper as for type A, and the solids rate device is used only for measurement rather than for control. In operation, solids circulation rate is determined essentially by gas velocity. Therefore, gas velocity and solids circulation rate could not be adjusted independently. Solids circulation rate follows what an operating velocity produces at whatever solids inventory the system was initially filled with. Owing to the absence of an intermediate hopper, the initial solids inventory could influence directly axial voidage profile, that is, the position of the inflection point changes with the initial solids inventory in the system.

3. Type C, Restrained System

As Fig. 5c (Yerushalmi et al., 1978) shows, this type has a large-diameter downcomer, and a solids rate controlling valve mounted at the lower end of the downcomer, resulting in a "restrained system." When the solids rate valve is fully open, its operating characteristics are essentially the same as those for Type B; while opening of the valve is small, it operates like Type A. In other words, Type C operates in an intermediate mode between Type A and Type B. Solids circulation rate is influenced jointly by solids inventory, gas velocity, and the opening of the controlling valve.

As regards inlet location, experiments showed that if the port for solids return is located far above the gas distributor, as for Type A (Li and Kwauk, 1980a), the solid particles would be immediately suspended after they enter, thus minimizing the inlet effect due to solids acceleration, with the result of a uniform axial voidage profile near the solids inlet at the bottom. If this port is close to the gas distributor, as for Type C, particles acceleration, especially for large/heavy particles, would bring about the non-uniform axial voidage profile shown for Type C apparatus.

The structure of the inlet (including the solids return device), such as V-shaped, L-shaped, U-shaped, pneumatically actuated, butterfly or slide, possesses each its unique effect on gas–solid flow.

The outlet may also have different configurations: right-angle bend, right-angle bend with guiding baffles, right-angle bend with a constricted horizontal tube. Experiments showed that as long as the outlet structure causes a dense layer at the top of the fast column, additional solids back-flow will occur, leading to increased solids concentration in the upper section of the fast fluidized bed.

If the effect of boundaries on gas–solid flow in fast fluidized beds were overlooked, gas/solids interaction might well be misunderstood.

B. TEST MATERIALS

The following test materials have often been used: FCC catalysts, aluminium oxide, silica gel, glass beads, silica or quartz sand, sea sand, coal and coal ash, petroleum coke, metal powders, resin particles, boric acid, and magnesite powder. Mean particle size ranges from 11 μm to 1,041 μm, and particle density, from 384 kg/m^3 to 7,970 kg/m^3. According to Geldart's classification (1973), most of these materials belongs to Class A, some to Class B, and a few to Class D or C, as listed in Table II.

C. MEASUREMENT METHODS AND INSTRUMENTATION

1. Average Solids Concentration

Traditionally, average solids concentration, or cross-sectional mean voidage, is determined by pressure gradient (Yerushalmi et al., 1976, Li et al., 1979; Hartge et al., 1986; Bai et al., 1987; Mori et al., 1988; Horio et al., 1988; Son et al., 1988; Schnitzlein and Weinstein, 1988; Glicksman et al., 1991; Ishii and Murakami, 1991; and others). Taps are installed along the fast fluidized column at given intervals and connected to U-tube manometers or to pressure transducers for monitoring pressure drops of the sections. In fast fluidization, pressure drops resulting from friction between the gas–solids mixture and the wall are quite small as compared to static solids pressure

TABLE II
Test Materials and Their Physical Properties

Materials	d_p μm	ρ_s kg/m³	Class.	D_t mm	H m	Investigators
FCC catalyst	49	1,070	A	152	8	Yerushalmi, 1979
Dicalite	33	1,670	A	152	8	Yerushalmi, 1979
HFZ-20	49	1,450	A	152	8	Yerushalmi, 1979
FCC catalyst	58	1,780	A	90	8	Li, 1979, 1980
FCC catalyst	59	900	A	144	8	Monceaux, 1986
FCC catalyst	94	1,546	A	186	8	Bai, 1987
FCC catalyst	60	1,000	A	50	2.8	Horio, 1988
FCC catalyst	76	1,714	A	305	12.2	Bader, 1988
FCC catalyst	85	1,500	A	400	8.5	Hartge, 1988
FCC catalyst	59	1,460	A	152	8	Schnitzlein, 1988
FCC catalyst	70	1,700	A	120	5.8	Arena, 1988
FCC catalyst	54	929	A	90	10	Tung, 1988
Vericulite	200	384	A	150	5.5	Bolton, 1988
Synclyst	60	1,000	A	150	5.5	Bolton, 1988
FCC catalyst	69	1,474	A	186	8	Yang, 1991
FCC catalyst	75	1,600	A	190	11.7	Azzi, 1991
FCC catalyst	46	1,780	A	50		Ishii, 1991
FCC catalyst	61	1,780	A	200		Ishii, 1991
FCC catalyst	57	930	A	100	5.5	Takeuchi, 1991
FCC catalyst	74	1,770	A	150	3	Kato, 1991
FCC catalyst	46	2,300		205	6.7	Nowak, 1991
FCC catalyst	56	729	A	50	4.6	Mori, 1991
FCC catalyst	64	(880)		200	6	Nakajima, 1991
Al(OH)$_3$	103	2,460	A–B	152	8	Yerushalmi, 1979
Alumina	54	3,160	A	90	8	Li, 1979
Alumina	81	3,090	A–B	90	8	Li, 1979
Alumina	64	1,800	A	152	6.4	Rhodes, 1986, 1988
Alumina-25	42	1,020	A	152	6.4	Rhodes, 1986, 1988
CBZ-1	38	1,310	A	152	6.4	Rhodes, 1988
Alumina	70	2,430	A	350	7	Rhodes, 1991
Alumina	43	2,003	A	90	10	Tung, 1988
Alumina	121	3,460	B	100	6.4	Chesonis, 1991
Silica gel	56			400	8	Hartge, 1986
Silica gel	187	703	A	186	8	Bai, 1987
Silica gel	603	789	B	186	8	Bai, 1987
Silica gel	1,041	1,303	B–D	186	8	Bai, 1987
Silica gel	165	711	A	186	8	Jin, 1988
Silica gel	80	706	A	187	8	Jin, 1988
Glass beads	157	2,420	A–B	152	8	Yerushalmi, 1979
Glass beads	90	2,543	A	400	10.5	Arena, 1986, 1988
Ballotini	88	2,600	A	41	6.4	Arena, 1991
Glass beads	240	2,450	B–D	150	3	Mori, 1991
No. 85 sand	268	2,650	D	152	8	Yerushalmi, 1979

TABLE II (continued)

Materials	d_p μm	ρ_s kg/m³	Class.	D_t mm	H m	Investigators
Sand	970	2,630	D	300 × 400	4	Brereton, 1986
Sand	250	2,630	D	300 × 400	4	Brereton, 1986
Sand	550	2,630	D	300 × 400	4	Brereton, 1986
Quartz sand	78	2,666	A	186	8	Bai, 1987
Quartz sand	31	2,672	A	186	8	Jin, 1988
Quartz	650	2,666	D	186	8	Jin, 1988
Silica sand	87	(1,200)	A	200	6	Nakajima, 1991
Sand-A	180	(1,536)	A–B	80	2.8	Wang, 1991
Sand-B	248	(1,641)	B–D	80	2.8	Wang, 1991
Sand-C	256	(1,634)	B–D	80	2.8	Wang, 1991
Sand	750	2,630	D	200	4	Cen, 1988
Sand	430	2,630	D	380	9.1	Son, 1988
Ottawa sand	185	2,700	B	152 × 152	7.3	Glicksman, 1991
Silica sand	140	2,522	B	400 × 300	4.5	Zheng, 1991
Raw coal	65	1,670	A	150	5.5	Li, 1984
Raw coal	4,000	1,700	D	200	4	Cen, 1988
Coal ash	939	2,200	D	186	8	Bai, 1987
Petrol cake	168	2,000	A–B	100	6.4	Chesonic, 1991
Pyri. cinder	56	3,050	A	90	8	Li, 1980
Magnesite	11	3,072	C	150	5.5	Li, 1983
Boric acid	162	1,435	A	50	3	Li, 1990
Iron ore	105	4,510	B	90	8	Li, 1980
Iron ore	125	5,250	B	150	3	Mori, 1991
Steel powder	52	7,300	A–B	33.6 × 33.6	1.6	Glicksman, 1991
Copper powder	35	7,970	A–B	33.6 × 33.6	1.6	Glicksman, 1991
Resin parti.	443	1,393	B	400 × 300	4.5	Zheng, 1991

heads, and hence could be neglected. The average solids concentration may be described as

$$\rho_b = \frac{dP}{dz} = \rho_s(1 - \varepsilon)g, \qquad (8)$$

or

$$\varepsilon = 1 - \frac{dP}{\rho_s g dz}. \qquad (9)$$

Arena *et al.* (1988) used a set of quick-closing valves to determine bed voidage, and by comparing their results with those determined by pressure drop, they confirmed that the two techniques agree with each other.

2. Local Solids Concentration

The optical fiber probe is an effective tool for measuring local voidage in fluidized beds. Qin and Liu (1982) developed a probe with a 2 × 2 mm² surface, and composed of two optical fiber groups arranged in alternate layers, one for illumination and the other for receiving reflected signals from solid particles, as shown in Fig. 6. Intensity of the reflected light is proportional to solids concentration, and the following relation was established between voidage and output signal voltage:

$$V = 4.836k \left\{ (1-\varepsilon)^{2/3} + \frac{[1-(1-\varepsilon)^{2/3}](1-\varepsilon)^{4/3}}{2.599 d_p^2} \right\}, \qquad (10)$$

in which k is an overall probe constant relative to size, shape and surface reflectance of particles. The probe was used in measuring radial voidage and its probability density function in fast fluidization (Li et al., 1980b), exhibiting rapid and sensitive characteristics. This technique has been widely adopted in studying fluidization (Tung et al., 1988; Horio et al., 1988; Hartge et al., 1988; Ishii and Murakami, 1991; Kato et al., 1991). Recently, Reh and Li (1991) have developed a crossed optical fiber probe which provides strong back-scattering signals from a well-defined small bed volume and possesses improved linear response to bed density.

3. Mass Flux

Mass flux of solids in the fast column can usually be measured by using internal sampling probes (Monceaux et al., 1986b; Bader et al., 1988; Rhodes

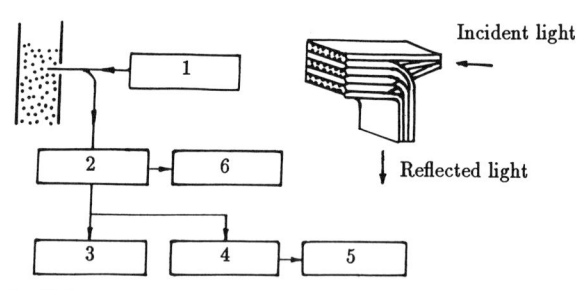

1—Light source;   2—Photomultiplier;   3—High voltage source
4—Probability density analyzer;   5—X-Y recorder
6—Timing digital integraph

FIG. 6. Instrumentation for optical-fiber measurement and alternate-layer arrangement of fibers (after Qin and Liu, 1982).

et al., 1988), from which it can be calculated from the amount of solid collected, divided by time of collection and area of the sampling tube. Isokinetic sampling is difficult to accomplish since gas velocity in the fast column varies significantly over its cross-section. Recently, Azzi et al. (1991) have developed a momentum probe composed of adjacent stainless steel tubes, as shown in Fig. 7. As the two branches are in opposition, the differential pressure DP between the tubes is a measure of the upward minus the downward momenta, that is,

$$(DP)_{local} = k_m(C_s v_s^2 + \rho_g \varepsilon_g u_g^2). \quad (11)$$

When the momentum of the gas is neglected,

$$(DP)_{local} = k_m C_s v_s^2. \quad (12)$$

Then the mass flux Φ_s is

$$\Phi_s = (C_s DP/k_m)^{1/2}, \quad (13)$$

in which k_m is a probe constant. When local solids concentration is known, mass flux can be calculated from the preceding momentum measurement.

4. Particle Velocity

If the solids mass flux Φ_s is measured as shown earlier, then the particle velocity can be calculated:

$$v_p = \Phi_s/C_s, \quad (14)$$

or

$$v_p = (DP/k_m C_s)^{1/2} \quad (15)$$

Particle momentum DP can be measured by a pitot tube (Bader et al., 1988), as well as by the momentum probe described earlier.

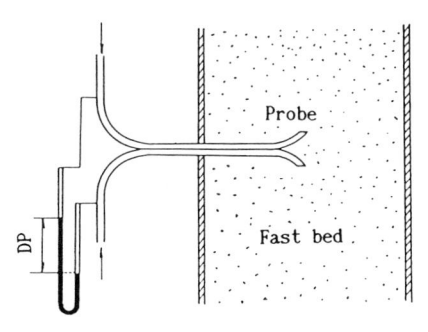

FIG. 7. Momentum probe measurement (after Azzi et al., 1991).

Another method of particle velocity measurement is the double tip optical fiber, which consists of two groups of optical fibers with a certain distance between each other, as shown in Fig. 6. When a solid particle passes the probe tip, two signals with a certain time interval in between would be detected. This time lag can be computed by means of the cross-correlation function (Qin and Liu, 1982):

$$R_{ab}(\tau) = \lim_{T \to \infty} \frac{1}{T} \int_0^T A(t)B(t+\tau)\,dt. \qquad (16)$$

From the distance between the two groups of optic fibers, and the time lag between the signals, both the particle velocity and probability velocity density function can be computed (Li *et al.*, 1980; Horio *et al.*, 1988; Hartge *et al.*, 1988; Nowak *et al.*, 1991).

The fiber-optic laser doppler velocimeter (LDV) is also used to measure particle velocity in gas–solid two-phase flow (Yang *et al.*, 1991; Bai, 1991). Figure 8 shows a schematic diagram of the working principle. The

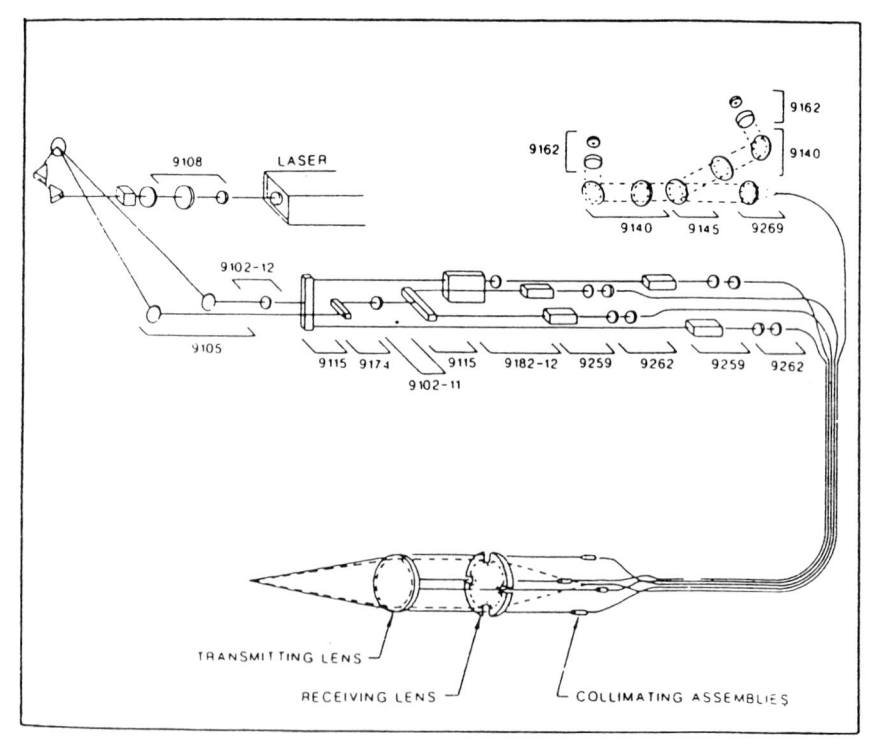

FIG. 8. Schematic diagram of light beam arrangement for TSI-LDV (after Yang *et al.*, 1991).

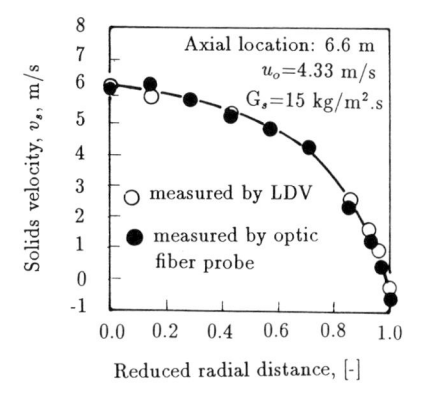

FIG. 9. Comparison of measured results by LDV and by optical fiber (after Yang et al., 1991).

LDV was operated in the backward scatter mode to obtain simultaneous measurement of two-dimensional velocity components. The resultant signals are transferred to a data acquisition system and processed on-line by a computer. A frequency shifter is employed to measure vertical particle velocity in both directions. As shown in Fig. 9, the LDV measured local particle velocity agree well with those from the optical fiber probe (Yang et al., 1991).

5. Flow Behavior Detection

In gas–solids fluidization non-uniformities always exist in space as well as time, leading to variations of local solids concentration and momentum. These differences as can be monitored by pressure fluctuation analysis (Yerushalmi et al., 1979; Wisecarver et al., 1986; Cai, 1989), needle capacitance probe (Li and Kwauk, 1980a) and optical fiber probe (Li and Kwauk, 1980a; Li et al., 1990), have been used to identify the transition of flow regimes. For that purpose, the random signals from these probes were processed through average amplitude, auto-correlation, cross-correlation and energy spectrum to judge transformation of fluidization regimes and the degree of gas/solids contacting.

6. Flow Structure Visualization

In early studies on fast fluidization (Yerushalmi et al., 1979; Li and Kwauk, 1980a), clustering of particles was recognized as a fundamental phenomenon, and much attention has since been given to the visualization of bed structure. Qin and Liu (1982) developed a fluorescent particle visualization technique, as shown in Fig. 10. Particles covered with a long-decay fluorescent powder

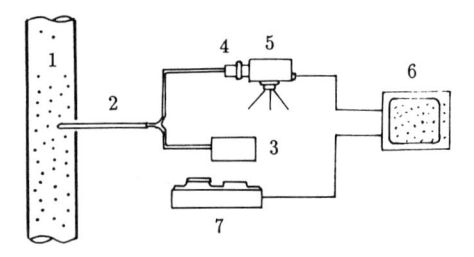

FIG. 10. Connection diagram for video camera to optical fiber probe (after Qin and Liu, 1982). 1, fluidized bed; 2, optical fiber probe; 3, focusing light source; 4, lens system; 5, video camera; 6, TV monitor: 7, video recorder.

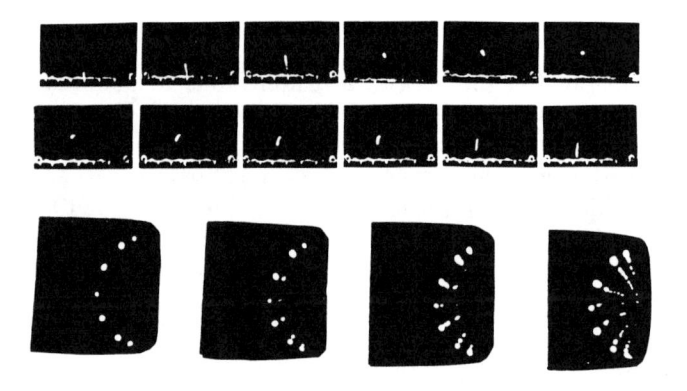

FIG. 11. Track of motion for particles flowing out from a side hole of a liquid-fluidized bed (after Qin and Liu, 1982).

were excited by illumination with ultraviolet from a point source of an optical fiber, to become fluorescent particles. The movement of these fluorescent particles can be tracked by visual observation in the dark and their tracks could also be recorded by a night-vision video camera after adequate photomultiplication, as shown in Fig. 11.

The bed structure of a two-dimensional fluidized bed, 550 mm in width, 15 mm in thickness and 2,600 mm in height (Bai et al., 1991), was studied through back lighting by observing and photography in the front. The photographs showed that the solid aggregates are U-shaped in the lower dense region, strand-shaped in the upper dilute region, as shown in Fig. 12.

Li et al. (1991) measured the size of clusters by using a video camera provided with a special optical fiber micrograph probe, as shown in Fig. 13.

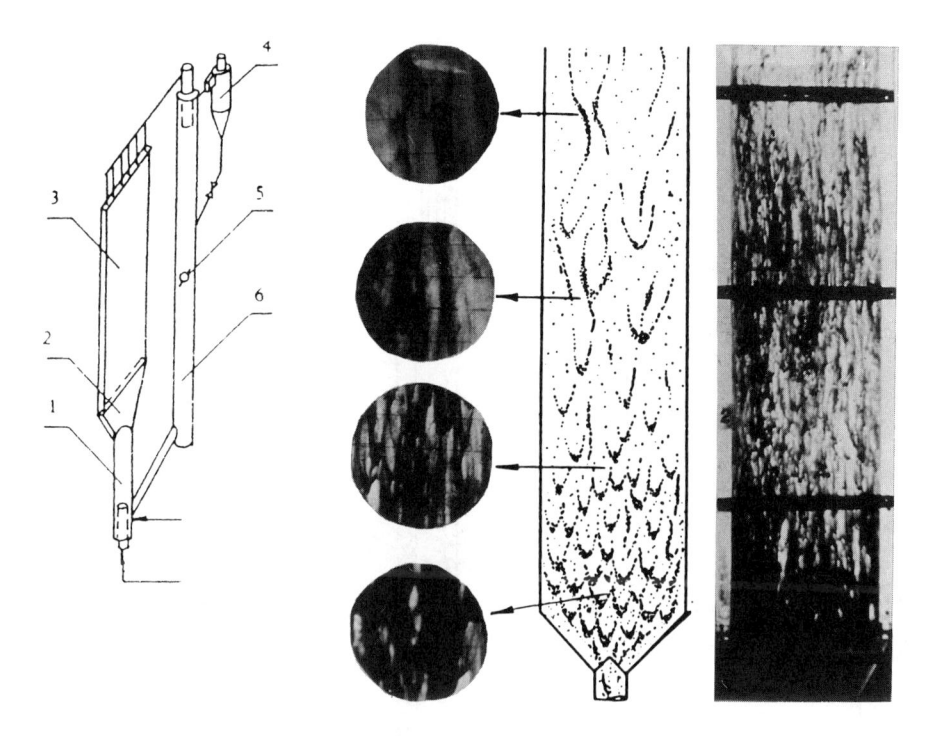

FIG. 12. Axial variation of appearance for clusters in a two-dimensional fast fluidized bed (after Bai et al., 1991). 1, riser; 2, distributor; 3, two-dimensional bed; 4, cyclone; 5, butterfly valve; 6, downcomer.

This probe consists of a set of lenses and an optical fiber flash light transmitter. The video camera is connected to an image collection plate and a TV monitor. The micrograph can also be recorded on a computer disk for future reproduction and data analysis. Experimental results indicated that the center of the bed, the shape of most clusters is strandlike, while near the wall, they are essentially spherical. A similar measurement system was presented by Takeuchi and Hirama (1991).

A holographic imaging technique, shown in Fig. 14, for studying clusters was developed and tested by Qu et al. (1985), consisting of an optical system where two lenses of high resolution are arranged with the same focus so that a collimated beam into the measuring system remains collimated when it leaves. The particles being photographed are transferred to become their images on the recording film. The holograms show the size, shape and number of clusters of test materials such as FCC catalyst, silica gel and glass beads.

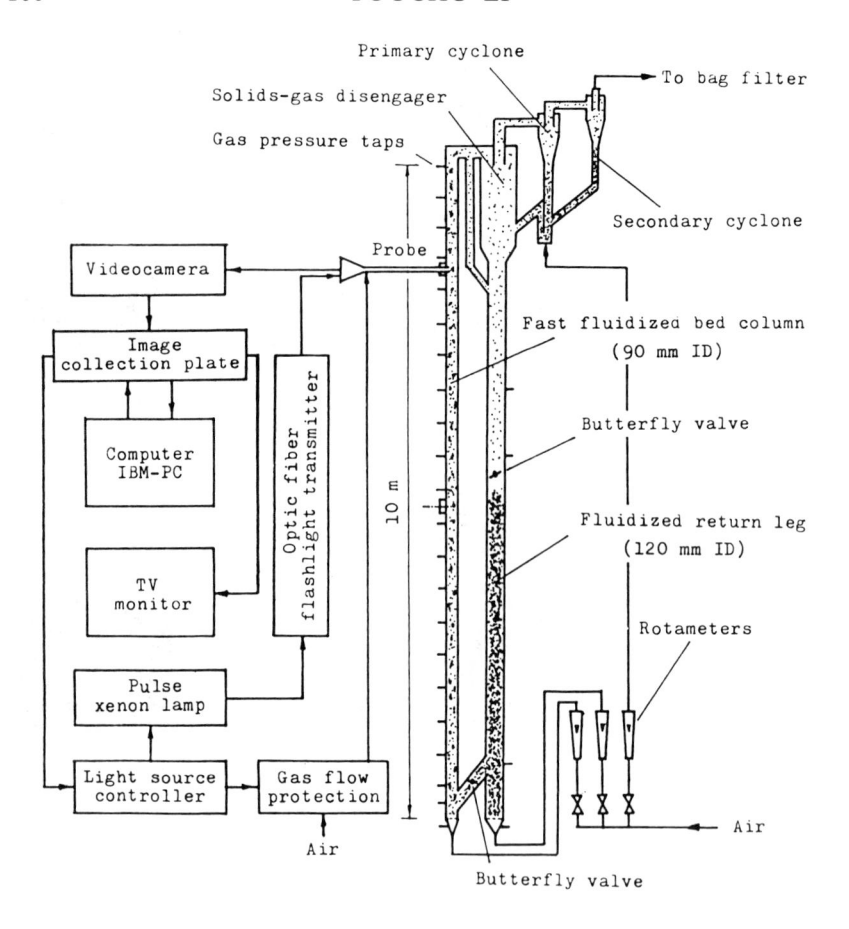

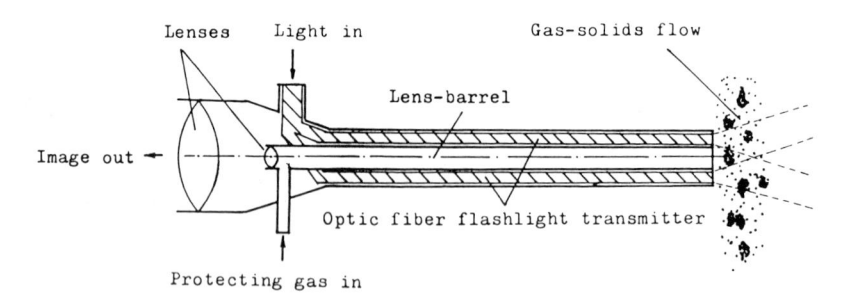

FIG. 13. Optical fiber micrograph probe and experimental apparatus (after Li et al., 1991).

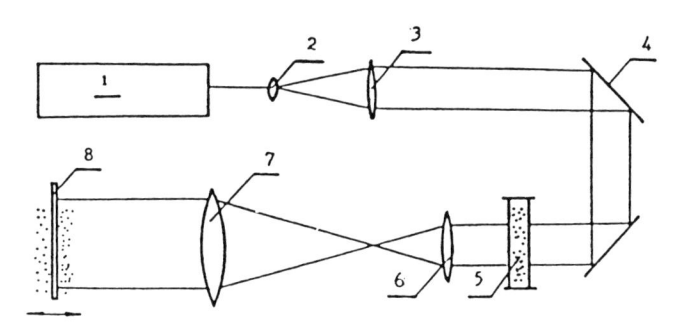

FIG. 14. Optical arrangement for holographic experiment (after Qu et al., 1985). 1, ruby laser; 2, beam expander; 3, collimating lens; 4, mirror; 5, particle field; 6, lens 1; 7, lens 2; 8, recording film.

III. Experimental Findings

A. VOIDAGE PROFILES

Voidage profiles represent one of the most important aspects of the flow structure of fast fluidization, which play an important role in gas and solids mixing, mass and heat transfer, and conversion in a chemical reactor. Considerable efforts have been given to studying the axial and radial variation of solids concentration: axially, dilute at the top and dense at the bottom, and radially, dilute in the center and dense in the vicinity of the wall. As already mentioned in Section II, these variations depend mainly on gas velocity and solids circulation rate and are also influenced by the configuration of the apparatus.

1. Axial Voidage Profile

Figure 15 shows typical results of axial voidage profile measurements (Li and Kwauk, 1980; Li et al., 1981). Experiments were conducted by using different solids materials in a Type A apparatus, 90 mm i.d. and 8 m high. Voidages were calculated from pressure gradient measurements. Solids circulation rate was controlled by a pneumatically actuated pulse feeder (see Section III in Chapter 7).

It can be seen from Fig. 15 that for any given solids circulation rate, most of the voidage profiles display a sigmoidal distribution, and the position of the inflection point becomes lower as gas velocity increases, indicating that solids inventory in the fast column diminishes with increasing gas velocity.

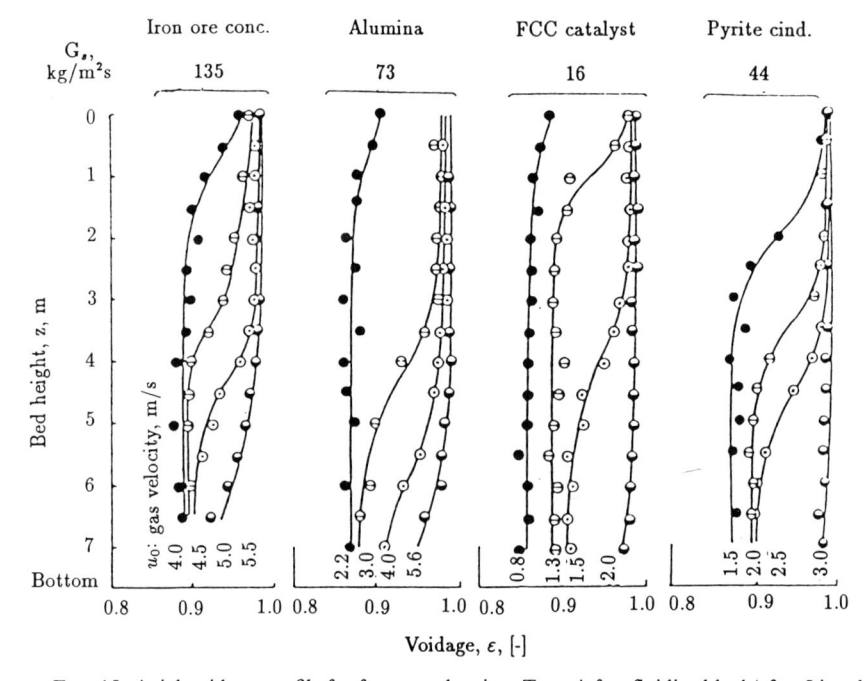

FIG. 15. Axial voidage profile for four powders in a Type A fast fluidized bed (after Li and Kwauk, 1980a).

The intermediate hopper acts as a storage balancing the inventory change. Evidently there is essentially no inlet and outlet effect, inasmuch as the voidage curves taper off into their respective asymptotic values for both the top and bottom sections of the fast column. This simple boundary condition facilitates flow modeling and the solution of the hydrodynamic equations. It should also be noted that the acceleration zone for the materials tested is too short to call for consideration.

Figure 16 shows the axial voidage profiles for pulverized coal in a Type B apparatus, with 150 mm i.d. and 6 m high (Li et al., 1984). Similar results have been obtained for calcined magnesite powder (Li et al., 1983). Measurement techniques remain the same as before, except for solids circulation rate. For any given initial solids inventory, when gas velocity is relatively low, the co-existence of a dilute phase region at the top and a dense phase region at the bottom appears, while as gas velocity increases beyond U_{pt}, this two phase structure disappears into an almost axially uniform voidage distribution. It should be pointed out that unlike Fig. 15, each of these curves stands for a different solids circulation rate which is determined by the corresponding gas velocity, that is, unlike Type A apparatus, Type B

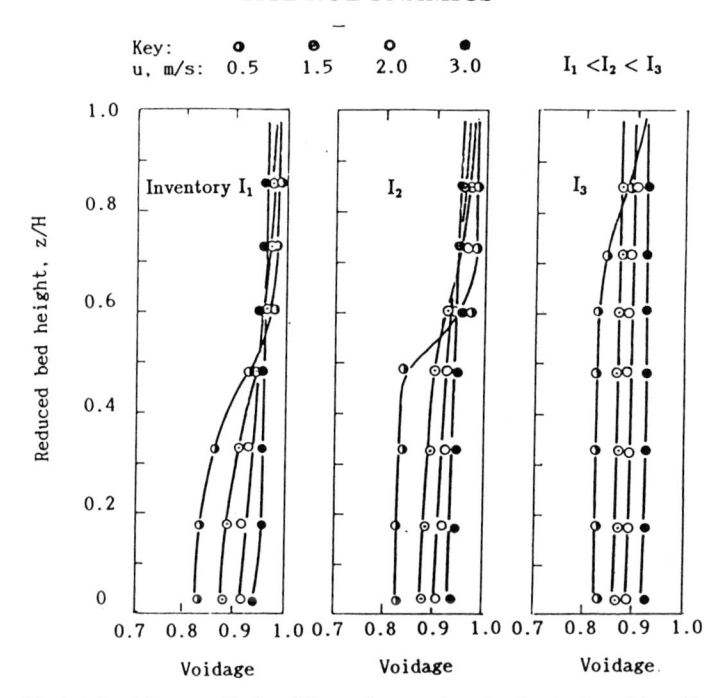

FIG. 16. Axial voidage profile for different inventories of pulverized coal in a Type B fast fluidized bed (after Li et al., 1984).

apparatus maintains different equilibrium solids inventory in the fast fluidized column, depending on the initial solids inventory, though for the same gas velocity. When the initial inventory is insufficient to give the required dense-phase region in the fast column, dilute phase region prevails throughout the fast column. Otherwise, a two-region voidage profile appears in the fast column, and by increasing initial solids inventory, the inflection point could be raised in the fast column, up to the very top, as shown in Fig. 17.

As regards Type C apparatus, typical of CCNY's setup (Yerushalmi et al., 1976), when the solid rate controlling valve opens but slightly, and the initial solids inventory is high, the voidage profile will be close to that for Type A apparatus, while when this valve is mostly open, the voidage profile will be similar to that for Type B apparatus. In most casts, its axial voidage variation will be affected by the initial solids inventory. The effect of solids inventory is reflected in what Weinstein et al. (1984) called "imposed pressure drop across the fast bed."

To describe the axial voidage profile, a one-dimensional model for steady-state operation has been developed as follows.

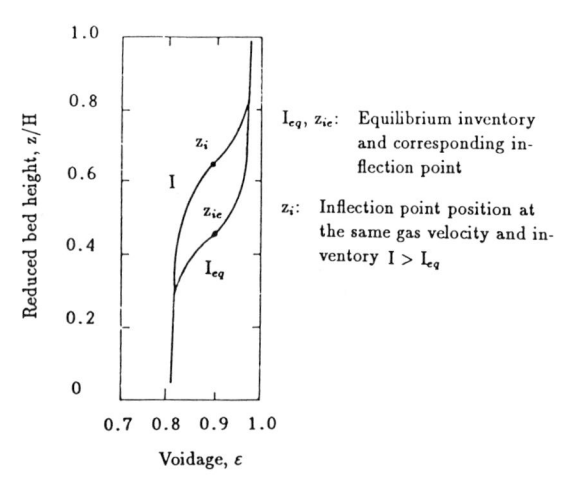

FIG. 17. Effect of solids inventory on axial voidage profile in a Type B fast fluidized bed (after Li et al., 1984).

I_{eq}, z_{ie}: Equilibrium inventory and corresponding inflection point

z_i: Inflection point position at the same gas velocity and inventory $I > I_{eq}$

Mass conservation:
for gas,

$$\frac{d}{dz}[\varepsilon \rho_g u_g] = 0; \qquad (17)$$

for solids,

$$\frac{d}{dz}[(1-\varepsilon)\rho_s v_s] = 0. \qquad (18)$$

Momentum conservation:
for gas,

$$\frac{d}{dz}[\varepsilon \rho_g u_g^2] = -\frac{dP}{dz} - F_d - F_{gg} - F_{gf}; \qquad (19)$$

for solids,

$$\frac{d}{dz}[(1-\varepsilon)\rho_s v_s^2] = F_d + F_b - F_{sg} - F_{sf}; \qquad (20)$$

for gas–solids mixture,

$$\frac{d}{dz}[\varepsilon \rho_g u_g^2 + (1-\varepsilon)\rho_s v_s^2]$$
$$= -\frac{dP}{dz} - (1-\varepsilon)(\rho_s - \rho_g)g - \varepsilon \rho_g g - F_{gf} - F_{sf}, \qquad (21)$$

in which F_b is buoyancy on the particles,

$$F_b = (1 - \varepsilon)\rho_g g; \tag{22}$$

F_d is drag force on the particles,

$$F_d = \tfrac{3}{4}(1 - \varepsilon)\rho_g \frac{\varepsilon C_d}{d_p}[u_g - v_s]^2; \tag{23}$$

F_{gg} is gravity of gas,

$$F_{gg} = \varepsilon \rho_g g; \tag{24}$$

F_{sg} is gravity of solids,

$$F_{sg} = (1 - \varepsilon)\rho_s g; \tag{25}$$

F_{gf}, F_{sf} are friction forces on the wall for gas and solids, respectively.

If friction between the particle and the wall can be neglected, from Eq. (18) and Eq. (20), we have

$$G_s = (1 - \varepsilon)\rho_s v_s, \tag{26}$$

$$\frac{dv_s}{dz} = \frac{G_s}{(1 - \varepsilon)^2 \rho_s} \frac{d\varepsilon}{dz}. \tag{27}$$

Then,

$$\frac{G_s^2}{(1 - \varepsilon)^3 (\rho_s - \rho_g)\rho_s g} \frac{d\varepsilon}{dz} \approx \frac{3}{4} C_d \frac{\mathrm{Re}_s^2}{\mathrm{Ar}} - 1, \tag{28}$$

where C_d denotes the apparent drag coefficient, which is related to the drag coefficient C_{ds} for the single particle through the voidage-dependent correction function $f(\varepsilon)$:

$$C_d = C_{ds} f(\varepsilon), \tag{29}$$

$$C_{ds} = \frac{24}{\mathrm{Re}_s} + \frac{3.6}{\mathrm{Re}_s^{0.313}},$$

and

$$\frac{d\varepsilon}{dz} = 0. \tag{30}$$

Therefore,

$$f(\varepsilon) = \frac{\mathrm{Ar}}{18\mathrm{Re}_s + 2.7\mathrm{Re}_s^{1.687}}. \tag{31}$$

When the inlet and outlet effects are minimal to the extent of being negligible, a set of correlations of the state parameters on the axial voidage profiles, as

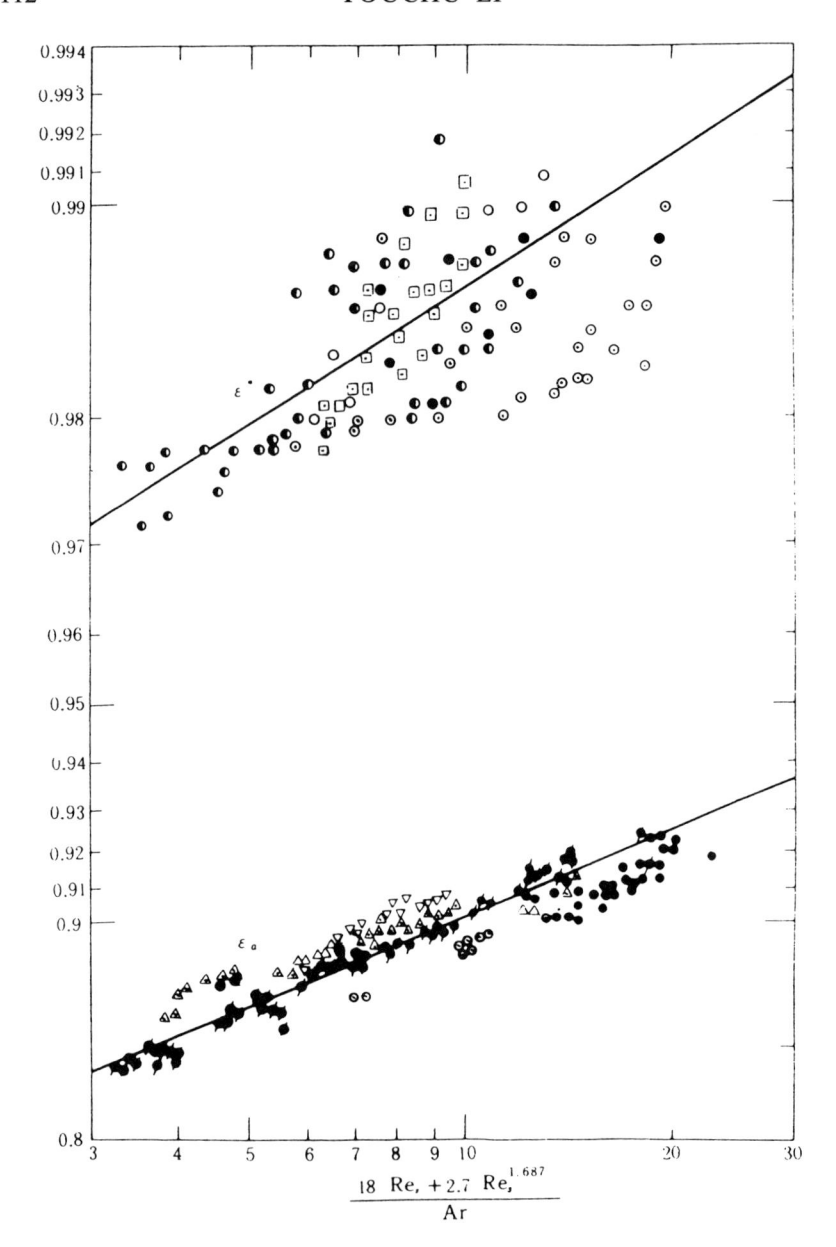

FIG. 18. Empirical correlation for voidages (after Li et al., 1982; Kwauk et al., 1986).

shown in Fig. 18, has been established (Li et al., 1982; Kwauk et al., 1986) in terms of the asymptotic voidages ε^* and ε_a:

top of bed:

$$1 - \varepsilon^* = 0.05547 \left[\frac{18\text{Re}_s^* + 2.7\text{Re}_s^{*1.687}}{\text{Ar}} \right]^{-0.6222}, \quad (32)$$

where

$$\text{Re}_s^* = \frac{d_p \rho_s}{\mu_g} \left[u_0 - u_d \left(\frac{\varepsilon^*}{1 - \varepsilon^*} \right) \right], \quad (33)$$

$$\text{Ar} = \frac{d_p^3 \rho_g g (\rho_s - \rho_g)}{\mu_g^2}; \quad (34)$$

bottom of bed:

$$\varepsilon_a = 0.7564 \left[\frac{18\text{Re}_{sa} + 2.7\text{Re}_{sa}^{1.687}}{\text{Ar}} \right]^{0.0741}, \quad (35)$$

where

$$\text{Re}_{sa} = \frac{d_p \rho_g}{\mu_g} \left[u_0 - u_d \left(\frac{\varepsilon_a}{1 - \varepsilon_a} \right) \right]. \quad (36)$$

The S-shaped axial voidage profile can be described by using the diffusion-segregation model (Li and Kwauk, 1980) as follows:

$$\ln \left(\frac{\varepsilon - \varepsilon_a}{\varepsilon^* - \varepsilon} \right) = \frac{1}{Z_0} (z - z_i). \quad (37)$$

The position of the inflection point or the height of the dense region for Type A apparatus can be estimated by the following empirical correlation:

$$z_i = H - 175.4 u_d \left[\frac{d_p g (\rho_s - \rho_g)}{\rho_g} \right]^{1.922} (u_0 - u_d)^{-3.844}. \quad (38)$$

The characteristic length, Z_0, representing the width of transitional zone between the dense and dilute regions at the inflection point, can be expressed as

$$Z_0 = 500 \exp[69(\varepsilon^* - \varepsilon_a)]. \quad (39)$$

For single-region voidage profiles, as can be seen in Fig. 16, the inflection point may be understood to lie beyond the top or the bottom of the fast column. In this case, the voidage profile would be described by Eq. (35) or Eq. (32) alone.

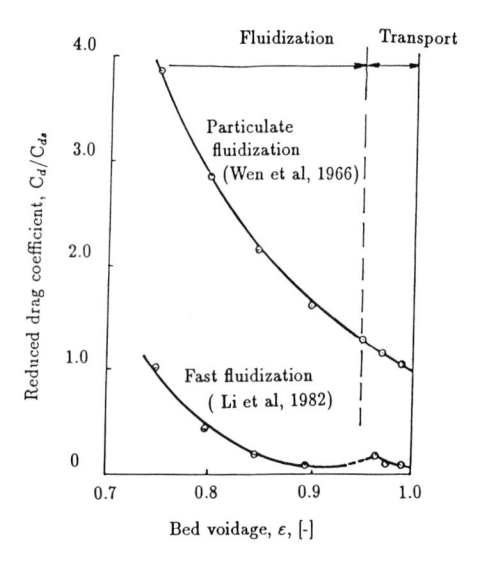

FIG. 19. Drag coefficient for different fluidization regimes (after Li et al., 1980b).

From experimental results, the correlations of the apparent drag coefficient can be obtained as follows:

$$C_d = [0.0232 \varepsilon_a^{-13.49}] C_{ds}; \tag{40}$$

for pneumatic transport:

$$C_d = [0.0633 \varepsilon^{*-35.01}] C_{ds}. \tag{41}$$

Figure 19 compares the apparent drag coefficients for different fluidization regimes, showing in particular that for fast fluidization is considerably less than that for particulate fluidization, mainly because of particles aggregation.

2. Radial Voidage Profile

Local voidages for FCC catalyst at various radial positions were measured with an optical fiber probe in a Type A apparatus, from which radial voildage profiles and their probability density functions were computed by Li et al. (1980b), as shown in Figs 20 and 21. When gas velocity is less than the incipient fast fluidization velocity of 1.25 m/s, the radial voidage profile is relatively flat; when gas velocity increases further, this profile becomes steeper: high in the center. As flow is transformed into pneumatic transport, the

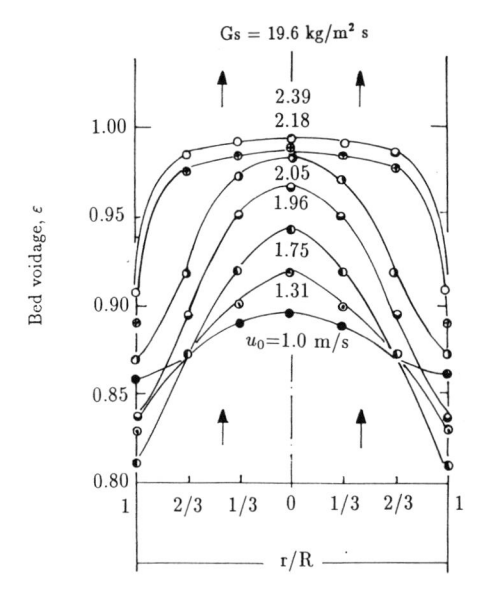

FIG. 20. Radial voidage profile for fast fluidized beds (after Li et al., 1980, 1985).

variation of the radial voidage in the core is very small, while most of the solids aggregate in the vicinity of the wall. This core–annulus distribution is, however, different from that for fast fluidization, for which there is considerable variation even in the core region.

Radial heterogeneity of solids distribution was also investigated by Weinstein et al. (1984, 1986a, 1986b), Tung et al. (1988, 1989), Horio et al. (1988), Hartge et al. (1988), and Azzi et al. (1991). Weinstein et al. and Azzi et al. used a γ-ray densitometer technique. Tung et al. (1988) developed a method of multiple regression for calibrating their optical fiber probe, with which they proposed a simple correlation to fit their experimental data (Tung et al., 1989), as shown in Fig. 22:

$$\varepsilon = \bar{\varepsilon}(0.191 + \phi^{2.5} + 3\phi^{11}), \tag{42}$$

where

$$\phi = \frac{r}{R}.$$

The preceding equation shows that as long as the cross-sectional average voidage $\bar{\varepsilon}$ is known, then the three-dimensional voidage profile could be calculated.

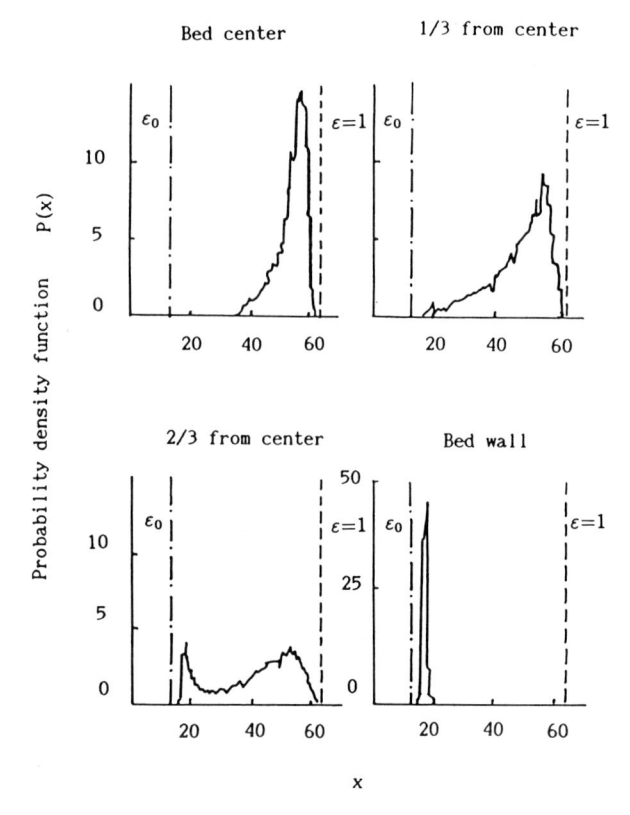

FIG. 21. Radial variation of probability density function for bed voidage (after Li et al., 1980b).

B. Bed Structure

In gas–solid fluidization, as gas velocity increases, the state of fluidization traverses in succession from particulate expansion, bubbling, turbulent, and fast fluidization to pneumatic transport. In this process of regime transition, the major gas flow is transformed from discontinuous bubbles into a continuous broth, and the solids, from a continuous dense emulsion phase into discontinuous clusters or strands. These bed structures have been observed visually and measured with capacitance probe and optical fiber probes (Li and Kwauk, 1980a). Both these probes produce high-frequency and low-amplitude random signals for fast fluidization, shown in Fig. 1b, implying absence of gas bubbles and improved gas–solid contacting. Analysis of the radial profiles of probability density function of these signals, shown

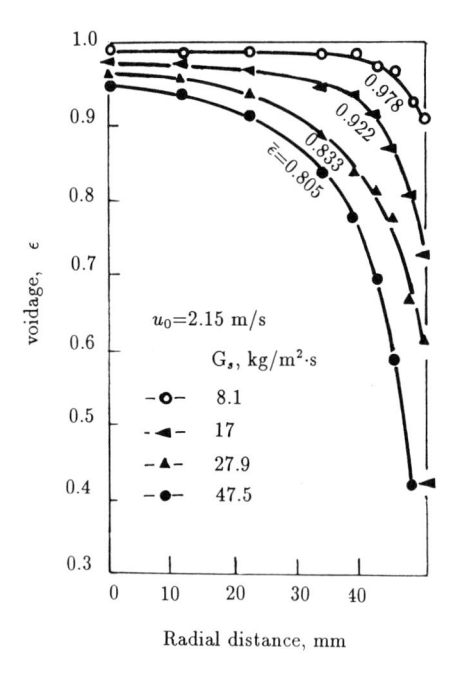

FIG. 22. Effect of solids flow rate on radial voidage profile (after Tung et al., 1988).

in Fig. 21, indicates the existence of heterogeneity (Li et al., 1980b; Hartge et al., 1988). Such dynamic signals for all the fluidization regimes, as shown in Fig. 23, were analyzed statistically and then compared with their respective flow structures, in the light of energy minimization (Li et al., 1990). These results show that in the range of low gas velocities, the auto-correlation function changes with large periodicity, corresponding to the motion of discrete gas bubbles or slugs, presumably because of insufficient momentum of the gas stream to break down the solids aggregates further. Here, bubbling represents a stage of energy minimization. As gas velocity increases, the periodicity of the auto-correlation functions acquires much higher frequency. In this case, the energy of the flowing gas has become high enough to dissolve the emulsion phase into clusters. Radially, minimal energy calls for the clusters to be carried up in the center of the fast column, and then they descend along the wall. This core–annulus distribution leads to radial variations of voidage, particle velocity and mass flux. Such a configuration of parameter distribution has also been elucidated from statistical analysis by Hartge et al. (1988).

It has been widely accepted that the phenomenon of fast fluidization is attributed to "cluster" formation. Photomicrography of the fast fluidization

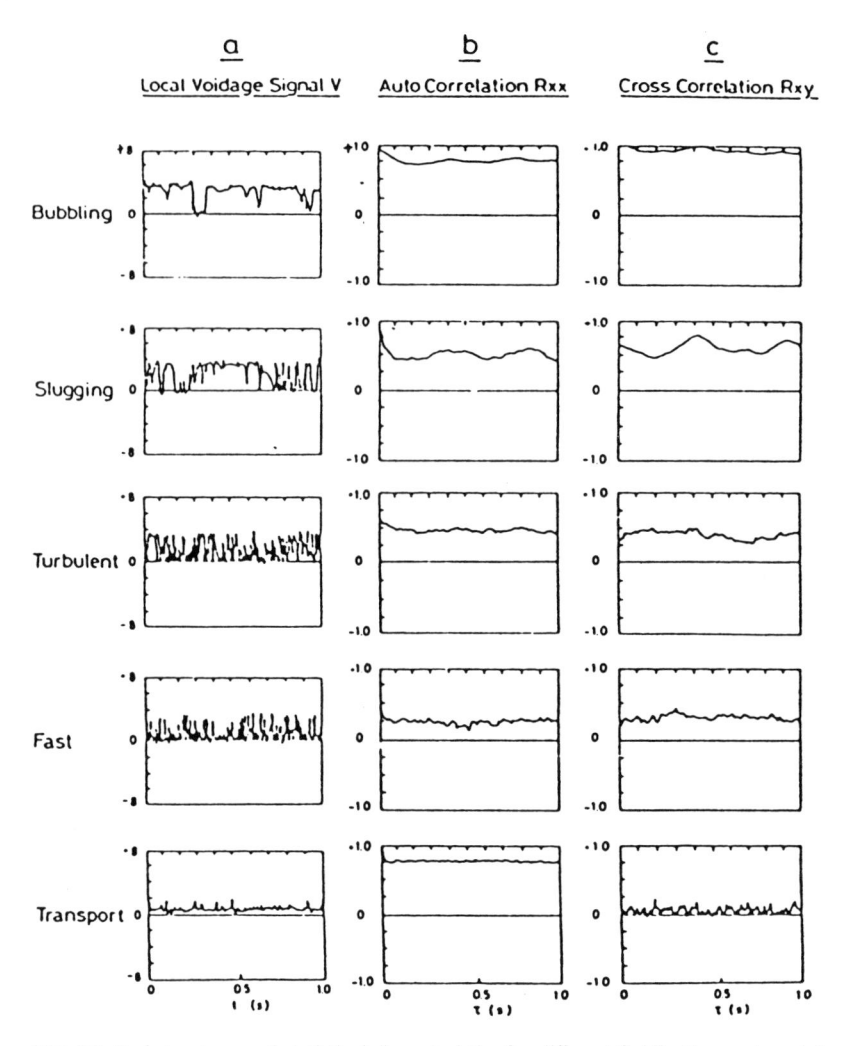

FIG. 23. Bed structure and statistical characteristics for different fluidization regimes (after Li et al., 1990).

taken with a video camera connected to an optical fiber probe (Li et al., 1991) clearly distinguishes two phases: a dilute dispersed phase and a dense cluster phase, as shown in Fig. 24. In the dispersed phase the solid particles are noted to be present essentially individually, while in the cluster phase they agglomerate with one another. In the center of the bed the shape of the clusters is mostly strandlike, while near the wall, it is essentially spherical. The radial distribution of cluster concentration is associated with that of

H (m)	U_o (m/s)	G_s (kg/(m²·s))	r (mm) 0	22.5	40
8.40	1.31	9.25			
8.40	2.62	51.35			
8.40	3.49	64.65			
4.77	1.31	7.32			
4.77	1.75	8.30			
4.77	3.49	30.05			

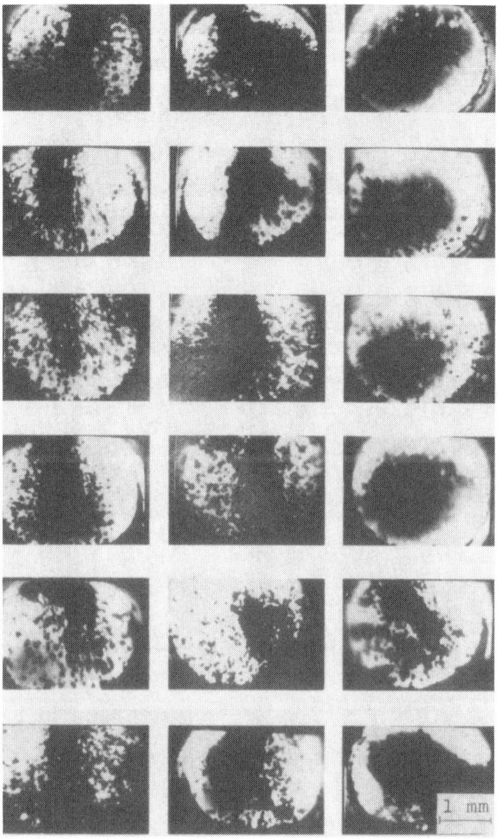

FIG. 24. Variation of two-phase structure with operating conditions (after Li *et al.*, 1991). Note: Dark color in circular visual field represents solid phase.

solids concentration. Clusters have also been studied visually by Takeuchi and Hiruma (1991) by using a video camera system. The photographs show that in the lower section of a fast fluidized bed, both relatively small and large packets (clusters) exist and move rapidly. The flow state is quite unstable. Qu *et al.* (1985) studies cluster in a large visual field using a holographic imaging technique, as shown in Figs 25 and 26. These holograms display the size, shape and number of clusters which are related to the operating conditions.

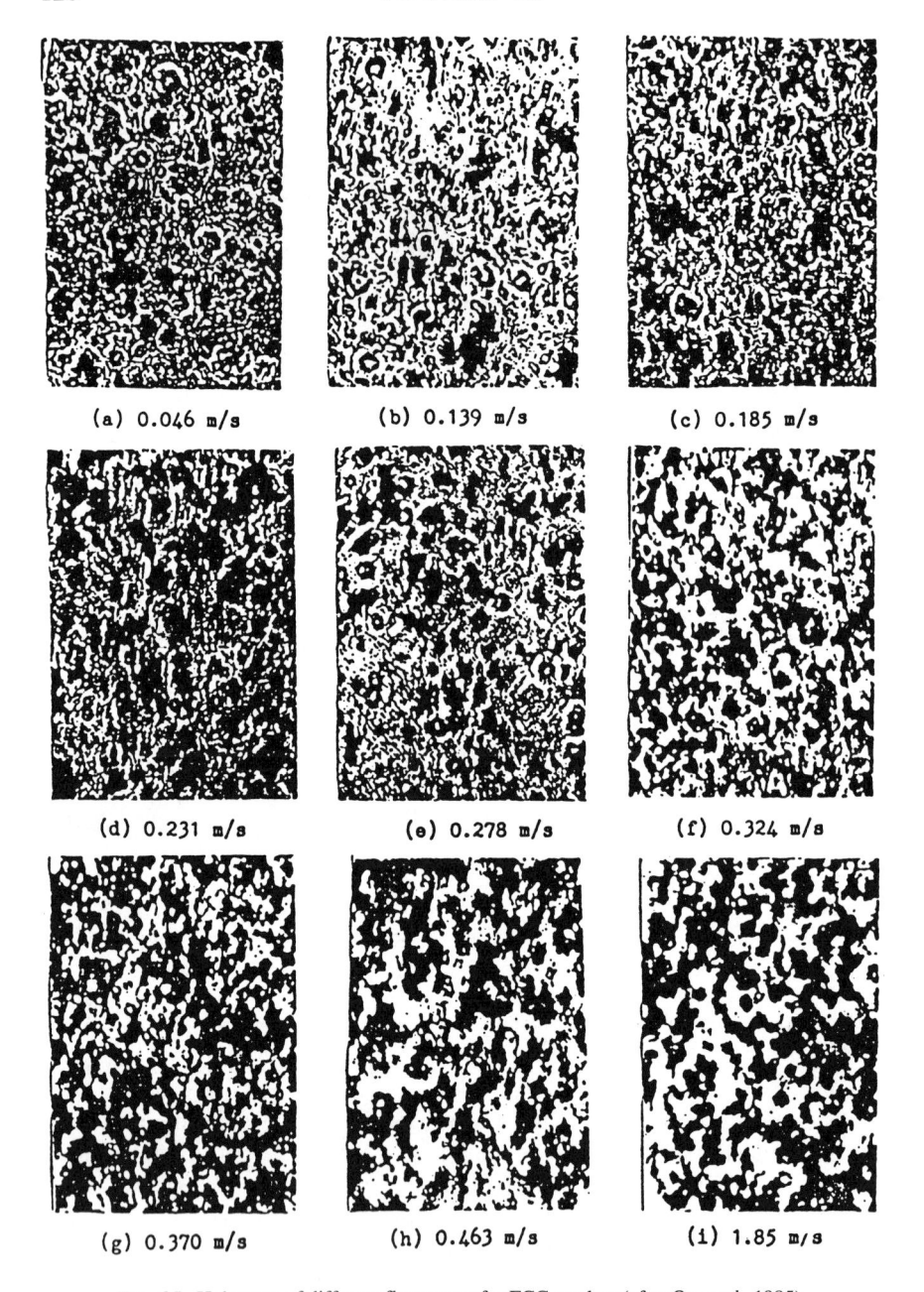

(a) 0.046 m/s
(b) 0.139 m/s
(c) 0.185 m/s
(d) 0.231 m/s
(e) 0.278 m/s
(f) 0.324 m/s
(g) 0.370 m/s
(h) 0.463 m/s
(i) 1.85 m/s

FIG. 25. Hologram of different flow states for FCC catalyst (after Qu *et al.*, 1985).

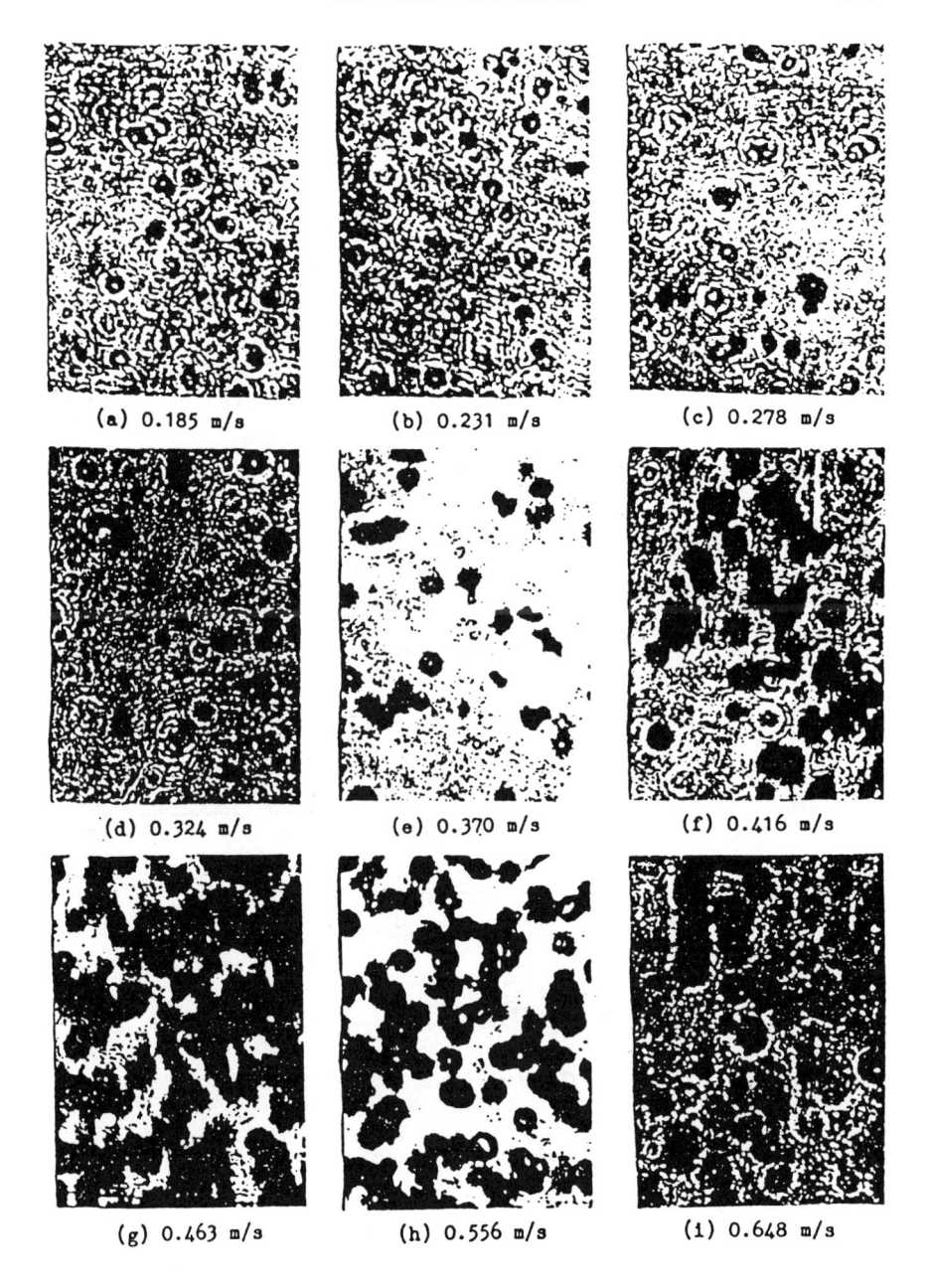

(a) 0.185 m/s
(b) 0.231 m/s
(c) 0.278 m/s
(d) 0.324 m/s
(e) 0.370 m/s
(f) 0.416 m/s
(g) 0.463 m/s
(h) 0.556 m/s
(i) 0.648 m/s

FIG. 26. Hologram of different flow states for silica gel particles (after Qu *et al.*, 1985).

C. Particle Velocity

Figure 27 shows local particle velocities in the radial direction measured by laser Doppler velocimeter at different levels of the fast bed (Yang et al., 1991). It can be seen from these diagrams that there is in general a maximum of particle velocity in the center of the bed, which gradually decreases along the radial direction to reach some minimum next to the wall. The radial profile of particle velocity depends on the overall flow state: relatively flat in the lower dense region, and much steeper in the upper dilute region. The influence of solids circulation rate and gas velocity is shown in Figs 28 and 29, respectively. In the center of the bed, local particle velocity increases with gas velocity and solid circulation rate, and decreases, in the vicinity of the wall.

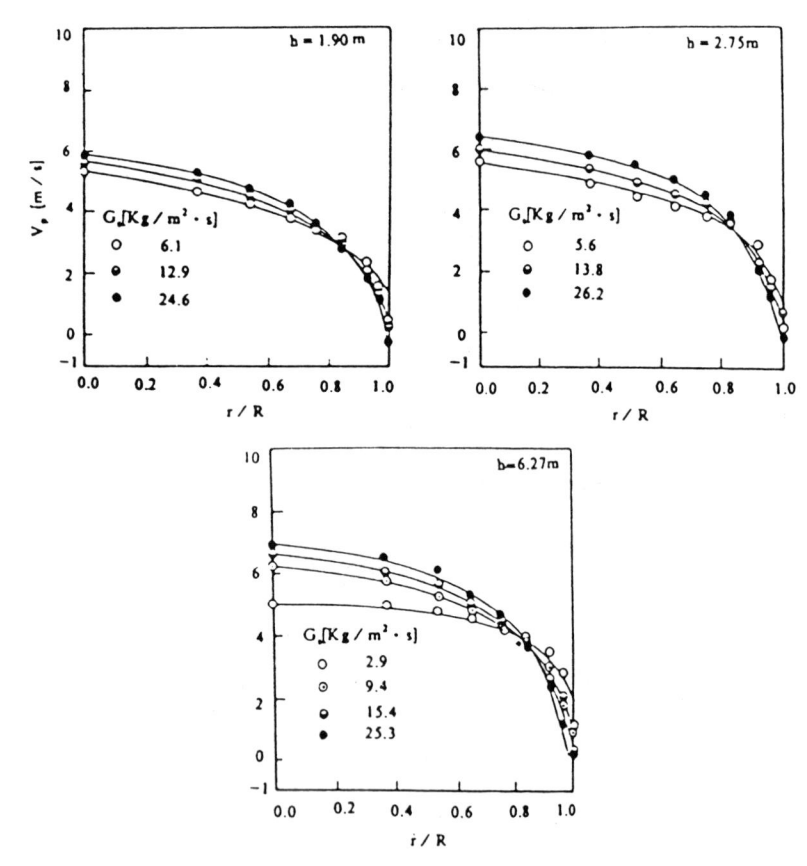

FIG. 27. Effect of solids circulation rate on radial distribution of local particle velocity at different bed heights (after Yang et al., 1991).

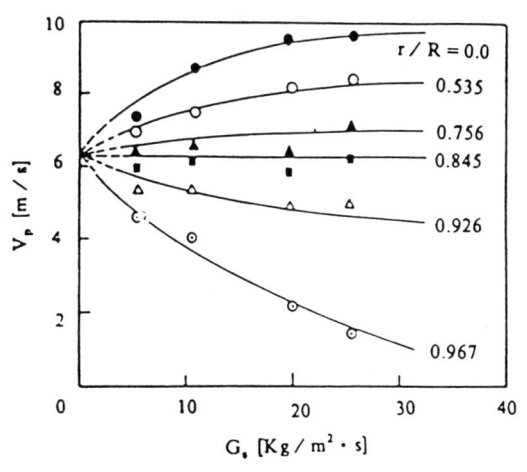

FIG. 28. Effect of solids circulation rate on radial local particle velocity (after Yang et al., 1991).

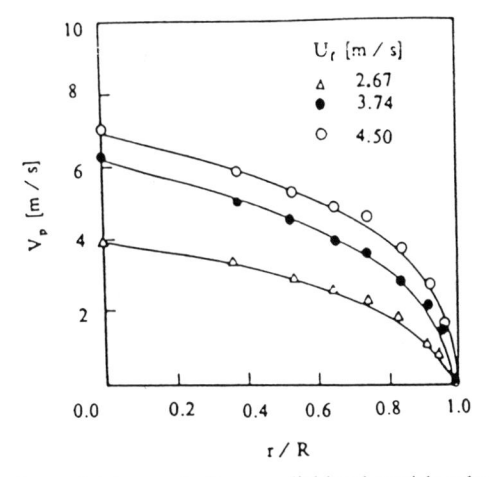

FIG. 29. Effect of superficial gas velocity on radial local particle velocity (after Yang et al., 1991).

D. GAS VELOCITY

For gas–solids two-phase flow, owing to the presence of solids, measurement of gas velocity is difficult. Zhang (1990) designed an isokinetic momentum probe to measure gas velocity with reduced experimental error caused by the presence of the solids. Figure 30 shows measured results of local gas velocity under various operating conditions.

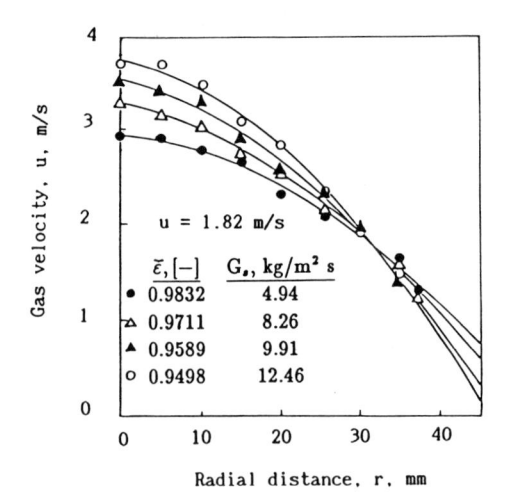

FIG. 30. Effect of average voidage on radial gas velocity profile (after Zhang, 1990).

From these diagrams, it can be seen that gas velocity in the fast fluidized bed diminishes along the radial direction from the center to the wall. Local gas velocities in the center are much higher than the corresponding superficial gas velocities, and radial heterogeneity of gas velocity increases with increase in superficial gas velocity and solids circulation rate. The local gas velocity can be estimated by using the following equations:

$$u_g = \left[\left(\frac{k_1}{2k_2}\right)^2 + \frac{k_v}{2k_2}\right]^{0.5} - \frac{k_1}{2k_2}, \tag{43}$$

$$k_1 = \frac{150\mu_g(1-\bar{\varepsilon})^2}{D_{eq}^2 \varepsilon^3}, \tag{44}$$

$$k_2 = \frac{1.75\rho_g(1-\bar{\varepsilon})}{D_{eq}\varepsilon^3}, \tag{45}$$

$$D_{eq} = d_p + 0.0004(u_0 - 0.2)(1-\bar{\varepsilon})^{28(1-\bar{\varepsilon})}, \tag{46}$$

$$u_0 = 2\int_0^1 u_g \phi \, d\phi. \tag{47}$$

E. MIXING

The efficiency of a gas–solids reactor can be greatly improved by controlling the degree of mixing, especially for non-zeroth order complex reactions approaching high conversion and/or high selectivity. Understanding of gas and solids mixing for fast fluidization is thus vitally significant for practical design. An earlier study concerning gas mixing in a fast fluidized bed (Cankurt and Yerushalmi, 1978) showed that the extent of gas backmixing is quite small. Recent research (Brereton et al., 1988; Bader et al., 1988; Wu, 1988; Dry and White, 1989; Li and Weinstein, 1989; Lou and Yang, 1990; Li and Wu, 1991; Werther et al., 1991) demonstrated that a certain degree of gas mixing exists, which is related to the heterogeneity of flow structure. Results concerning mixing of solids (Ambler et al., 1990; Chesonis et al., 1991; Patience et al., 1991; Bai, 1991) indicated that it seems to approach perfect mixing.

1. Gas Backmixing

In experimental studies of gas backmixing (Wu, 1988; Li and Wu, 1991), a given amount of a tracer gas, such as hydrogen or helium, is injected through a solenoid valve for a short period of time into the fast fluidized bed through a nozzle. Gas samples are withdrawn through a sampling probe inserted into the bed at a short distance of, say, 40 mm upstream from the tracer nozzle. The content of the tracer gas in the sample gas is normally determined by a thermal conductivity cell. Figure 31 shows that for a given solids circulation rate, the degree of gas backmixing decreases with increase in gas velocity. It is worthy to note that when gas velocity is raised from 1.0

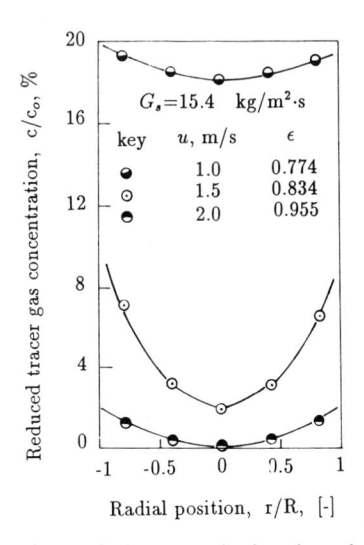

FIG. 31. Influence of gas velocity on gas backmixing (after Li and Wu, 1991).

m/s to 1.5 m/s (for FCC catalyst), gas backmixing is sharply reduced, primarily because of the transition of the flow regime from turbulent to fast fluidization at a gas velocity of about 1.25 m/s. In this case, the radial profile of the tracer content is quite steep, implying that high backflow of gas takes place near the wall. For both dense phase fluidization ($\varepsilon = 0.774$, $u = 1.0$ m/s), and dilute phase transport ($\varepsilon = 0.955$, $u = 2.0$ m/s), the radial profiles of tracer concentration is relatively flat, implying that backmixing due to backflow and transverse mixing are of the same order.

Experimental results by Li and Weinstein (1989) also indicated that in the bubbling and slugging regimes, radial concentration profiles are essentially flat, and that the axial dispersion coefficient is proportional to gas velocity; for the turbulent regime, a very sharp radial concentration profile exists, and gas backmixing mainly results from the downflow of solid particles near the wall; in the fast regime, backmixing of gas in the central core of the bed is negligible, while intensive backmixing appears near the wall, which decreases, however, with increasing gas velocity.

Figure 32 shows that for gas velocity of 1.5 m/s, gas backmixing increases with solids circulation rate, inasmuch as solids circulation causes backflow of gas. All of the curves in Figs 31 and 32 (Wu, 1988; Li and Wu, 1991) possess the common feature that gas backmixing gradually increases along the radial direction, in a manner analogous to radial solids concentration profiles. These are attributed to heterogeneous gas–solids flow, with

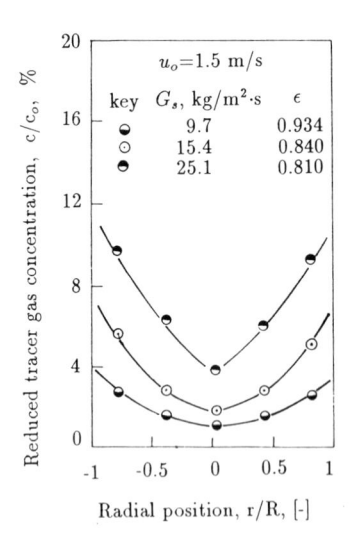

FIG. 32. Influence of solids circulation rate on gas backmixing at a velocity higher than the incipient fast fluidization velocity (after Li and Wu, 1991).

HYDRODYNAMICS

solids moving up in the center region and down in the vicinity of the wall, thus leading to increased gas backflow near the wall.

2. Axial Gas Dispersion

The extent of gas dispersion can usually be computed from experimentally measured gas residence time distribution. The dual probe detection method followed by least square regression of data in the time domain is effective in eliminating error introduced from the usual pulse technique which could not produce an ideal Delta function input (Wu, 1988). By this method, tracer is injected at a point in the fast bed, and tracer concentration is monitored downstream of the injection point by two sampling probes spaced a given distance apart, which are connected to two individual thermal conductivity cells. The response signal produced by the first probe is taken as the input to the second probe. The difference between the concentration-versus-time curves is used to describe gas mixing.

The mechanism of gas mixing in fast fluidization probable involves

1. gas diffusion caused by radial gas velocity distribution;
2. eddy diffusion caused by local solids circulation;
3. gas mixing caused by interchange of solids between clusters and their surroundings with rapid succession of cluster formation and dissolution.

However, contrary to bubbling fluidization, major gas flow in fast fluidization has changed from the discontinuous bubble phase into a continuous phase, with the dominant characteristics of bubbleless gas–solids contacting. Thus, it is reasonable to treat the gas mixing process in terms of a pseudo-homogeneous dispersion model. On the other hand, based on experimental findings (Adams, 1988; Brereton et al., 1988), the radial dispersion coefficient is small compared to axial dispersion. Thus, neglecting radial dispersion is probably acceptable within the accuracy of experimental measurement. Consequently, gas mixing for fast fluidized bed can be described by the following one-dimensional pseudo-homogenous diffusion model:

$$\frac{dC}{d\theta} = \frac{1}{\text{Pe}}\left(\frac{d^2C}{d\xi^2} - \frac{dC}{d\xi}\right), \tag{48}$$

in which

$$\theta = \frac{t}{\bar{t}}; \quad \xi = \frac{z}{L}; \quad C = \frac{c}{c_0}; \quad \text{Pe} = \frac{UL}{D_a}. \tag{49}$$

Initial and boundary conditions for this open system are

$$\theta = 0, \quad C(\xi, 0) = 0;$$
$$\xi = 0, \quad C(0, \theta) = C_1(0, \theta); \quad (50)$$
$$\xi = \infty, \quad C(\infty, \theta) \neq \infty.$$

The gas dispersion number Pe can be obtained by means of least square analysis in the time domain to fit the experimentally determined response curves, that is,

$$\frac{d\psi}{d\,\text{Pe}} = 0, \quad (51)$$

where

$$\psi = \sum_{i=1}^{N} [c_2(t_i) - c_2^*(t_i)]^2, \quad (52)$$

$$c_2^* = \int_0^t g(t - t') c_1(t') \, dt. \quad (53)$$

$g(t)$ is the transfer function for Delta-function input.

Figure 33 gives experimental results to show the influence of bed voidage

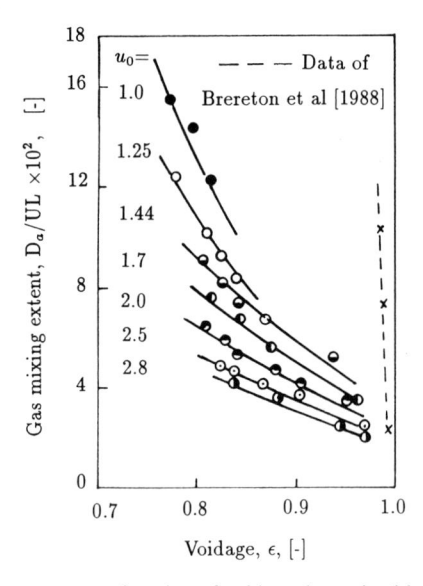

FIG. 33. Gas mixing extent as a function of voidage determined by gas velocity and solids circulation rate (after Li and Wu, 1991).

on D_a/UL for FCC catalyst. It can be seen that D_a/UL decreases with increase in bed voidage and superficial gas velocity, and with decrease in solids circulation rate. In the ranges of gas velocities of less than about 1.25 m/s, corresponding to bubbling fluidization, the influence of gas velocity on D_a/UL is very strong, while beyond this gas velocity, this variation becomes flat, indicating that raising gas velocity is much more favorable for controlling gas mixing. Figure 34 gives dispersion coefficients for different flow regimes, identified roughly by their flow behaviors associated with gas velocity and solids circulation rate. The values of dispersion coefficients for the turbulent regime are above 0.4 m²/s; for the fast regime roughly from 0.3 to 0.5 m²/s; and for pneumatic transport close to 0.2 m²/s. These figures indicate that axial dispersion coefficient are two to three orders of magnitude greater than those for radial dispersion, which range from 0.0003 to 0.007 m²/s, as given by Adams (1988).

Based on these results, the following generalized correlation, spanning the entire fluidization regime, from bubbling to pneumatic transport, as shown in Fig. 35, has been established (Li and Wu, 1991):

$$D_a = 0.1953\varepsilon^{-4.1197} \tag{54}$$

According to bed voidage, which can be calculated from gas velocity and solids circulation rate by using Eqs. (32) to (38), the axial gas dispersion coefficient can thus be easily obtained.

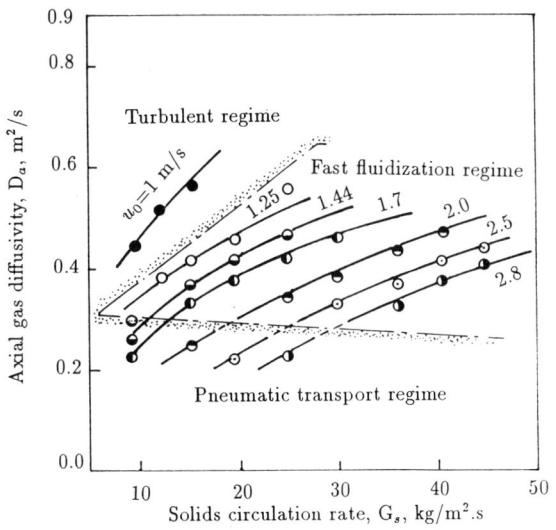

FIG. 34. Influence of gas velocity and solids circulation rate on gas diffusivity (after Li and Wu, 1991).

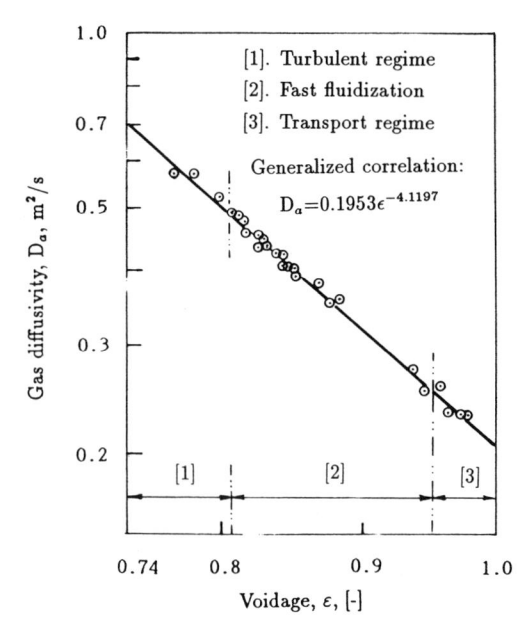

FIG. 35. Correlation of gas diffusivity with bed voidage (after Li and Wu, 1991).

Chart legend and annotations:
- [1]. Turbulent regime
- [2]. Fast fluidization
- [3]. Transport regime
- Generalized correlation: $D_a = 0.1953\epsilon^{-4.1197}$
- Y-axis: Gas diffusivity, D_a, m²/s
- X-axis: Voidage, ϵ, [-]

Further experiments on axial gas mixing were conducted by Lou and Yang (1990) in a 31 mm i.d. by 6 m high fast bed, using as a tracer gas helium injected in the dense-phase region and sampled in the dilute-phase region. The test materials used were iron catalyst ($d_p = 90$ μm; $\rho_s = 3{,}100$ kg/m³), silica gel ($d_p = 167$ μm; $\rho_s = 794$ kg/m³) and sand ($d_p = 202$ μm; $\rho_s = 2{,}570$ kg/m³). For gas velocities between 3 and 6 m/s and solids circulation rates between 50 and 90 kg/m²s, the values of D_a range from 0.1 to 1.8 m²/s. From their experimental data, these authors developed the following correlation:

$$\frac{D_a \rho_g}{\mu_g} = 0.11 (\text{Re})^{1.6} \left(\frac{G_s}{\rho_s U_t}\right)^{0.62} \left(\frac{\rho_s}{\rho_g}\right)^{1.2}. \tag{55}$$

This equation indicates that axial gas dispersion coefficient increases with gas velocity and solids circulation rate. This dependence of axial dispersion coefficient on gas velocity is at variance in the available literature.

3. Radial Gas Dispersion

In most measurements on radial gas dispersion (Yang et al., 1984; Bader et al., 1988; Adams, 1988; Werther et al., 1991), helium, methane or carbon

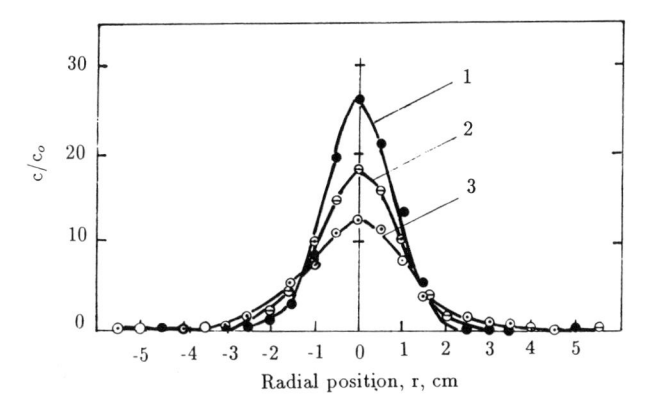

FIG. 36. Radial profile of tracer concentration at $u_0 = 4.1$ m/s; G_s, kg/m^{2}s: (1) 43.5; (2) 74; (3) 106. (After Yang et al., 1984.)

dioxide was employed as the tracer gas injected continuously at the centerline of the fast bed and sampled downstream by means of a radially traversing sampling probe. Typical results measured in a 115 mm i.d. by 8 m high fast bed are presented in Fig. 36 (Yang et al., 1984), and by assuming axial diffusion negligible and radial diffusion coefficient constant, the variation of tracer concentration can be described by

$$u\frac{dc}{dz} = D_r\left(\frac{d^2c}{dr^2} + \frac{1}{r}\frac{dc}{dr}\right). \tag{56}$$

The boundary conditions are

$$z = 0, \quad r = 0 \quad c = c_i \tag{57}$$
$$r \neq 0 \quad c = 0;$$
$$z = \infty, \quad c \to c_0$$
$$r = R \quad \frac{dc}{dr} = 0 \tag{58}$$

The solution of Eq. (56) is

$$\frac{c}{c_0} = 1 + \sum_{n=1}^{n=\infty} \frac{J_0(\beta_n r)}{J_0^2(\beta_n R)} e^{\left(-\beta_n^2 D_r \frac{z}{u}\right)}, \tag{59}$$

where the radial dispersion coefficient D_r can be obtained by regression analysis of experimental data. Dependence of D_r on the operating parameters is shown in Fig. 37, from which Yang et al. (1984) developed the following

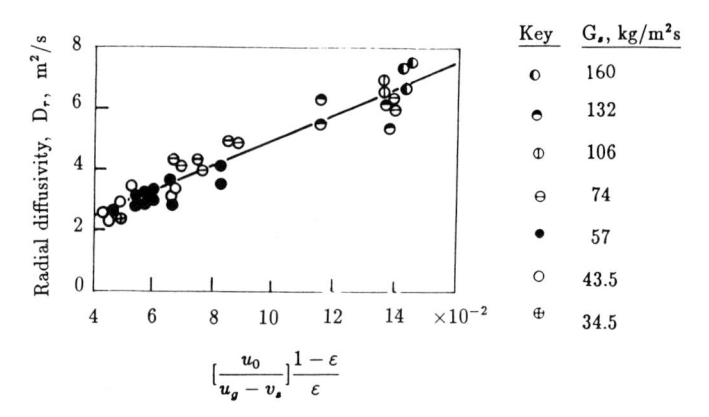

FIG. 37. Correlation of radial diffusivity with experimental parameters (after Yang et al., 1984).

TABLE III
MEASURED VALUES OF RADIAL DISPERSION COEFFICIENT

Investigators	Tracer	d_p μm	D_t mm	H m	u m/s	ρ_b kg/m³	D_r ×10⁴ m²/s
V. Zoonen, 1962	H₂	20–150	50	10	2.5–12	150	3.2–13
Yang, 1984	He	220	115	8	3.5–5.3		2.6–7.1
Adams, 1988	CH₄	76	300 × 400	4	3.5–4.5	55	50–180
Bader, 1988	He	76	305	12.2	3.7–6.1	300	26–65
Werther, 1991	CO₂	130	400	8.5	3.0–6.2		25–55

correlation:

$$D_r = 43.4\left(\frac{u_0}{u_g - u_p}\right)\left(\frac{1 - \varepsilon}{\varepsilon}\right) + 0.7. \tag{60}$$

Experimental results show that the radial dispersion coefficient increases with solids circulation rate and with bed density, and decreases, though slightly, with gas velocity. Experiments are yet inconclusive as to the dependence of D_r on column diameter.

Values of radial dispersion coefficients from literature sources are listed in Table III.

Table III shows that the difference between values of D_r can be as high as an order of magnitude, although for nearly the same range of gas velocity. That is probably because experiments were performed in different-diameter apparatus and experimental data were treated by different models.

4. Solids Mixing

Measurement of solids mixing is even more difficult than gas mixing. Since considerable bed material recycles continuously following any solids tracer injection, the background tracer concentration in the circulating fluidized bed keeps increasing, thus often leading to unacceptable error. For this, a solid tracer with a short life seems to be highly desirable.

Bader et al. (1988) used common salt as a solid tracer, which was injected into a flowing catalyst bed. Solids samples, withdrawn downstream of tracer injection, were leached with water and the salt concentration determined by electrical conductivity of the solution. Their results indicated substantial solids backmixing. Li et al. (1991) observed solids mixing in a fast fluidized bed combustor by using raw coal as a tracer, which was injected into the ash bed. Their results also showed that near-perfect mixing prevailed. Similar experiments was also conducted by Chesonis et al. (1991) in a cold model.

Computation of axial dispersion coefficients for solids was provided by Patience et al. (1991), in the following simple one-dimensional model using the data obtained from RTD measurements of a radioactive solid tracer in a 82.8 mm i.d. by 5 m high circulating fluidized bed, that is,

$$\frac{\partial c}{\partial t} + \frac{\partial(v_p c)}{\partial z} = D_s \frac{\partial^2 c}{\partial z^2}, \tag{61}$$

$$v_p = \frac{L}{\mu - t_{inj}/2}, \tag{62}$$

$$D_s = \left(\sigma^2 - \frac{t_{inj}^2}{24}\right)\frac{v_p^3}{2L}. \tag{63}$$

For gas velocities ranging 4.1 to 6.3 m/s, the measured axial solids dispersion coefficients D_s varied from 0.39 to 0.91 m^2/s for dilute solid suspension, while for the whole fast bed, in the same gas velocity range, D_s was 0.22 to 1.67 m^2/s. Axial solid dispersion coefficient generally increases with both gas velocity and solids circulation rate. Milne and Berruti (1991) have also given their own model for solids mixing.

Bai (1991) developed a solids tracing technique for studying both gas and solid mixing at the same time, in which solid particles soaked with a volatile organic reagent were used as the tracer. When fluidized, the organic compound on the surface of the particles is transferred rapidly into the gas phase. The solids tracer was injected at the bottom of the fast bed, and samples were collected at the outlet at the top. The tracer concentration was determined by gas chromatography. The axial dispersion of the gas and the solids can be described by the following equations:

gas:

$$\frac{\partial c_{sg}}{\partial t} + \frac{\partial (u_g c_{sg})}{\partial z} - D_{ag}\frac{\partial^2 c_{sg}}{\partial z^2} = \frac{1-\varepsilon}{\varepsilon}c_s f(t), \qquad (64)$$

solids:

$$\frac{\partial c_s}{\partial t} + \frac{\partial (v_s c_s)}{\partial z} - D_{as}\frac{\partial^2 c_s}{\partial z^2} = 0. \qquad (65)$$

The axial dispersion coefficients D_{ag} and D_{as} can thus be obtained from their respective RTD, typical values of which are shown in Fig. 38. Figure 39 shows further that D_{ag} increases with solids circulation rate but decreases with increasing gas velocity. Figure 40 shows that solids mixing is greater than gas, and Fig. 41 shows that fine particles mix better than coarse.

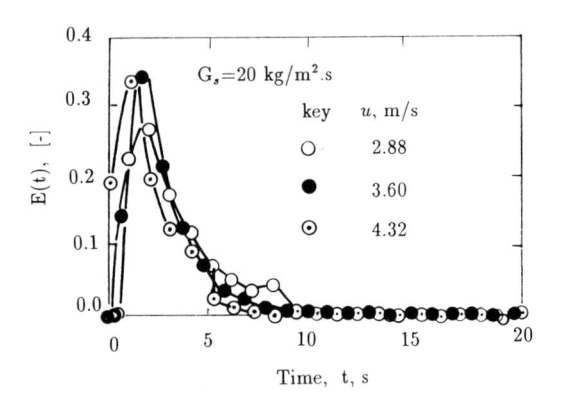

FIG. 38. Typical RTD in a fast fluidized bed, measured by solids tracer (after Bai, 1991).

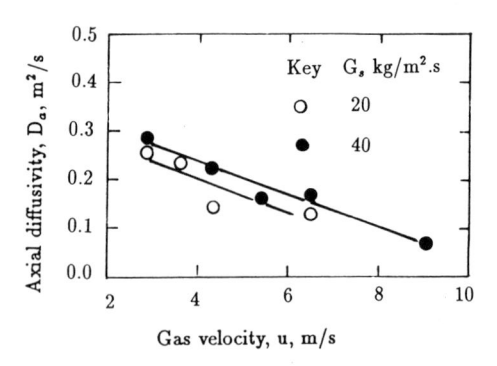

FIG. 39. Influence of operating conditions on axial gas diffusivity (after Bai, 1991).

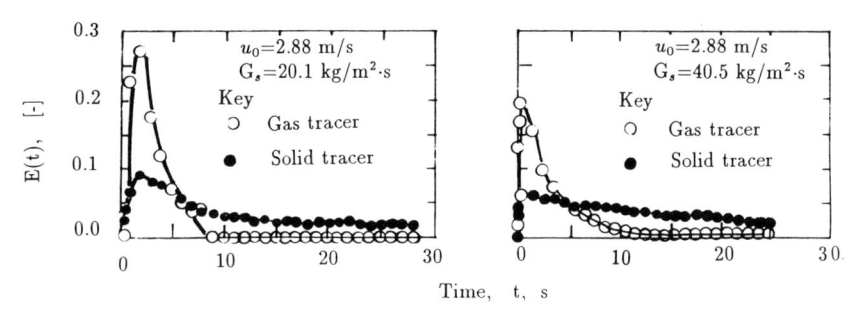

FIG. 40. Comparison of solids mixing with gas mixing in a fast fluidized bed (after Bai, 1991).

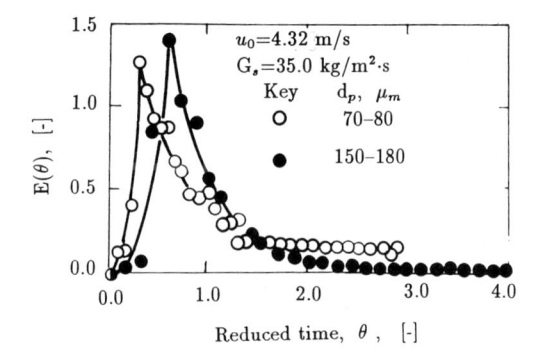

FIG. 41. Mixing for different particle sizes (after Bai, 1991).

F. Effect of Boundaries

Boundaries in fast fluidization refer mainly to the column wall as well as the inlet and outlet. Effect of the wall on pressure drop due to friction between the fluidized solids and the wall surface is minimal (Li *et al.*, 1978), although it is the very cause of radial distribution of parameters. The configuration of the inlet and the outlet often strongly affect gas–solids flow, especially with regard to axial voidage profile.

1. Effect of Outlet

Outlet design generally consists of the four types shown in Fig. 42:

1. right-angle bend;
2. right-angle bend with a guiding baffle;
3. right-angle bend with dead space at the top;
4. contracted horizontal side tube.

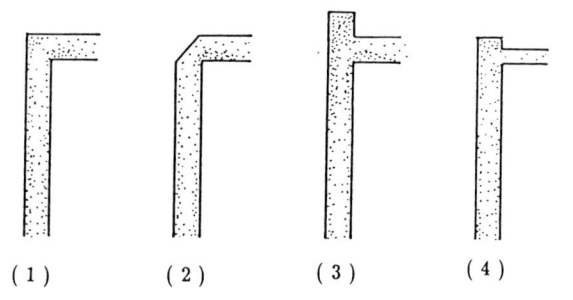

FIG. 42. Different types of outlet geometry.

When a gas–solid mixture moves upward and passes through these bend-shaped outlets, solids tend to separate from the gas stream by virtue of their greater inertia. As a result, part of the solids accumulates in the bottom of the horizontal tube section for cases (1) and (2), and forms a dense pocket at the top of the fast column for cases (3) and (4). Some of the accumulated solids descend into the fast column, densifying its upper section, often eccentrically as shown in Fig. 43 (Li et al., 1978). In the particular apparatus used, the height of the dead space is in the range of 0 to 100 mm when measured from the upper edge of the horizontal exit tube (equivalent to $H/D = 0$ to 1). In this H/D range, the effect of dead height on axial voidage profile is accentuated, while for $H/D > 1$, this effect does not increase much. Solids accumulation extends downward to a distance, determined by both gas velocity and solids circulation rate, up to some upper limits. Based on their experiments, Jin et al. (1988) proposed the following correlation to estimate the height of this zone of increased solids concentration:

$$\frac{z}{D_t} = 85.7 - 8.43 \times 10^{-4}\,\text{Re}, \qquad \text{as } 3 \times 10^4 < \text{Re} < 8 \times 10^4, \tag{66}$$

in which

$$\text{Re} = \frac{u_g D_t \rho_g}{\mu_g}.$$

The effect of the dead space in densifying solids at the top of the fast column has been utilized to increase its solids inventory as well as to protect the bend from severe erosion.

2. *Effect of Inlet*

As mentioned in Section II.A, Type C apparatus, or the "restrained system," possesses strong inlet effect. Figure 44 gives several typical configurations

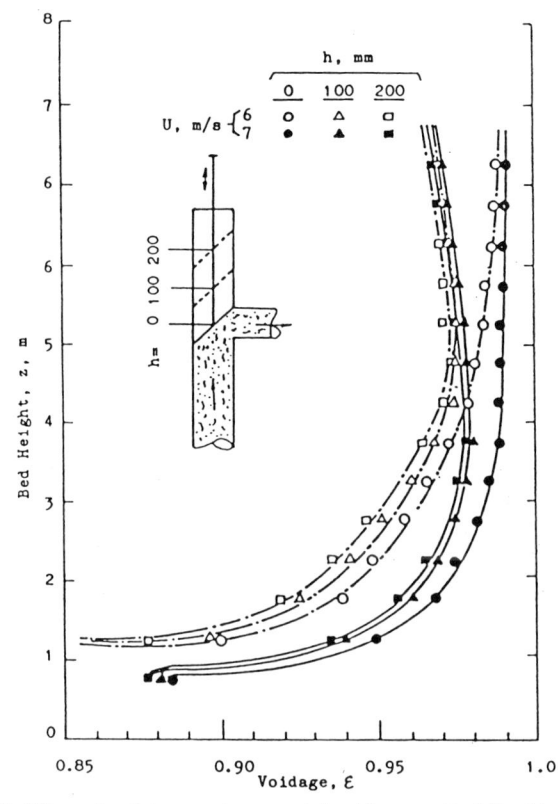

FIG. 43. Effect of outlet geometry on axial voidage profile (after Li et al., 1978).

for this type of apparatus. Measurements showed that they give each its own unique flow structure. Effect of the Type (1) inlet of Fig. 44 is minimal, as shown in Fig. 15 (Li and Kwauk, 1980a). Among recent studies on inlets (Xia and Tung, 1989; Bai et al., 1990; Mori et al., 1991), Xia and Tung (1989) demonstrated that for Type 4 in Fig. 44 with primary aeration only, axial voidage profiles are normal S-shaped curves (see Fig. 45). But for Type (4) of Fig. 44 with primary and secondary aerations, abnormal voidage profiles could appear as shown in Fig. 46. Bai et al. (1990) investigated the effect of the degree of the opening of the butterfly valve on flow structure, and indicated that for small openings, voidage is mostly greater than 0.95 and varies exponentially with bed height, while for large openings, axial voidage profiles display the coexistence of a dilute-phase region at the top and a dense-phase region at the bottom, or even a single dense-phase region with voidages less than 0.95, provided the solids inventory is sufficiently high. The latter

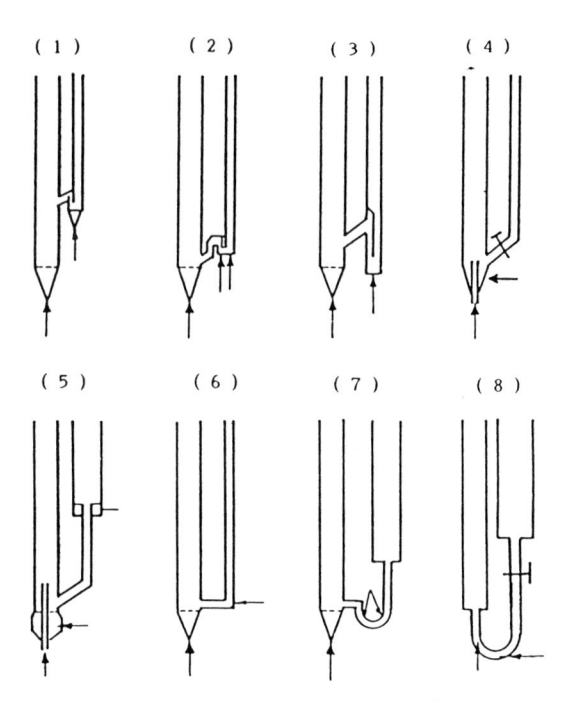

FIG. 44. Typical inlet configurations.

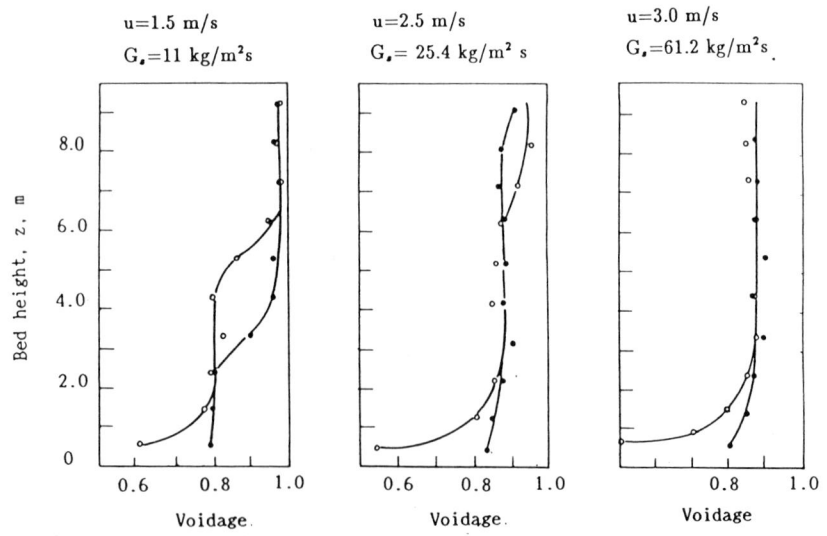

FIG. 45. Comparison of axial voidage profiles for FCC catalyst with one (○) or two (●) gas inlets—inlet (4) (after Xia and Tung, 1989).

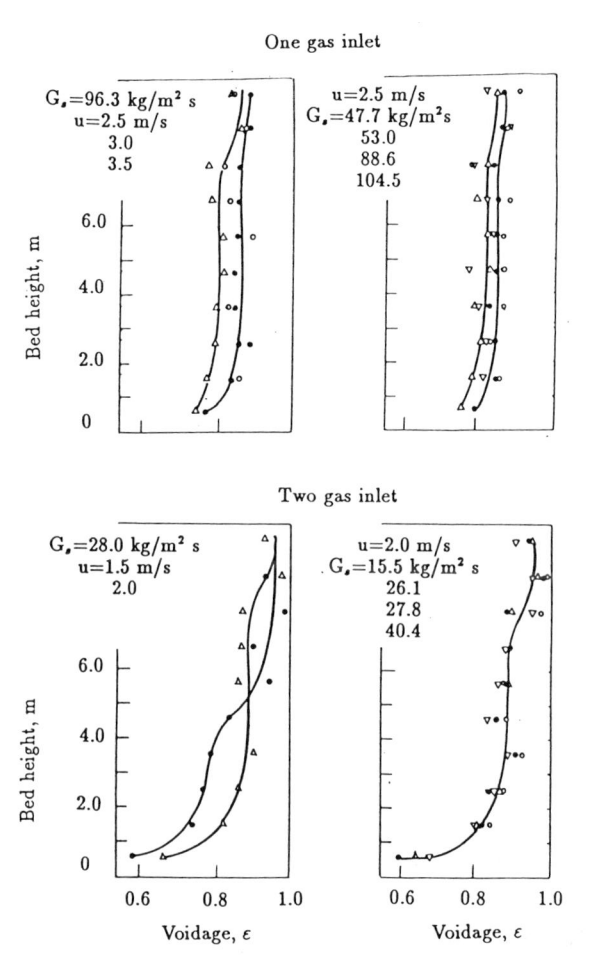

FIG. 46. Dependence of axial voidage profile on inlet geometry and operating conditions—inlet (4) (after Xia and Tung, 1989).

operation is similar to that happens in Type B apparatus, as already noted in Section II.B, while the former operation is analogous to pneumatic transport. Type (5) of Fig. 44 (Hartge et al., 1986) has performance similar to that of Type (4). Mori et al. (1991) showed that for the Type (7) inlet of Fig. 44, bed voidage and its axial distribution are dependent on solids properties, operating conditions and the geometric dimension of the equipment. It is expected that Types (6) and (8) inlets of Fig. 44 would have similar characteristics.

Notation

Ar	Archimedes number, dimensionless	G_s	solids circulation rate, kg/m²s
c	concentration, (v)%	G_{sm}	minimum solids circulation rate, kg/m²s
c_i	concentration of tracer at a injection point, (v)%	H	height of fast fluidized bed, m
c_0	integrating average concentration of tracer, (v)%	I	solids inventory, kg
		I_{eq}	equilibrium inventory for fast fluidization, kg
c_s	solids concentration, kg/m³	k	k_1, k_2, k_m experimental constant
C_d	drag coefficient, dimensionless	L	length or distance, m
C_{ds}	drag coefficient for single particle, dimensionless	ϕ	reduced radius, dimensionless
D_a	axial gas diffusivity, m²/s	Φ_s	solids flux, kg/m²s
D_r	radial gas diffusivity, m²/s	P	pressure, kg-wt/m²
D_s	solid particle diffusivity, m²/s	Pe	Peclet number, dimensionless
D_t	diameter of fluidized bed, m	R	radius of column, m
		r	radial position, m
D_{eq}	particle equivalent diameter, m	Re	Reynolds number, dimensionless
DP	momentum difference, or pressure drop, Pa	Re_c	Reynolds number at incipient fast point, dimensionless
d_p	particle diameter, m		
$f(\varepsilon)$	voidage-dependent correction function, dimensionless	Re_m	Reynolds number defined by Eq. (5), dimensionless
		Re_s	relative Reynolds number, defined by Eq. (33) or (36), dimensionless
F_b	buoyance on particles, kg-wt		
F_a	drag force on particles kg-wt	t	time, s
		t_{inj}	injection of tracer, s
F_g	gravity of gas, kg-wt	$\bar{t}$	mean residence time, s
F_s	gravity of solids, kw-wt	U_{pt}	pneumatic transport velocity, m/s
F_{gf}	friction force on the wall for gas, kg-wt	U_t	terminal velocity, m/s
F_{sf}	friction force on the wall for solids, kg-wt	U_{tf}	incipient fast fluidization velocity, m/s
g	acceleration of gravity, m/s²	u	velocity, m/s
		u_0	superficial gas velocity, m/s

u_d	superficial solid velocity, m/s	ε_a	asymptotic voidage for dense-phase region of fast bed, dimensionless
u_g	actual gas velocity, m/s		
v_s	actual solids velocity, m/s	ε^*	asymptotic voidage for dilute-phase region of fast bed, dimensionless
v_p	actual particle velocity, m/s		
V	output signal voltage, V		
z	axial coordinate, m	ζ	dimensionless distance
z_i	location of point of inflection for fast fluidization, m	θ	dimensionless time
		μ_g	viscosity of gas, kg/ms
		ρ_b	density of bed, kg/m^3
Z_0	characteristic length for fast fluidization voidage profile, m	ρ_g	density of gas, kg/m^3
		ρ_s	density of solids, kg/m^3
		σ^2	second moment
ε	voidage, dimensionless	ψ	numerical function

References

Adams, C. K. "Gas mixing in a fast fluidized bed," *in* "Circulating Fluidized Bed Technology-II" (P. Basu and J. F. Large, eds.), pp. 299–306. Pergamon Press, 1988.

Ambler, P. A., Milne, B. J., Berruti, F., and Scott, D. S. "Residence time distribution of solids in a circulating fluidized bed: Experimental and modelling studies," *Chem. Eng. Sci.*, **45**(8), 2179–2186 (1990).

Arena, U., Cammarota, A., and Pistane, L. "High velocity fluidization behavior of solids in a laboratory scale circulating fluidized bed," *in* "Circulating Fluidized Bed Technology" (P. Basu, ed.), pp. 119–128. Pergamon Press, 1986.

Arena, U., Cammarota, A., Massimilla, L., and Pirozzi, D. "The hydrodynamic behavior of two circulating fluidized bed units of different sizes," *in* "Circulating Fluidized Bed Technology II" (P. Basu and J. F. Large, eds.), pp. 223–230. Pergamon Press, 1988.

Arena, U., Malandrini, A., Marzocchella, A., and Massimilla, L. "Flow structures in the risers of laboratory and pilot CFB units," *in* "Circulating Fluidized Bed Technology III" (P. Basu, M. Horio and M. Hasatani, eds.), pp. 137–144. Pergamon Press, 1991.

Avidan, A. A., and Yerushalmi, J. "Bed expansion in high velocity fluidization," *Powder Techn.* **32**, 223–232 (1982).

Azzi, M., Turlier, P., Large, J. F., and Bernard, J. R. "Use of a momentum probe and gammadensity to study local properties of fast fluidized beds," *in* "Circulating Fluidized Bed Technology III" (P. Basu, M. Horio and M. Hasatani, eds.), pp. 189–194. Pergamon Press, 1991.

Bader, R., Findlay, J., and Knowlton, T. M. "Gas/solids flow patterns in a 30.5-cm-diameter circulating fluidized bed," *in* "Circulating Fluidized Bed Technology II" (P. Basu and J. F. Large, eds.), pp. 123–128. Pergamon Press, 1988.

Bai, D. Ph. D. dissertation, Tsinghua University, Beijing, China, pp. 131–149 (1991).

Bai, D., Jin, Y., Yu, Z., and Yao, W. "A study on the performance characteristics of the circulating fluidized beds," *Chem. Reaction Engineering and Technology* (in Chinese) **3**(1), 24–32 (1987).

Bai, D., Jin, Y., Yu, Z., and Yang, Q. "The axial distribution of the cross-sectionally averaged voidage in fast fluidized bed," *Chem. Reaction Eng. and Technology* (in Chinese) **6**(1), 63–71 (1990a).

Bai, D., Jin, Y., and Yang, C. "Influence of the inlet structures on flow in fast fluidized bed," *Fifth National Conference on Fluidization (in Chinese)*, Beijing, China, April 9–14, pp. 94–97 (1990b).

Bai, D., Jin, Y., and Yu, Z. "Clusters observation in a two dimensional fast fluidized bed," *in* "Fluidization '91: Science and Technology," (M. Kwauk and M. Hasatani, eds.), pp. 110–115. Science Press, Beijing, 1991.

Bolton, L. W., and Davidson, J. F. "Recirculation of particles in fast fluidized risers," *in* "Circulating Fluidized Bed Technology" (P. Basu and J. F. Large, eds.), pp. 139–146. Pergamon Press, 1988.

Brereton, C. M. H., and Stromberg, L. "Some aspects of the fluid dynamic behavior of fast fluidized beds," *in* "Circulating Fluidized Bed Technology (P. Basu, ed.), pp. 133–144. Pergamon Press, 1986.

Brereton, C. M. H., Grace, J. R., and Yu, J. "Axial gas mixing in a circulating fluidized bed," *in* "Circulating Fluidized Bed Technology II" (P. Basu and J. F. Large, eds.), pp. 307–314. Pergamon Press, 1988.

Cai, P. "The transition of flow regime in dense phase gas–solid fluidized bed," Ph. D. Thesis, Tsinghua University (China) and Ohio State University (U.S.A.) (1989).

Cankurt, N. T., and Yerushalmi, J. "Gas mixing in high velocity fluidized beds," *in* "Fluidization" (J. Davidson and D. Kearins, eds.), pp. 387–392. Cambridge University Press, 1978.

Cen, K., Fan, J., Luo, Z., Yan, J., and Ni, M. "The prediction and measurement of particle behavior in circulating fluidized bed," *in* "Circulating Fluidized Bed Technology II" (P. Basu and J. F. Large, eds.), pp. 213–222. Pergamon Press, 1988.

Chesonis, D. C., Klinzing, G. E., Shah, Y. T., and Dassori, C. G. "Solid mixing in a circulating fluidized bed," *in* "Circulating Fluidized Bed Technology III" (P. Basu, M. Jorio and M. Hasatani, eds.), pp. 587–592. Pergamon Press, 1991.

Choi, J. H., Yi, C. K., and Son, J. E. "Axial voidage profile in a cold mode circulating fluidized bed," *in* "Circulating Fluidized Bed Technology III" (P. Basu, M. Horio and M. Hasatani, eds.), pp. 131–136. Pergamon Press, 1991.

Dry, R. J. "Radial concentration profiles in a fast fluidized bed," *Powder Technol.* **49**, 37–44 (1987).

Dry, R. J., and White, C. C. "Gas residence time characteristics in a high-velocity circulating fluidized bed of FCC catalyst," *Powder Techn.* **58**, 17 (1989).

Fujima, Y., Tagashira, K., Takahashi, Y., Ohme, S., Ichimura, S., and Arakawa, Y. "Conceptual study on fast fluidization formation," *in* "Circulating Fluidized Bed Technology III" (P. Basu, M. Horio, and M. Hasatani, eds.), pp. 85–90. Pergamon Press, 1991.

Fusey, I., Lim, C. J., and Grace, J. R. "Fast fluidization in a concentric circulating fluidized bed," *in* "Circulating Fluidized Bed Technology" (P. Basu, ed.), pp. 409–416. Pergamon Press, 1986.

Geldart, D. "Types of gas fluidization," *Powder Technol.* **7**(5), 285–290 (1973).

Glicksman, L. R., Westphalen, D., Brereton, C., and Grace, J. "Verification of the scaling laws for circulating fluidized beds," *in* "Circulating Fluidized Bed Technology III" (P. Basu, M. Horio and M. Hasatani, eds.), pp. 119–124. Pergamon Press, 1991.

Hartge, E. U., Li, Y., and Werther, J. "Analysis of the local structure of the two-phase in a fast fluidized bed," *in* "Circulating Fluidized Bed Technology" (P. Basu, ed.), pp. 153–160. Pergamon Press, 1986.

Hartge, E. U., Rensner, D., and Werther, J. "Solids concentration and velocity patterns in circulating fluidized beds," *in* "Circulating Fluidized Bed Technology II" (P. Basu and J. F. Large, eds.), p. 165. Pergamon Press, 1988.

Horio, M. "Hydrodynamics of circulating fluidization—present status and research needs," *in* "Circulating Fluidized Bed Technology III" (P. Basu, M. Horio, and M. Hasatani, eds.), pp. 3–14. Pergamon Press, 1991.

Horio, M., Morishita, K., Tachibana, O., and Murata, N. "Solid distribution and movement in circulating fluidized beds," *in* "Circulating Fluidized Bed Technology II" (P. Basu and J. F. Large, eds.), p. 147. Pergamon Press, 1988.

Ishii, H., and Murakami, I. "Evaluation of the scaling law of circulating fluidized beds in regard to cluster behavior," *in* "Circulating Fluidized Bed Technology III" (P. Basu, M. Horio and M. Hasatani, eds.), pp. 125–130. Pergamon Press, 1991.

Jin, Y., Yu, Z., Chi, C., and Bai, D. "The influence of exit structures on the axial distribution of voidage in fast fluidized bed," *in* "Fluidization '88: Science and Technology" (M. Kwauk and M. Hasatani, eds.), pp. 165–173. Science Press, Beijing, 1988.

Kato, K., Takarada, T., Tamura, T., and Nishino, K. "Particle hold-up distribution in a circulating fluidized bed," *in* "Circulating Fluidized Bed Technology III" (P. Basu, M. Horio and M. Hasatani, eds.), pp. 145–150. Pergamon Press, 1991.

Kwauk, M. "Generalized fluidization II: Accelerative motion," *Scientia Sinica* **13**(9), 1477 (1964).

Kwauk, M., Wang, N., Li, Y., Chen, B., and Shen, Z. "Fast fluidization at ICM," *in* "Circulating Fluidized Bed Technology" (P. Basu, ed.), pp. 33–62. Pergamon Press, 1986.

Lewis, W. K., Gilliland, F. D., and Bauer, W. C. "Characteristics of fluidized particles," *Ind. Eng. Chem.* **41**, 1104 (1949).

Li, H., Xia, Y., Tung, Y., and Kwauk, M. "Micro-visualization of two-phase structure in a fast fluidized bed," *in* "Circulating Fluidized Bed Technology III" (P. Basu, M. Horio and M. Hasatani, eds.), pp. 183–188. Pergamon Press, 1991.

Li, J., and Weinstein, H. "An experimental comparison of gas backmixing in fluidized beds across the regime spectrum," *Chem. Eng. Sci.* **44**, 1697–1705 (1989).

Li, J., Tung, Y., and Kwauk, M. "Axial voidage profiles of fast fluidized beds," *in* "Circulating Fluidized Bed Technology II" (P. Basu and J. F. Large, eds.), p. 193. Pergamon Press, 1988.

Li, J., Reh, L., Tung, Y., and Kwauk, M. "Comparison of various behaviors for different fluidization regimes" (in Chinese), *Proceedings of Fifth National Conference on Fluidization, Beijing, China, April 9–13*, pp. 110–113 (1990).

Li, Y., and Kwauk, M. "The dynamics of fast fluidization," *in* "Fluidization" (J. R. Grace and J. M. Matsen, eds.), pp. 537–544. Plenum Press, New York, 1980a.

Li, Y., and Wang, F. "Experiments of hydrodynamics for roasted magnesite in a fast fluidized bed," Research report (in Chinese), Institute of Chemical Metallurgy, Academia Sinica (1983).

Li, Y., and Wang, F. "Flow behavior of fast fluidization," *CREEM Seminar, Institute of Chemical Metallurgy, Academia Sinica*, p. 183 (1985).

Li, Y., and Wu, P. "A study on axial gas mixing in a fast fluidized bed," *in* "Circulating Fluidized Bed Technology III" (P. Basu, M. Horio and M. Hasatani, eds.), pp. 581–586. Pergamon Press, 1991.

Li, Y., Wang, F., Dai, D., Wang, Y., and Zhu, Q. "Fast fluidization—Features of flow for iron ore concentrate," Research report (in Chinese), Institute of Chemical Metallurgy, Academia Sinica (1978).

Li, Y., Chen, B., Wang, F., and Kwauk, M. "Gas–solid flow in a fast fluidized bed," *J. Chem. Ind. and Eng.* (in Chinese) **30**(2), 143–152 (1979).

Li, Y., Chen, B., Wang, F., Wang, Y., and Kwauk, M. "Experimental studies on fast fluidization," *Chemical Metallurgy* (in Chinese), No. 4, pp. 20–45 (1980b).

Li, Y., Chen, B., Wang, F., and Kwauk, M. "Rapid fluidization," *Int. Chem. Eng.* **21**, 670–678 (1981).

Li, Y., Chen, B., Wang, F., and Wang, Y. "Hydrodynamic correlations for fast fluidization," *in* "Fluidization: Science and Technology" (M. Kwauk and D. Kunii, eds.), pp. 124–134. Science Press, Beijing, China, 1982.

Li, Y., Wang, F., and Sun, K. "Measurements of fast fluidization for calcined magnesite powder," Research report (in Chinese), Institute of Chemical Metallurgy, Academia Sinica (1983).

Li, Y., Wang, F., and Shen, T. "Characteristics of operation for pulverized coal in a cold mode fast fluidized bed combustor," Research report (in Chinese), Institute of Chemical Metallurgy, Academia Sinica (1984).

Li, Y., Wang, F., and Zeng, Q. "A new process of preparing anhydrous boric oxide by dehydration of boric acid in a fast fluidized bed," *Chem. Reaction Eng. and Techn.* (in Chinese) **6**(2), 43–49 (1990).

Lou, G., and Yang, G. "Axial gas dispersion in a fast fluidized bed" (in Chinese), *Proceedings of Fifth National Conference on Fluidization, Beijing, China, April 9–13*, pp. 155–158 (1990).

Milne, B. J., and Berruti, F. "Modelling the mixing of solids in circulating fluidized beds," *in* "Circulating Fluidized Bed Technology III" (P. Basu, M. Horio and M. Hasatani, eds.), pp. 575–580. Pergamon Press, 1991.

Monceaux, L., Azzi, M., Molodtsof, Y., and Large, J. F. "Overall and local characterization of flow regimes in a circulating fluidized bed," *in* "Circulating Fluidized Bed Technology" (P. Basu, ed.), p. 147. Pergamon Press, 1986a.

Monceaux, L., Azzi, M., Molodtsof, Y., and Large, J. F. "Particle mass flux profiles and flow regime characterization in a pilot-scale fast fluidized bed unit," *in* "Fluidization V" (K. Østergaard and A. Sørensen, eds.), pp. 337–344. Eng. Found., New York, 1986b.

Mori, S., Yan, Y., Kato, K., Matubara, K., and Lui, D. "Hydrodynamics of circulating fluidized bed," *in* "Circulating Fluidized Bed Technology III" (P. Basu, M. Horio and M. Hasatani, eds.), pp. 113—118. Pergamon Press, 1991.

Mutsers, S. M. P., and Rietema, K. "The effect of interparticle forces on the expansion of a homogeneous gas–solid fluidized bed," *Powder Technol.* **18**, 239–248 (1977).

Nakajima, M., Harada, M., Asai, M., Yamazaki, R., and Jimbo, G. "Bubble fraction and voidage in an emulsion phase in the transition to a turbulent fluidized bed," *in* "Circulating Fluidized Bed Technology III" (P. Basu, M. Horio and M. Hasatani, eds.), pp. 79–84. Pergamon Press, 1991.

Nowak, W., Mineo, H., Yamazaki, R., and Yoshida, K. "Behavior of particles in a circulating fluidized bed of a mixture of two different size particles," *in* "Circulating Fluidized Bed Technology III" (P. Basu, M. Horio and M. Hasatani, eds.), pp. 219–224. Pergamon Press, 1991.

Patience, G. S., Chaouki, J., and Kennedy, G. "Solids residence time distribution in CFB reactors," *in* "Circulating Fluidized Bed Technology III" (P. Basu, M. Horio and M. Hasatani, eds.), pp. 599–604. Pergamon Press, 1991.

Perales, J. F., Coll, T., Llop, M. F., Puigjaner, L., Arnaldos, J., and Casal, J. "On the transition from bubbling to fast fluidization regimes," *in* "Circulating Fluidized Bed Technology III" (P. Basu, M. Horio and M. Hasatani, eds.), pp. 73–78. Pergamon Press, 1991.

Qin, S., and Liu, G. "Application of optical fibers to measurement and display of fluidized systems," *in* "Fluidization: Science and Technology" (M. Kwauk and D. Kunii, eds.), pp. 258–267. Science Press, Beijing, 1982.

Qu, Z., Wei, J., and Kwauk, M. "Holographic imaging of clusters in dilute fluidized suspensions," Research Report, Institute of Chemical Metallurgy, Academia Sinica (1985).

Reddy Karri, S. B., and Knowlton, T. M. "A practical definition of the fast fluidization regime," *in* "Circulating Fluidized Bed Technology III" (P. Basu, M. Horio and M. Hasatani, eds.), pp. 67–72. Pergamon Press, 1991.

Reh, L. "Fluidized bed processing," *Chem. Eng. Progr.* **67**(2), 58 (1971).

Reh, L., and Li, J. "Measurement of voidage in fluidized beds by optical probes," *in* "Circulating Fluidized Bed Technology III" (P. Basu, M. Horio and M. Hasatani, eds.), pp. 163–170. Pergamon Press, 1991.

Rhodes, M. J., and Geldart, D. "The hydrodynamics of recirculating fluidized beds," *in* "Circulating Fluidized Bed Technology" (P. Basu, ed.), pp. 193–200. Pergamon Press, 1986.

Rhodes, M. J., and Geldart, D. *Powder Technol.* **53**, 155–162 (1987).

Rhodes, M. J., Laussmann, P., Villain, F., and Geldart, D. "Measurement of radial and axial solids flux variations in the riser of a circulating fluidized bed," *in* "Circulating Fluidized Bed Technology II" (P. Basu and J. F. Large, eds.), p. 155. Pergamon Press, 1988.

Rhodes, M. J., Mineo, O., and Hirama, T. "Particle motion at the wall of the 305 mm diameter riser of a cold model circulating fluidized bed," *in* "Circulating Fluidized Bed Technology III" (P. Basu, M. Horio and M. Hasatani, eds.), pp. 171–176. Pergamon Press, 1991.

Schnitzlein, M., and Weinstein, H. "Flow characterization in high velocity fluidized beds using pressure fluctuations," *Chem. Eng. Sci.* **43**, 2605–2614 (1988).

Son, J. E., Choi, J. H., and Lee, C. K. "Hydrodynamics in a large circulating fluidized bed," *in* "Circulating Fluidized Bed Technology II" (P. Basu and J. F. Large, eds.), p. 113. Pergamon Press, 1988.

Takeuchi, H., and Hirama, T. "Flow visualization in the riser of a circulating fluidized bed," *in* "Circulating Fluidized Bed Technology III" (P. Basu, M. Horio and M. Hasatani, eds.), pp. 177–182. Pergamon Press, 1991.

Takeuchi, H., Hirama, T., Chiba, T., Biswas, J., and Leung, L. S. "A quantitative definition and flow regime diagram for fast fluidization," *Powder Techn.* **47**(2), 195–199 (1986).

Tung, Y., Li, J., and Kwauk, M. "Radial voidage profiles in a fast fluidized bed," *in* "Fluidization: Science and Technology" (M. Kwauk and D. Kunii, eds.), pp. 139–145. Science Press, Beijing, 1988.

Tung, Y., Zhang, W., Wang, Z., Tiu, X., and Ching, X. "Further study of radial voidage profiles in fast fluidized beds," *Eng. Chemistry and Metallurgy* **10**(2), 18–23 (1989).

Van Deemter, J. "Mixing patterns in large-scale fluidized beds," *in* "Fluidization" (J. R. Grace and J. M. Matsen, eds.), p. 69. Plenum Press, New York, 1980.

Van Zoonen, D. "Measurements of diffusional phenomena and velocity profiles in a vertical riser," *in* "Proceedings of the Symposium on the Interaction between Fluids and Particles" (P. A. Rottenburg, ed.), pp. 64–71. Institution of Chemical Engineering, 1962.

Wang, X. S., and Gibbs, B. M. "Hydrodynamics of a circulating fluidized bed with secondary air injection," *in* "Circulating Fluidized Bed Technology III" (P. Basu, M. Horio and M. Hasatani, eds.), pp. 225–234. Pergamon Press, 1991.

Weinstein, H., Graff, R., Meller, N., and Shao, M. "The influence of the imposed pressure drop across a fast fluidized bed," *in* "Fluidization" (D. Kunii and R. Toei, eds.), pp. 299–306. Engineering Foundation, 1984.

Weinstein, H., Shao, M., and Schnitzlein, M. "Radial variation in solid density in high velocity fluidization," *in* "Circulating Fluidized Bed Technology" (P. Basu, ed.), pp. 201–206. Pergamon Press, 1986a.

Weinstein, H., Graff, R. A., Meller, M., and Shao, M. "Radial variation in void fraction in a fast fluidized bed," *in* "Fluidization 5" (K. Østergaad and A. Sørensen, eds.), pp. 329–336. Eng. Found., New York, 1986b.

Weinstein, H., Li, J., Bandlamundi, E., Feindt, H. J., and Graff, R. A. "Gas backmixing of fluidized beds in different regimes and different regions," *in* "Fluidization" (J. R. Grace, L. W. Schemilt and M. A. Bergougnou, eds.), p. 57. Eng. Found., New York, 1989.

Werther, J., Hartge, E. U., Kruse, M., and Nowak, W. "Radial mixing of gas in the core zone of a pilot scale CFB," *in* "Circulating Fluidized Bed Technology III" (P. Basu, M. Horio and M. Hasatani, eds.), pp. 593–598. Pergamon Press, 1991.

Wilhelm, R. H., and Kwauk, M. "Fluidization of solid particles," *Chem. Eng. Progr.* **44**, 201 (1948).

Wisecarver, K. D., Kitano, K., and Fan, L. S. "Pressure fluctuation in a multi-solid pneumatic transport bed," *in* "Circulating Fluidized Bed Technology" (P. Basu, ed.), pp. 145–152. Pergamon Press, 1986.

Wu, P. "Behavior of gas mixing in a fast fluidized bed," M. S. Dissertation, Institute of Chemical Metallurgy, Academia Sinica (1988).

Xia, Y., and Tung, Y. "Effect of inlet and exit structures on fast fluidization," *in* "MPR '87–89" (Y. Tung, Y. Xia and M. Kwauk, eds.), pp. 163–179. Institute of Chemical Metallurgy, Academia Sinica, 1989.

Yang, G., Huang, Z., and Zhao, L. "Radial gas dispersion in a fast fluidized bed," *in* "Fluidization" (D. Kunii and R. Toei, eds.), pp. 145–152. Eng. Found., New York, 1984.

Yang, Y., Jin, Y., Yu, Z., Wang, Z., and Bai, D. "The radial distribution of local particle velocity in a dilute circulating fluidized bed," *in* "Circulating Fluidized Bed Technology III" (P. Basu, M. Horio and M. Hasatani, eds.), pp. 145–152. Pergamon Press, 1991.

Yerushalmi, J., and Cankurt, N. T. "Further studies of the regimes of fluidization," *Powder Techn.* **24**, 187 (1979).

Yerushalmi, J., Turner, D., and Squires, A. M. "The fast fluidized bed," *Ind. Eng. Chem. Proc. Des. Dev.* **15**, 47–53 (1976).

Yerushalmi, J., Cankurt, N. T., Geldart, D., and Liss, B. "Flow regimes in vertical gas–solid contact systems," *AIChE Sympos. Series* **74**(176), 1–13 (1978).

Zhang, B. "A study on the radial distribution of gas velocity in a fast fluidized bed," M. S. Dissertation, Institute of Chemical Metallurgy, Academia Sinica (1990).

Zheng, Q., Wang, X., and Li, X. "Heat transfer in circulating fluidized beds," *in* "Circulating Fluidized Bed Technology III" (P. Basu, M. Horio and M. Hasatani, eds.), pp. 263–268. Pergamon Press, 1991.

4. MODELING

LI JINGHAI

INSTITUTE OF CHEMICAL METALLURGY
ACADEMIA SINICA
ZHONGGUAN VILLAGE, BEIJING, PEOPLE'S REPUBLIC OF CHINA

I. Basic Concepts	148
A. System Designation	148
B. Parameter Definition	150
II. Methodology for Modeling Fast Fluidization	156
A. Pseudo-fluid Approach	156
B. Two-Phase Approach	157
III. Pseudo-fluid Models	157
A. k–ε Turbulence Model	157
B. Two-Channel Model	158
C. Two-Dimensional Diffusion Model	158
IV. Two-Phase Model—Energy-Minimization Multi-scale (EMMS) Model	159
A. Three Scales of Interaction	160
B. Energy Analysis and System Resolution	161
C. Momentum and Mass Conservation for the ST Subsystem	165
D. Energy Minimization and the PD, PFC and FD Regimes	168
E. The EMMS Model	171
V. Local Hydrodynamics—Phases	171
A. Definition of Fast Fluidization and Its Three Operating Modes	172
B. Hydrodynamic States	177
C. Gas/Solid Contacting	184
VI. Overall Hydrodynamics—Regions	188
A. Axial Hydrodynamics	189
B. Radial Hydrodynamics	190
Notation	197
References	199

This chapter is devoted to formulating the hydrodynamic behavior of fast fluidized beds for understanding both local and overall heterogeneous flow structures in systems which were described experimentally in the last chapter. With a brief review on approaches to modeling fast fluidization and on a generalized system designation, the so-called energy-minimization multi-scale (EMMS) model will be presented and then applied to a quantitative analysis of gas/solid contacting to account for both local two-phase structure and overall two-region coexistence in the axial and the radial directions.

I. Basic Concepts

A. System Designation

Figure 1 graphically summarizes the flow structure of fast fluidization introduced in the last chapter, as characterized by

1. local two-phase structure consisting of a solid-rich dilute phase and a gas-rich dense phase;

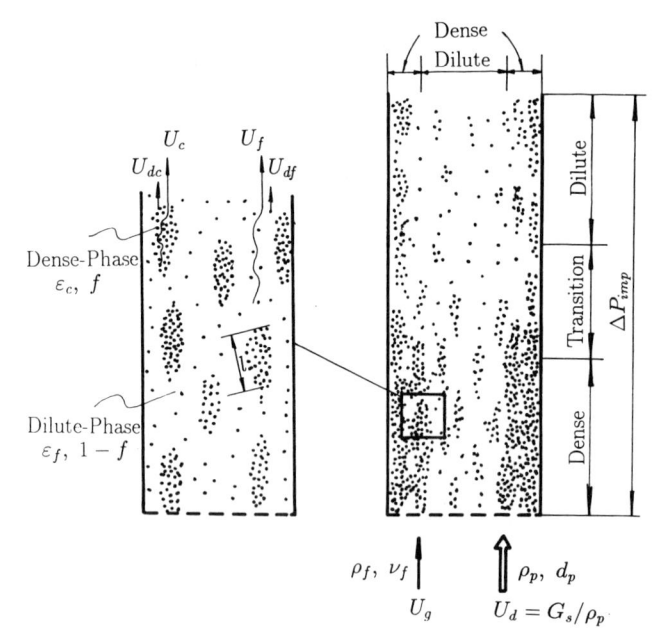

FIG. 1. Heterogeneous flow structure in particle–fluid systems and related parameters.

2. two-region coexistence in the axial direction—a dilute region at the top and a dense region at the bottom;
3. radial heterogeneous distribution with a dilute region in the core and a dense region near the wall;
4. both local and overall heterogeneities are subject to operating conditions, material properties and boundary conditions.

To describe the complicated variation of the flow structure in fast fluidization with operating parameters, material properties and boundary conditions, the following system of designation is proposed (MK-90-10-30).

The term *meso-heterogeneity* denotes local heterogeneity describing localized coexisting of *phases*, the solid-rich dense phase and the fluid-rich dilute phase, due to the *primary instability* inherent in fluid–particle systems. The configurations of phase combination which depend on operating parameters, notably fluid velocity, are described by a spectrum of *regimes*: particulate expansion, bubbling fluidization, turbulent fluidization, fast fluidization and dilute transport, as occurs in gas fluidization at Geldart's Class A powders. The constitution of the regime spectrum is subject to variations in material properties, that is, not all the regimes mentioned would show up for systems composed of different materials. For instance, particulate expansion is seldom detected for Geldart's Class D particles fluidized by a gas, and L/S systems always fluidize particulately. Such a material-dependent change of regime spectrum is described as *patterns*.

The term *macro-heterogeneity* denotes overall heterogeneity describing segregation into dilute and dense *regions* in different parts of the equipment due to *secondary instability* of fluid–particle systems caused by equipment configuration. For fast fluidization in a cylindrical vessel, for instance, macro-heterogeneity is resolved in two dimensions: *axial*, a dense region at the bottom surmounted by a dilute region at the top, and *radial*, a wall region of low voidage surrounding a core region of high voidage.

In summary, the proposed system of designation comprises the following four categories:

Phase—state of particle aggregation (meso-heterogeneity) (see table).

	Continuous		Discontinuous
Dense:	emulsion	or	clusters
Dilute:	broth	or	bubbles

Regime—configuration of phase combination dependent on operating parameters: bubbling, turbulent, fast, transport.

Pattern—constitution of regime spectrum dependent on material properties: bubbling/transport for coarse G/S systems; particulate/bubbling/turbulent/fast/transport for FCC catalyst/air systems; particulate only for most L/S systems.

Region—spatial distribution of phases dependent on boundary conditions (macro-heterogeneity): top and bottom; core and wall.

B. PARAMETER DEFINITION

To describe the complex phase structures in fast fluidization as shown in Fig. 1, a system of parameters is shown in Table I. These parameters can be grouped into three types: independent, dependent and characteristic, as will be described in this section.

1. Independent Parameters

Independent parameters are those which can be changed at will by design and in operation, and the dynamic states of particle–fluid systems will follow. They usually consist of material properties, operating conditions and boundary conditions.

a. Material Properties. Material properties include fluid density ρ_f, fluid viscosity u_f, particle density ρ_p and particle diameter d_p, which are process-independent.

Since most particles handled in engineering are multisized, a characteristic mean diameter has to be defined. The most commonly used mean diameter is the so-called surface-to-volume mean diameter defined as

$$\bar{d}_s = \left(\sum_{i=1}^{N} x_i/d_i \right)^{-1},$$

which is dependent only on size distribution without consideration of the interaction between the particles and the fluid, and therefore causes deviation in flow calculations. In order to reasonably represent the hydrodynamic equivalence between multisized particles and monosized particles, the interaction between particles and fluid, in addition to geometry, was considered, leading to the following hydrodynamic mean diameter (Li et al., 1991c; Li and Kwauk, 1994):

$$\bar{d} = \bar{C}_D \bigg/ \sum_{i=1}^{N} C_{D_i} x_i/d_i.$$

TABLE I
PARAMETERS AND FORMULAS

Parameters	Dense Phase	Dilute Phase	Inter Phase	Global
Fluid density	ρ_f	ρ_f	ρ_f	
Particle density	ρ_p	ρ_p	$\rho_p(1-\varepsilon_c)$	
Solid-phase scale	d_p	d_p	l (cluster)	
Voidage	ε_c	ε_f	$1-f$	$\varepsilon = \varepsilon_f(1-f)+\varepsilon_c f$
Volume fraction	f	$1-f$		1
Superficial velocity: gas	U_c	U_f	$U_f(1-f)$	$U_c = U_f(1-f) + U_c f$
solid	U_{dc}	U_{df}		$U_d = U_{df}(1-f) + U_{dc} f$
slip	$U_{sc} = U_c - \dfrac{U_{dc}\varepsilon_s}{1-\varepsilon_s}$	$U_{sf} = U_f - \dfrac{U_{df}\varepsilon_f}{1-\varepsilon_f}$	$U_{si} = \left(U_f - \dfrac{U_{dc}\varepsilon_c}{1-\varepsilon_f}\right)(1-f)$	$U_s = U_g - \dfrac{U_d \varepsilon}{1-\varepsilon}$
Characteristic Reynolds number	$\mathrm{Re}_c = (1-\varepsilon_s)\left|\dfrac{U_{dc}\varepsilon_s}{1-\varepsilon_s}\right| d_p/\nu_f$	$\mathrm{Re}_f = \left(U_f - \dfrac{U_{df}\varepsilon_f}{1-\varepsilon_f}\right) d_p/\nu_f$	$\mathrm{Re}_i = \dfrac{U_{si} l}{\nu_f}$	
Drag coefficient: single sphere	$C_{D_{0c}} = \dfrac{24}{\mathrm{Re}_c} + \dfrac{3.6}{\mathrm{Re}_c^{0.313}}$	$C_{D_{0f}} = \dfrac{24}{\mathrm{Re}_f} + \dfrac{3.6}{\mathrm{Re}_f^{0.313}}$	$C_{D_{0i}} = \dfrac{24}{\mathrm{Re}_i} + \dfrac{3.6}{\mathrm{Re}_i^{0.313}}$	
fluidized spheres	$C_{D_c} = C_{D_{0c}} \varepsilon_c^{-4.7}$	$C_{D_f} = C_{D_{0f}} \varepsilon_f^{-4.7}$	$C_{D_i} = C_{D_{0i}}(1-f)^{-4.7}$	
Number of particles or clusters in unit volume	$m_c = (1-\varepsilon_c)\left/\dfrac{\pi d_p^3}{6}\right.$	$m_f = (1-\varepsilon_f)\left/\dfrac{\pi d_p^3}{6}\right.$	$m_i = f\left/\dfrac{\pi l^3}{6}\right.$	
Force acting on each particle or cluster	$F_{dense} = C_{D_c} \dfrac{\pi d_p^2}{4} \dfrac{\rho_f}{2} U_{sc}^2$	$F_{dilute} = C_{D_f} \dfrac{\pi d_p^2}{4} \dfrac{\rho_f}{2} U_{sf}^2$	$F_{bulk} = C_{D_i} \dfrac{\pi l^2}{4} \dfrac{\rho_f}{2} U_{si}^2$	
Total force in unit volume or pressure drop	$\Delta P_{dense} = m_c F_{dense}$	$\Delta P_{dilute} = m_f F_{dilute}$	$\Delta P_{inter} = F_{inter} = m_i F_{bulk}$	
Power in unit volume	$(W_{st})_{dense} = \Delta P_{dense} U_c f$	$(W_{st})_{dilute} = \Delta P_{dilute} U_f (1-f)$	$(W_{st})_{inter} = \Delta P_{inter} U_f (1-f)$	W_{st}
for unit mass of particles	$(N_{st})_{dense} = (W_{st})_{dense}/(1-\varepsilon)\rho_p$	$(N_{st})_{dilute} = (W_{st})_{dilute}/(1-\varepsilon)\rho_p$	$(N_{st})_{inter} = (W_{st})_{inter}/(1-\varepsilon)\rho_p$	N_{st}

b. Operating Conditions. Configuration of phase combination in particle–fluid two-phase flow depends on operating conditions which include superficial fluid velocity and superficial particle velocity, defined respectively as

$$U_g = \frac{\text{total fluid mass flow rate over the overall cross section}}{\text{(cross-sectional area)(fluid density)}},$$

$$U_d = \frac{\text{total solid mass flow rate over the overall cross section}}{\text{(cross-sectional area)(solid density)}}.$$

As defined here, U_g and U_d are independent of each other, and they differ from their corresponding actual velocities U_g/ε and $U_d/(1-\varepsilon)$, which vary with the dependent parameter ε.

c. Boundary Conditions. Boundary conditions affecting the distribution of regions in space are equipment-related external constraints, which generally consist of the retaining wall, entrance and exit configurations and the imposed pressure drop ΔP_{imp} over the particle–fluid two-phase system.

Wall effect gives rise to radial heterogeneity. Although Soo (1989), Ding et al. (1992) and Ding and Gidaspow (1990) proposed boundary conditions for their models, it is still considered necessary to measure relevant parameters near the wall, such as fluid velocity, particle velocity, and void fraction. Inlet and outlet effects are even more difficult to generalize, and will therefore be discussed only in a qualitative sense.

The imposed pressure drop ΔP_{imp} is a global factor related to the effects of the solids inventory, the resistance of the solids circulation loop and the *configuration of the system*. It plays an important role in determining axial voidage profiles (Weinstein et al., 1983; Li et al., 1988a), which, according to operation, fall into three modes: dilute region only, dense region only and both regions coexisting in a unit. Its effect will be discussed in Section V.A.

2. Dependent Parameters

Figure 1 shows the eight principal dependent parameters in a generalized particle–fluid two-phase system, the local state of which can be described by some functional relationship between a dependent parameter vector $\mathbf{X}(r)$ in terms of these eight parameters:

$$\mathbf{X}(r) = \{\varepsilon_f(r), \varepsilon_c(r), f(r), U_f(r), U_c(r), l(r), U_{d_f}(r), U_{d_c}(r)\}.$$

These eight dependent parameters will be described next.

a. Particle and Fluid Velocity. Various average velocities can be derived as shown in Table I:

Local average fluid velocity:
$$U_g(r) = U_f(r)(1 - f(r)) + U_c(r)f(r).$$

Local average particle velocity:
$$U_d(r) = U_{d_f}(r)(1 - f(r)) + U_{d_c}(r)f(r).$$

From these, the corresponding cross-sectional average velocities can be defined:

$$U_g = \frac{2}{R^2} \int_0^R U_g(r) r \, dr,$$

$$U_d = \frac{2}{R^2} \int_0^R U_d(r) r \, dr.$$

b. Voidage. Voidage represents the degree of expansion of particle–fluid systems, defined as

$$\text{Voidage} = \frac{(\text{total volume}) - (\text{volume occupied by particles})}{\text{total volume}}.$$

For heterogeneous particle–fluid systems, a variety of voidages, with respect to phases and regions, are employed for describing their complicated structures. Local voidage in the dense-phase $\varepsilon_c(r)$ and local voidage in the dilute-phase $\varepsilon_f(r)$ are the two basic voidages from which other voidages are derived:

Local average voidage:
$$\varepsilon(r) = \varepsilon_c(r)f(r) + \varepsilon_f(r)(1 - f(r)).$$

Cross-sectional average voidage:
$$\bar{\varepsilon} = \frac{2}{R^2} \int_0^R \varepsilon(r) r \, dr.$$

Global average voidage:
$$\bar{\bar{\varepsilon}} = \frac{1}{H} \int_0^H \bar{\varepsilon}(h) \, dh.$$

Usually $\bar{\bar{\varepsilon}}$ and $\bar{\varepsilon}$ are calculated from measured pressure drops, while measurements of $\varepsilon(r)$, $\varepsilon_c(r)$ and $\varepsilon_f(r)$ are much more delicate because of difficulty in calibration.

c. *Phase Structure.* The dependent parameters, $l(r)$, for the dimension of a particle cluster, and $f(r)$, for the volume fraction occupied by the dense phase, are designed to describe the phase structure of particle-fluid systems together with $\varepsilon_c(r)$ and $\varepsilon_f(r)$. Among others (Yerushalmi and Cankurt, 1979; Subbarao, 1986), a correlation for the size of a particle aggregate $l(r)$ was proposed by Li (1987) and Li *et al.* (1988c) with the assumption that $l(r)$ is inversely proportional to the input energy. Details will be introduced in Section IV.C.

It should be understood that $l(r)$ is used fictitiously as an equivalent dimension for calculating the interaction between the dense and dilute phases, rather than a parameter to represent the real size of a cluster.

3. Characteristic Parameters

Characteristic parameters are those which are not related directly to operating and boundary conditions, but depend only on material properties.

a. *Drag Coefficient.* Because of the heterogeneous structure in particle–fluid two-phase flow, three drag coefficients are needed for calculating multi-scale interactions between particles and fluid, as listed in Table I. Drag coefficient for single particles can be calculated for $Re_p < 1{,}000$ (Flemmer and Banks, 1986) as

$$C_{D_o} = 24/Re_p + 3.6/Re_p^{0.313},$$

and for particles in homogeneous suspensions (Wallis, 1969) as

$$C_D = C_{D_o} \varepsilon^{-4.7}.$$

Although particles in two-phase flow are not uniformly distributed, the dense and the dilute phases can be considered, each in its own, as uniform suspensions, and the global system can thus be regarded as consisting of dense clusters dispersed in a broth of separately distributed discrete particles, as shown in Fig. 1. The preceding correlations will therefore be used respectively in the dilute and the dense phases, for calculating micro-scale fluid–particle interaction, and also for evaluating meso-scale interphase interaction between clusters and the broth, as shown in Table I, for C_{D_c}, C_{D_f} and C_{D_i}.

b. *Terminal Velocity.* The terminal velocity U_t of a particle is also called its free falling velocity. It is the maximum velocity which a single particle eventually reaches while falling down freely in a still fluid, or the minimum velocity for a flowing fluid to carry this single particle upward. The value of U_t can be derived from force balance between the gravity of a particle and the drag exerted by the fluid, that is,

$$\frac{\pi d_p^3}{6}\rho_p g = C_{D_o}\frac{\pi d_p^2 \rho_f U_t^2}{4 \cdot 2};$$

therefore,

$$U_t = \left(\frac{4gd_p(\rho_p - \rho_f)}{3\rho_f C_{D_o}}\right)^{0.5}.$$

c. Minimum Fluidization Velocity U_{mf}. Minimum fluidization velocity is the lowest fluid velocity at which a particle–fluid system starts to fluidize, which can be calculated from (Kunii and Levenspiel, 1969)

$$\frac{1.75}{\phi_s \varepsilon_{mf}^3}\left(\frac{d_p U_{mf}\rho_f}{\mu_f}\right)^2 + \frac{150(1-\varepsilon_{mf})}{\phi_s^2 \varepsilon_{mf}^3}\frac{d_p U_{mf}\rho_f}{\mu_f} = \frac{d_p^3 \rho_f(\rho_p - \rho_f)g}{\mu_f^2},$$

where ε_{mf} is the bed voidage at minimum fluidization which depends on the physical properties of the particles and their mode of packing, and ϕ_s is the shape factor. Without knowing ϕ_s and ε_{mf}, U_{mf} can be approximated as (Kunii and Levenspiel, 1969):

for small particles ($Re_p < 20$):

$$U_{mf} = \frac{d_p^2(\rho_p - \rho_f)g}{1{,}650 u_f};$$

for large particles ($Re_p > 1{,}000$):

$$U_{mf}^2 = \frac{d_p(\rho_p - \rho_f)g}{245\rho_f}.$$

A extensive review on U_{mf} was made by Couderc (1985), listing a variety of correlations for U_{mf}.

d. Saturation Carrying Capacity. Saturation carrying capacity K^* is the critical flux of solids flow corresponding to the transition from dense-phase fluidization to dilute-phase transport, at which these two hydrodynamic states coexist. This is often called "choking." Though much attention has been paid to measuring K^*, resulting in a number of correlations as reviewed by Briens and Bergougnou (1986), Satija *et al.* (1985) and Yang (1983), for example (Knowlton and Bachovchin, 1975),

$$\left(\frac{U_{pt}}{gd_p}\right)^{0.5} = 9.07\left(\frac{\rho_p}{\rho_f}\right)^{0.347}\left(\frac{K^* d_p}{\mu_f}\right)^{0.214}\left(\frac{d_p}{d_B}\right)^{0.246},$$

its underlying mechanism is still not understood, as will be discussed in Section V.A.

4. Derived Parameters

Other parameters in Table I, exclusive of those discussed here, can be formulated as functions of the preceding parameters and are therefore called derived parameters, which will be dealt with wherever they arise.

II. Methodology for Modeling Fast Fluidization

There are two principal approaches to formulate particle–fluid two-phase flow in fluidized beds: the *pseudo-fluid* model and the *two-phase* model.

A. Pseudo-fluid Approach

The pseudo-fluid approach inherits the methodology for analyzing single-phase fluid flow and considers a fluid–particle system to consist of k distinct fictitious fluids with interaction between each other.

The simplest case is called the *one-fluid model*, that is, $k = 1$, assuming that particles are distributed in the fluid discretely. It was used for modeling voidage distributions in fast fluidized beds (Li and Kwauk, 1980; Bai et al., 1988; Zhang et al., 1990).

A more popular pseudo-fluid model is the so-called *two-fluid model*, that is, $k = 2$, in which the particle phase is considered as a fictious fluid, and separate hydrodynamic equations are established for both the particle phase and the fluid phase (Soo, 1967; Jackson, 1963; Grace and Tuot, 1979; Gidaspow et al., 1986, 1989; Arastoopour and Gidaspow, 1979; Yang, 1988; Louge and Chang, 1990; Ding and Gidaspow, 1990; Bai et al., 1991b).

For considering radial heterogeneity in a two-phase system, the pseudo-fluid model is simultaneously applied to the core dilute region and the wall dense region, resulting in the so-called two-channel model (Nakamura and Capes, 1973; Bai et al., 1988; Yang, 1988; Ishii et al., 1989; Berruti, 1989; Rhodes, 1990).

By analyzing the turbulent kinetic energy of the fluid, the two-fluid model was modified to the so-called k–ε model (Militzer, 1986; Zhou and Huang, 1990) for dilute two-phase flow, which has been extended to calculating radial distributions in fast fluidized beds (Yang, 1992).

Although the pseudo-fluid approach can be strictly formulated, it is hardly capable of describing the flow structure difference between the dilute phase and the dense phase, that is, the scale cannot reach that of individual particles, and regime transitions, related to inflective changes of flow structure, cannot yet be treated.

B. Two-Phase Approach

The two-phase approach is based on the phenomenological nature of particle–fluid flow, which is two-phase in structure. It considers a heterogeneous system to consist of two different phases, a solid-rich dense phase and a gas-rich dilute phase, for each of which mass and momentum conservation is analyzed with consideration of interaction and exchange between the two phases. The two-phase model was proposed for bubbling fluidization (Davidson, 1961; Grace and Clift, 1974) to analyze the bubble phenomenon, and was then used for calculating cluster diameter (Yerushalmi et al., 1978; Subbarao, 1986) and analyzing local flow structure (Hartge, 1988) in fast fluidized beds. To account for the stability of two-phase structure in fast fluidization, Li (1987) and Li et al. (1988c) proposed the energy-minimization multi-scale (EMMS) model, considering that mass and momentum conservation alone is not sufficient for determining the flow structure in fast fluidization, and that an additional condition is needed.

The two-phase approach can be used to analyze the interaction between the fluid and individual particles, though it is not convenient for analyzing turbulent and time-dependent behavior of the system. Generally speaking, and in the author's opinion, the pseudo-fluid approach is more suitable for dilute two-phase flow, while the two-phase approach is more suitable for dense fluidization.

Considering the variety of the pseudo-fluid models established by international colleagues, as reviewed earlier, this chapter will mainly discuss the two-phase approach, but with a brief introduction to several pseudo-fluid models developed in China.

III. Pseudo-fluid Models

A. k–ε Turbulence Model

A k–ε turbulence model was proposed by Yang (1992) to simulate the flow field in circulating fluidized beds, consisting of the following equations:
1. conservation for gas with respect to mass, momentum, turbulent kinetic energy and dissipation rate;

2. conservation for particles with respect to mass and momentum;
3. turbulence for both gas and particles;
4. boundary conditions.

His model paid special attention to the influence of particles on gas turbulence and the collision between particles, and was solved numerically to calculate radial profiles of voidage, gas velocity and particle velocity, showing good agreement with experiments.

B. Two-Channel Model

For simulating radial heterogeneity in fast fluidization, Bai et al. (1988) formulated the so-called two-channel model with consideration of the interaction and exchange between these two regions through their interface, from which axial voidage profiles and voidages in these two regions were calculated. This model was further improved (Bai et al., 1991b) by applying the two-fluid model to both the core and the annular regions for calculating radial voidage profiles in fast fluidized beds.

C. Two-Dimensional Diffusion Model

Zhang et al. (1990) developed a more general diffusion model for calculating axial and radial voidage distributions in fast fluidized beds by extending Li and Kwauk's (1980) model.

Combining Brownian movement and random walk with particle motion in a fast fluidized bed, they derived the following Fokker–Planck equation with respect to voidage:

$$\frac{\partial \varepsilon}{\partial t} + \frac{\partial (u_z \varepsilon)}{\partial z} = D_a \frac{\partial^2 \varepsilon}{\partial z^2} + \frac{D_r}{r} \cdot \frac{\partial}{\partial r}\left(r \cdot \frac{\partial \varepsilon}{\partial r}\right).$$

Assuming that radial and axial diffusion coefficients are equal, that is, $D_a = D_r$, Eq. (1) then becomes

$$\frac{1}{A} \cdot \frac{\partial \varepsilon}{\partial z} = \frac{\partial^2 \varepsilon}{\partial z^2} + \frac{\partial^2 \varepsilon}{\partial r^2} + \frac{1}{r} \cdot \frac{\partial \varepsilon}{\partial r},$$

where $A = G_s/\rho_p f'$; f' is called the frequency factor, taken to be 0.02 by optimization (Zhang et al., 1991). By separating variables with respect to h and r, the solution of Eq. (2) can be written as

$$\varepsilon(r, z) = (C_1 \cdot e^{\alpha_1 z} + C_2 \cdot e^{\alpha_2 z}) \cdot [C_3 \cdot r^{1.5} \cdot J_0(kr) + C_4 \cdot r^{1.5} \cdot Y_0(kr)],$$

where α_1 and α_2 are the characteristic roots of the separated equation with respect to h and can be formulated as

$$\alpha_1 = \frac{1/A + ((1/A)^2 + 4k^2)^{1/2}}{2},$$

$$\alpha_2 = \frac{1/A - ((1/A)^2 + 4k^2)^{1/2}}{2},$$

and $J_0(kr)$ and $Y_0(kr)$ are the Bessel function and the Neumann function, respectively.

According to Li and Kwauk's model, Zhang and Xie derived $C_i(i = 1$–$4)$, resulting in the final correlations for spatial voidage distribution (Zhang et al., 1991):

For $z \leq Z_i$,

$$\varepsilon(r, z) = \left[\varepsilon_a + \frac{\varepsilon^* - \varepsilon_a}{2} \cdot e^{\alpha_1(z - Z_i)}\right] \cdot \left[1 - \left(1 - \frac{\varepsilon_{mf}}{\varepsilon^*}\right)\left(\frac{r}{R}\right)^{3.5} \frac{4 - (kr)^2}{4 - (kR)^2}\right],$$

for $z \geq Z_i$,

$$\varepsilon(r, z) = \left[\varepsilon^* - \frac{\varepsilon^* - \varepsilon_a}{2} \cdot e^{\alpha_2(z - Z_i)}\right] \cdot \left[1 - \left(1 - \frac{q\varepsilon_{mf}}{\varepsilon^*}\right)\left(\frac{r}{R}\right)^{3.5} \frac{4 - (kr)^2}{4 - (kR)^2}\right],$$

where q and k depend on material properties, and $q = 1.75$, $k = 0.01$ for FCC particles.

IV. Two-Phase Model—Energy-Minimization Multi-scale (EMMS) Model

For very dilute particle–fluid systems in which particles are discretely distributed, invoking the conditions for the conservation of mass and momentum normally suffices to define its hydrodynamic states, because, as the traditional theory expects, it always operates in a *single* state. The methodology of dealing with dilute systems has been inherited, though regrettably, in studying dense particle–fluid systems which are characterized by structural heterogeneity and regime multiplicity, for which some additional constraints need to be specified.

An appropriate understanding of particle–fluid two-phase flow rests on an adequate analysis of the local hydrodynamics corresponding to the scale of bubbles or clusters, as well as the overall hydrodynamics corresponding to the scale of the retaining vessel. Local hydrodynamics, designated as phases, is rooted in the intrinsic characteristics of particle–fluid systems, whereas overall hydrodynamics, designated as regions, depends on geometrical boundary conditions. Any specific local hydrodynamic state in a system is subject to the constraints of overall hydrodynamics, whereas every point of the overall system has to satisfy the constraints of local hydrodynamics.

This section will first deal with the *phases* in particle–fluid two-phase flow by developing a mathematical model to quantify local hydrodynamic states. This analysis will reveal the insufficiency of the conditions for the conservation of mass and momentum alone in determining the hydrodynamic states of heterogeneous particle–fluid systems, and calls for a methodology different from what is used in analyzing dilute uniform flow. For this purpose the concept of *multi-scale interaction* between particles and fluid and the principle of *energy minimization* are proposed.

Then, overall hydrodynamics of fast fluidization—*region*—will be discussed by extending the EMMS model to both axial and radial directions. Other two aspects of local hydrodynamics—*regime and pattern*—will not be involved as this book is limited to the fast fluidization regime.

A. Three Scales of Interaction

To account for the intrinsic characteristics of particle–fluid two-phase flow in fast fluidization, the particles and the fluid are considered to interact with each other on both a micro-scale and meso-scale level to produce local or meso-scale heterogeneity (phases), and the overall fluid–particle system interacts with the equipment boundaries on a much larger scale to produce macro-scale heterogeneity (regions).

1. Micro-scale

Micro-scale interaction is the smallest scale of interaction between the particles and fluid in the system, corresponding to the size of the constituent particles and prevailing in both the dense phase and the dilute phase. This interaction, expressed as force acting on a single particle, can be written for the dense phase:

$$F_{\text{dense}} = C_{D_c} \frac{\pi d_p^2}{4} \frac{\rho_f U_{sc}^2}{2}$$

and for the dilute phase:

$$F_{\text{dilute}} = C_{D_f} \frac{\pi d_p^2}{4} \frac{\rho_f U_{sf}^2}{2}.$$

2. Meso-scale

Meso-scale interaction is concerned with the interaction between clusters and the dilute-phase broth surrounding them, or the interaction between bubbles and the emulsion in which they exist. For the former, this interaction is expressed as force acting on a cluster by the broth through the so-called interphase

$$F_{\text{bulk}} = C_{D_i} \frac{\pi l^2}{4} \frac{\rho_f U_{si}^2}{2},$$

where U_{si} is the superficial relative velocity between the clusters and the dilute-phase broth as defined in Table I. Here, for simplicity, the interaction between the particles in the clusters and the particles in the broth is neglected.

3. Macro-scale

Macro-scale interaction occurs between the global particle–fluid system and its boundaries, resulting in macro-heterogeneity. Macro-scale interaction generally extends in both the radial (lateral) and axial directions. Radial macro-scale interaction is caused by wall effect leading to radial distribution of parameters. Axial macro-scale interaction originates primarily from ΔP_{imp} and inlet and outlet effects. The imposed pressure drop ΔP_{imp} also governs the shape of axial profiles. Macro-scale interaction underlies the dependence of meso-scale hydrodynamics on location, and it will be discussed in Section VI in connection with overall hydrodynamics.

In effect, such a multi-scale analysis resolves a macro-scale heterogeneous system into three meso- to micro-scale subsystems—dense-phase, dilute-phase and inter-phase. Thus, modeling a heterogeneous particle–fluid two-phase system is reduced to calculations for the three lower-scale subsystems, making possible the application of the much simpler theory of particulate fluidization to aggregative fluidization and the formulation of energy consumptions with respect to phases (dense, dilute and inter) and processes (transport, suspension and dissipation).

B. Energy Analysis and System Resolution

1. Specific Energy Consumption

a. Per Unit Mass of Particles. The *total* energy associated with a flowing particle–fluid system, expressed as power per *unit mass* of solids, N_T, is considered to consist of the sum of two portions, one used in suspending and transporting the particles N_{st}, and one purely dissipated in particle collision, circulation, acceleration, etc., N_d. N_{st} can be further split into that for particles suspension N_s and for transport N_t, and it can also be apportioned between the dense cluster phase, the surrounding dilute phase and interaction between the two, that is,

$$N_T = N_{st} + N_d,$$
$$N_{st} = N_s + N_t$$
$$= (N_{st})_{dense} + (N_{st})_{dilute} + (N_{st})_{inter}.$$

In fact, N_s, which results from the slip between the fluid and the particles, is also dissipated because it does not contribute to the upward motion of the particles, making the total dissipated energy equal to $N_s + N_d$. However, this portion of dissipated energy is responsible for retaining the potential energy of the particles which are suspended in the system, that is, keeping the system expanded, and is therefore different from the purely dissipated energy N_d.

The average voidage of the system ε is formulated additively from the dense and dilute phases as

$$\varepsilon = \varepsilon_c f + \varepsilon_f (1 - f).$$

If wall friction is neglected, then

$$N_T = \frac{\Delta P U_g}{(1 - \varepsilon)\rho_p} = \frac{(1 - \varepsilon)\rho_p g U_g}{(1 - \varepsilon)\rho_p} = U_g g.$$

The three components of N_{st} in the three subsystems can thus be calculated as follows:

$$(N_{st})_{dense} = \frac{1}{(1 - \varepsilon)\rho_p} \Delta P_{dense} U_c f,$$

$$(N_{st})_{dilute} = \frac{1}{(1 - \varepsilon)\rho_p} \Delta P_{dilute} U_f (1 - f),$$

$$(N_{st})_{inter} = \frac{1}{(1 - \varepsilon)\rho_p} \Delta P_{inter} U_f (1 - f).$$

Substituting the foregoing and from the expressions for ΔP_{dense}, ΔP_{dilute} and ΔP_{inter} in Table I, we get

$$N_{st} = \frac{1}{(1 - \varepsilon)\rho_p} [\Delta P_{dense} U_c f + \Delta P_{dilute} U_f (1 - f) + \Delta P_{inter} U_f (1 - f)]$$

$$= \frac{3}{4(1-\varepsilon)\rho_p}\left(C_{D_c}\frac{1-\varepsilon_c}{d_p}\rho_f U_{sc}^2 U_c f + C_{D_f}\frac{1-\varepsilon_f}{d_p}\rho_f U_{sf}^2 U_f(1-f)\right.$$
$$\left. + C_{D_i}\frac{f}{l}\rho_f U_{si}^2 U_f(1-f)\right).$$

By difference, the dissipated energy N_d can be deduced from N_{st} and N_T:

$$N_d = N_T - N_{st} = N_T - (N_s + N_t),$$

where N_t, the energy for transporting the particles at a flow rate of G_s through a distance of $1/(1-\varepsilon)\rho_p$ and related to unit mass of particles in unit area, can be written as

$$N_t = \frac{G_s g}{(1-\varepsilon)\rho_p},$$

and then, again by difference, the energy for suspending the particles can be calculated:

$$N_s = N_{st} - N_t.$$

The total energy consumption N_T has thus been resolved into its components according to the constituent processes involved in particle–fluid interaction.

b. Per Unit Volume of Vessel. By multiplying the preceding energy terms by $(1-\varepsilon)\rho_p$, which is the total mass of particles in unit volume, they are converted into the corresponding energy consumptions with respect to *unit volume* of the retaining vessel, that is,

$$W_T = N_T(1-\varepsilon)\rho_p = \Delta P U_g,$$
$$W_{st} = N_{st}(1-\varepsilon)\rho_p,$$
$$W_s = N_s(1-\varepsilon)\rho_p,$$
$$W_d = N_d(1-\varepsilon)\rho_p,$$
$$W_t = N_t(1-\varepsilon)\rho_p.$$

c. Per Unit Mass of Fluid. Corresponding to N, energy consumption per unit mass of fluid can be calculated by dividing W with $\varepsilon\rho_f$, which is the total mass of fluid in unit volume. For instance, the energy consumption per unit mass of fluid for suspending and transporting particles is

$$\frac{W_{st}}{\varepsilon\rho_f}.$$

2. Characterization of Energy Consumptions

Obviously, the different energy terms characterize different aspects of particle–fluid two-phase flow, among which those related to particle suspension and transport are of importance in analyzing fluid–particle interaction, and, in particular, for identifying system stability:

N_{st} characterizes the intrinsic tendency of particles toward an array of the lowest interaction with the fluid;

W_{st} or $W_{st}/\varepsilon\rho_f$ represents the intrinsic tendency of the fluid to seek flow paths for the lowest resistance to flow.

3. System Resolution

According to the preceding energy analysis and using the symbols defined in Fig. 1, Fig. 2 shows graphically how a particle–fluid two-phase system can be resolved into the suspension and transport (ST) subsystem corresponding to N_{st}, and the energy dissipation (ED) subsystem corresponding to N_d.

The beginning of this section has already shown that both N_d and N_{st} are dependent on vector **X**, that is, related to flow structure, while N_T is only subject to the independent parameter U_g, and therefore independent of flow structure as long as the particles are suspended.

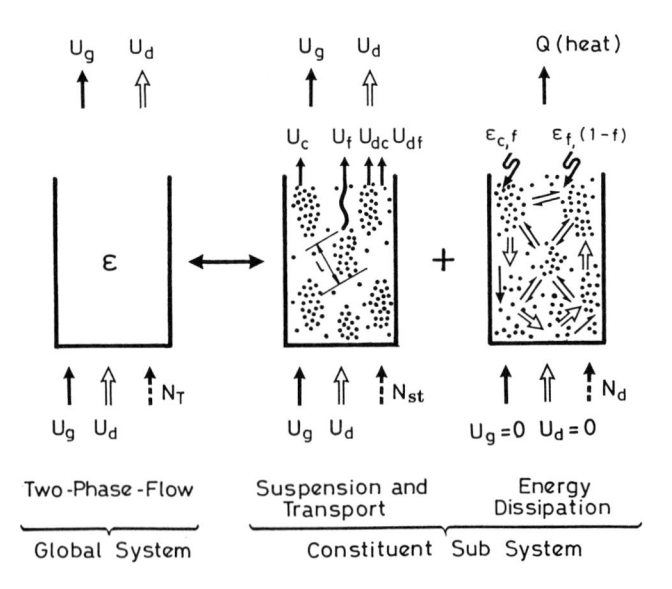

FIG. 2. System resolution for particle–fluid systems.

C. Momentum and Mass Conservation for the ST Subsystem

Following Fig. 2, Fig. 3 shows the two steps used in analyzing the ST subsystem. The first step shows the whole ST subsystem as a combination of the two phases interacting with each other through an interface—a dense phase with voidage ε_c and volume fraction f, and a dilute phase with voidage ε_f and volume fraction $(1-f)$. The overall fluid flow is split into two streams—$U_c f$ through the dense phase and $U_f(1-f)$ through the dilute phase, and the pressure drops due to fluid flow in the two phases should be equal. The second step depicts the interaction between the dense and the dilute phases as taking place through an independent fictitious interphase between the clusters and the surrounding broth. The energy consumed in the ST subsystem N_{st} is thereby resolved into the three constituent terms given in Section IV.B:

$$N_{st} = (N_{st})_{dense} + (N_{st})_{dilute} + (N_{st})_{inter}.$$

This resolution of N_{st} makes possible the formulation of the following six equations on mass and momentum conservation:

1. Force Balance for Clusters in Unit Bed Volume

The number of clusters in a unit bed volume is m_1, and the number of

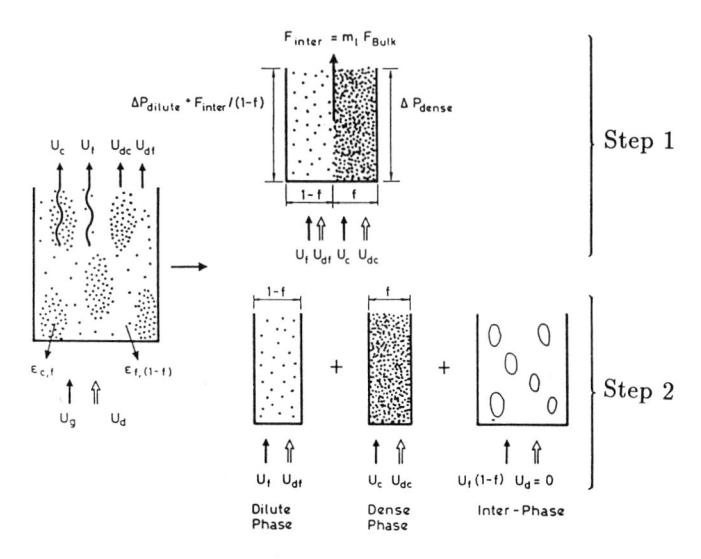

Fig. 3. Two-step resolution for ST subsystem.

particles in the dense phase in this volume is $m_c f$. The hydrodynamic forces acting on these clusters can be equated to their weight:

$$m_c f F_{\text{dense}} + F_{\text{bulk}} m_1 = F_1 m_1,$$

where F_{dense} and F_{dilute} were defined in Section I, and F_1 is the weight of a single cluster, defined as

$$F_1 = \frac{\pi l^3}{6}(1 - \varepsilon_c)\rho_p g.$$

Referring to Table I, we get

$$F_1(\mathbf{X}) = \frac{3}{4} C_{D_c} \frac{1 - \varepsilon_c}{d_p} \rho_f U_{sc}^2 f + \frac{3}{4} C_{D_i} \frac{f}{l} \rho_f U_{si}^2 - (1 - \varepsilon_c)(\rho_p - \rho_f)gf = 0.$$

2. Equal Pressure Drop

As shown in Step 1 of Fig. 3, the fluid flowing in the dilute phase has to support the discrete particles it contains as well as the clusters in suspension, and the combined forces result in a pressure drop equal to that of the parallel dense-phase fluid flow:

$$\begin{array}{ccc} \Delta P_{\text{dilute}} & + \quad \Delta P_{\text{inter}}/(1 - f) & = \quad \Delta P_{\text{dense}} \\ \| & \| & \| \\ m_f F_{\text{dilute}} & + \quad F_{\text{bulk}} m_1/(1 - f) & = \quad m_c F_{\text{dense}} \\ \underbrace{\text{particles} \qquad \qquad \text{clusters}} & & \underbrace{} \\ \text{dilute phase} & & \text{dense phase} \end{array}$$

from which,

$$F_2(\mathbf{X}) = C_{D_f} \frac{1 - \varepsilon_f}{d_p} \rho_f U_{sf}^2 + \frac{f}{1 - f} C_{D_i} \frac{1}{l} \rho_f U_{si}^2 - C_{D_c} \frac{1 - \varepsilon_c}{d_p} \rho_f U_{sc}^2 = 0.$$

3. Particle Suspension for Dilute Phase

The weight of particles in the dilute phase is

$$m_f F_{\text{dilute}} = (1 - \varepsilon_f)(\rho_p - \rho_f)g,$$

from which

$$F_3(\mathbf{X}) = \frac{3}{4} \frac{1 - \varepsilon_f}{d_p} C_{D_f} \rho_f U_{sf}^2 - (1 - \varepsilon_f)(\rho_p - \rho_f)g = 0.$$

4. Continuity for the Fluid

$$F_4(\mathbf{X}) = U_g - U_f(1 - f) - U_c f = 0.$$

5. Continuity for the Particles

$$F_5(\mathbf{X}) = U_d - U_{d_f}(1 - f) - U_{d_c} f = 0.$$

6. Cluster Diameter

The diameter of a cluster l is assumed to be inversely proportional to the energy input according to Chavant (1984):

$$l = \frac{K}{\text{energy input}},$$

where the input energy refers to that which is responsible for breaking up the dense phase into clusters. At minimum fluidization, $l \to \infty$. Therefore, the energy input $(N_{st})_{mf}$ at minimum fluidization should be subtracted from the total input N_{st}, thus leading to the expression

$$l = \frac{K}{N_{st} - (N_{st})_{mf}},$$

where

$$(N_{st})_{mf} = (U_g)_{mf} g = \left(U_{mf} + \frac{U_d \varepsilon_{mf}}{1 - \varepsilon_{mf}}\right) g.$$

As Matsen (1982) mentioned for very fine particles, when the fluid velocity is high enough for the voidage to reach 0.9997, all particles are discretely distributed in the fluid, that is, clusters disappear, or $l = d_p$. On the other hand, Wu et al. (1991) found that any system possesses its own unique maximum voidage ε_{max} beyond which clusters do not exist. When voidage reaches ε_{max}, N_{st} is almost all consumed in transporting particles at a solids flow rate of G_s with hardly any energy dissipated, that is, N_s and N_d can both be neglected. Therefore,

$$N_{st} \doteq N_t \text{ or } U_g g \doteq \frac{G_s g}{(1 - \varepsilon_{max})\rho_p},$$

that is,

$$(N_{st})|_{\varepsilon = \varepsilon_{max}} = U_g g = \frac{U_d g}{1 - \varepsilon_{max}}.$$

Thus, we get

$$d_p = \frac{K}{\dfrac{U_d g}{1 - \varepsilon_{max}} - (N_{st})_{mf}},$$

from which

$$l = \frac{d_p\left(\dfrac{U_d g}{1 - \varepsilon_{max}} - (N_{st})_{mf}\right)}{N_{st} - (N_{st})_{mf}},$$

leading to

$$F_6(\mathbf{X}) = l - \frac{d_p\left(\dfrac{U_d g}{1 - \varepsilon_{max}} - \left(U_{mf} + \dfrac{U_d \varepsilon_{mf}}{1 - \varepsilon_{mf}}\right)g\right)}{N_{st} - \left(U_{mf} + \dfrac{U_d \varepsilon_{mf}}{1 - \varepsilon_{mf}}\right)g} = 0.$$

This equation shows that ε_{max} depends not only on operating conditions but also on material properties. For instance, its value for L/S systems can be much lower than those for G/S systems. Such a dependency was analyzed by Chen and Li (1994).

The six equations in $F_i(\mathbf{X})(i = 1, 2, \ldots, 6)$ just described in terms of force balance and continuity, are, however, not sufficient to determine the hydrodynamic state of a heterogeneous particle–fluid system, because with these, multiple solutions would result among which only one is valid to represent the stable state. An additional condition is needed to define this solution.

D. Energy Minimization and the PD, PFC and FD Regimes

Minimizing the energy consumption for transporting and suspending particles is considered to be this missing condition (Li, 1987; Li et al., 1988b). For steady *single-fluid* flow, Helmboltz first proposed the so-called minimum energy dissipation theorem for incompressible Newtonian fluids with constant viscosity and negligible interior and conservative volumetric forces. This theorem states that the integral of the energy dissipation over the whole flow field tends toward a minimum under unchanged boundary conditions (Lamb, 1945). However, it has not been extended to particle–fluid *two-phase* flow.

Azbel and Liapis (1983) analyzed gas/liquid systems with the assumption that the available energy at steady state is at a minimum. Reh (1971) mentioned the concept of the lowest resistance to fluid flow, and in a somewhat alternate way, the so-called minimum pressure drop was used by Nakamura and Capes (1973) in analyzing the annular structure in dilute transport risers. The instability of a uniform particle-fluid suspension was analyzed by introducing small disturbances into the system (Jackson, 1963; Grace and Tuot, 1979; Batchelar, 1988).

In cocurrent-up fluid–particle two-phase flow, the fluid tends to choose an upward path with minimal resistance, while the particles tend to array themselves with minimal potential energy. Stability of the two-phase system calls for mutual coordination, as much as possible, between the fluid and the particles in following their respective tendencies. This applies for all the three broad regimes of operation: fixed bed, fluidization and transport. When neither the fluid nor the particles can dominate the system, either has to compromise and yield its intrinsic tendencies to those of the other to reach a stable state. However, if the system is fully dominated by either the fluid or the particles, the intrinsic tendency of the dominant one will be satisfied exclusively, with full suppression of that of the other.

If the fluid velocity is lower than U_{mf}, that is, for the fixed bed, the fluid–particle system is totally dominated by the particles, inasmuch as the flowing fluid cannot change the geometric state of the system, which is therefore fully particle-dependent, or *particle-dominated* (PD). In fact, this regime belongs to single-fluid flow through a maze of complicated channels composed of particles packed under gravity, and the theory of single-phase flow can be used.

When fluidized, at velocities between U_{mf} and U_{pt}, neither the particles nor the fluid can dominate the other in displaying either's tendency exclusively: They have to compromise with each other in such a way that the particles seek as much as possible minimal potential energy and the fluid flows through them as much as possible with minimal resistance. In the lower-velocity end of this range, the particles tend to aggregate into a continuous dense-phase emulsion to admit excess gas to flow in the form of bubbles, and as velocity increases toward the higher-velocity end of this range, the particles in the emulsion tend to shred themselves into strands, which are discontinuous and dense-phase, distributed in a broth of sparsely dispersed individual particles. On a macro scale, particles tend to aggregate next to the wall of the retaining vessel, leaving a core of dilute solids concentration in the center of the vessel. Such two-phase structures in this velocity range are but irrefutable phenomena of nature. Evidently the particles, made mobile by the flowing fluid, react in turn upon the fluid to affect its flow and velocity distribution. The fluid with its intrinsic tendency to choose an upward path with minimal resistance, and the particles with their intrinsic tendency to array themselves with minimal potential energy, seem to accommodate each other to form this *particle–fluid-compromising* (PFC) regime. This regime is characterized by minimal energy per unit mass of the particles, N_{st}, for which particles aggregate into clusters or emulsion, and also by conditionally minimizing W_{st} and $W_{st}/\varepsilon\rho_f$, for which the fluid flows with relatively low resistance.

At higher velocities, beyond U_{pt}, all particles fed into the system are transported in dilute phase out of the system without prolonged residence, and without being able to aggregate themselves to the same extent as the U_{mf}–U_{pt} range. Inasmuch as particle arrangement is suppressed by the high-velocity fluid, this dilute single-phase regime is *fluid-dominated* (FD) and is characterized by minimal energy either per unit mass of the fluid, $W_{st}/(\varepsilon\rho_f)$, or per unit volume W_{st}. Under this condition, the fluid disperses the particles as discretely as possible in order to achieve the lowest resistance to fluid flow. Maximal particles dispersion corresponds to maximal energy expenditure per unit mass of particles, that is, N_{st} = max. Therefore, the stability condition for this regime can be represented by W_{st} = min, $W_{st}/(\varepsilon\rho_f)$ = min and N_{st} = max. The condition N_{st} = max represents the ideal case of uniform, or particulate expansion of particles for which N_{st} = max = $U_g g$ = constant.

Other relationships between the extrema of N_{st}, W_{st} and $W_{st}/\varepsilon\rho_f$ were further discussed by Li and Kwauk (1994).

The choice of the minimal energy to characterize different regimes—N_{st} for fluidization, and W_{st} for dilute-phase transport—may sound *a priori* or hypothetical, but as any hypothesis goes, its justification lies in how well it is subsequently corroborated by fact, as will be demonstrated in the later parts of this chapter.

Transitions between these three regimes are distinct. PD/PFC transition occurs at the minimum fluidization velocity U_{mf} at which the fluid has acquired enough velocity to force the particles to move, thus enabling the particles to compromise with the fluid. Transition from the PFC to the FD regime takes place at a fluid velocity corresponding to its saturation carrying capacity K^* (choking), at which the fluid begins to dominate the particles to realize the lowest resistance to flow. Details about these transitions will be further discussed in Section V.A.

With the transition from the PFC regime to the FD regime, N_{st}, characterizing the intrinsic tendency of the particles, jumps from a minimum to a maximum, while W_{st} and $W_{st}/\varepsilon\rho_f$, both characterizing the intrinsic tendency of the fluid, change from a conditional minimum to an absolute minimum, implying that the particles have lost their dominance over the system and have surrendered to the fluid at the choking point.

E. THE EMMS MODEL

If we consider the set of six equations on mass and momentum conservation on the one hand, and the characteristic energy extrema for stability of the three broad regimes of operation on the other, mathematical modeling for local hydrodynamics of particle–fluid two-phase flow beyond minimum fluidization needs therefore to satisfy the following constraints:

(a) N_{st} = extreme(min. for $G_s \geq K^*$, max for $G_s \leq K^*$)
(b) $F_i(\mathbf{X}) = 0 \, (i = 1, 2, \ldots, 6)$ } Model LG (Local–General),
(c) $U_{sc} \geq 0, U_{sf} \geq 0, U_{si} \geq 0$

where U_{sc}, U_{sf} and U_{si} stand for the slip velocities in the dense phase, the dilute phase and the interphase, respectively:

$$U_{sc} = U_c - \frac{U_{dc}\varepsilon_c}{1 - \varepsilon_c},$$

$$U_{sf} = U_f - \frac{U_{df}\varepsilon_f}{1 - \varepsilon_f},$$

$$U_{si} = \left(U_f - \frac{U_{dc}\varepsilon_f}{1 - \varepsilon_c}\right)(1 - f).$$

This is a nonlinear optimization problem with eight parameters and nine constraints, called energy-minimization multi-scale (EMMS) modeling, from which the parameter vector $\mathbf{X}$ and various energy consumptions can be calculated.

For solving the EMMS model, the saturation carrying capacity K^* has to be determined to ascertain whether N_{st} should be minimized or maximized in Model LG, that is, to identify the transition from the PFC regime to the FD regime. This transition is characterized by the following equality (Li et al., 1992):

$$(W_{st})_{PFC} \equiv (W_{st})_{(N_{st})_{min}} = (W_{st})_{(N_{st})_{max}|\varepsilon_c = \varepsilon_{mf}} \equiv (W_{st})_{FD}$$

The physical implication of this transition will be explained with computation in the next section.

V. Local Hydrodynamics—Phases

The EMMS model is a nonlinear optimization problem involving eight parameters and nine constraints consisting of both equalities and inequalities, and can be solved by using the so-called GRG-2 algorithm developed by Lasdon et al. (1978) for coding the generalized reduced gradient method. This section presents the computation results for system FCC/air ($d_p = 54$ μm, $\rho_p = 929.5$ kg/m^3) without introducing the relevant computation techniques which were elucidated in detail by the authors in their recent book (Li and Kwauk, 1994). All computation results satisfy the Kuhn–Tucker condition.

A. Definition of Fast Fluidization and Its Three Operating Modes

Figure 4 shows the changes of W_{st} calculated with respect to N_{st} = min for the PFC regime and to N_{st} = max for the FD regime, respectively. The two curves cross each other at point B:

$$(W_{st})_{(N_{st})_{min}} = (W_{st})_{(N_{st})_{max}|\varepsilon_c = \varepsilon_{mf}},$$

which defines the critical condition for breaking up of the two-phase PFC regime of dense fluidization into a uniform FD regime of dilute transport, hence, the saturation carrying capacity K^*. There is a maximum of W_{st}

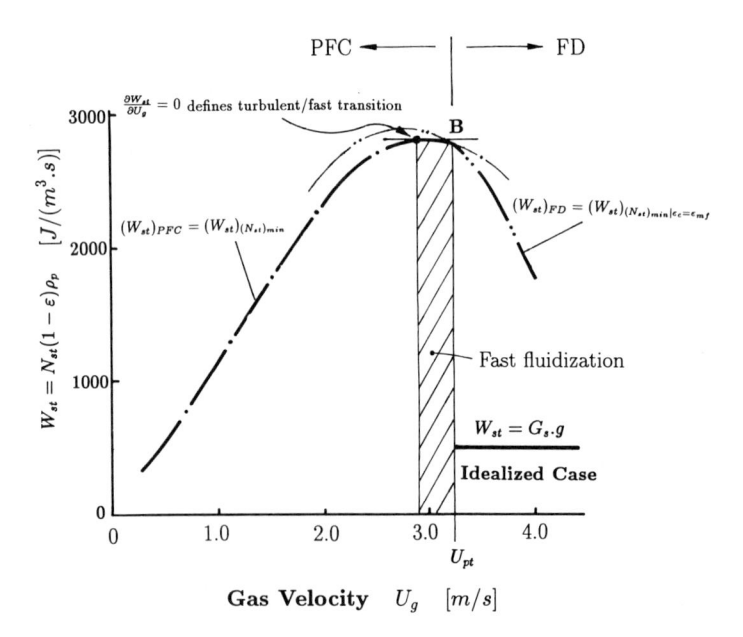

FIG. 4. Definition of saturation carrying capacity and transition from PFC to FD regime (system FCC/air).

marking the onset of fast fluidization (Li et al., 1988b), that is, fast fluidization starts with

$$\frac{\delta W_{st}}{\delta U_g} = 0$$

and terminates at point B as just defined.

According to the preceding definition, the relationship between the saturation carrying capacity K^* and fluid velocity U_g can be calculated, as shown in Fig. 5, which defines the transition from the PFC regime to the FD regime, that is, from fast fluidization to dilute transport, as corroborated by the experimental points (the data at high velocities was transported from Fig. 7).

For instance, if the solids flow rate is specified at $G_s = 50$ kg/(m²s), choking will take place at $U_g = 3.21$ m/s for system FCC/air as indicated in the figure. Throughout the entire regime spectrum, only at this unique point (U_{pt}, K^*) can both dense-phase fluidization and dilute-phase transport coexist. At velocities higher than U_{pt}, only dilute transport can exist, shown as Mode FD in Fig. 4; at velocities lower than U_{pt}, only dense-phase fluidization can take place, shown as Mode PFC in Fig. 4. The transition point at U_{pt} identifies the unique Mode PFC/FD on the curve of Fig. 5 for the coexistence of both modes, the relative proportion of which depends on other external conditions such as the imposed pressure ΔP_{imp} as reported by Weinstein et al. (1983).

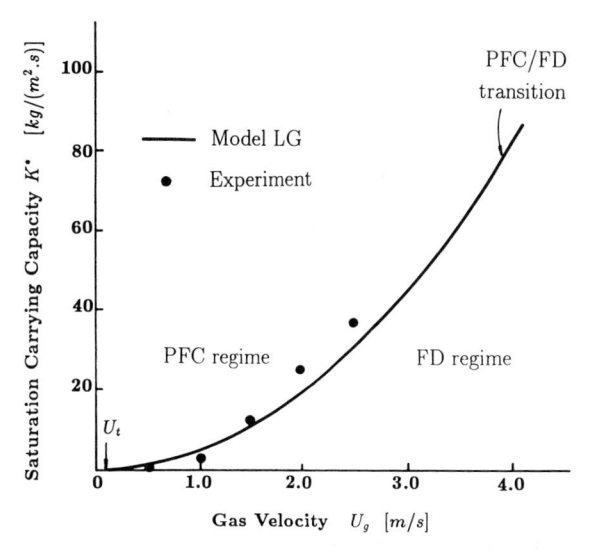

FIG. 5. Definition of carrying capacity K^* and its change with gas velocity U_g (FCC/air).

Experiments conducted by Li et al. (1988a) confirmed the three operating modes in a circulating fluidized bed of 90 mm i.d. by using a scanning transducer-valved system. The imposed pressure drop ΔP_{imp} was altered by changing the solids inventory I, or by manipulating the solid rate-control valve.

Figure 6 shows the dependence of axial voidage profiles on gas velocity. If gas velocity increases without changing solids inventory, the difference between the bottom dense region and the top dilute region will be reduced, accompanied by an increase of solids flow rate, until the difference disappears. It is evident that solids flow rate is dependent on gas velocity.

Figure 7 shows the dependence of axial voidage profiles on solids inventory, which was first recognized by Weinstein et al. (1983), for system FCC/air ($\rho_p = 1,450$ kg/m^3, $d_p = 59$ μm). Two kinds of voidage profile curves are shown. The curves for $I = 15, 20$ kg in Fig. 7a and $I = 15, 20, 22$ kg in Fig. 7b are S-shaped, with the co-existence of two regions, and transition occurring inside the unit, while solids can be fed at the bottom of the bed at saturation flow rates. Variation of solids inventory I does not result in any change of

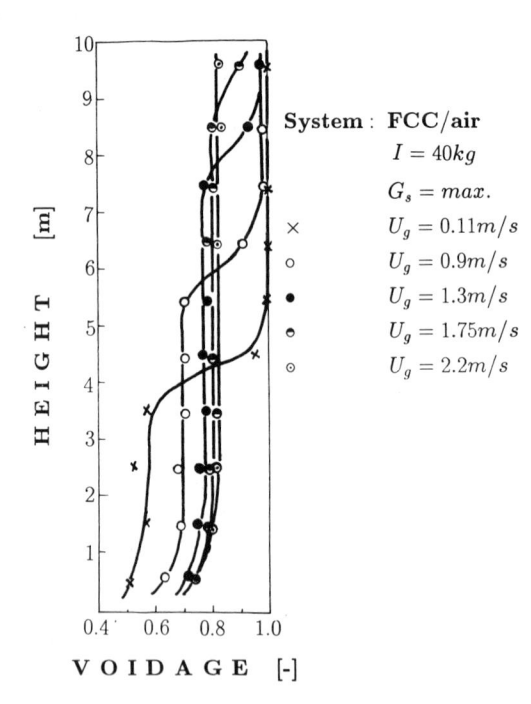

FIG. 6. Dependence of axial voidage profiles on gas velocity at a constant solids inventory (FCC/air).

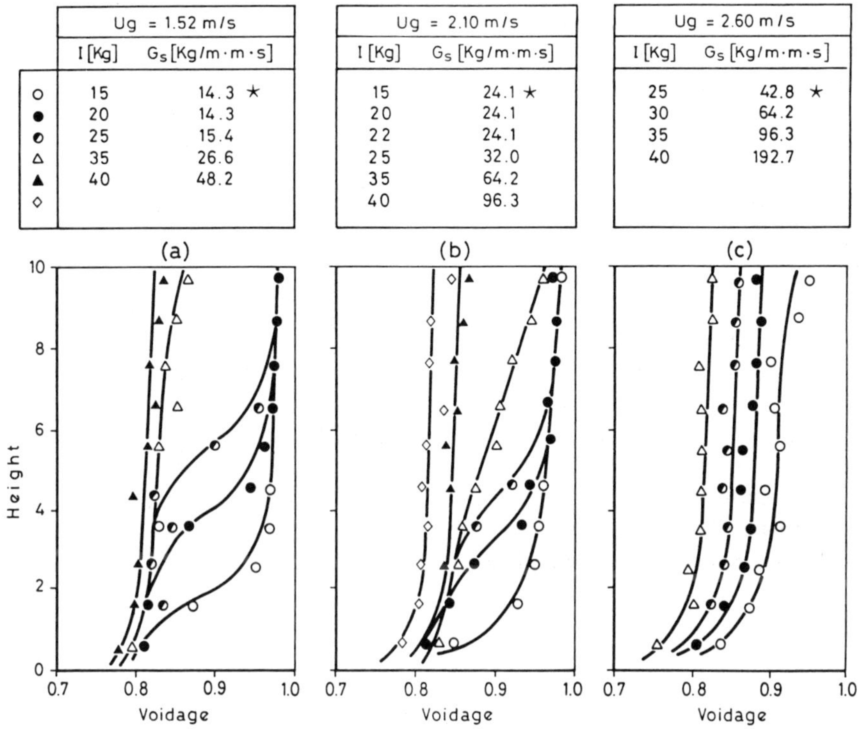

FIG. 7. Axial voidage profiles for system FCC/air. (*) Data used in Fig. 5.

G_s, ε^* and ε_a, but only displaces the position of the inflection point Z_i. Solids rate G_s is always equal to the saturation carrying capacity at the corresponding gas velocity ($K^* = 14.3$ kg/m²s for $U_g = 1.52$ m/s; $K^* = 24.1$ kg/m²s for $U_g = 2.1$ m/s). Asymptotic voidages for the dilute and dense phases, ε^* and ε_a respectively, are only functions of U_g and relevant material properties.

At $G_s = K^*$, any change in the opening of the solid-rate control valve can only upset the initial equilibrium temporarily. For instance, if the opening is decreased, the input solid rate becomes smaller than K^*, temporarily. However, the output solid rate still remains at K^*, thus depleting the fast bed of solids. Meanwhile the voidage of the dilute phase at the top ε^* also remains constant. As a result, the dilute-phase region extends toward the bottom of the bed. The excess solids which have thus far been removed from the fast bed have accumulated in the slow bed, thus increasing the imposed pressure at the bottom, forcing more solids flow into the fast bed. This dynamic process continues until G_s equals K^* again, but at a new equilibrium position of the point of inflection Z_i, which is lower than before.

On increasing I, the inflection point in the axial voidage profile will move upward until it passes beyond the top of the bed. In the latter case, $G_s > K^*$, and a new operating region is reached, in which the axial voidage profiles are no longer S-shaped, as shown by all the curves in Fig. 7c and by the curves for $I = 25$, 35 and 40 kg in Figs 7a and 7b. On the other hand, decreasing I will make the inflection point Z_i move downward until it passes beyond the bottom of the bed, with $G_s < K^*$. In this operating region, the flow pattern becomes dilute-phase transport, and neither is the voidage profile S-shaped. In the two operating regions for $G_s < K^*$ and $G_s > K^*$, G_s is independent of U_g and can be changed independently. Therefore, the bed voidage would be related to both U_g and G_s; axial voidage profiles are not S-shaped; and any change in I would result in a corresponding change in G_s.

If the solids control valve is fully closed at the state of an S-shaped profile, the local states of both regions will not change; however, the top dilute region will extend toward the bottom with a constant solid flow rate K^* until the dense region disappears, the whole bed becomes dilute, the particles in the bed start to be blown out, and finally the bed becomes empty.

Figure 8 shows the dependence of axial voidage profile on gas velocity

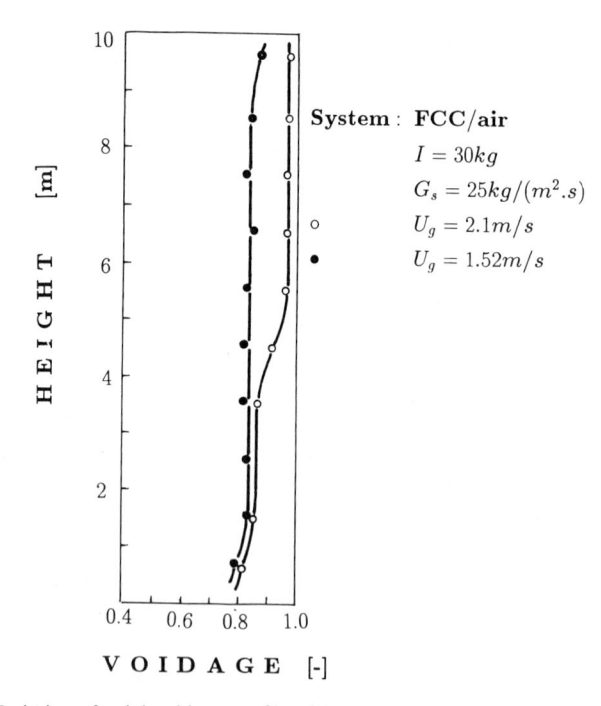

FIG. 8. Variation of axial voidage profile with gas velocity at constant solids flow rate.

at a fixed solids flow rate. If we reduce the gas velocity of the state with an S-shaped axial voidage profile, the inflection point will immediately disappear, showing the unique relation between U_g and K^*. Figure 9 gives experimental results with hollow glass beads, showing changes similar to Fig. 7 for FCC particles.

Figure 10 outlines the characteristics of axial voidage profiles, showing three operating modes of fast fluidized beds on the basis of the three variables, U_g, G_s and ΔP_{imp}. Zone PFC/FD in the saddle area of Fig. 10a is that for the co-existence of two flow regions with $G_s = K^*$, for which the constant-U_g curves are horizontal lines. Zone PFC toward the right is that for the dense single-phase region with $G_s > K^*$, and zone FD toward the left, for the dilute-single-phase region with $G_s < K^*$. Point I_{min} represents the minimum solids inventory for the development of two flow regions. Point D is the critical point for S-shaped axial voidage profiles. If U_g is larger than the gas velocity corresponding to this point, the S-shaped axial voidage profile lapses into dilute flow. On the other hand, if G_s is larger than the solid flow rate corresponding to point D, the S-shaped axial voidage profile is replaced by dense flow.

It is easy to see that zone PFC/FD in Fig. 10a corresponds to operation at choking, that is, $U_g = U_{pt}$ and $G_s = K^*$, and the right-hand side, or zone PFC, corresponds to the operation of $U_g < U_{pt}$ and $G_s > K^*$ in the PFC regime, and the left-hand side, or zone FD, to the operation of $U_g > U_{pt}$ and $G_s < K^*$ in the FD regime.

Experiments also indicated that at any constant gas velocity, S-shaped profiles can exist only within certain limits of I, as represented by the boundary of zone PFC/FD in Fig. 10a. Thus, to realize a given mode of operation, the necessary global conditions need to be satisfied in addition to intrinsic hydrodynamics. Bed height can also affect axial voidage profiles, as already noted. The smaller the bed height is ($Z_i < Z_2$), the shorter the horizontal section of $G_s = f(U_g, I)$ is in Fig. 10a, and the more difficult it is to develop the S-shaped profile. Additional data in Fig. 10b for hollow glass beads further identify the three operating modes just enumerated.

B. Hydrodynamic States

Figure 11 shows the flow structures of the circulating fluidized bed (CFB) for all the preceding three operating modes. The curves in the lower diagram give computation results of ε_f, ε_c and ε, while the operating conditions and flow structures are summarized at the top.

With increasing fluid velocity, a particle–fluid system starts with the particle-dominated fixed bed terminating at U_{mf}, spans the particle–fluid-compromising regimes of particulate, bubbling, turbulent and fast fluidization,

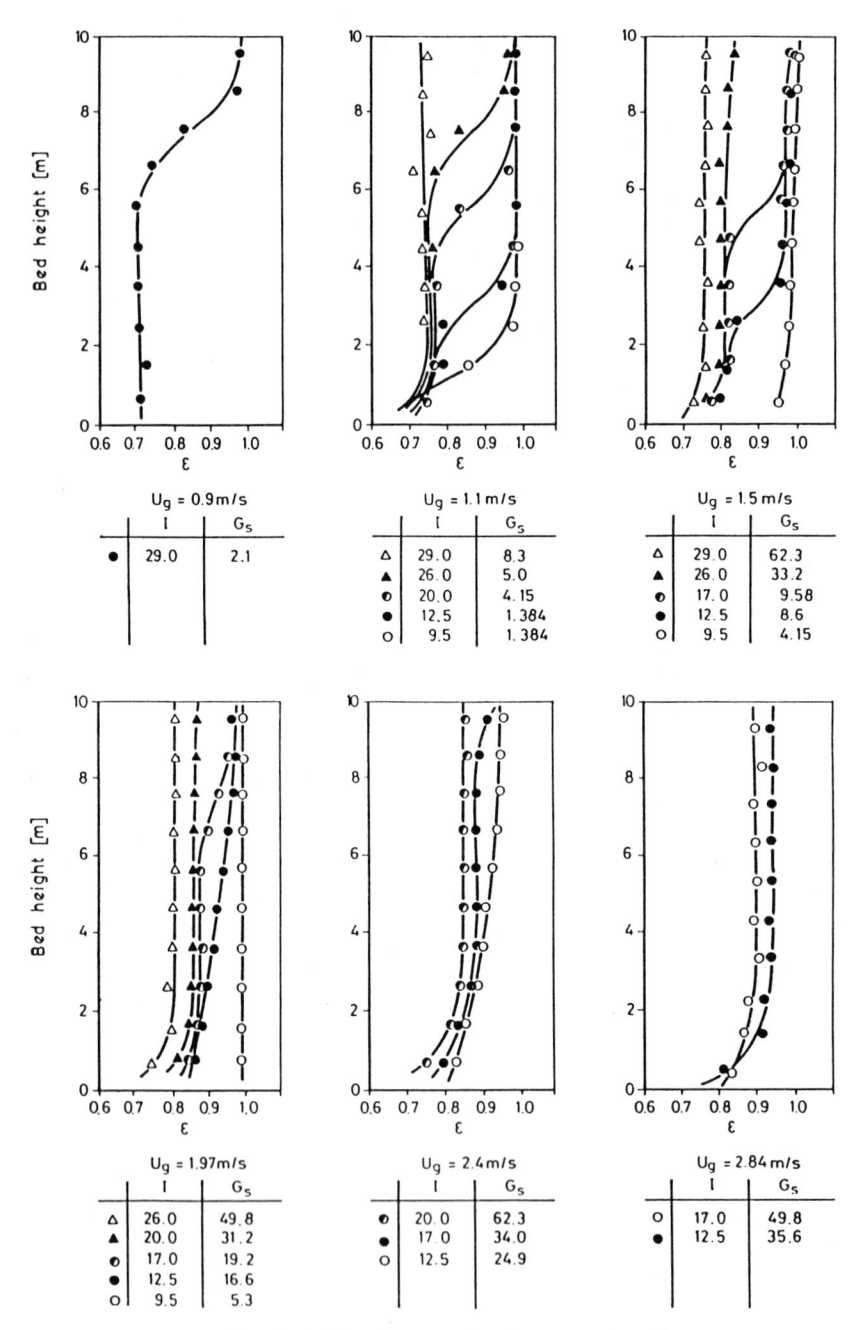

Fig. 9. Axial voidage profiles for system glass/air.

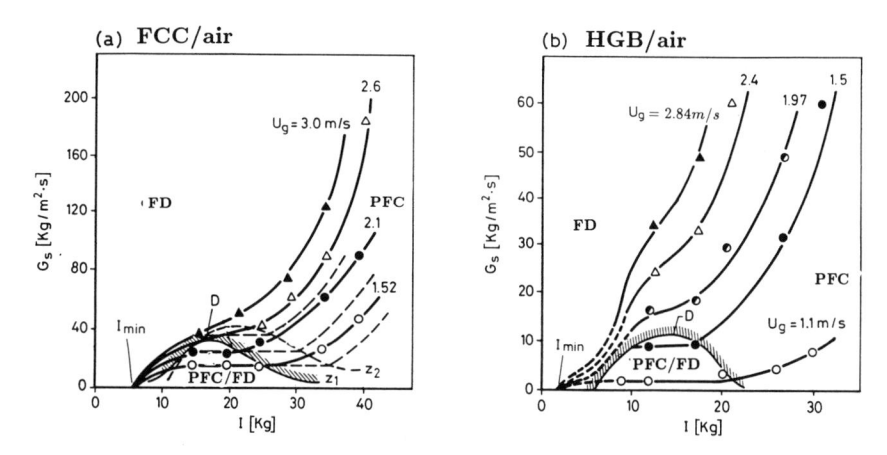

FIG. 10. Relationship between U_g, G_s and ΔP_{imp} for different operating modes.

and finally becomes fluid-dominated at the so-called choking velocity U_{pt} with the onset of dilute transport. In actual reality transport is yet possessed with vestigial heterogeneity, which disappears, however, at U_{uni} while lapsing into idealized transport. Below U_{mf} and beyond U_{uni}, the particle–fluid two-phase flow is single-phased, and therefore the traditional theories apply. In the range from U_{mf} to U_{uni}, Model LG has two solutions corresponding to $(N_{st}) = \min$ and $(N_{st}) = \max$, of which only one, depending on conditions, is stable.

At gas velocities below U_{pt}, the system is stabilized at $N_{st} = \min$, since $(W_{st})_{(N_{st})_{min}} < (W_{st})_{(N_{st})_{max}|\varepsilon_c = \varepsilon_{mf}}$. Therefore, Mode PFC prevails, showing a dense region only with an average voidage of ε_a, operating in the PFC regime at $G_s > K^*$ with a two-phase structure—a dense phase with voidage ε_c, which is close to ε_{mf}, and a dilute phase with voidage ε_f, which approaches unity. These characteristic voidages are shown as optical voidage measurements at the top of the figure. For Mode PFC, the external condition $\Delta P_{imp} \geq (1 - \varepsilon_a)\rho_p g H$ must be satisfied.

With increasing gas velocity, $(W_{st})_{(N_{st})_{min}}$ approaches $(W_{st})_{(N_{st})_{max}|\varepsilon_c = \varepsilon_{mf}}$ gradually, and reaches it finally at U_{pt}, corresponding to $G_s = K^*$, that is, $(W_{st})_{(N_{st})_{min}} = (W_{st})_{(N_{st})_{max}|\varepsilon_c = \varepsilon_{mf}}$, implying that both the PFC regime and FD regime are stable, that is, Mode PFC/FD characterized by the coexistence of these two regimes with a dense region at the bottom having average voidage ε_a and a dilute region at the top having average voidage ε^* for the idealized case, or ε_1^* for real nonidealized cases with some degree of particle aggregation. Mode PFC/FD shows an S-shaped axial voidage profile with an inflection point Z_i, which changes with ΔP_{imp} in the range from

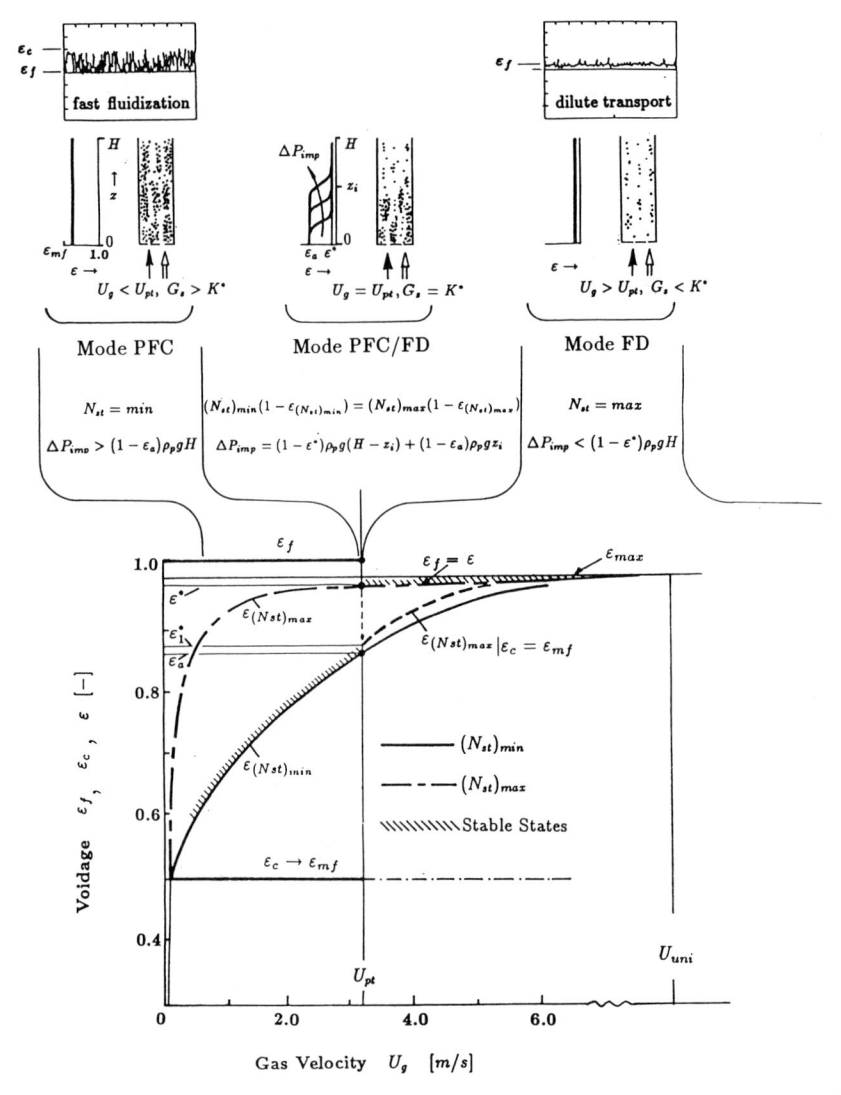

FIG. 11. Calculation showing essential aspects of hydrodynamics of circulating fluidized beds ($G_s = 50$ kg/m^{2}s).

$(1 - \varepsilon^*)\rho_p g H$ to $(1 - \varepsilon_a)\rho_p g H$ and is stabilized at $\Delta P_{imp} = (1 - \varepsilon^*)\rho_p g Z_i + (1 - \varepsilon_a)\rho_p g(H - Z_i)$. In real units, Mode PFC/FD sometimes may not appear because of limited vessel height.

Beyond U_{pt}, the FD regime replaces the PFC regime which becomes unstable at $G_s < K^*$, resulting in Mode FC, showing a dilute region only, and characterized by $N_{st} = $ max and $\Delta P_{imp} \leq (1 - \varepsilon^*)\rho_p g H$. In this mode, the dilute phase voidage ε_f deviates from values close to unity as shown by calculated results and corrborated by optical measurements shown at the top of the figure. For the idealized case, a single-phase uniform structure prevails with $\varepsilon = \varepsilon_f$ but for any real case, a certain extent of particle aggregation exists, until with increase in gas velocity it approaches the idealized homogeneous condition. Here, only two extreme cases can be calculated and discussed because of the lack of understanding of the structure of clusters existing in any real nonidealized FD regime: One is the idealized case with uniform structure as discussed earlier, for which average voidage at the choking point jumps from ε_a to ε^*; the other is the case with the assumption of nonideality, that is, clusters still exist with voidage $\varepsilon_c = \varepsilon_{mf}$ as in the PFC regime. For the latter, the average voidage calculated from Model LG will change as shown by the dashed line for $\varepsilon_{(N_{st})_{max}|\varepsilon_c = \varepsilon_{mf}}$, and the voidage at the choking point would jump from ε_a to ε_1^*. The structural difference of particle aggregates between Mode PFC and Mode FD (much smaller for FD) warrants further study.

The high frequency of cluster formation and dissolution in fast fluidization is reflected in high-frequency random voidage fluctuations as shown at the top subfigure of Fig. 11. Such a change in two-phase behavior promotes efficient gas/solids contacting.

For Mode PFC and Mode FD, any change in ΔP_{imp} certainly causes variation of solid flow rate G_s, and hence, local voidage in the whole unit, while in Mode PFC/FD it only affects the location of the inflection point Z_i.

As soon as gas velocity reaches U_{uni}, the corresponding bed voidage ε reaches ε_{max}, at which the particle concentration is so low that clusters can no longer exist, particles are dispersed discretely and the system becomes really uniform. At this point, the two solutions characterized by $(N_{st})_{max}$ and $(N_{st})_{min}$ approach each other, and idealized transport prevails.

Figure 12 shows the computed results for the local hydrodynamic states of the system FCC/air to illustrate the change of the status parameters with gas flow rate U_g for two solids rates $G_s = 50$ and 75 kg/(m²s).

Figure 12a shows that the voidages for the dense cluster phase and the surrounding dilute phase remain essentially constant at $\varepsilon_c \to \varepsilon_{mf} = 0.5$ and $\varepsilon_f \to 1.0$, respectively, indicating that stability calls for this heterogeneity of the coexistence of the two phases, which leads to bubbling at small gas velocities and clustering at large gas velocities.

As gas velocity U_g increases, the average voidage ε in Fig. 12a increases while the cluster fraction f in Fig. 12c decreases, until at some gas velocity a sudden change takes place with f dropping to zero for the idealized case, denoting the disappearance of the cluster phase. At this point single-phase solids transport begins to take over the two-phase structure which has hitherto

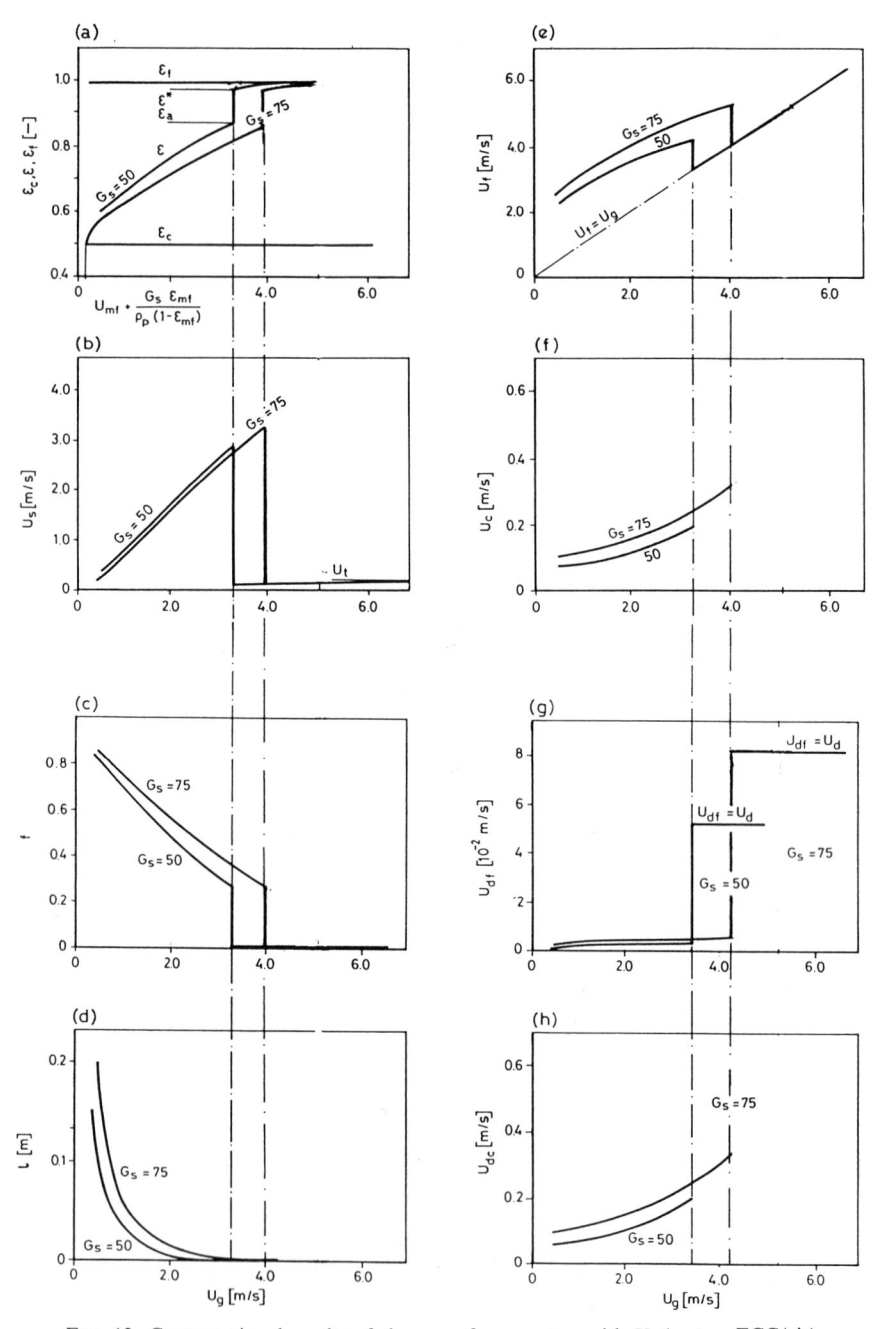

FIG. 12. Computational results of changes of parameters with U_g (system FCC/air).

existed, with a new voidage of ε^* as Fig. 12a shows, from which the average voidage increases gradually with further increase in gas velocity.

Figure 11 has already shown that only at this singular point of sudden change can two regions coexist in a reactor, a dilute transport region with voidage ε^* surmounting a dense region with voidage ε_a, as is the case for the S-shaped voidage distribution curve in fast fluidization shown for Mode PFC/FD. The position of the point of inflection in this S-shaped curve can vary and is determined by the imposed pressure ΔP_{imp} across the fast fluidized bed, or the solids inventory of the entire calculating fluid–bed system, as explained in connection with Fig. 11. At gas velocities less than that at this singular point, only one heterogeneous region of a two-phase structure can exist, and at gas velocities greater than that at this singular point, only a single-phase region for transport can exist.

Figure 12b shows that before this sudden change, in this PFC regime, the slip velocity U_s between solid particles and the surrounding fluid is far greater than the terminal velocity U_t of the particles. After this sudden jump, in he FD regime, however, U_s drops to a much lower value close to U_t as one approaches the idealized case. However, for nonidealized cases, U_s in the FD regime is still somewhat higher than U_t because of residual particle aggregation, as shown by Li et al. (1991b) experimentally.

Figure 12d shows that the dimension of clusters l is large at low gas velocities (down to dense phase continuous for bubbling fluidization), decreasing with gas velocity down to zero at the point of this sudden change, at which clusters disappear as noted already.

Figure 12e shows that gas velocity in the dilute phase U_f increases with the entering gas velocity U_g until it drops at the point of sudden change when clusters disappear, to a value identical to that of U_g. Figure 12f shows that gas velocity in the dense cluster phase U_c, which is approximately an order of magnitude smaller than the dilute phase velocity U_f, also increases with the entering gas velocity U_g until U_c terminates with the disappearance of clusters at the point of sudden change.

Figure 12g shows that solids velocity U_{df} in the dilute phase is small when clusters exist, implying that major solids flow occurs through the clusters, only to jump to a high value at the point of sudden change. Figure 12h indicates that solids velocity U_{dc} in the dense cluster phase increases steadily with the entering gas velocity U_g and terminates at the point when clusters cease to exist all of a sudden.

The dominant mechanism of particle aggregation is related to the stability of the system: In the PD regime, particles are held together by gravity; in the PFC regime, the fluid and the particles compromise with each other to seek an energy-minimized state, resulting in the aggregation of particles to form a two-phase structure; in the FD regime, the high fluid velocity works against this tendency of particle aggregation, thus disrupting the two-phase structure, with a dramatic drop in cluster diameter l.

C. GAS/SOLID CONTACTING

Figure 13 supplements Fig. 12 by comparing the changes in the different velocities and the different drag coefficients in the dilute, dense and inter phases. This figure shows that U_{sf} and C_{Df} for the dilute phase, broth, keep essentially constant, while U_{sc} for the dense phase decreases with increasing fluid velocity U_g, resulting in an increasing C_{Dc} which stays at a much higher level. However, the slip velocity U_{si} between the two phases is high, and the corresponding drag coefficient C_{Di} is low. Inasmuch as a high drag coefficient signifies efficient particle–fluid contacting, it can be deduced that the high slip velocity U_s in heterogeneous two-phase flow, such as a circulating fluidized bed, does not contribute much to gas/solid contacting. Intensive

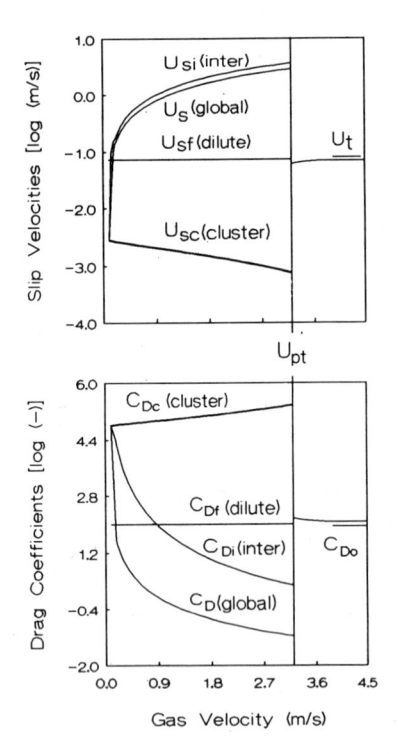

FIG. 13. Dependences of slip velocities and drag coefficients on gas velocity in different phases.

gas/solid contacting can therefore be considered to result mainly from incessant exchanges between the two phases caused by frequent dissolution and reformation of clusters.

The two-phase structure in a particle–fluid system is a localized meso-scale phenomenon of heterogeneity and has been shown to result from the tendency of the system to seek a minimal energy for suspending and transporting the particles, N_{st} = min. Figure 14a shows that if the fluid–particle system were uniform, N_{st} would be $U_g g$, which is higher than the corresponding N_{st} for the heterogeneous structure composed of the sum of $(N_{st})_{dense}$ for the dense phase, and $(N_{st})_{dilute} + (N_{st})_{inter}$ for the dilute and inter phases. The lower value of N_{st} for the heterogeneous structure is due to the low value of the $(N_{st})_{dense}$ component, as shown in the lowest curve in Fig. 14a. It was shown that N_{st} increases with increase in ε_c and decrease in ε_f. Therefore, in minimizing N_{st}, ε_c tends toward the minimal value of ε_{mf} and ε_f towards the maximal value of unity, as shown in Fig. 11 for the PFC regime.

The FD region at the top is characterized by the dominance of the fluid over the movement of particles, as already shown in Fig. 11. When the fluid-dominated FD regime is first formed, the clusters of fast fluidization are disintegrated to form an essentially one-phase structure in which the particles are, however, not completely discretely suspended, that is, at a much higher concentration as compared to the ε_f computed for the broth before U_{pt}. This is shown by the fluctuating voidage considerably above zero, as can be seen in the upper right-hand side of Fig. 11.

According to the degree of uniformity of the system, the fluid-dominated FD-regime can be divided into two subregions: dilute-phase transport for real systems and idealized dilute-phase transport. The transition between the two subregimes occurring at the value of ε_{max}.

To elucidate further the mechanism of local heterogeneity as evidenced by the division of fluid flow into the two phases, a term called mass-specific intensity of particle–fluid interaction I_m is proposed:

$$I_m = \frac{N_{st}}{U_g \rho_f},$$

which provides a criterion for evaluating fluid–particle contacting between unit mass of fluid and unit mass of particles.

For the dense and the dilute phases, the corresponding mass-specific intensities of particle–fluid interaction are

$$(I_m)_{dense} = \frac{(N_{st})_{dense}}{U_c f \rho_f},$$

$$(I_m)_{dilute} = \frac{(N_{st})_{dilute} + (N_{st})_{inter}}{U_f (1-f) \rho_f}.$$

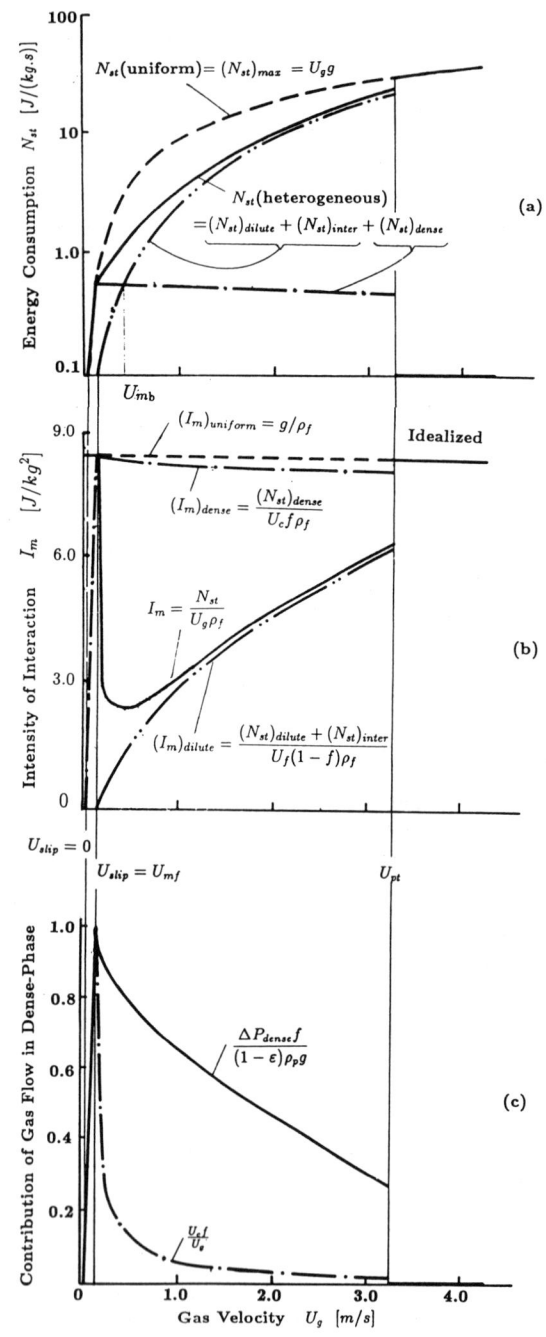

FIG. 14. Mass-specific intensity of G/S interaction in different phases (FCC/air, $G = 50$ kg/m^{2}s).

Figure 14b shows that $(I_m)_{dense}$ is much higher than $(I_m)_{dilute}$, implying that a great part of the total weight of the solid particles $(1 - \varepsilon)\rho_p g$ is borne mainly by a small portion of upflowing fluid in the dense phase $U_c f/U_g$, shown as $\Delta P_{dense} f/(1 - \varepsilon)\rho_p g$ in Fig. 14c, while the major portion of the fluid flows through the dilute phase with hardly any infiltration into the dense phase. In uniform suspension, from its definition, I_m reaches the maximum of g/ρ_f. From Fig. 14b, we can understand that the formation of the two-phase heterogeneous structure leads to a dramatic decrease of the mass-specific intensity of fluid/particle interaction. Although its value in the dense phase is close to that in the uniform suspension, its contribution to the average intensity is not so much due to the little fluid flow $U_c f/Ug$ in the dense phase, as can be seen in 14b. Figure 14b shows that the local average mass-specific intensity of fluid/particle interaction I_m is critically dependent on flow regimes. Starting from incipient fluidizing, I_m drops sharply with bubbling, then increases with U_g, until at the choking velocity U_{pt} it jumps to its maximum g/ρ_f, meaning that maximal mass-specific interaction between the fluid and particles occurs in the state of uniform structure.

Therefore, we define relative contacting intensity

$$I'_m = \frac{I_m}{(I_m)_{max}} = \frac{I_m \rho_f}{g},$$

which evaluates the role of heterogeneity in reducing particle–fluid contacting intensity. The value of I'_m changes from zero to a maximum of unity in the state of uniform structure.

In addition to mass-specific intensity of particle–fluid interaction, volume-specific intensity I_v is another important criterion for evaluating particle–fluid contacting in a system. From Section IV.B, W_{st} is itself a measure of volume-specific intensity of particle–fluid interaction, that is,

$$I_v = N_{st}(1 - \varepsilon)\rho_p = W_{st}.$$

Figure 15 maps the variation of this volume-specific intensity with gas velocity U_g and solids flow rate G_s. The subfigure at the top shows the shadowed cross-section for $G_s = 80$ kg/(m²s). Maximal I_v corresponds to the most efficient particle–fluid contacting per unit volume, and W_{st} should be integrated volumetrically to yield the global effectiveness of particle–fluid contacting in a reactor.

Mass or heat transfer between fluid and particles is related not only to fluid–particle contacting, but also to the exchange between the dense and the dilute phases. Therefore, further efforts are needed to unravel how the latter effect, which is characterized by repeated dissolution and reformation of particle aggregates, should be incorporated in order to evaluate the overall mass or heat transfer process.

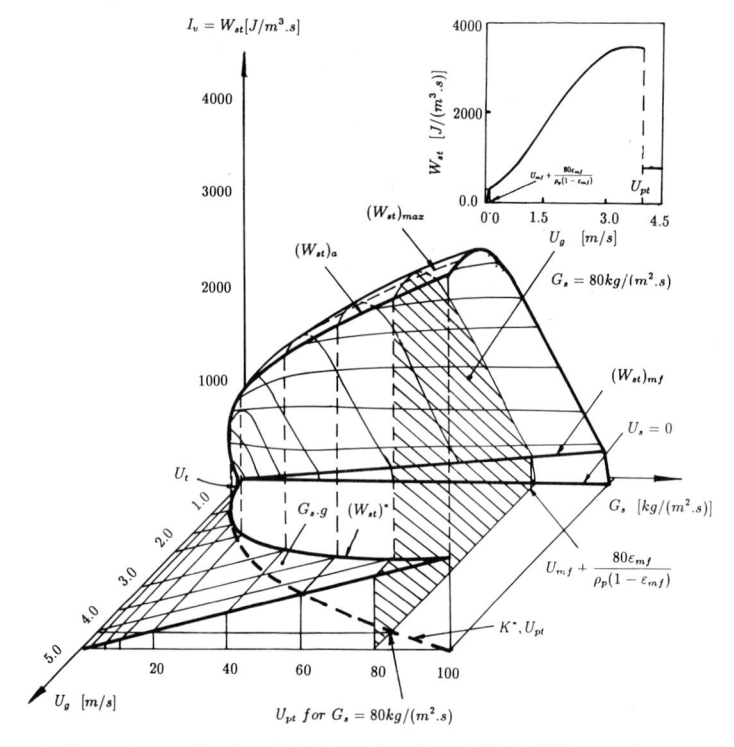

FIG. 15. Dependence of volumetric intensity of particle–fluid interaction on operating conditions (FCC/air).

The preceding consideration seem to imply that for efficient fluid–particle contacting in a reactor of a limited volume, high I_v should be chosen, while if the number of particles in a reactor is limited, the system should be operated at as high a value of I_m as possible.

VI. Overall Hydrodynamics—Regions

While local hydrodynamics of particle–fluid two-phase flow—phases, regimes and patterns—is concerned with its intrinsic stability, of direct engineering significance is its overall hydrodynamics, which deals with its space-dependent characteristics subject to the boundary conditions set by the retaining vessel. For the axisymmetric equipment generally employed in engineering, overall hydrodynamics is resolved into the axial and the radial directions: top and bottom regions for the former, and core and wall regions for the latter. On this macro scale, therefore, the principle of energy minimization has to be applied on an extended scale, in order to quantify the heterogeneous flow field.

A. Axial Hydrodynamics

1. Axial Heterogeneity

Axial heterogeneity is mainly related to the coexistence of two different regions in terms of local hydrodynamics—a bottom dense region with average voidage ε_a and a top dilute region with average voidage ε^*, bridged together by a transition region, so as to form an S-shaped profile, such was first treated by Li and Kwauk (1980) for fast fluidized beds:

$$\frac{\varepsilon(z) - \varepsilon_a}{\varepsilon^* - \varepsilon(z)} = \exp\left(-\frac{z - Z_i}{Z_0}\right),$$

in which Z_i is the location of the point of inflection in the profile, and Z_0 is called the characteristic length, which governs how quickly the top and bottom regions merge into each other and is related to operating conditions and material properties.

According to the analysis of local hydrodynamics in Section V.A, the top dilute region operates in the FD regime, while the bottom dense region operates in the PFC regime. Therefore, whenever the S-shapd profile appears, the corresponding solid flow rate is bound to be equal to the saturation carrying capacity K^*. Such axial heterogeneity depends not only on the local hydrodynamics given by Model LG, but also on the pressure drop imposed across the unit ΔP_{imp}, which affects the position of the inflection point Z_i, and as shown in Fig. 11, can operate in three modes (see table).

	Local Conditions	Overall Conditions	Voidage Regions
Mode PFC	$U_g < U_{pt}$, $G_s > K^*$	$\Delta P_{imp} > (1 - \varepsilon_a)\rho_p gH$	dense only
Mode PFC/FD	$U_g = U_{pt}$, $G_s = K^*$	$(1 - \varepsilon^*)\rho_p gH < \Delta P_{imp} < (1 - \varepsilon_a)\rho_p gH$	dilute/dense
Mode FD	$U_g > U_{pt}$, $G_s < K^*$	$\Delta P_{imp} < (1 - \varepsilon^*)\rho_p gH$	dilute only

Therefore, to realize any desired mode of operation, it is necessary to control both the local and the overall conditions.

2. Mathematical Modeling

The three modes of operation discussed in Section V.A are identified as follows:

Mode PFC: 1-region $(W_{st})_{(N_{st})_{min}} < (W_{st})_{(N_{st})_{max}}|_{\varepsilon_c = \varepsilon_{mf}}$,

Mode PFC/FD: 2-region $(W_{st})_{(N_{st})_{min}} = (W_{st})_{(N_{st})_{max}}|_{\varepsilon_c = \varepsilon_{mf}}$,

Mode FD: 1-region $(W_{st})_{(N_{st})_{min}} > (W_{st})_{(N_{st})_{max}}|_{\varepsilon_c = \varepsilon_{mf}}$.

To calculate the two-regioned axial parameter profile in Mode PFC/FD, ε_a and ε^* are determined as follows.

For the one-dimensional profile now under consideration, the hydrodynamics of the top dilute region can be described by

$$\left. \begin{array}{l} (N_{st})^* = \max, \\ F_i(\mathbf{X}^*) = 0 (i = 1, 2, \ldots,), \\ U_{sc}^* \geq 0, U_{sf}^* \geq 0, U_{si}^* \geq 0. \end{array} \right\} \text{Model LG–FD}$$

Solution of this model gives $\mathbf{X}^*$ and ε^*. Of course, it is only possible to calculate the two extreme cases—the uniform case with $N_{st} = \max$ and the completely nonidealized case with $N_{st} = \max|_{\varepsilon_c = \varepsilon_{mf}}$.

The hydrodynamics of the bottom dense region is governed by

$$\left. \begin{array}{l} (N_{st})_a = \min, \\ F_i(\mathbf{X}_a) = 0 (i = 1, 2, \ldots, 6), \\ U_{sc_a} \geq 0, U_{sf_a} \geq 0, U_{si_a} \geq 0, \end{array} \right\} \text{Model LG–PFC}$$

which gives $\mathbf{X}_a$ and ε_a.

From the values of ε_a and ε^* thus calculated, and ΔP_{imp} computed from solids inventory and the geometry of the equipment, Z_i can be deduced from pressure balance:

$$Z_i = \frac{\Delta P_{imp} - (1 - \varepsilon_a)\rho_p gH}{(\varepsilon_a - \varepsilon^*)\rho_p gH}.$$

Then the entire axial voidage profile can be calculated from the model mentioned earlier in the present section.

B. Radial Hydrodynamics

1. Radial Heterogeneity

Radial heterogeneity, showing a dilute core region surrounded by a dense annular region next to the wall, is attributed to the role of the wall in promoting minimization of fluid–particle interaction by inducing a dense region near wall with low N_{st}, thus resulting in the radial distributions of all parameters. Such a heterogeneous flow structure affects the performance of a chemical reactor critically, in a positive way sometimes, but likely in a negative way, because the formation of radial distribution results in considerable decrease in fluid–particle contacting and leads to backmixing of both fluid and particles (Li and Weinstein, 1989).

2. Mathematical Modeling

Model LG applies just as well to radial heterogeneity, though in a somewhat different functional form, to describe the fluid dynamics at any radial position r:

(a) $N_{st}(r) = \text{extreme}(\min \text{ for } G_s(r) \geq K^*, \max \text{ for } G_s(r) \leq K^*),$
(b) $F_i(X(r)) = 0 \, (i = 1, 2, \ldots, 6),$ } Model LR
(c) $U_{sc}(r) \geq 0, U_{sf}(r) \geq 0, U_{si}(r) \geq 0.$

This model, Model LR (Local–Radial), describes the dependence of $X(r)$ on gas velocity $U_g(r)$ and particle velocity $U_d(r)$, which are, however, not manipulatable operating parameters, but should be correlated with the average superficial gas velocity U_g and the average superficial particle velocity $U_d \, (= G_s/\rho_p)$ as well as boundary conditions.

In order to calculate $U_g(r)$ and $U_d(r)$ from the specified operating parameters U_g and G_s, it is necessary to know the dominant factor defining radial heterogeneity. Extending the energy minimization method to overall hydrodynamics, a stable radial profile calls for not only Model LR for local hydrodynamics at every point, but also the minimization of the cross-sectional average $\bar{N}_{st}$ for overall stability, which is defined by

$$\bar{N}_{st} = \frac{2}{R^2(1-\bar{\varepsilon})} \int_0^R N_{st}(r)[1 - \varepsilon(r)] r \, dr, \tag{1}$$

where the cross-sectional average bed voidage $\bar{\varepsilon}$ is defined as

$$\bar{\varepsilon} = \frac{1}{R^2} \int_0^R \varepsilon(r) r \, dr. \tag{2}$$

To fulfill both local and overall stability, all parameters would adjust themselves radially in such a way that not only is $N_{st}(r)$ at any radial position minimized for the PFC regime, or maximized for the FD regime, but also $\bar{N}_{st}$ for the *whole* cross-section is minimized, yielding *radial* profiles as governed by the following Overall-Radial model:

(a) $\bar{N}_{st}$ = extreme(min for $G_s > K^*$, max for $G_s < K^*$),

(b) Model LR(r),

(c) $U_g = \dfrac{2}{R^2} \int_0^R U_g(r) r \, dr$,

(d) $U_d = G_s/\rho_p = \dfrac{2}{R^2} \int_0^R U_d(r) r \, dr$,

(e) $\Delta P_{st}(r)$ = constant,

(f) boundary conditions.

⎫ Model OR

The parameter vector **X**(r), describing this twofold optimization problem, depends on both local and overall fluid dynamics. For a system without wall effects, Model OR simplifies to Model LG for local hydrodynamics.

3. Solution of Model OR

In theory, given the required boundary conditions, it is possible to calculate the radial profiles of all parameters with Model OR from the specified operating variables, U_g and G_s. An approximate solution to the simplest case of the PFC regime, for example, that is, N_{st} = min, is possible by simplifying Model OR with experimental results, for calculating $U_d(r)$ from $U_g(r)$ and $\varepsilon(r)$.

Radial heterogeneity is assumed to be distributed in terms of a heterogeneity factor $K(r)$, defined as the ratio of the equivalent cluster diameter $l(r)$ to the particle diameter d_p, that is, $K(r) = k(r)/d_p$. When the voidage profile $\varepsilon(r)$ is known, $K(r)$ can also be expressed as $K(\varepsilon(r))$. At minimum fluidization, the cluster diameter is evidently equal to the diameter of the unit, and at infinite bed expansion, $\varepsilon = 1.0$ and $l = d_p$, that is,

$$K(\varepsilon_{mf}) = d_B/d_p \quad \text{and} \quad K(\varepsilon = 1.0) = 1.0.$$

At any radial position r, $U_d(r)$ can be calculated from $U_g(r)$ and $\varepsilon(r)$ by using Model LR through a trial $f(\mathbf{A}, r)$, which may be approximated by polynomials or other adequate functions with the parameter vector **A**, to be optimized with respect to $\bar{N}_{st}$ = min. Now, the problem becomes how to find $K(r)$, that is, how to determine **A**, in order to satisfy energy minimization, force balance, continuity and the given boundary conditions. Therefore, Model OR is simplified for the PFC regime as the K-Radial (KR) model:

To find $\quad K(r) = f(\mathbf{A}, r)$

Satisfying (a) $\bar{N}_{st} = \min,$
(b) Model LR(r),
(c) $K(\varepsilon_{mf}) = d_B/d_p,$
(d) $K(\varepsilon = 1.0) = 1.0,$
(e) $\bar{U}_d = \dfrac{2}{R^2}\displaystyle\int_0^R U_d(r)r\,dr.$

} ModelKR–PFC

For calculating the radial profiles in the FD regime, for instance, in the top dilute region, $\bar{N}_{st}$ should be maximized to form the counterpart Model KR-FD.

Model KR (PFC or FD) can also be solved by using the GRG-2 algorithm for both local and overall energy minimization (Li et al., 1991a).

Figure 16 shows the radial profiles of parameters involving the solid velocity $U_d(r)$ calculated from Model KR using the data provided by Bader et al. (1988) on radial distribution of voidage and gas throughput as shown in Fig. 16f.

Figure 16a shows the calculated radial distribution of the heterogeneity factor $K(r)$. In the dilute core region, the bed structure is shown to be close to homogeneous with small particle clusters. Beyond $r = 12$ cm near the wall, the heterogeneity factor increases dramatically, indicating a much more aggregated bed structure. Energy consumption $N_{st}(r)$ depends strongly on bed structure, as Fig. 16b shows—very high in the dilute core region, but extremely low in the dense wall region. Such a distribution of $N_{st}(r)$ and particle population leads to minimization of $\bar{N}_{st}$, and hence to a stable radial profile.

The real solids velocity profile calculated from Model KR, and the real gas velocity profile deduced from Fig. 16f, are shown in Fig. 16c for comparison. Because of the near-homogeneous bed structure in the dilute core region, solids velocity reaches high values approaching the magnitude of the gas velocity. In the dense wall region, low and even negative solids velocity exists, which causes local solids and gas backmixing. A comparison of the calculated and the measured (Bader et al., 1988) solids velocity profiles, which are independent from each other because the latter was not used for simulation, is shown in Fig. 16d. A reasonable agreement is achieved.

The calculated radial profile of slip velocity between gas and solids is shown in Fig. 16e. The lowest slip velocity occurs in the center of the unit because of uniformity in this region, the highest not far from the wall. It is worth to note that the local slip velocity shown in this figure is lower than the cross-sectional average slip velocity $\bar{U}_s$, or $U_g/\bar{\varepsilon} - G_s/\rho_p(1 - \bar{\varepsilon}) = 3.5$ m/s.

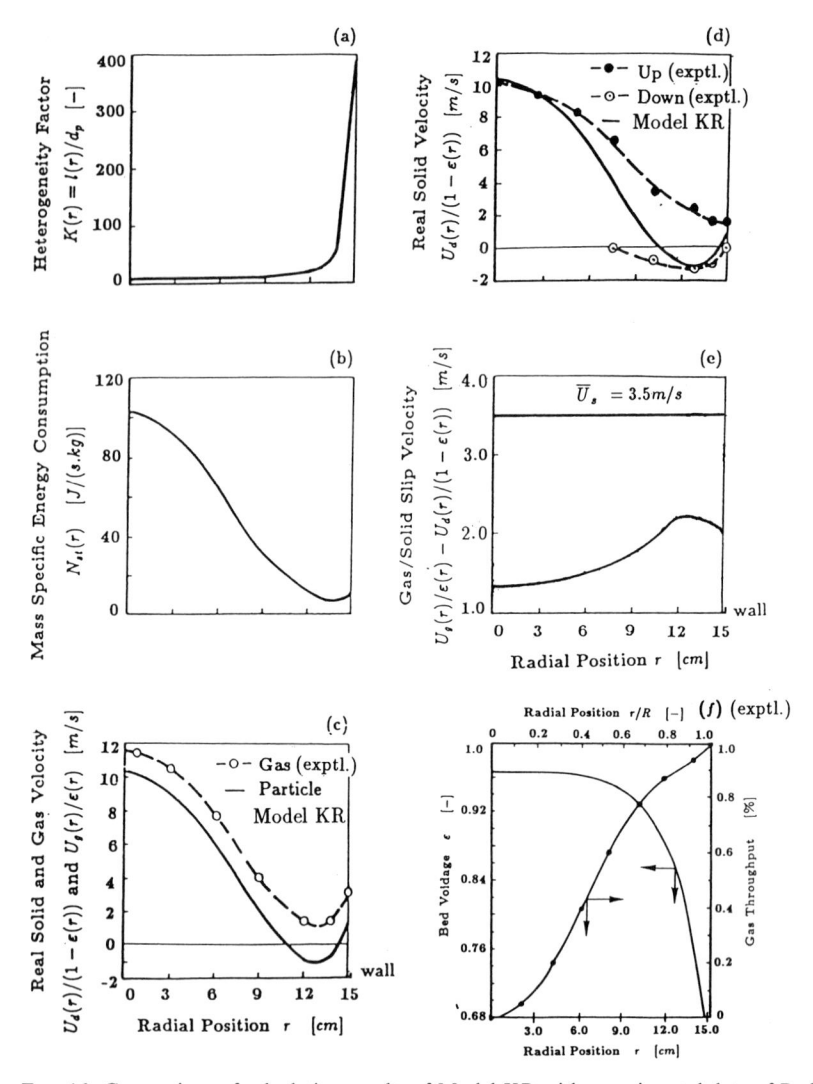

FIG. 16. Comparison of calculation results of Model KR with experimental data of Bader et al. (1988). ($d_p = 76$ μm, $\rho_p = 1,714$ kg/m^3, $d_B = 30.5$ cm, $U_g = 3.7$ m/s, $G_s = 98$ kg/m^{2}s.)

This indicates that the high slip velocity for the overall circulating fluidized bed reactor should be attributed not only to local heterogeneity (aggregation of particles), but also to overall heterogeneity—radial profile, as reported by Rhodes and Geldart (1985), and axial profile. Therefore, the global average slip velocity is not adequate for correlating particle/fluid contacting in the system, but local slip velocity, or even more intrinsically, local fluid–particle interaction intensity, must be used as shown below.

With the preceding discussion on radial modeling, we are now in a position to re-examine the definition of fluid–particle interaction intensity of Section V.C. Figure 17 compares radial distributions of slip velocity U_s/ε, fluid–particle interaction intensity I_m and local volumetric intensity of fluid–particle interaction I_v. It is noted here that there are external values for all these three parameters with respect to radial position, each characterizing a different aspect of fluid–particle contacting. Maximum slip velocity corresponds to minimum I_v and a peak of I_m, indicating that though slip velocity denotes the relative motion between fluid and particles, it does not really characterize the efficiency of fluid–particle contacting. Therefore, evaluation of both I_m and I_v would be necessary for understanding fluid–particle contacting in a heterogeneous system.

Evidently, an efficient reactor should be operated at both high I_m and high I_v, which usually prevail in the uniform dense-phase state. For the case shown in Fig. 17, fluid–particle contacting in the core region is much more intensive than in the wall region. If the core region becomes more dilute, the opposite case might occur because of decreasing I_v in the core.

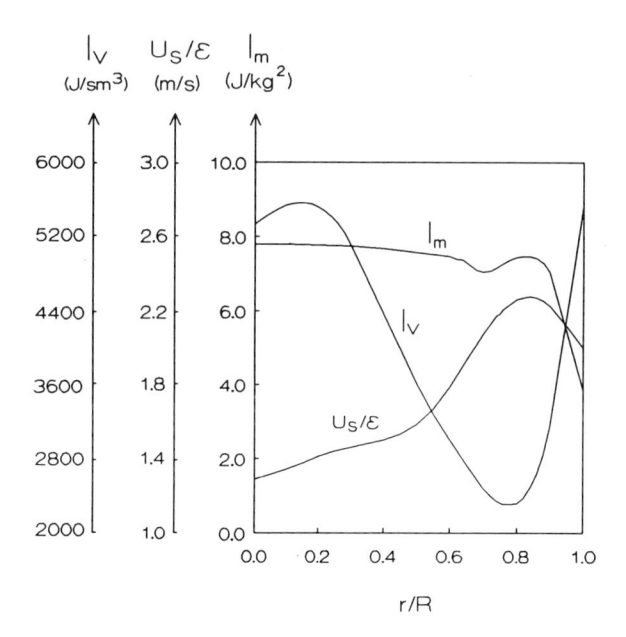

FIG. 17. Radial distributions of I_v, I_m and U_s/ε from experimental data of Bader *et al.* (1988).

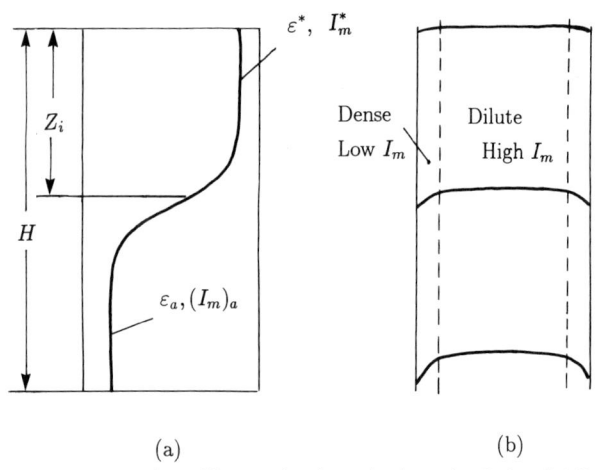

FIG. 18. Distribution of interaction intensity in a circulating fluidized bed.

In considering particle–fluid contacting for any global systems, the local values of $I'_m(h, r)$ need to be averaged:

$$\overline{I'_m} = \frac{\int_0^H \int_0^R I'_m(h, r)[1 - \varepsilon(h, r)]\rho_p U_g(h, r)\rho_f 2\pi r \, dr \, dh}{\int_0^H \int_0^R [1 - \varepsilon(h, r)]\rho_p U_g(h, r)\rho_f 2\pi r \, dr \, dh}.$$

For the CFB system shown in Fig. 18a, such an average could be approximated by

$$\overline{I'_m} = \frac{(1 - \varepsilon^*)Z_i(I'_m)^* + (1 - \varepsilon_a)(H - Z_i)(I'_m)_a}{(1 - \varepsilon^*)Z_i + (1 - \varepsilon_a)(H - Z_i)}.$$

In principle, $\overline{I'_m}$ can be optimized with respect to operating conditions, both radially and axially, as shown in Fig. 18.

Similarly, the global averaged $I_v(h, r)$ can be expressed as

$$\overline{I_v} = \frac{\int_0^H \int_0^R I_v(h, r) 2\pi r \, dr \, dh}{\int_0^H \int_0^R 2\pi r \, dr \, dh}.$$

Notation

A	parameter vector to be optimized	I	solids inventory in a unit, [kg]
C_D	drag coefficient of a single particle in suspension, [-]	I_m	mass-specific intensity of particle–fluid interaction, [J/kg²]
C_{D_0}	drag coefficient of a single particle in air, [-]	I'_m	dimensionless mass-specific intensity of particle–fluid interaction
$\bar{C}_D$	average C_D with respect to $\bar{d}$, [-]		
$\bar{d}$	hydrodynamic mean particle diameter, [m]	I_v	volume-specific intensity of particle–fluid interaction, [J/s m³]
$\bar{d}_s$	volume surface mean diameter, [m]		
d_B	unit diameter, [m]	K	factor
d_p	particle diameter, [m]	K^*	saturation carrying capacity, [kg/(m² s)]
D_a	axial diffusion coefficient, [m²/s]	$K(r)$	radial heterogeneity factor, [-]
D_r	radial diffusion coefficient, [m²/s]	l	equivalent diameter of particle clusters, [m]
f	volume fraction of dense phase, [-]		
$F_i(\mathbf{X})$	equations for mass and momentum conservation	m	particle or cluster number in unit volume, [-]
F	force acting on each particle or cluster, [(kg m)/s²] or objective function in GRG-2	N	energy consumption with respect to unit mass, [J/(s kg)] or total number of particle fractions
g	gravity acceleration, [m/s²]	n_i	particle number for fraction i
G_s	solids flow rate, [kg/(m²s)]	ΔP	pressure gradient or total force in unit volume, [kg/(m² s²)]
H	total height of unit, [m]	ΔP_{imp}	imposed pressure drop across a unit [kg/(m s²)]
h	height coordinate, [m]		

R	radius of unit [m]		profile, [m]
R_e	Reynolds number		
R_{xx}	auto-correlation coefficient	**SUBSCRIPTS**	
R_{xy}	cross-correlation coefficient	a	bottom dense region
r	radial coordinate, [m]	c or dense	dense phase
T	time interval, [s]	d	dissipation
t	time, [s]	f or dilute	dilute phase, fluid
U_c	superficial gas velocity in dense phase, [m/s]	i or inter	inter phase
		min	minimum
		mf	minimum fluidization
U_d	superficial solids velocity, [m/s]	mb	minimum bubbling
U_f	superficial gas velocity in dilute phase, [m/s]	max	maximum
		pt	value for choking point
U_g	superficial gas velocity, [m/s]	p	particles
		s	suspension
U_s	slip velocity, [m/s]	st	suspension and transport
U_t	terminal velocity of particles, [m/s]	t	transport
U_{uni}	minimum U_g for dilute uniform suspension, [m/s]	T	total
		FD	fluid-dominating regime
u_p	real solid velocity, [m/s]	PFC	particle–fluid-compromising regime
$\bar{u}_d$	average solid velocity, [m/s]	z	parameters with respect to axial direction
u_f	real fluid velocity		
u_s	real slip velocity	*	top dilute region
W	energy consumptions with respect to unit volume, [J/(m³ s)]		
		GREEK LETTERS	
$\mathbf{X}$	variable vector	ε	local average voidage, [-]
x	mass fraction		
z	axial coordinate with the origin at the top, [m]	ε_1^*	average voidage in the dilute region at the top in real case with the
Z_i	inflection point of axial voidage		

ε_{max}	assumption of the existence of clusters maximum voidage for occurrence of clusters	ϕ_s OVERLINE	shape factor of particles	
ρ	density, [kg/m³]	—	time average or cross-sectional average	
v	kinematic viscosity, [m²/s]	=	global average over the whole limit	
μ	viscosity, [kg/(m s)]			

References

Arastoopour, H., and Gidaspow, D. "Analysis of IGT pneumatic conveying data and fast fluidization using a thermohydrodynamic model," *Powder Technology* **22**, 77 (1979).

Azbel, D., and Liapis, A. I. Chapter 16, "Hydrodynamics of gas solids flow," *in* "Handbook of Fluids in Motion" (Cheremisinoff, N. P., ed.), p. 427–452. Ramesh Gupta, 1983.

Bader, R., Findlay, J., and Knowlton, T. M. "Gas/solid flow patterns in a 30.5 cm diameter circulating fluidized Bed," *in* "Circulating Fluidized Bed Technology II" (P. Basu and J. F. Large, eds.), p. 123. Pergamon Press, 1988.

Bai, D., Jin, Y., and Yu, Z. "The two-channel model for fast fluidization," *in* "Fluidization '88—Science and Technology" (Kwauk, M., and Hasatani, M., eds.), pp. 155–164. Science Press, Beijing, 1988.

Bai, D., Jin, Y., and Yu, Z. "Acceleration of particles and momentum exchange between gas and solids in fast fluidized beds," *in* "Fluidization '91—Science and Technology" (Kwauk, M., and Hasatani, M., eds.), pp. 46–55. Science Press, Beijing, 1991a.

Bai, D., Jin, Y., Yu, Z., and Gan, N. "Radial profiles of local solid concentration and velocity in a concurrent downflow fast fluidized bed," *in* "Circulating Fluidized Bed Technology III" (P. Basu, M. Horio and M. Hasatani, eds.), p. 157. Pergamon Press, 1991b.

Batchelar, G. K. "A new theory of the instability of a uniform fluidized bed," *J. Fluid Mech.* **193**, 75–110 (1988).

Berruti, F., and Kalogerakis, N. "Modeling the internal flow structure of circulating fluidized beds," *Can. J. of Chem. Eng.* **67**, 1010–1014 (1989).

Briens, C. L., and Bergougnou, M. A. "New model to calculate the choking velocity to monosize and multisize solids in vertical pneumatic transport lines," *Can. J. of Chem. Eng.* **64**, 196 (1986).

Chavant, V. V. "Physical principles in suspension and emulsion processing," *in* "Advances in Transport Processes" (A. S. Mujumdar and R. A. Mashelkar, eds.), p. 1. John Wiley & Sons, New York, 1984.

Chen, A., and Li, J. "Particle aggregation in particle-fluid two-phase flow," *in* "Fluidization '94—Science and Technology" (the organizing committee of CJF-5, ed.), p. 254. Chemical Industry Press, Beijing, 1994.

Couderc, J.-P. Chapter 1, "Incipient fluidization and particulate systems," *in* "Fluidization" (J. F. Davidson, R. Clift and D. Harrison, eds.), p. 1. Academic Press, London, 1985.

Davidson, J. F. "Discussion," *Trans. Inst. Chem. Eng.* **39**, 230–232 (1961).

Ding, J., and Gidaspow, D. "A bubbling fluidization model using theories of granular flow," *AIChE J.* **36**, 523 (1990).

Ding, J., Lyczkowski, R. W., Sha, W. T., Altobelli, S. A., and Fukushima, E. "Analysis of liquid–solids suspension velocities and concentrations obtained by NMR imaging," *NSF/DlE Workshop on Flow of Particles and Fluids, Sep. 17–18, 1992, Gaithersburg, Maryland*.

Flemmer, R. L. C., and Banks, C. L. "On the drag coefficient of a sphere," *Powder Technology* **48**, 217 (1986).

Gidaspow, D. "Hydrodynamics of fluidization and heat transfer: Supercomputer modeling," *Appl. Mech. Rev.* **39**, 1–22 (1986).

Gidaspow, D., Tsuo, Y. P., and Luo, K. M. *In* "Fluidization VI" (Grace, J. R., Shemilt, L. W., and Bergougnou, M. A., eds.), pp. 81–88. Engineering Foundation, New York, 1989.

Grace, J. R., and Clift, R. "On the two-phase theory of fluidization," *Chem. Eng. Sci.* **29**, 327 (1974).

Grace, J. R., and Tuot, J. "A theory for cluster formation in vertically conveyed suspensions of intermediate density," *Trans. IChemE.* **57**, 49 (1979).

Hartge, E.-U., Rensner, D., and Werther, J. "Solids concentration and velocity patterns in circulating fluidized beds," *in* "Circulating Fluidized Bed Technology III" (P. Basu, J. F. Large, eds.), p. 165. Pergamon Press, 1988.

Ishii, H., Nakajima, T., and Horio, M. "The annular flow model of circulating fluidized beds," *J. of Chem. Eng. of Japan* **22**, 484–490 (1989).

Jackson, R. "The mechanics of fluidized beds, I. The stability of the state of uniform fluidization," *Trans. Inst. Chem. Engrs.* **44**, 13–21 (1963).

Knowlton, T. M., and Bachovchin, D. M. "The determination of gas–solids pressure drop and choking velocity as a function of gas density in a vertical pneumatic conveying line," *International Conference on Fluidization, Pacific Grove, June 15–20*, p. 253 (1975).

Kunii, D., and Levenspiel, O. "Fluidization Engineering," Wiley, New York, 1969.

Lamb, H. "Hydrodynamics," pp. 617–619. Dover, New York (1945).

Lasdon, L. S., Waren, A. D., Jain, A., and Ratner, M. "Design and testing of a GRG code for nonlinear optimization," *in* "ACM Transactions on Mathematical Software," Vol. 4, No. 1, pp. 34–50 (1978).

Li, J. "Multi-scale modeling and method of energy minimization for particle–fluid two-phase flow," Ph. D. Thesis, Institute of Chemical Metallurgy, Academia Sinica, Beijing (1987).

Li, J., and Kwauk, M. "Particle–fluid two-phase flow," to be published by Metallurgical Industry Press, Beijing, 1994 (in press).

Li, J., and Weinstein, H. "An experimental comparison of gas backmixing in fluidized beds across the regime spectrum," *Chem. Eng. Sci.* **44**, 1697 (1989).

Li, J., Tung, Y., and Kwauk, M. "Axial voidage profiles in fast fluidized beds," *in* "Circulating Fluidized Bed Technology II" (P. Basu and J. F. Large, eds.), p. 193. Pergamon Press, 1988a.

Li, J., Tung, Y., and Kwauk, M. "Energy transport and regime transition of particle fluid two phase flow," *in* "Circulating Fluidized Bed Technology II" (P. Basu, J. F. Large, eds.), p. 75. Pergamon Press (1988b).

Li, J., Tung, Y., and Kwauk, M. "Multi-scale modeling and method of energy minimization in particle–fluid two-phase flow," *in* "Circulating Fluidized Bed Technology II" (P. Basu and J. F. Large, eds.), p. 89. Pergamon Press, 1988c.

Li, Jinghai, Reh, Lothar, and Kwauk, Mooson. "Application of energy minimization principle to hydrodynamics of circulating fluidized beds," *in* "Circulating Fluidized Bed Technology III" (P. Basu, M. Horio and M. Hasatani, eds.), p. 163. Pergamon Press, 1991a.

Li, J., Reh, Li, Tung, Y., and Kwauk, M. "Multi-aspect behavior of fluidization in different regimes," *in* "Fluidization '91—Science and Technology" (M. Kwauk and M. Hasatani, eds.), p. 11. Science Press, Beijing, 1991b.

Li, J. Z., Yang, L., and Liu, W. "Hydrodynamic mean diameter of multisized particles," *in* "Proceeding of CJChEC," p. 754. Tianjin University Press, 1991c.

Li, J., Kwauk, M., and Reh, L. "Role of energy minimization in gas/solid fluidization," *in*

"Fluidization VII" (O. E. Potter and D. G. Nicklin, eds.), p. 83. Engineering Foundation, New York, 1992.
Li, Y., and Kwauk, M. "Hydrodynamics of fast fluidization," in "Fluidization" (J. R. Grace, ed.), p. 537. Plenum Press, 1980.
Louge, M., and Chang, H. *Powder Technology* **60**, 197–201 (1970).
Matsen, J. M. "Mechanisms of choking and entrainment," *Powder Technology* **32**, 21–33 (1982).
Militzer, J. "Numerical prediction of the fully developed two-phase (air–solids) flow in a pipe," in "Circulating Fluidized Bed Technology" (P. Basu, ed.), pp. 173–184. Pergamon Press, 1986.
Nakamura, K., and Capes, C. E. "Vertical pneumatic conveying: A theoretical study of uniform and annular particle flow models," *Can. J. of Chem. Eng.* **51**, 39–46 (1973).
Reh, L. "Fluidized bed processing," *Chem. Eng. Prog.* **67**(2), 58 (1971).
Rhodes, M. J. *Powder Technology* **60**, 27–38 (1990).
Rhodes, M. J., and Geldart, D. "From minimum fluidization to pneumatic transport—a critical review of the hydrodynamics," in "Circulating Fluidized Bed Technology" (P. Basu, ed.), p. 21. Pergamon Press, 1985.
Satija, S., Young, J. B., and Fan, L. S. "Pressure fluctuation and choking criterion for vertical pneumatic conveying of fine particles," *Powder Technology* **43**, 257 (1985).
Soo, S. L. "Fluid Dynamics of Multiphase Systems," Blaisdell, Waltham, Massachusetts, 1967.
Soo, S. L. "Particulates and Continuum: Multiphase Fluid Dynamics," Hemisphere, New York, 1989.
Subbarao, D. "Chester and lean-phase behavior," *Powder Technology* **46**, 101–107 (1986).
Wallis, G. B. "One-Dimensional Two-Phase Flow," p. 182. McGraw-Hill, New York, 1969.
Weinstein, H., Graff, R. A., Meller, M., and Shao, M. "The influence of the imposed pressure drop across a fast fluidized bed," *Proc. 4th Intern. Conf. Fluidization, Kashikojima, Japan,* pp. 299–306 (1983).
Wu, W., Li, J., Yang, L., and Kwauk, M. "Critical condition for particle aggregation in particle-fluid two-phase flow," *Journal of Engineering Thermophysics* (in Chinese) **13**, 324 (1991).
Yang, W. C. "Criteria for choking in vertical pneumatic conveying lines," *Powder Technology* **35**, 143 (1983).
Yang, W. C. "A model for the dynamics of a circulating fluidized bed loop," *In* "Circulating Fluidized Bed Technology II" (Basu, P., and Large, J. F., eds.), pp. 181–191. Pergamon Press, 1988.
Yang, Y. "Study of concurrent up- and down-flow in circulating fluidized beds," Ph.D. Thesis, Tsing Hua University, Beijing, 1992.
Yerushalmi, J., and Cankurt, N. J. "Further studies of the regimes of fluidization," *Powder Technology* **24**, 187 (1979).
Yerushalmi, J. R., Cankurt, N. T., Geldart, D., and Liss, B. "Flow regimes in gas–solids contact systems," *AIChE Symp. Ser.* **76**, 1 (1978).
Zhang, H., Xie, Y. S., Chen, Y., and Hasatani, M. *In* "Circulating Fluidized Bed Technology III" (P. Basu, M. Horio and M. Hasatani, eds.), p. 151. Pergamon Press, 1990.
Zhang, H., Xie, Y. S., and Hasatani, M. "A new two dimensional correlation for voidage distribution of fast fluidized beds," in "Proc. of 4th China–Japan Symposium Fluidization, Beijing, China," p. 21. 1991.
Zhou, L., and Huang, X. *Science in China (Series A)* **33**, 52–59 (1990).

5. HEAT AND MASS TRANSFER

YU ZHIQING AND JIN YONG

DEPARTMENT OF CHEMICAL ENGINEERING
TSING HUA UNIVERSITY
BEIJING, PEOPLE'S REPUBLIC OF CHINA

I. Introduction	203
II. Experimental Studies of Heat Transfer	204
A. Effect of Variables	205
B. Radial Distribution of Heat Transfer Coefficients	210
C. Axial Distribution of Heat Transfer Coefficients	216
D. Variation of Local Heat Transfer Coefficient along the Surface	216
E. Heat Transfer Enhancement	219
III. Theoretical Analysis and Models for Heat Transfer	223
A. Gas Convective Transfer	223
B. Radiative Transfer	224
C. Particle Convective Transfer	224
D. Validation of the Model	230
IV. Heat Transfer between Gas and Particles	232
V. Mass Transfer	235
Notation	235
References	236

I. Introduction

Most of the reactions that proceed in FFB are strongly exothermic or endothermic, and heat must be removed from or supplied into the bed in order to control the bed temperature. So heat transfer in FFB is of great importance for design and operation.

One of the remarkable features of the FFB is its very high heat transfer rate. Because of extensive solids mixing in the FFB, the heat transfer rate between particles is several orders of magnitude higher than that of silver by conductive heat transfer. The high heat capacity and thermal conductivity and large surface area of the solids particles make it possible to heat up the gas quickly to reach an equilibrium state.

It has long been realized that the heat transfer rate between a wall and a gas flowing through a fluidized bed of granules is at least one order of magnitude higher than that would be expected for the same flow conditions in an empty column. But when the gas velocity is increased to reach the fast fluidized regime, the rate of transfer will decrease because of decreasing bed density.

For years a number of studies, both experimental and theoretical, have been conducted on heat transfer in FFB, though very few have been made on mass transfer.

In this chapter, emphasis will be given to heat transfer in fast fluidized beds between suspension and immersed surfaces to demonstrate how heat transfer depends on gas velocity, solids circulation rate, gas/solid properties, and temperature, as well as on the geometry and size of the heat transfer surfaces. Both radial and axial profiles of heat transfer coefficients are presented to reveal the relations between hydrodynamic features and heat transfer behavior. For the design of commercial equipment, the influence of the length of heat transfer surface and the variation of heat transfer coefficient along the surface will be discussed. These will be followed by a description of current mechanistic models and methods for enhancing heat transfer on large heat transfer surfaces in fast fluidized beds. Heat and mass transfer between gas and solids in fast fluidized beds will then be briefly discussed.

II. Experimental Studies of Heat Transfer

Most heat transfer studies were carried out under ambient conditions and with fine particles in small experimental sets. Though exhaustive reviews have been presented (e.g., Grace, 1986; Glicksman, 1988; Leckner, 1990), the data obtained at Tsing Hua University (Bi and Jin, 1989, 1990; Bai and Jin, 1992) will be the focus of this discussion and will be compared to the results of other researchers.

Heat transfer processes occurring in fast fluidized bed generally include

- heat transfer between particles;
- heat transfer between gas and solid particles;
- heat transfer between particle suspension and heat transfer surfaces (wall or otherwise immersed in the bed).

A. Effect of Variables

The dependence of heat transfer on operating conditions, gas/solid properties and the geometries of both the bed and the heat transfer surface, is illustrated conceptually in Fig. 1.

1. Influence of Suspension Density

The measured heat transfer coefficients are often represented as a function of solids concentration, as shown in the experimental data of Fig. 2 for a wide range of operating conditions, bed temperatures and solids sizes. It has been found that there exists the relation $h \propto (1 - \varepsilon)^n$, in which the values of exponent n generally ranges from 0.5 to 1.0. The strong dependence of heat transfer coefficient on solids concentration implies that heat transfer between suspension and surfaces is dominated by a convective particle transfer process.

2. Influence of Particle Properties

Figure 3 presents the heat transfer coefficients of four kinds of materials (FCC catalyst, quartz sand and two kinds of silica gel of different mean diameters). The larger the particle size is, the smaller the heat transfer coefficients are, much in the same manner as for bubbling fluidized beds. Figure 3 shows that the greater the density ρ_p, the higher the heat transfer

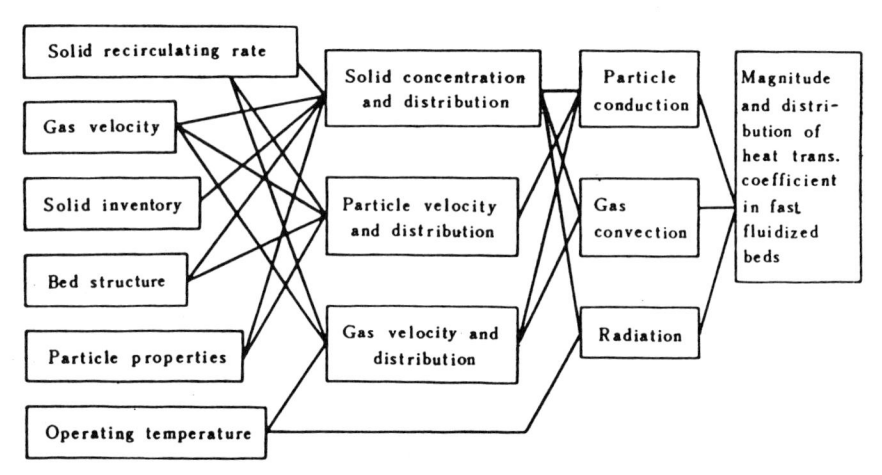

FIG. 1. Conceptual illustration of the dependence of heat transfer on operating conditions, gas/solid properties and the geometries of the bed and the heat transfer surface.

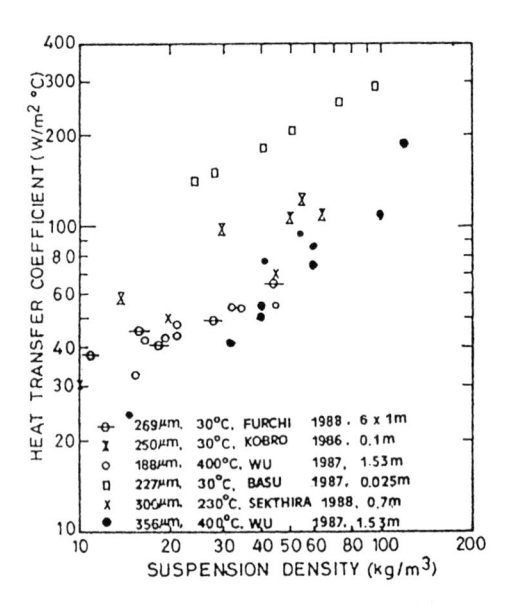

FIG. 2. Effect of suspension density on heat transfer coefficients (Basu, 1990).

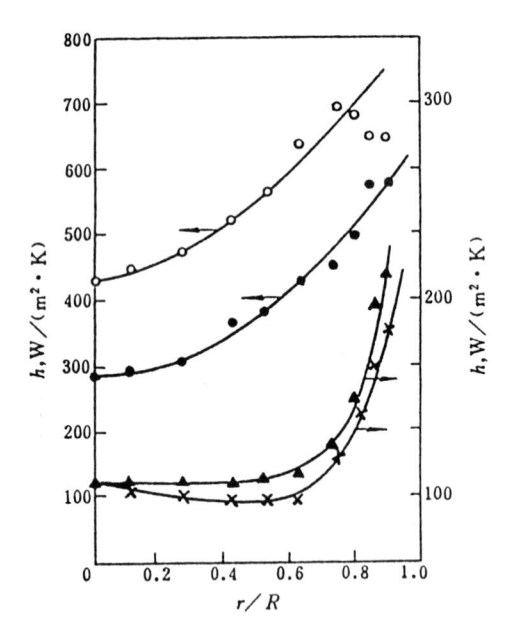

Particle	ρ_s, kg/m³	d_p, mm
Silicagel A	706	0.280
Silicagel B	710	0.140
FCC catalyst	1,473	0.048
Quartz sand	2,643	0.31

FIG. 3. Effect of particle properties on heat transfer coefficients (Bi *et al.*, 1989). $U=2.35$ m/s, $\varepsilon=0.88$, ○, quartz, ●, FCC. $U=4.81$ m/s, $\varepsilon=0.948$, ×, Silicagel A, △, Silicagel B.

coefficients. Apparently smaller particles have lower thermal resistance, thus yielding higher heat transfer coefficients. However, as shown in Fig. 3, the effect of particle size decreases when gas velocity and the length of heat transfer surface increase.

3. Influence of Bed Temperature

Figure 4 plots, against suspension density, the heat transfer coefficients measured by Basu (1990) over a wide range of bed temperature for 296 μm sand, by Kobro and Brereton (1986) at a temperature of 850°C for 250 μm sand and by Grace and Lim (1989) at 880°C for 250–300 μm sand. The overall heat transfer coefficient is shown to increase with bed temperature. Before radiation becomes dominant in heat transfer, the observed rise in heat transfer coefficient with bed temperature may be explained as follows. The gas convective component is expected to decrease mainly because of the inverse dependence of gas density on temperature. On the other hand, the particles convective component will increase with temperature, thus leading to an increase in gas conductivity, because the latter is dominant for

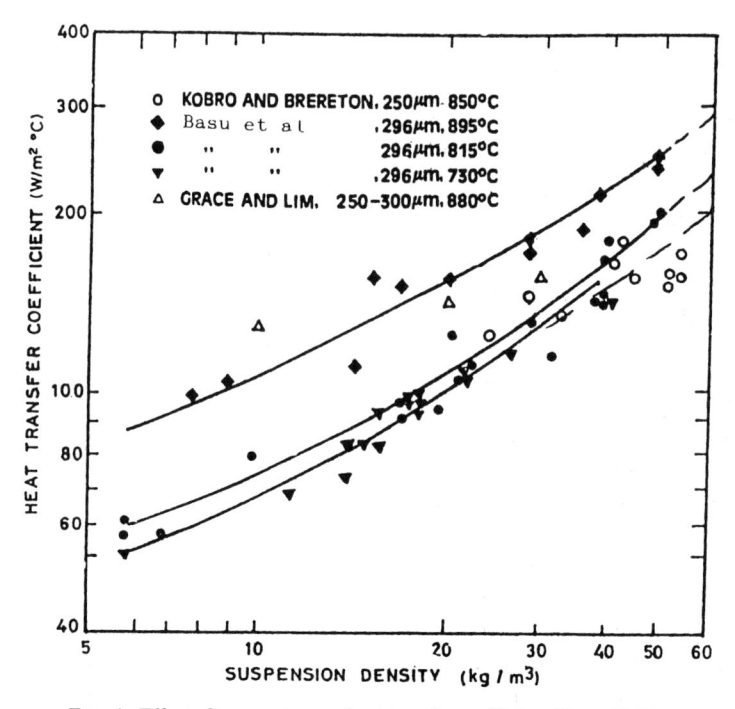

FIG. 4. Effect of temperature on heat transfer coefficient (Basu, 1990).

fine-particle beds, and the net effect is a rise in heat transfer rate. At even higher temperatures, radiation is dominant, thus favoring a further rise in heat transfer coefficients.

4. Influence of Surface Orientation

It is easily understood that local heat transfer coefficient is closely related to the local behaviors of gas and particles. When the heat transfer surface is suspended in the column for heat transfer coefficient measurements, a certain connecting stick is usually used. The use of the connecting stick will disturb the flow pattern of the particles around the probes, so that local heat transfer coefficients measured for probes pointing upward is expected to be different from those obtained for downward probes.

Experimental studies have been conducted by Bi *et al.* (1990) in a fast fluidized bed 186 mm in inside diameter at ambient temperature. FCC particles with a mean size of 48 μm were employed, and three cylindrical heat probes, 10 mm in diameter and 40, 80 and 160 mm in length, respectively, were used. The probes, made of copper-sheathed inner heating elements, were instrumented with thermocouples 40 mm apart on the surface for the measurement of surface temperature. Located 2.9 m above the distributor, the probe was installed either upward or downward (see Fig. 5), and was moved along the radial direction by two connecting sticks for the measurement of heat transfer coefficients for different orientations and at different radial positions.

Figure 6 shows the radial profiles of heat transfer coefficients for both the downward and the upward probes under the same operating conditions. Lower heat transfer coefficient for the upward probe is observed at the central region and at regions near the bed wall where particles fall downward. Because of high particle concentration near the wall, the difference is, however, not significant in this region.

The horizontal sticks were far from the heat transfer surface, so their influence on the gas–solids flow would be considered minimal. If the vertical stick is taken as an extended part of the probe, the upper line in Fig. 6 can be considered as the result of a short probe with no stick, and the bottom line as the result of a certain long probe with no stick. The distance between the two lines can be regarded as the variation caused by the influence of the front part of the probe on heat transfer of the back surface.

When the heat transfer surface is located horizontally in a bed, the heat transfer coefficients obtained windward and leeward are also quite different. Experimental measurements conducted by Liu *et al.* (1990), where a Ni-Cr band 0.1 mm thin, 5.7 mm wide and 1376 mm long was wound on a Bakelite rod 25 mm in diameter as the heat transfer surface, have well demonstrated this variation, as shown in Fig. 7.

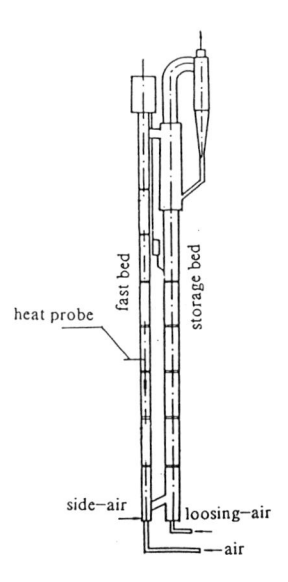

FIG. 5. Schematics of experimental apparatus and probes (Bi et al., 1990).

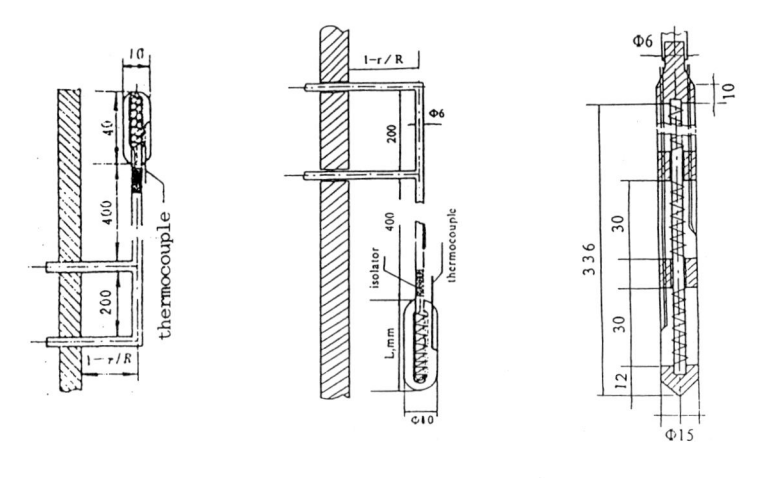

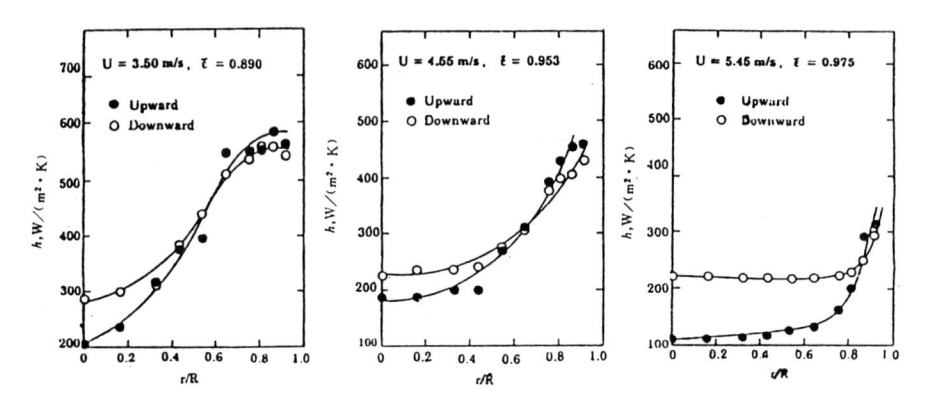

FIG. 6. Effect of surface orientation on heat transfer coefficients (Bi et al., 1990).

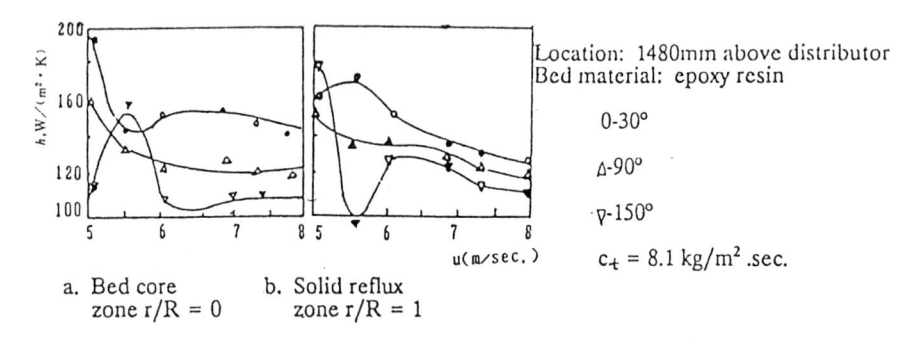

Location: 1480mm above distributor
Bed material: epoxy resin

○ 0-30°
△-90°
▽-150°

a. Bed core zone r/R = 0
b. Solid reflux zone r/R = 1

$c_+ = 8.1$ kg/m^2 .sec.

FIG. 7. Effect of windward and leeward facing on heat transfer coefficient (Liu et al., 1990).

B. RADIAL DISTRIBUTION OF HEAT TRANSFER COEFFICIENTS

1. Influence of Operating Conditions

Non-uniform distribution of gas and solids velocities and solids concentration in FFB leads to non-uniform distribution of heat transfer coefficients. Figure 8 shows the radial distribution of heat transfer coefficients, measured at four bed sections for different solids circulating rates. For all sections, heat transfer coefficients, just like solids concentrations, increase with increasing solids circulating rate. In general, heat transfer coefficients change directly with solids concentrations. Local heat transfer coefficients are comparatively low and the profiles are flat in the center region, but they increase sharply near the bed wall. Comparison of different bed sections shows that the effect of solids recirculation rate is highly significant for the lower bed sections but less significant for the higher bed sections.

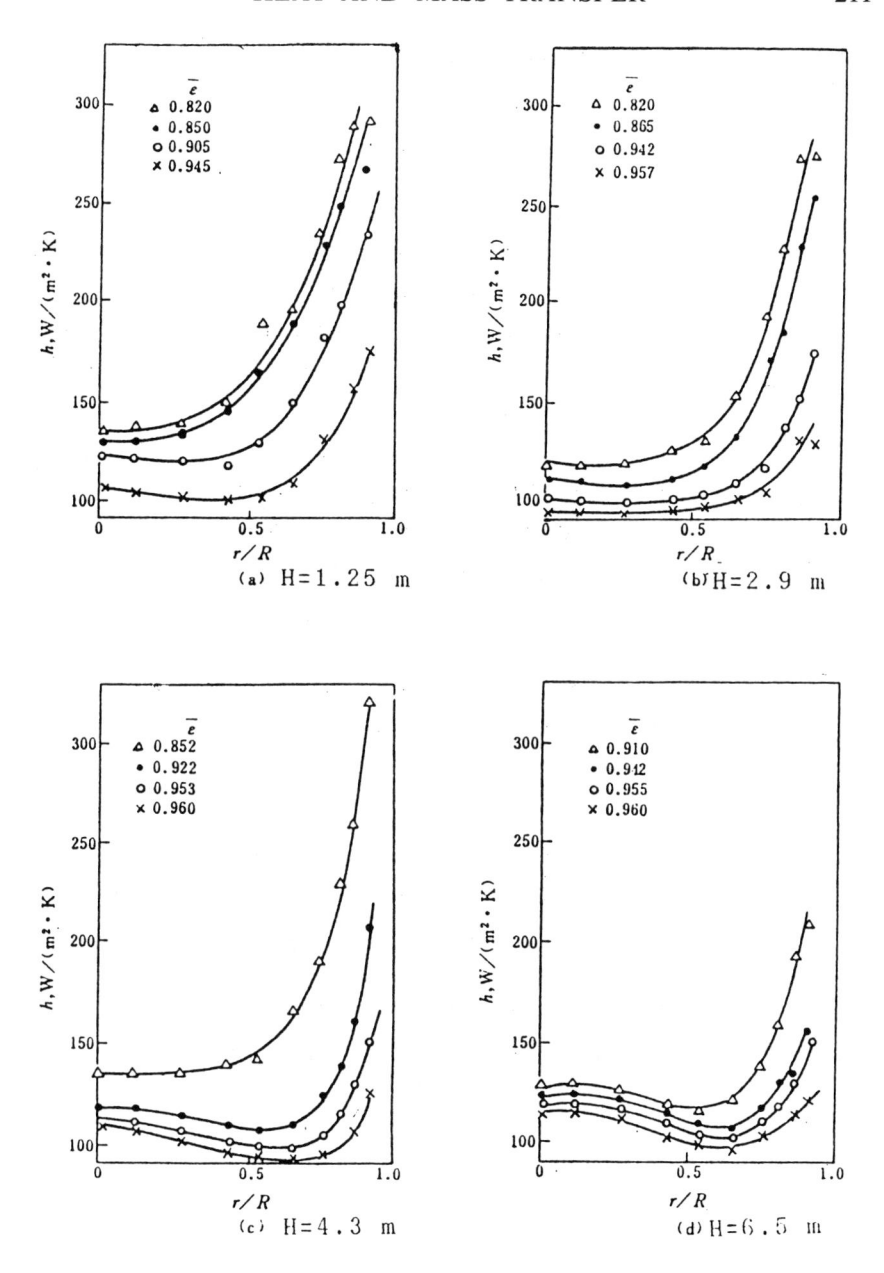

FIG. 8. Effect of solid recirculating on radial distribution of heat transfer coefficient at $U = 3.7$ m/s for silica gel ($d_p = 280$ μm) (Bi et al., 1990). G_s: △, 84.4 kg/(m²s); ●, 68.2 kg/(m²s); ○, 56.3 kg/(m²s); ×, 42.6 kg/(m²s).

As for the higher bed sections (Figs. 8c and 8d), where solids concentrations are low, the radial profile of heat transfer coefficients has a minimum point in the region of r/R from 0.5 to 0.8, showing an inconsistency in the trend of variation with solids concentration; in other words, at this particular region heat transfer becomes more complex.

The effect of gas velocity on radial distribution of heat transfer coefficient is shown in Figs 9 and 10. With increasing gas velocity the heat transfer coefficients decrease. For the lower bed sections (see Fig. 10) the radial distributions are mainly affected by solids concentrations, and for the higher bed sections this trend changes significantly.

2. Empirical Correlations and Predictions

To predict the radial profile of heat transfer coefficient in fast fluidized beds, an empirical correlation has been proposed by Bi et al. (1989) in dimensionless form as follows:

$$\frac{hd_p}{k_g} = a(1 - \bar{\varepsilon})^{n_1} \left(\frac{U_g d_g \rho_g}{\mu_g}\right)^{n_2} \left(\frac{d_p}{D}\right)^{0.53} \left(\frac{C_{pp}\rho_p}{C_{pg}\rho_g}\right)^{0.26}, \qquad (1)$$

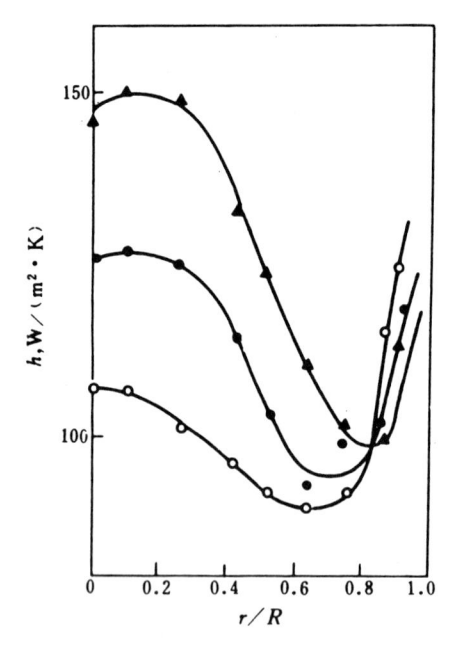

FIG. 9. Effect of superficial gas velocity on heat transfer coefficient at constant bed density for Silicagel A (Bai et al., 1990). $\varepsilon = 0.960$, $H = 6.5$ m. U: ○, 3.65 m/s; ●, 6.00 m/s; ▲, 7.33 m/s.

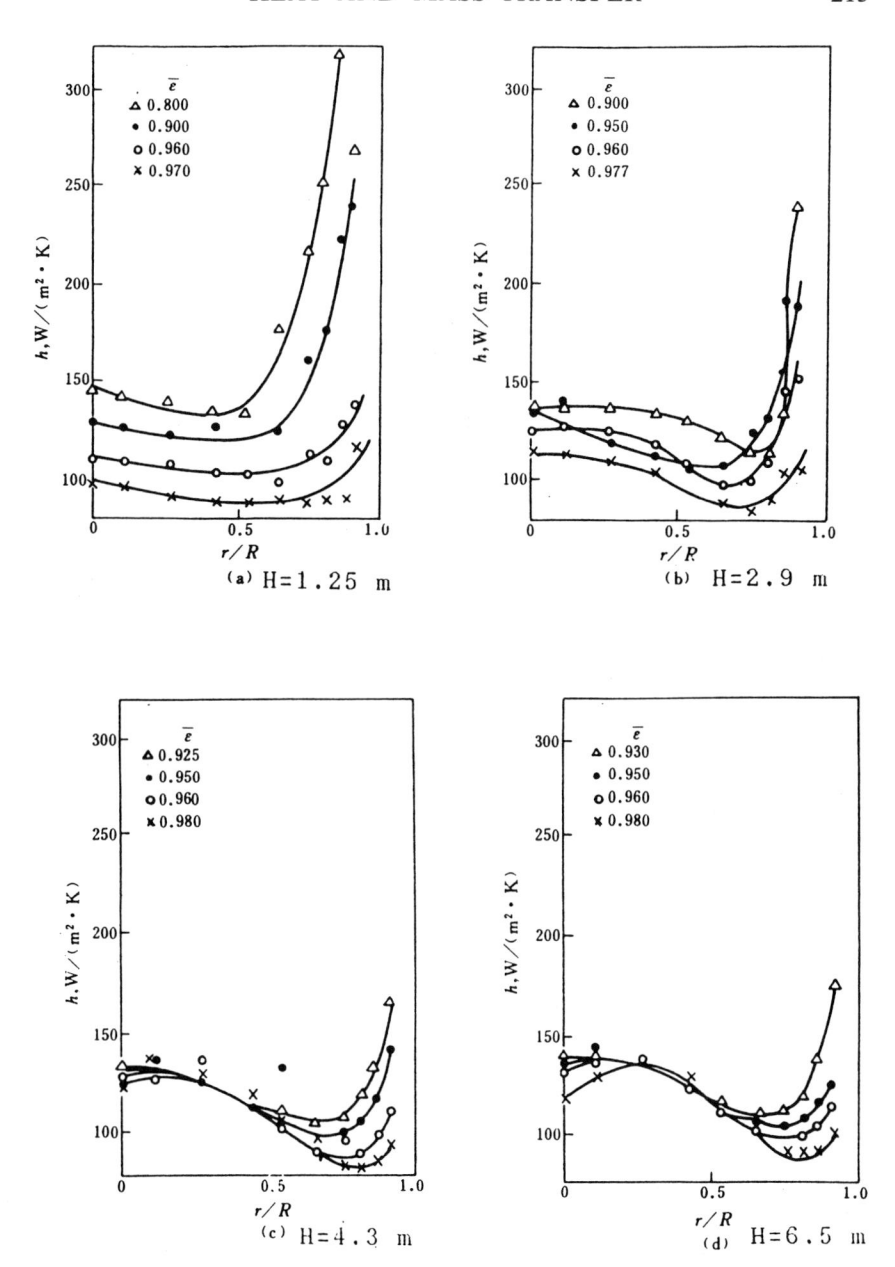

FIG. 10. Effect of superficial gas velocity on radial distribution of heat transfer coefficient at $U = 6.0$ m/s for Silicagel A (Bi *et al.*, 1990). G_s: △, 133.8 kg/(m²s); ●, 93.9 kg/(m²s); ○, 73.1 kg/(m²s); ×, 42.1 kg/(m²s).

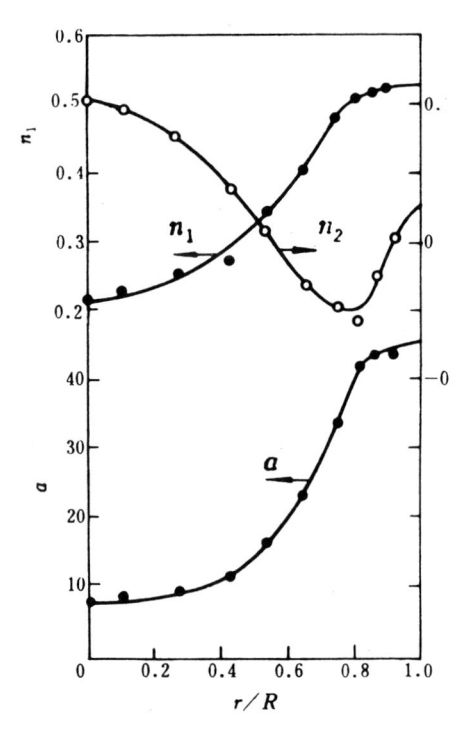

FIG. 11. Effect of radial position on constants a, n_1, n_2, in Eq. (1) (Bi et al., 1990).

where a, n_1 and n_2 are empirical functions of radial positions (as shown in Fig. 11).

When $r/R < 0.80$:

$$a = 7.695 + 13.54(r/R) - 61.23(r/R)^2 + 120.4(r/R)^3,$$

$$n_1 = 0.220 - 0.0262(r/R) + 0.413(r/R)^2 + 0.097(r/R)^3,$$

$$n_2 = 0.0995 + 0.0753(r/R) - 0.738(r/R)^2 + 0.496(r/R)^3.$$

When $r/R > 0.80$:

$$a = -220.6 + 607.7(r/R) - 349.5(r/R)^2,$$

$$n_1 = 0.409 + 0.154(r/R) - 0.035(r/R)^2,$$

$$n_2 = -1.607 + 3.113(r/R) - 1.480(r/R)^2.$$

The characteristic axial and radial distributions of heat transfer coefficients computed from the preceding formula are shown in Fig. 12. In the calculation, the sectional average voidage $\bar{\varepsilon}$ can be established from any of the known

HEAT AND MASS TRANSFER 215

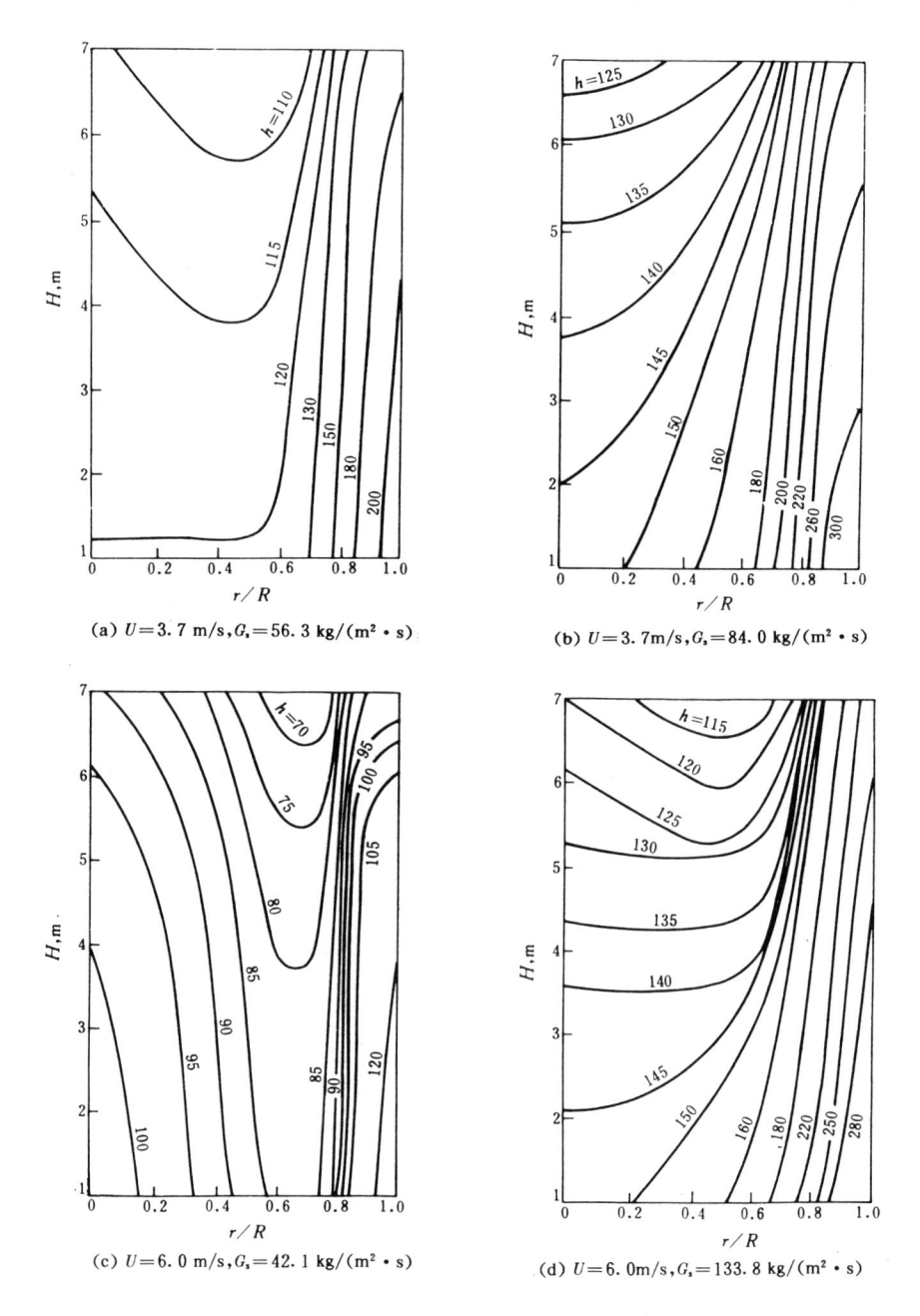

(a) $U=3.7$ m/s, $G_s=56.3$ kg/(m² · s)

(b) $U=3.7$ m/s, $G_s=84.0$ kg/(m² · s)

(c) $U=6.0$ m/s, $G_s=42.1$ kg/(m² · s)

(d) $U=6.0$ m/s, $G_s=133.8$ kg/(m² · s)

FIG. 12. Computed distribution of heat transfer coefficient in fast fluidized bed for Silicagel A (Bi et al., 1990).

methods, such as the two-channel model proposed by the authors, or the Li–Kwauk model of Chapter 4, Section III.D.

From the preceding discussion the radial distribution of heat transfer coefficient in fast fluidized beds can be distinguished as three types:

1. When solids concentration is high ($\bar{\varepsilon} < 0.9$), the heat transfer coefficient changes directly with solids concentration.
2. At low solids concentration ($0.9 < \bar{\varepsilon} < 0.96$), a minimum point of heat transfer coefficient appears in the region of radial position r/R from 0.5 to 0.8.
3. At very low solids concentration ($0.96 < \bar{\varepsilon}$) and high gas velocity, gas convection becomes significant, and the heat transfer coefficient profile becomes more complex, with values higher in the region near the axis than those near the bed wall.

C. Axial Distribution of Heat Transfer Coefficients

Usually, the feature of the axial distribution of solids concentration is dilute at the top sections and dense at the lower sections. Consequently, the heat transfer coefficient decreases gradually along the bed height (Zhang et al., 1987; Anderson et al., 1987; Bi et al., 1990). Heat transfer coefficients at any axial position of the bed increase with increasing solids concentration (see Fig. 13). With increasing gas velocity the axial distribution of heat transfer coefficient becomes more and more uniform. Based on a large number of experimental data for four kinds of particles, Bi et al. (1990) proposed the following empirical correlation:

$$\frac{\bar{h}d_p}{k_g} = 22.5(1 - \bar{\varepsilon})^{0.41} \left(\frac{d_p}{D}\right)^{0.53} \left(\frac{C_{pp}\rho_p}{C_{pg}\rho_g}\right)^{0.26}. \qquad (2)$$

Inasmuch as heat transfer depends on the hydrodynamic features of fast fluidization, if the fast fluidized bed is equipped with an abrupt exit, the axial distribution of solids concentration will have a C-shaped curve (Jin et al., 1988; Bai et al., 1992; Glicksman et al., 1991. See Chapter 3, Section III.F.1). The heat transfer coefficient will consequently increase in the region near the exit, as reported by Wu et al. (1987).

D. Variation of Local Heat Transfer Coefficient along the Surface

As mentioned earlier, increasing the length of the heat transfer surface will decrease the overall heat transfer coefficient. In order to elucidate the

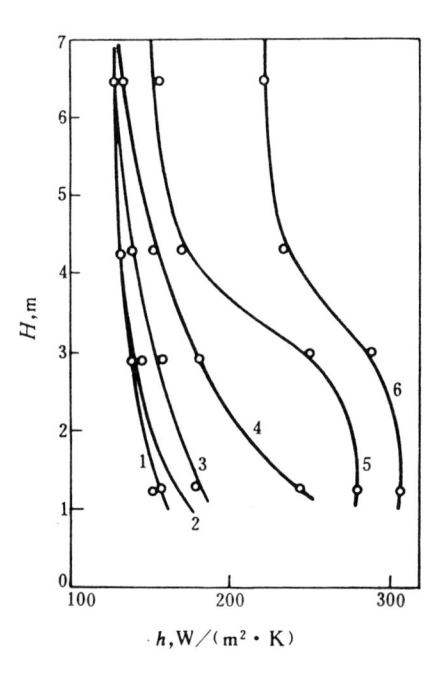

FIG. 13. Axial distribution of heat transfer coefficient at different radial positions for Silicagel A (Bi et al., 1990). r/R: (1) 0.0, (2) 0.22, (3) 0.43, (4) 0.65, (5), 0.75, (6) 0.85.

underlying mechanism, experimental studies have been made independently by Wu et al. (1989) and Bai et al. (1990). Wu and co-workers employed a water-cooled membrane, 1.59 m long, located at the bed wall, and heat was transferred from the bed at 400–800°C to the wall maintained at 40–50°C. Their data as well as those of Kobro and Brereton (1986) are plotted in Fig. 14. In the experiments of Bai and co-workers, a specially designed heat transfer probe, 15 mm in o.d. and 336 mm long, was installed vertically to measure the profile of local heat transfer coefficients along the probe. Typical results are shown in Fig. 15. Local heat transfer coefficient along the heat transfer surface is shown to decrease from top to bottom of the probe, regardless of where the probe is located. Such decrease in heat transfer coefficient near the bed wall is greater than that near the bed center. These results suggest that when a tall heat transfer surface is immersed at any radial position of the bed, the solids particles tend to form a cluster layer next to the surface. Heat transfer between the cluster layer and the surface is mainly caused by unsteady particle convection. In the course of the movement of the clusters in the layer, the driving force of heat transfer decreases as the particles are progressively heated by the surface.

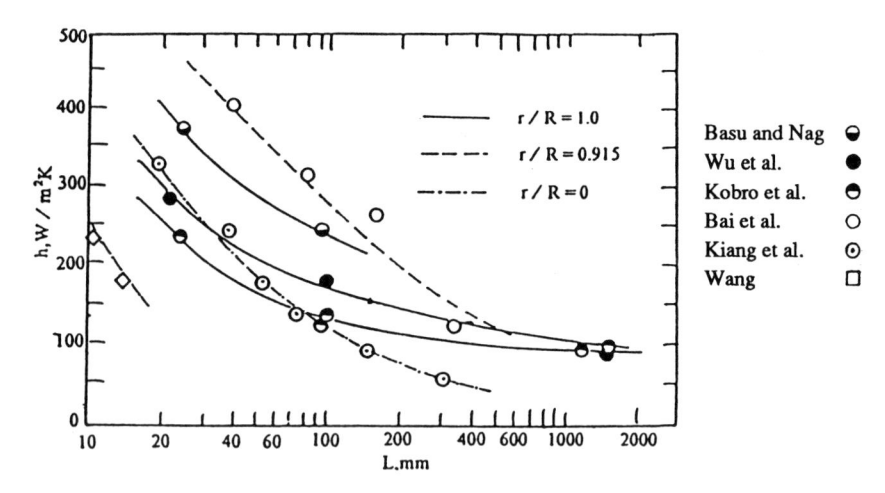

FIG. 14. Relationship between local heat transfer coefficients and probe length (Bai et al., 1992). $\rho_m = 60$ kg/m^3.

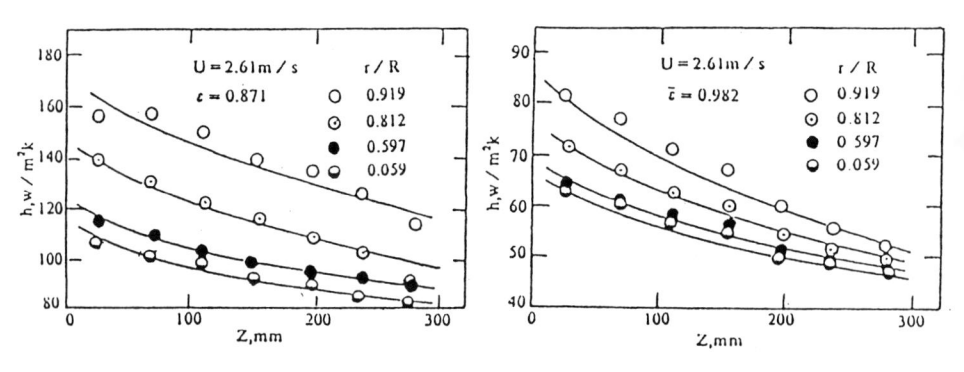

FIG. 15. Profiles of local heat transfer coefficient along the probe (Bai et al., 1992).

Figure 16 shows that the profiles of local heat transfer coefficients along the probe become flatter when the voidage is higher. Because heat transfer involves predominantly a transient process between the particles and the surface, local heat transfer coefficients at any radial position increase with increase in particle concentration. At the upper part of the probe, fresh clusters come into contact with the surface where the driving force of heat transfer reaches a maximum, and the effect of voidage on heat transfer coefficient is expected to be greater than that at the lower part of the probe. Because of the large solids concentration near the bed wall, the effect of voidage on heat transfer coefficients along the probe is larger here than that at the bed center (see Fig. 16).

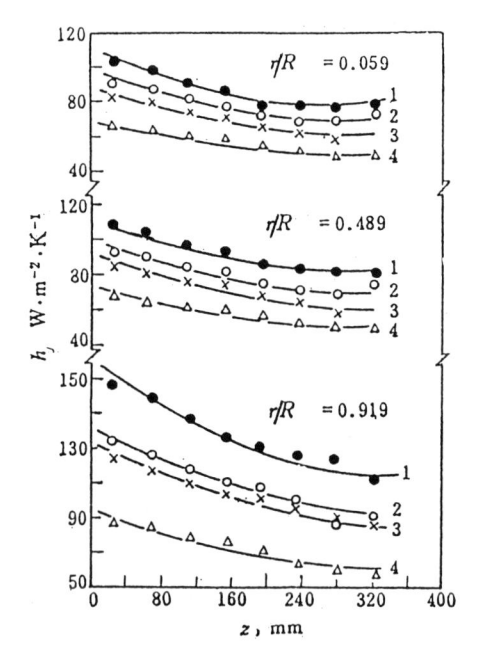

FIG. 16. Effect of bed voidage on profiles of heat transfer coefficient along the probe (Bai et al., 1992). ε: (1) 0.8866, (2) 0.9261, (3) 0.9513, (4) 0.9782.

Figure 17 shows that the effect of gas velocity on the profiles of local heat transfer coefficients is larger at the upper part of the probe than at the lower part.

E. HEAT TRANSFER ENHANCEMENT

Decrease in overall heat transfer coefficient with the length of heat transfer surface is detrimental in the commercial use of CFB reactors. The key to enhanced heat transfer is the frequent renewal of the cluster layer at the surface by minimizing its continued development. Based on this idea, a method to enhance heat transfer was proposed (Bai, 1991) to install special rings, shown in Fig. 18 as the A (without ring) and B (with rings) probes, each composed of eight identical sub-probes made of aluminium alloy, each 30 mm long and 15 mm in i.d. These sub-probes are connected to each other with Teflon isolators, each 12 mm long. The total length of the probe is 336 mm. Axial heat flux along the surface is interrupted by these Teflon thermal isolators. The probes are heated by internal electric heaters, and the surface temperatures are measured with thermocouples.

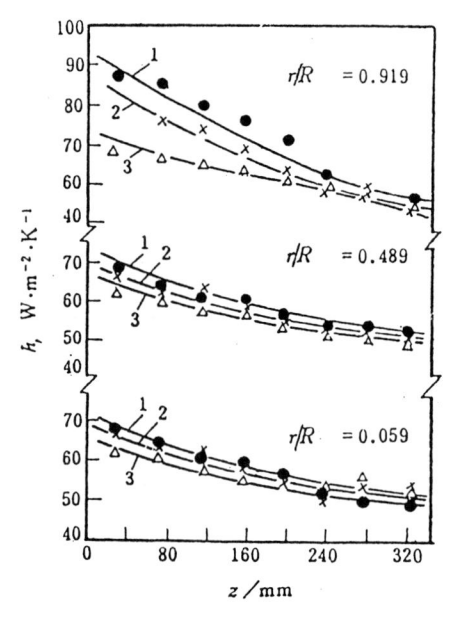

FIG. 17. Effect of gas velocity on profiles of heat transfer coefficient along the probe (Bai et al., 1992). u, m/s: (1) 3.62, (2) 4.50, (3) 5.50.

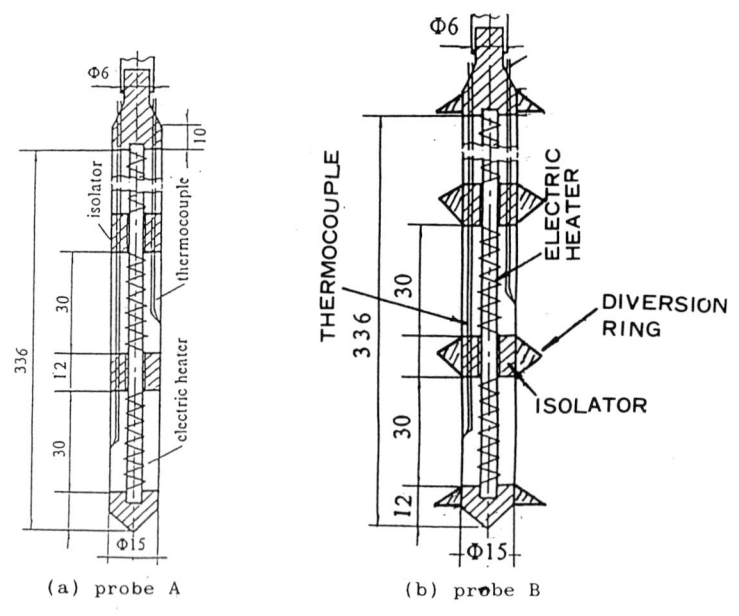

FIG. 18. Schematics of heat transfer probe (Bai et al., 1992). (a) Probe A; (b) probe B.

The local heat transfer coefficients measured by both probe B and probe A are compared in Fig. 19. The local heat transfer coefficient for probe B is evidently larger than that for probe A. This difference in local heat transfer coefficients measured by the two probes gradually decreases from top to bottom of the probes at any radial position in the bed. Such enhancement in heat transfer is attributed to the promotion of cluster layer renewal surrounding the probe, as has been well recognized (Glicksman, 1988; Wu et al., 1989, 1990; Basu, 1990; Leckner, 1991; and Bai et al., 1990).

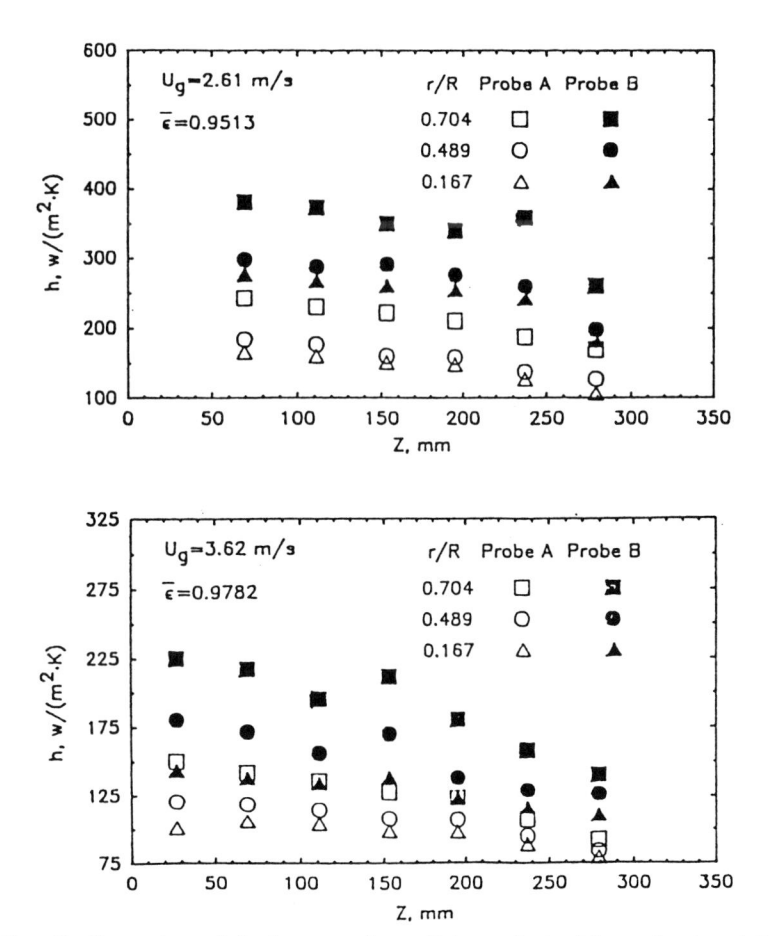

FIG. 19. Comparison of the heat transfer coefficients obtained by probe A and B (Bai et al., 1991).

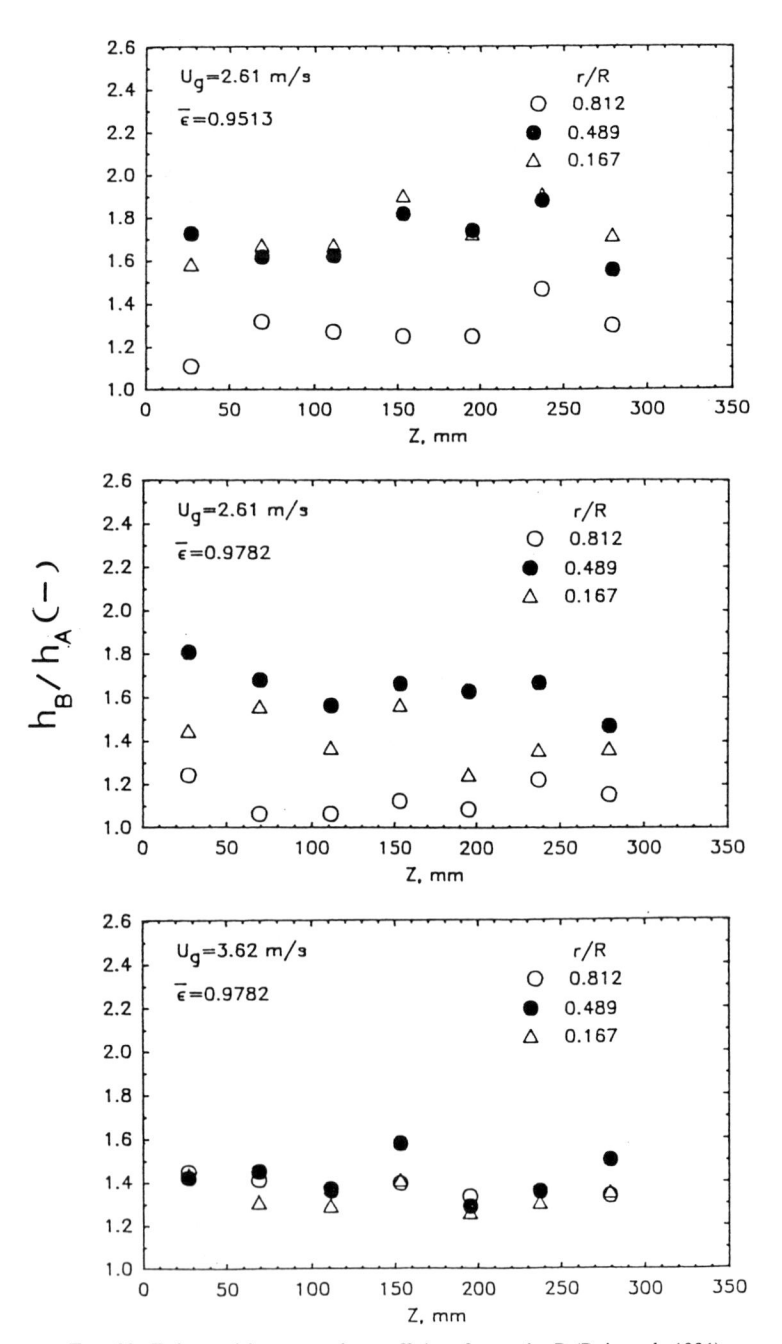

FIG. 20. Enhanced heat transfer coefficient for probe B (Bai et al., 1991).

Figure 20 indicates that the ratio of h_B/h_A, between probes B and A, is in the range of 1.1 to 1.8 and keeps almost constant along the axial direction of the probe. The ratio of h_B/h_A is large at low gas velocities, and when the probes are located at the center of the bed (see Fig. 20). This implies that the distribution of heat transfer coefficients, even with rings, along the radial direction of the bed becomes flatter at a higher bed level where solids concentration is low. Stated alternatively, with increasing gas velocity the influence of the rings on heat transfer coefficient diminishes.

III. Theoretical Analysis and Models for Heat Transfer

The heat transfer coefficient h_C between the suspension and an immersed surface may be considered to be composed of three additive components:

1. the particle convective transfer component h_{pc}, which represents the heat transfer due to particle motion within the bed;
2. the gas convective transfer component h_{gc}, which represents the heat transfer to or from the gas percolating through the bed;
3. the radient transfer component h_{rad}, which represents the radiative heat transfer between FFB and the heat transfer surface.

The second and the third components become significant only at high temperatures ($>700°C$) and low solids concentrations (<30 kg/m^3). In fast fluidized beds, the motion of the particles plays an overriding role in the heat transfer process, since the solids particles have larger heat capacity and higher thermal conductivity. Most of the heat transfer models reported in the literature give emphasis to particle convective transfer.

A. Gas Convective Transfer

In addition to direct contact with clusters, the wall of a fast bed is constantly exposed to the up-flowing gas, which contains dispersed solids (Li *et al.*, 1988). The gas convective component can be estimated on the basis of correlations for gas flow alone through the column, at the same superficial gas velocity and with the same physical properties. When a tall heat transfer surface is used or the bed is operated at high solids concentrations, errors caused by using different approaches will usually be small since h_{gc} is generally much less than h_{pc}, provided the solids concentration is low and temperature high.

B. Radiative Transfer

The radiative component h_{rad} can be estimated by treating the suspension as a grey body:

$$h_{rad} = \frac{\sigma(T_{susp}^4 - T_{surf}^4)}{\left[\dfrac{1}{e_{surf}} + \dfrac{1}{e_{susp}} - 1\right](T_{susp} - T_{surf})}. \tag{3}$$

In fast fluidized beds, a proper modification to the suspension emissivity is usually needed according to the gas–solids flow pattern, such as the following correlation originally proposed by Grace (1982):

$$e_{susp} = 0.5(1 + e_p). \tag{4}$$

C. Particle Convective Transfer

The particle convective heat transfer component is usually treated on the basis of the penetration or packet theory originally proposed by Mickley and Fairbanks (1955) assuming that the clusters are formed next to the immersed surface (e.g., Subbarao and Basu, 1986; Basu and Nag, 1987; Zhang et al., 1987; Liu et al., 1990). In that case, the clusters of solids and voids or dispersed phase are assumed to come into contact with the heat transfer surface alternatively, and the heat transfer coefficient can be given as follows:

$$h_{pc} = \frac{1}{\dfrac{d_p}{10k_g} + \left(\dfrac{t_c \pi}{4k_c C_{pc}\rho_c}\right)^{0.5}}. \tag{5}$$

Subbarao and Basu (1986), Basu and Nag (1987) and Basu (1990) derived the expression of cluster residence time t_c on the heat transfer surface based on Subbarao's (1986) cluster model, although the model is not widely accepted. Lu et al. (1990) and Zhang et al. (1987) have also obtained empirical correlations independently for predicting cluster residence time based on their heat transfer experiments. However, because of the lack of available and reliable information about the residence time of clusters at the surface and the fraction of the clusters in solids suspension, a significant discrepancy between the results predicted by the different approaches mentioned earlier has been observed. Besides, it should be pointed out that the major shortcoming in the earlier models is that they all take no account of the heat transfer surface length.

To account for the influence of the length of the heat transfer surface on heat transfer, several "cluster-layer" type models have been reported in the literature (Glicksman, 1988; Wu et al., 1990; Mahalingam et al., 1991; Basu, 1990). However, all these models can only be applied to heat transfer between the suspension and the walls.

To explain the underlying heat transfer mechanism, the heat transfer model proposed by the authors (Bai et al., 1992a) is illustrated in Fig. 21 based on the following assumptions:

1. When the heat transfer surface is immersed in the bed at any radial position, the solids particles form a downflowing cluster layer surrounding the surface. The thickness of the cluster layer grows up to reach a final constant value.
2. Heat transfer between the cluster layer and the surface is mainly caused by unsteady particle convection. The cluster layer is renewed from time to time by the moving particles and fluidizing gas.
3. The heat transfer process causes the radial temperature gradient to vanish at a certain distance away from the heat transfer surface, and at the outer boundary of this layer the derivative of radial temperature dT/dt is zero.
4. The thermo-physical properties of both gas and particles are constant.
5. Radial heat transfer is negligible.

According to these assumptions, the heat transfer process can be described

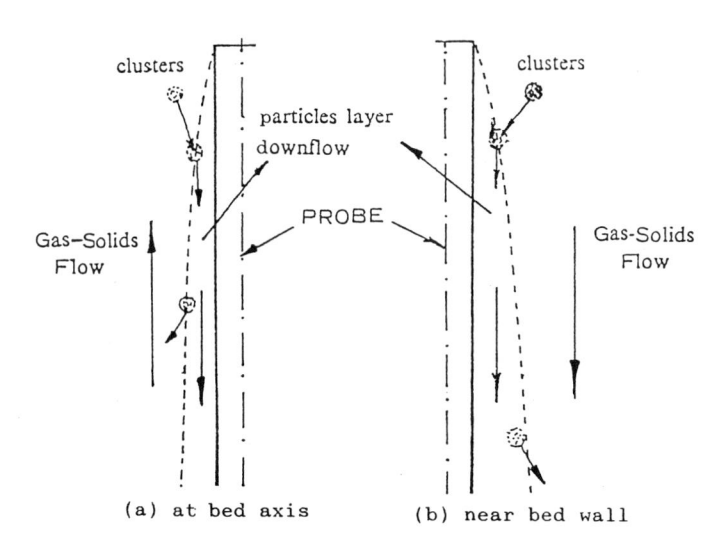

FIG. 21. Mechanism of cluster heat transfer (a) at bed axis, (b) near bed wall (Bai et al., 1991).

by the following differential equation and boundary conditions:

$$\nabla^2 T = \frac{1}{\alpha_c}\frac{\delta T}{\delta t}, \quad r_h < r < r_h + \delta$$

$$T(r_h, t) = T_{surf};$$

$$T(r, 0) = T_b; \tag{6}$$

$$\left.\frac{\delta T}{\delta r}\right|_{r_h + \delta \cdot t} = 0.$$

For an immersed cylindrical surface, the solution of Eq. (6) yields the following time-averaged heat transfer coefficient:

$$h = -\frac{1}{T_{surf} - T_b}\kappa_c \left.\frac{\delta \bar{T}}{\delta r}\right|_{r=r_h}$$

$$= \kappa_c \sqrt{\frac{s}{\alpha_c}} \frac{-I_1\left(\sqrt{\frac{s}{\alpha_c}}r_h\right)K_1\left[\sqrt{\frac{s}{\alpha_c}}(r_h+\delta)\right] + K_0\left(\sqrt{\frac{s}{\alpha_c}}(r_h+\delta)\right) + K_1\left(\sqrt{\frac{s}{\alpha_c}}r_h\right)I_1\left[\sqrt{\frac{s}{\alpha_c}}(r_h+\delta)\right]}{I_0\left(\sqrt{\frac{s}{\alpha_c}}r_h\right)K_1\left[\sqrt{\frac{s}{\alpha_c}}(r_h+\delta)\right] + K_0\left(\sqrt{\frac{s}{\alpha_c}}r_h\right)I_1\left[\sqrt{\frac{s}{\alpha_c}}(r_h+\delta)\right]}. \tag{7}$$

Equation (7) shows that the local heat transfer coefficient is a function of the thickness and the renewal frequency of the cluster layer. If the gas–solids flow is fully developed and $r_h \gg d_p$, Eq. (7) can be simplified to

$$h = \kappa_c\sqrt{\frac{s}{\alpha_c}}, \tag{8}$$

where the effective thermal conductivity of the cluster layer k_c can be estimated from the expression proposed by Gelperin and Einstein (1971):

$$k_c = k_g\left[1 + \frac{(1-\varepsilon_c)(1-k_g/k_p)}{k_g/k_p + 0.28\varepsilon_c^{0.63(k_p/k_g)^{0.18}}}\right] \tag{9}$$

for particles with diameter less than 0.5–0.7 mm abd $k_p/k_g < 5{,}000$. The thermal diffusivity of the cluster layer is calculated as follows:

$$\alpha_c = k_c/[\rho_p(1-\varepsilon_c)C_{pp} + \rho_g\varepsilon_c C_{pg}], \tag{10}$$

where the averaged voidage of the cluster layer is assumed to be

$$\varepsilon_c = (\varepsilon + \varepsilon_{mf})/2. \tag{11}$$

This is based on the assumption that the voidage at the heat transfer surface is ε_{mf}, and the voidage varies linearly with radial distance within the cluster layer. The local voidage is expressed as follows (Tung et al., 1989):

$$\varepsilon = \bar{\varepsilon}^{(0.191 + \phi^{2.5} + 3\phi^{11})}. \tag{12}$$

The renewal frequency of the cluster layer at the heat transfer surface is mainly dependent on the solids exchange rate between the layer and the other suspension, and usually assumed to be proportional to the ratio of the characteristic velocity to the characteristic length. Here we define

$$s = CV_c/Z, \tag{13}$$

where C is a proportionality constant and V_c is the descending velocity of the cluster layer.

In order to evaluate the descending velocity of the cluster layer, an alternative approach proposed originally by Glicksman (1988) is adopted. As shown in Fig. 22, clusters are assumed to sweep along the surface and then accelerate downwards under the action of gravity, until they reach a maximum velocity, V_{cmax}. According to an analysis of force balance, the cluster layer descending velocity can be expressed as

$$V_c = V_{cmax}[1 - \exp(-gZ/V_{cmax})], \tag{14}$$

where V_{cmax} can be determined directly or reduced indirectly from experimental measurements.

The parameter C represents the probability of contact between the cluster and the heat transfer surface when the cluster layer velocity remains constant.

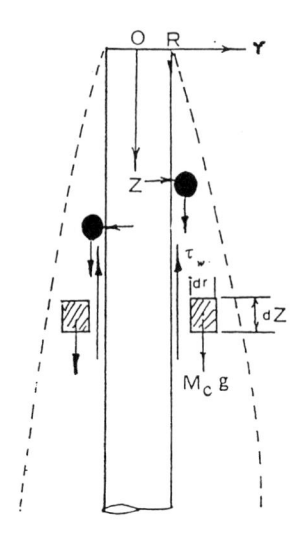

FIG. 22. Schematics of cluster layer structure (Bai et al., 1991).

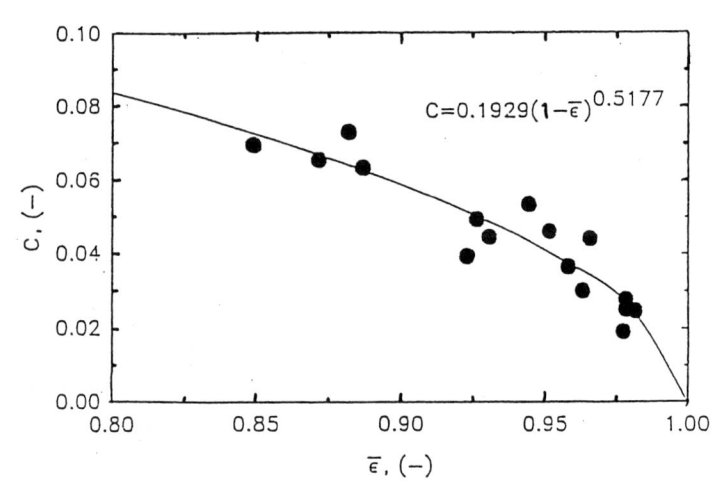

FIG. 23. Effect of bed voidage on parameter C (Bai et al., 1991).

The parameter C is directly related to the solids concentration of the bed. The results deduced from the heat transfer coefficients show that the value of the parameter C increases with increasing solids concentration of the bed, but it is almost independent of gas velocity, as shown in Fig. 23. The following empirical correlation (Bai, 1991) is proposed for the calculation of the parameter C:

$$C = 0.193(1 - \bar{\varepsilon})^{0.518}. \qquad (15)$$

Figure 24 shows the variation of the maximum descending velocity of the cluster layer, V_{cmax}, along the radial direction of the bed under different operating conditions. The radial distribution of V_{cmax}, which is in the range of 1.2 to 2.5 m/s, is fairly flat, except in the region near the wall where a sharp increase in V_{cmax} can be observed. Because of the significant effect of the difference between particle flow direction and cluster layer flow direction, i.e., concurrent in the region near the bed wall and countercurrent in the center region, V_{cmax} is larger in the region near the bed wall than that in the center region. Also, V_{cmax} increases with decreasing average voidage at any radial position of the bed (see Fig. 24). This results from increase both in the tendency of particle aggregation and in the solids concentration within the cluster layer with decreasing average voidage. It is expected that the heat transfer coefficient is smaller in the center region than that in the region near the wall, and that the decreasing tendency of heat transfer coefficient along the surface increases with radial position towards the wall.

If the gas velocity is increased and a constant average voidage is kept,

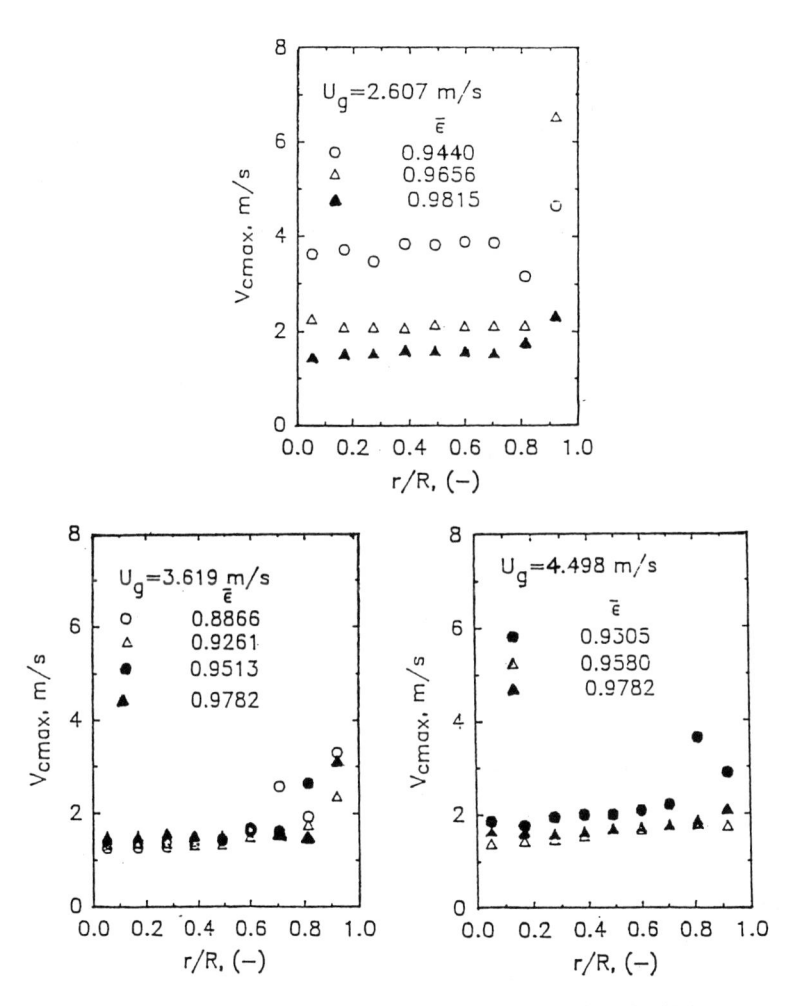

FIG. 24. Effect of radial position r/R on V_{max} (Bai et al., 1991).

then, because of the increase in particle velocities at all radial positions of the bed (Yang et al., 1991), V_{cmax} at any radial position will decrease. Consequently, there is a decrease in the renewal frequency of the cluster layer at the heat transfer surface. The local heat transfer coefficient will therefore decrease with increasing gas velocity (Bai et al., 1991).

The correlation of V_{cmax} can be expressed as follows:

$$V_{cmax} = 9.0(U_g/\bar{\varepsilon})^{-0.677}(1-\varepsilon)^{0.165}. \tag{16}$$

The average deviation of this correlation is less than 20%.

D. VALIDATION OF THE MODEL

The validity of the proposed model, Eq. (7), has been checked against experimental data (Bi *et al.*, 1990; Bai *et al.*, 1991, 1992; Basu *et al.*, 1987, 1990; Wu *et al.*, 1987, 1989) including local and overall heat transfer coefficients between the suspension and the immersed surfaces and the bed wall. These data cover a wide range of operating conditions ($U_g = 2$ to 8 m/s, $T_b = 25$ to 860°C), particle diameters ($d_p = 31$ to 500 μm) and surface sizes ($L = 40$ to 1,590 mm), which are summarized in the table.

References	Solids	d_p (μm)	T_b (°C)	L (mm)
Bai *et al.* (1991)	FCC	54	25	336
Basu and Nag (1987)	Sand	87	30–45	
		227	30–45	100
Basu (1990)	Sand	296	730	100
Bi *et al.* (1991)	FCC	18	25	40
				80
				160
Feugier *et al.* (1987)	Sand	95		
		215	400	95
Fraley *et al.* (1983)	Glass beads	37	30	150
Furchi *et al.* (1988)	Glass beads	269	30	1,000
Kobro and Brereton (1988)	Sand	250	30	100
Mickley and Trilling (1949)	Glass beads	70–451	150	650
Nag and Moral (1991)	Sand	310	80	300
Sekthira *et al.* (1988)	Sand	300	230	700
Subbarao and Basu (1986)	Sand	130	25–30	
		260	25–30	100
Wu *et al.* (1987)	Sand	188	227	1,530
Wu *et al.* (1989)	Sand	241	410–860	1,220
				1,590

1. Comparison with Profiles of Local Heat Transfer Coefficients along Heat Transfer Surface

The variation in local heat transfer coefficient with axial position of the surface is compared with model predictions in Fig. 25 to show good agreement for different radial positions and different operating conditions. The average deviation between model predictions and 1,100 experimental data points is less than 10%, as shown in Fig. 25.

Figure 26 compares the model predictions with the experimental data of Wu *et al.* (1989) for a membrane wall surface 1.59 m long at a suspension density of 54 kg/m³. Good agreement, especially at lower bed temperatures (Fig. 26a) is clearly seen. At higher temperatures, because of increase in

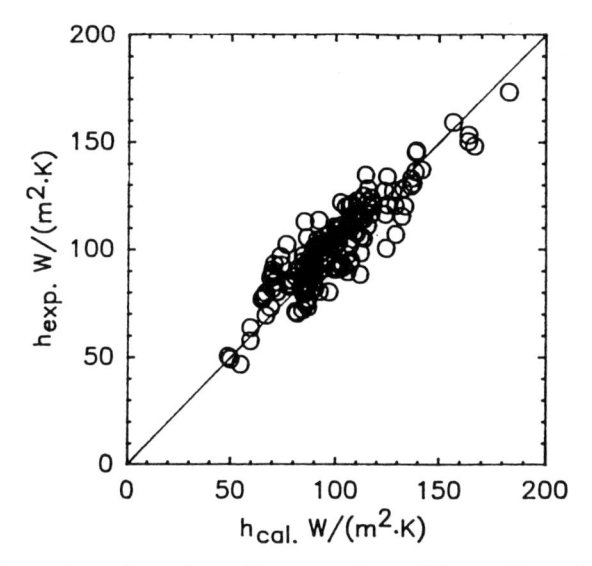

FIG. 25. Comparison of experimental heat transfer coefficients with model, Eq. (7) (Bai et al., 1991).

radiative heat transfer, the model predictions are relatively lower than experimental data. If the radiative component (h_r obtained from the model of Kolar et al., 1979, is indicated by broken line in Fig. 26) is considered, closer agreement is expected.

2. Comparison with Overall Heat Transfer Coefficient

In order to check the validity of the proposed model in a wide range of operating conditions, particle diameters and surface sizes, literature data on overall heat transfer coefficients have been collected. To obtain the overall heat transfer coefficients from the proposed model, Eq. (7) is integrated numerically over the length of the surface from $z = 0$ to $z = L$. Figure 27 shows a plot of h_{cal}/h_{exp} vs. L for a suspension density of around 60 kg/m^3. Most of the experimental data are interpolated from their original figures of 11 different groups of workers. Good agreement between the model predictions and the experimental data measured by the probes longer than 0.1 m is obtained (see Fig. 27). If the probe is short (<0.1 m), however, model predictions of the overall heat transfer coefficients are lower than those measured by experiments (i.e., $h_{cal}/h_{exp} < 1$), possibly because a shorter probe length reduces the thickness of the cluster layer as well as its thermal resistance. Apparently the cluster layer will disappear for a very short probe, and in this case the "cluster-layer" model would be invalid.

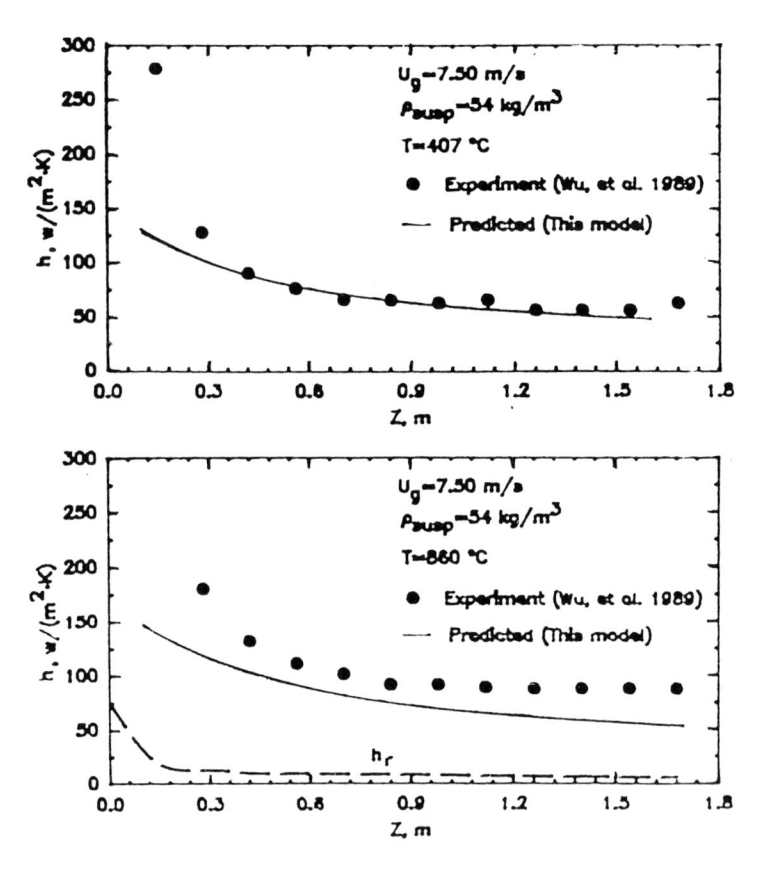

FIG. 26. Comparison of profiles of heat transfer coefficient along probe length with model, Eq. (7) (Bai et al., 1991).

IV. Heat Transfer between Gas and Particles

With respect to heat transfer between gas and particles in fast fluidized beds, there is one notable study by Watanabe *et al.* (1991) in a riser 21 mm in inside diameter and 1,800 mm in height using two kinds of particles with four different sizes. Particles heated up to about 340 K with heaters suspended in an external solids feeder are supplied to the bottom of the riser. While the particles flow upward in the riser, heat is transferred from the particles to the flowing air. By measuring both temperatures of the gas and the solids and incorporating some approximate assumptions (e.g., assuming that the gas and solids are in plug flow), the heat transfer coefficient h_{gp} can be obtained.

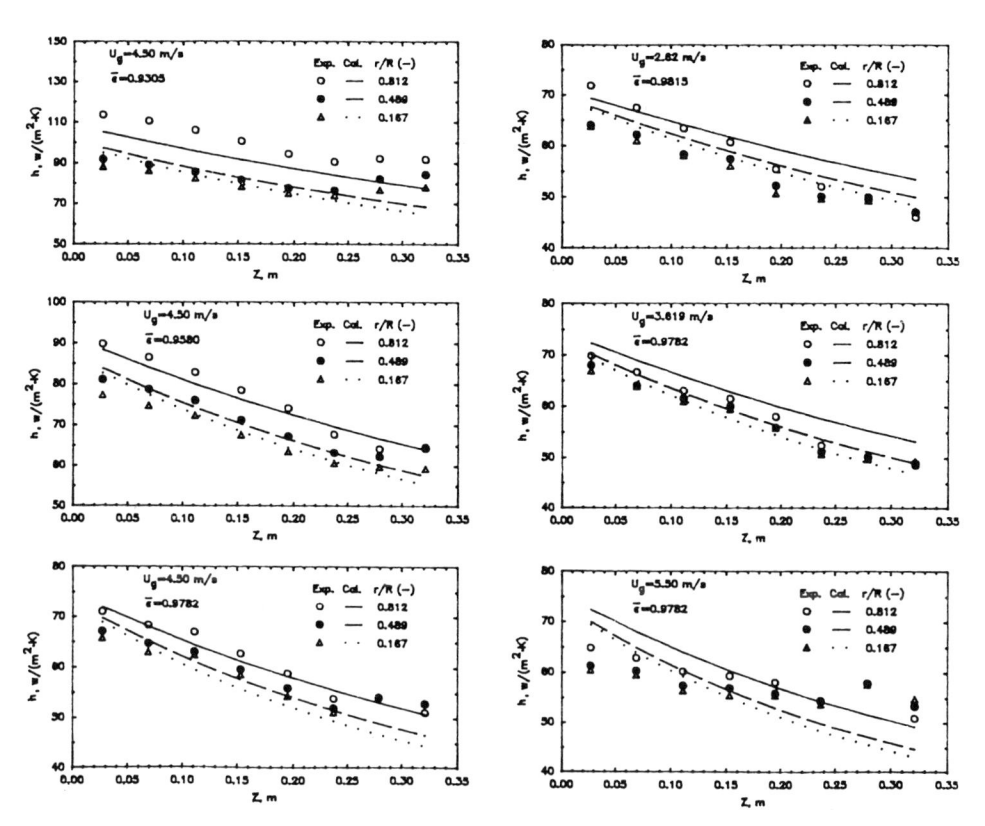

FIG. 27. Profiles of local heat transfer coefficient along the probe: comparison of model and experimental data (Bai et al., 1991).

Figure 28 shows typical profiles of gas and particle temperatures along the bed height. Temperature rises rapidly at the bed bottom section of the bed and gradually reaches a constant value. Because of the large heat capacity of the particles, the temperature of the particles decreases slightly at the lower section of the bed. Above 0.5 m from the distributor the temperature field attains thermal equilibrium.

The heat transfer coefficient h_{gp} along the axis, shown in Fig. 29, decreases sharply in the region near the bottom of the riser. Except in this region, h_{gp} gradually decreases with increasing distance from the distributor and finally approaches a constant value. The profile of the heat transfer coefficient is similar to that of the mean slip velocity, higher at the bottom and gradually decreasing towards the top of the bed.

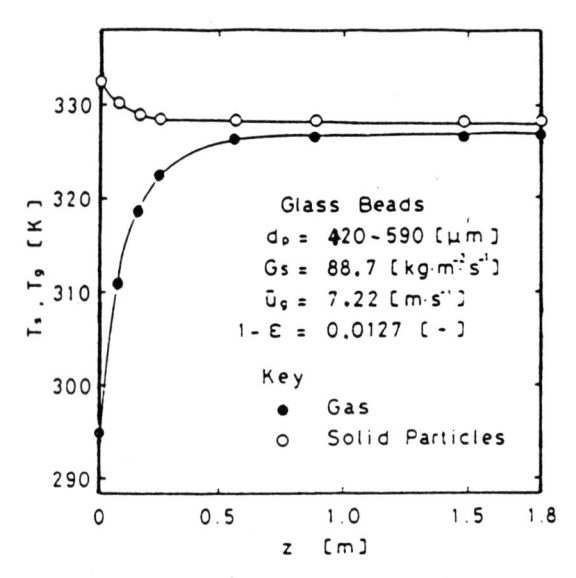

FIG. 28. Temperature profiles of gas and solids in the riser (Watanabe, 1991).

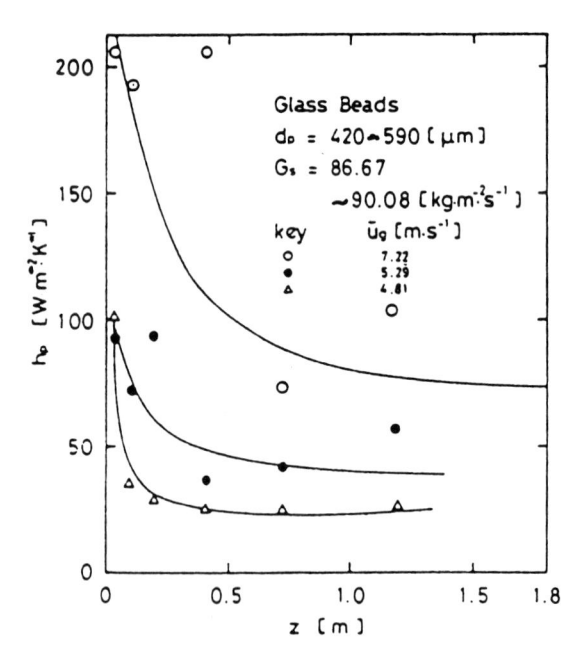

FIG. 29. Local heat transfer coefficient along the axis of the riser (Watanabe, 1991).

V. Mass Transfer

There are very few studies on mass transfer in fast fluidized beds. An exploratory experiment was carried out by Shen and Kwauk (1985) in a circulating fluidized bed 24 mm in i.d. using active carbon particles with a mean diameter of 0.57 mm and CCl_4 as a tracer. The mass transfer coefficient is calculated by incorporating some assumptions, e.g., that the concentration of the tracer at external surface of the particles is zero and the gas is in plug flow. It is of interest to see that h_m at the bottom increases with height, which suggests improved gas/solid contact. Further decrease in the value of h_m with height is attributed to the negligible tracer concentration at the external surface of the particle.

Notation

A	area of heat transfer surface, m²	s	renewal frequency of cluster layer, 1/s
d_p	mean diameter of particles, m	T	temperature, K
		$\bar{T}$	time-averaged temperature within cluster layer, K
C	parameter		
C_p	heat capacity, kJ/kg K		
h	local heat transfer coefficient, W/(m² K)	t	time, s
		U_g	superficial gas velocity, m/s
$\bar{h}$	overall heat transfer coefficient, W/(m² K)	V	solids velocity, m/s
k	heat conductivity, W/m K	Z	axial coordinate along probe, m
L	total length of heat transfer surface, m		
Q	power supplied by heater, W	**Greek Letters**	
R	bed radius, m	α	temperature conductivity, J/gK
r	radial coordinate of bed or distance from the center of heat transfer surface, m	ε	local voidage
		$\bar{\varepsilon}$	cross-sectionally average voidage
r_h	radius of heat transfer surface, m	ρ	density, kg/m³
		φ	radial position, r/R

Subscripts

		g	gas
		mf	minimum fluidization
A	probe A	max	maximum value
B	probe B	i	the ith subprobe
b	bed	p	particle
c	cluster layer	surf	heat transfer surface

References

Andersson, B. A., Johnsson, F., and Leckner, B., "Heat flow measurement in fluidized bed boilers," *Proc. Int. Conf. on Fluid. Bed Combustion* (Mustonen, J.R., ed.) pp. 592–598 ASME, (1987).
Bai, D. R. Ph. D. dissertation, Tsing Hua University, 1991.
Bai, D. R., Jin, Y., Yu, Z. Q., and Zhi, J. *Powder Technology* **71**, 51–58 (1991).
Bai, D. R., Jin, Y., and Yu, Z. Q. *Chem. Reaction Eng. and Technol.* (in Chinese) **8**(2), 224–235 (1992).
Bai, D. R., Jin, Y., Yu, Z. Q., and Cao, C. S. Paper presented at China–Japan Conference on Chem. Eng., p. 701, Tianjin University Press, Tianjin, 1990.
Basu, P. *Chem. Eng. Sci.* **45**, 3123–3136 (1992a).
Basu, P., and Nag, P. K. *Int. J. Heat Mass Transfer* **30**, 2399–2409 (1987).
Bi, H. T., Jin, Y., Yu, Z. Q., and Bai, D. R., The radial distribution of heat transfer coefficients in fast fluidized bed. *Proc. of 6th Int. Conf. on Fluid.* Banff, Canada, pp. 702. (1989).
Bi, H. T., Jin, Y., Yu, Z. Q., and Bai, D. R. *in* "Circulating Fluidized Bed Technology III" (Basu, P., Horio, M. and Hasatani, M., eds.), pp. 233–238. Pergamon Press, Oxford, 1990.
Feugier, A., Gaulier, C., and Martin, G. *Proc. 9th Int. Conf. on Fluid. Bed Combustion*, p. 613 (1987).
Fraley, L., Lin, Y. Y., Hsiao, K. H., and Solbakken, A. ASME Paper 83-HT-92, Seattle (1983).
Furchi, J. C. L., Goldstein, L., Lombardi, G., and Mosheni, M. *In* "Circulating Fluidized Bed Technology II" (Basu, P., and Large, J. F., eds.), pp. 263–270. Pergamon Press, Oxford, 1988.
Gelperin, N. I., and Einstein, V. G. *In* "Fluidization" (Davidson, J. F., and Harrison, D., eds.), pp. 471–450. Academic Press, London, 1971.
Glicksman, L. R. *In* "Circulating Fluidized Bed Technology II" (Basu, P., and Large, J. F., eds.), pp. 13–19. Pergamon Press, Oxford, 1988.
Glicksman, L. R., Westphalen, D., Breerton, C., and Grace, J. R. *in* "Circulating Fluidized Bed Technology III" (Basu, P., Horio, M., and Hasatani, M., eds.), pp. 119–124. Pergamon Press, Oxford, 1991.
Grace, J. R. Fluidized bed heat transfer. *In* "Handbook of Multiphase Flow" (Hestroni, G. ed.), pp. 9–70. McGraw-Hill Hemisphere, Washington, DC, 1982.
Grace, J. R. *In* "Circulating Fluidized Bed Technology" (Basu, P., ed.), pp. 63–80. Pergamon Press, New York, 1986.
Halder, D. K., and Basu, P. *AIChE Symp. Ser.* **84**, 262 (1988).
Kobro, H., and Brereton, C. *In* "Circulating Fluidized Bed Technology" (Basu, P., ed.), p. 263. Pergamon Press, Toronto, 1986.
Kobro, H., and Brereton, C. *In* "Circulating Fluidized Bed Technology" (Basu, P., ed.), pp. 263–272. Pergamon Press, Toronto, 1991.
Kolar, A. K., Grewal, N. S., and Saxena, S. C. *Int. J. Heat Mass Transfer* **22**, 1695–1703 (1979).
Leckner, B. *In* "Circulating Fluidized Bed Technology III" (Basu, P., Horio, M., and Hasatani, M., eds.), pp. 27–38. Pergamon Press, Oxford, 1990.

Li, J., Tung, Y., and Kwauk, M. *In* "Circulating Fluidized Bed Technology II" (Basu, P., and Large, P. eds.), pp. 193–204. Pergamon Press, Oxford, 1988.
Liu, D. C., Yang, H. P., Wang, Y. L., and Lin, Z. J. *In* "Circulating Fluidized Bed Technology III" (Basu, P., Horio, M., and Hasantani, M., eds.). pp. 275–281. Pergamon Press, Oxford, 1990.
Lu, H. L., Yang, L. D., Bao, Y. L., and Chen, L. Z. *In* "5th National Conference on Fluidization" (in Chinese), Beijing, China, pp. 164–167 (1990).
Mahalingam, M., and Kolar, A. K. *In* "Circulating Fluidized Bed Technology III" (Vasu, P., Horio, M., and Hasatani, M., eds.), pp. 239–248. Pergamon Press, Oxford, 1991.
Martin, H. *Chem. Ing. Tech.* **52**, 199–209 (1980).
Martin, H. *Chem. Eng. Prog.* **18**, 157–223 (1984).
Mickley, H., and Fairbanks, D. F. *AIChE J.* **1**, 374–384 (1955).
Mickley, H. S., and Trilling, C. A. *Ind. Eng. Chem.* **41**, 1135–1147 (1949).
Nag, P. K., and Moral, M. N. A. *In* "Circulating Fluidized Bed Technology III" (Basu, P., Horio, M., and Hasatani, M., eds.), pp. 253–262. Pergamon Press, Oxford, 1990.
Sekthira, A., Lee, Y. Y., and Genetti, W. E. *25th National Heat Transfer Conference, Houston,* 1988.
Shen Z., and Kwauk, M. "Mass transfer between gas and solid in fast fluidization," Institute of Chemical Metallurgy, Academia Sinica.
Subbarao, D., and Basu, P. *In* "Circulating Fluidized Bed Technology" (Basu, P., ed.), p. 281. Pergamon Press, Toronto, 1986.
Tung, Y., Zhang, W. L., Wang, Z. Z., Wen, X. L., and Qin, Y. *Chemical Engineering and Metallurgy* (in Chinese) **10**(2), 17–23 (1989).
Watanabe, T., Chen, Y., Hasatani, M., Xie, Y. S., and Naruse, I. *In* "Circulating Fluidized Bed Technology III" (Basu, P., Horio, M., and Hasatani, M. eds.), p. 283. Pergamon Press, Oxford, 1991.
Wu, R. J., Lim, J., Chaouki, J., and Grace, J. R. *AIChE J.* **33**, 1888–1893 (1987).
Wu, R. L., Grace, J. R., Lim, C. J., and Brereton, C. M. H. *AIChE J.* **35**, 1685–1691 (1989).
Wu, R. J., Grace, J. R., and Lim, C. J. *Chem. Eng. Sci.* **45**, 3389–3398 (1990).
Yang, Y. L., Jin, Y., Yu, Z., Wang, Z., and Bai, D. *In* "Circulating Fluidized Bed Technology III" (Basu, P., Horio, M., and Hasatani, M., eds.), pp. 201–206. Pergamon Press, Oxford, 1991.
Zhang, W. M., Wang, Z. M., and Chen, G. T. *Chinese National Conference on Chem. Eng.* (in Chinese) (1987).
Zigler, E. N., Koppel, L. B., and Brazelton, W. T. *Ind. Eng. Chem. Fundam.* **3**, 324–328 (1964).

6. POWDER ASSESSMENT

MOOSON KWAUK

INSTITUTE OF CHEMICAL METALLURGY
ACADEMIA SINICA
ZHONGGUAN VILLAGE, BEIJING, PEOPLE'S REPUBLIC OF CHINA

I. Geldart's Classification	241
II. Powder Characterization by Bed Collapsing	243
III. Modeling the Three-Stage Bed Collapsing Process	245
IV. Instrument for Automatic Surface Tracking and Data Processing	247
V. Qualitative Designation for Bed Collapsing	250
VI. Quantifying Fluidizing Characteristics of Powders	251
VII. Improving Fluidization by Particle Size Adjustment	253
VIII. Measure of Synergism for Binary Particle Mixtures	260
Notation	264
References	265

The characteristics of the powder being fluidized contribute as much as the design of the fast fluidized bed and the choice of its operating conditions to efficient particle–fluid contacting. While no tailor-made specifications are yet available to qualify a powder for superior operation in fast fluidization, a wide velocity range for the fast bed regime and a gradual, rather than abrupt, transition (such as choking for coarse powders) to transport are desirable features, as will be shown in Fig. 11 in this chapter. As a matter of fact, the need for such characteristics has been well known in fluid catalytic cracking, for which the microspheroidal catalyst particles have to be fabricated according to a strict regimen in order to ensure their performance. While such experience could well be adapted to other artificially prepared powder materials, it could also be extended to the choice of disintegration equipment and processes in the case of naturally occurring solids.

The efficiency of particle–fluid contacting in fluidization, popularly described in terms of the quality of fluidization, has its origin not only in the physical properties of the fluidizing medium and of the solid material of which the particles are composed, but also in the characteristics and in the group behavior of the particles while in motion. Particle characteristics include size, size distribution, shape, and surface roughness or texture, while particle group behavior refers to the interaction between a particle and its neighboring particles as well as the movement of the particles as a group. Particle group behavior could again be related to static parameters such as the angle of repose or the shear strength, as well as to dynamic parameters such as efflux time and flowability, many of which are understood only at a phenomenological–empirical level. Direct application of these readily available parameters to the description or characterization of fluidization is, however, difficult. What is needed are methods, both theoretical and instrumental, suited to particle–fluid systems, to properly correlate the quality of fluidization to these more basic, experimentally determinable parameters.

Early investigations were concerned mostly with physical properties, somewhat with particle characteristics but little with particle group behavior. Even so, significant results were obtained. For instance, the distinction between L/S fluidization and G/S fluidization, *viz.*, "particulate" and "aggregative," and the provision of criteria for such distinction (Wilhelm and Kwauk, 1948; Harrison *et al.*, 1961; Romero and Johanson, 1962), most of which were based on the Froude number. Other criteria were then proposed involving fluctuating parameters in fluidization (Rietema, 1967), for instance, pressure drop or voidage.

In all these correlations, however, parameters on particle characteristics and particle group behavior were hardly invoked. These investigations did lead, on the other hand, to the solutions of problems of considerable practical value, such as the careful specification of particle size distribution for artificially produced fluidizable materials, e.g., catalyst for fluid catalytic cracking, or graphite particles for heat treatment of metallic parts in fluid beds (Chen and Song, 1980; Jin, 1980; Jin and Zhao, 1984). Perfection of these artifacts guarantees a small velocity range above incipient fluidization where bubbling is essentially suppressed, and, within this small velocity range, the G/S system can be made to behave comparably to its L/S counterpart.

It was thus gradually recognized that the distinction between particulate and aggregative fluidization had ignored the finer details in regard to the not-so-good particulate fluidization for L/S systems, e.g., large and/or heavy particles such as lead in water for which liquid cavities similar to bubbles in G/S systems were observed, on the one hand, and the smoother type of aggregative fluidization for G/S systems, e.g., light and/or fine, especially size-graded particles which suppressed the formation of large bubbles, on the other. In other words, the particulate–aggregative transition should be viewed as continuous (Tung, 1981; Tung and Kwauk, 1982; Foscolo and Gibilaro, 1984; Geldart and Wong, 1985; Rowe, 1986) rather than abrupt.

Fluidization quality in terms of material properties, particle characteristics, and particle group behavior thus needs to be assessed on three scales: gross scale of the fluidized bed (macro scale), aggregate scale of gas bubbles, and particle clusters (meso scale), and scale of the discrete, individual particles (micro scale), as described in Chapter 4.

According to such methodology, powders can be classified in relation to the regimes in which they tend to perform in the fluidized state.

I. Geldart's Classification

Geldart (1972, 1973) classified powders with respect to their fluidizing characteristics into four groups, as already mentioned in Chapter 1, Section V, which are summarized in Table I in terms of typical particle size, particle characteristics, fluidizing behavior and bubble characteristics, and described again as follows.

Group A represents the best powders. They are fine and aeratable, such as cracking catalyst. They fluidize nicely, and expand particulately after reaching the point of incipient fluidization u_{mf}, until the first bubbles appear at a higher velocity, the minimum bubbling velocity u_{mb}. It is thus evident that the ratio of u_{mb}/u_{mf} is always greater than unity, and the greater is this value, the better the powder performs in fluidization.

Group B powders are of intermediate particle size, such as sand. They do not fluidize so smoothly as Group A, for bubbles form as soon as the incipient fluidization velocity is reached. It is thus evident that the ratio u_{mb}/u_{mf} is equal to unity. When the fluidizing gas is turned off suddenly, Group B powders would collapse immediately. In the fluidized state, the rising bubbles travel upward faster than the interstitial gas flow rate, and are therefore designated as "fast bubbles."

The next hardest to fluidize are the *Group C* powders. They are very fine powders, such as flour, for which cohesiveness due to interparticulate forces of one kind or another interferes with their fluidizing behavior. As a matter of fact, Group C powders channel rather than fluidize with gas flow. Consequently, the pressure drop is, more often than not, less than that required to support the weight of the solid particles. Fluidization of Group C powders may be improved by mechanical stirring or vibration, by the addition of small amounts of even finer particles such as fumed silica, or even of coarser particles, or by anti-electrification by conductive coatings, such as tin oxide or graphite.

The hardest to fluidize are the *Group D* powders. They are coarse particles such as wheat or soybeans. They can better be spouted than fluidized, and

TABLE I
GELDART'S CLASSIFICATION OF POWDERS

Group	Typical d_p	Particle Characteristics	Fluidizing Behavior	Bubble Characteristics

C	20	fine
		cohesive
		interparticle forces
		e.g., flour
			channel rather than fluidize
			pressure drop < particle weight
			fluidization improved by
			  stirring
			  vibration
			  addition of submicrons
			  anti-electrification	

A	30–100	fine aeratable e.g., cracking catalyst	appreciable range between u_{mf} and u_{mb}, $viz.$, $u_{mb}/u_{mf} > 1$ fluidize nicely	bubbles have clouds "fast bubbles"
B	100–800	intermediate size e.g., sand	bubble at incipient fluidization $viz.$, $u_{mb}/u_{mf} = 1$ collapse immediately upon shutting off gas flow	
D	1,000	coarse e.g., wheat	can be spouted mix poorly when fluidized appreciable particle attrition rapid elutriation of fines relatively sticky materials can be fluidized	bubbles cloudless "slow bubbles" bubble velocity less than interstitial gas velocity

when fluidized, the particles mix poorly. As with Group B powders, with Group D powders, bubbles form as soon as they start to fluidize, that is, $u_{mf}/u_{mf} = 1$. Bubbles formed in Group D powders move more slowly than the interstitial gas and are therefore known as "slow bubbles." With Group D powders, solids attrition is appreciable, and when mixed with fines, the smaller particles are elutriated rapidly. Because of their relatively large size, even sticky Group D materials may be fluidized.

Geldart found that the preceding classification could be charted quantitatively by two variables: the effective density of the particles ($\rho_s - \rho_f$) and their average surface–volume diameter d_{sv}, as shown in Fig. 1a. In Geldart's original chart for air, the boundary between Groups A and C is empirically determined and consists of a rather diffuse belt. The boundary between A and B is defined by the relation

$$(\rho_s - \rho_f)d_p = 0.225,$$

and the boundary between Groups B and D is

$$(\rho_s - \rho_f)d_p^2 = 1.$$

Grace (1986) improved Geldart's original diagram by incorporating data for gases other than air and for operations at pressures higher than atmospheric. The group boundaries proposed by Grace are based on the Archimedes number, as shown in Fig. 1b:

boundary C/A $Ar = 0.31–1.3$;

A/B $Ar = 1.03 \times 10^6 (\rho/\rho_f)^{-1.275}$;

B/D $Ar = 1.45 \times 10^6$.

II. Powder Characterization by Bed Collapsing (Relevant references include Tung, 1981; Tung and Kwauk, 1982; Tung et al., 1989.)

The spectrum of states of transition between particulate and aggregative fluidization calls on the one hand for devising experimental techniques for recording the related phenomenological manifestations and on the other for developing analytical procedures for quantifying these experimental findings in relation to the properties, characteristics and group behavior of the powders.

Broadly speaking, for G/S systems, three modes of particle–fluid contacting may be recognized to take place simultaneously, as shown in Fig. 2: bubbles containing sparsely disseminated particles, emulsion of densely suspended

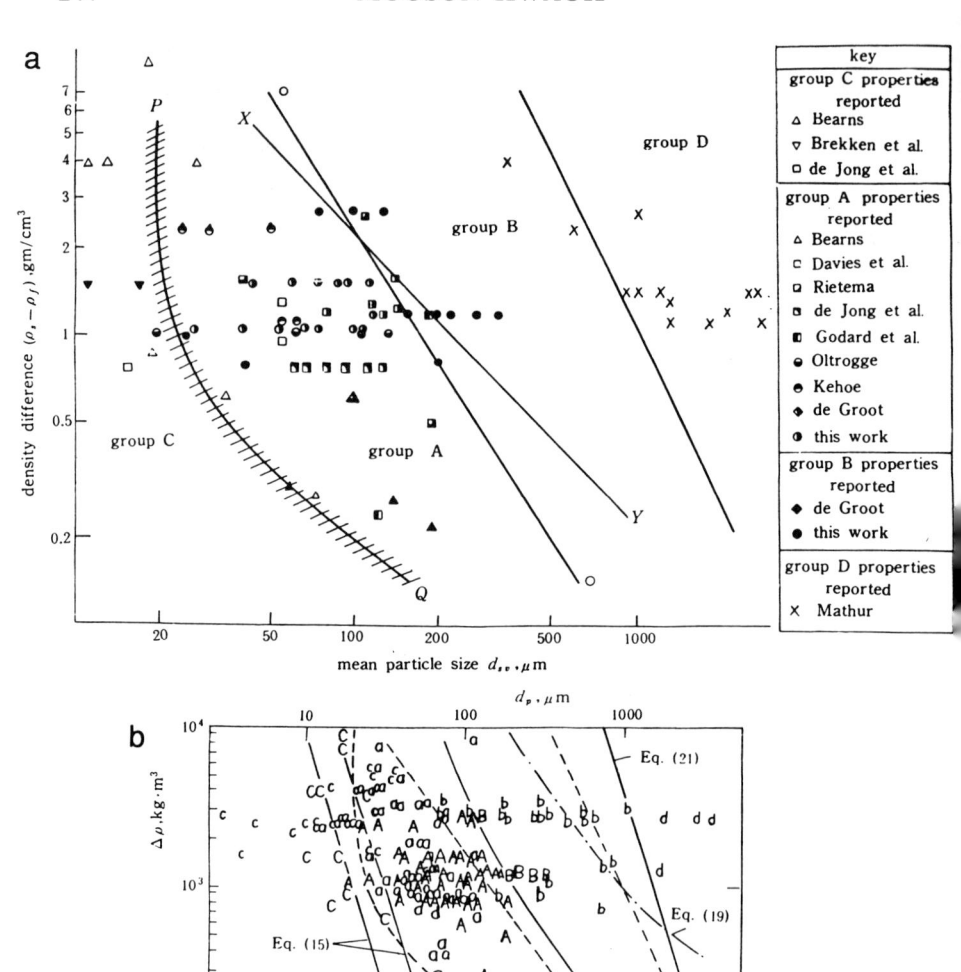

FIG. 1. Geldart's powder classification chart. (a) Geldart's original diagram for air (Geldart, 1973). (b) Grace's improvement on Geldart's diagram (Grace, 1986).

particles, and defluidized (transient as well as persistent) particles not fully suspended hydrodynamically by the flowing gas. When the gas fluidizing a powder, exhibiting all these modes of contacting, is turned off abruptly, the fluidized bed will collapse and subside in three consecutive stages:

1. a rapid initial stage for bubble escape,

POWDER ASSESSMENT 245

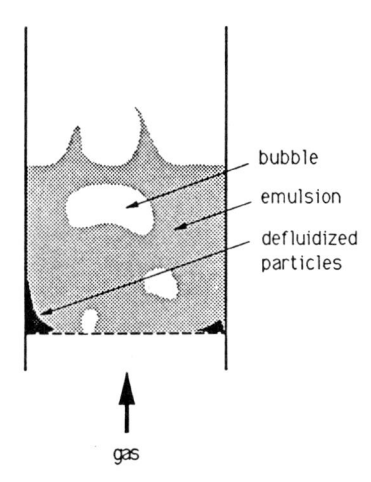

FIG. 2. The three modes of particle–fluid contacting in G/S fluidization.

2. an intermediate stage of hindered sedimentation with constant velocity of the dense emulsion of mobile particles, and
3. a final decelerating stage of solids consolidation for the incompletely suspended particles.

Thus, the random, spatial distribution of the three modes of particle–fluid contacting is transformed into the ordered, temporal sequence of the three stages of the sedigraph. The bed collapsing method has been used, too, for other objectives in studies on fluidization (Rietema, 1967; Morooka *et al.*, 1973).

III. Modeling the Three-Stage Bed Collapsing Process

Figure 3 shows the variation of the solids bed surface with time for the three-stage bed collapsing process. In this figure, (i) shows bubbles dispersed throughout the bed at $t < 0$. After the fluid is shut off, the bubbles ascend through the bed, until at $t = t_b$, a bubble-free dense phase B remains, as shown in (ii) and (iii). The initial stage of rapid bed collapse resulting from the depletion of bubbles is called the *bubble escape* stage. Thereafter, dense phase B collapse slowly at a constant rate. While layer B descends, the particles near the bottom of the bed begins to pile up, building up the accumulated layer D. This second stage will be called *hindered sedimentation*. The voidage throughout layer B remains constant during this stage, and at

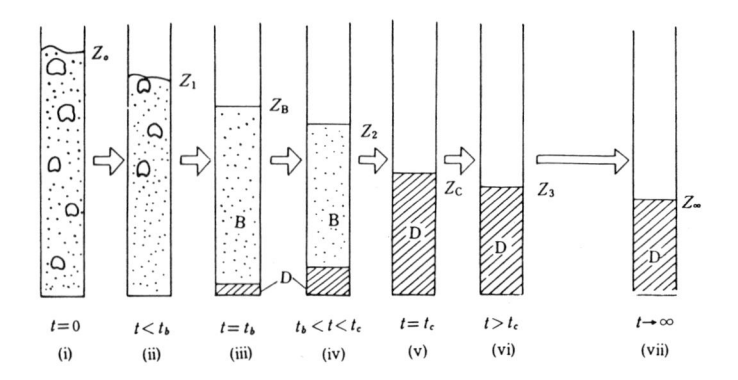

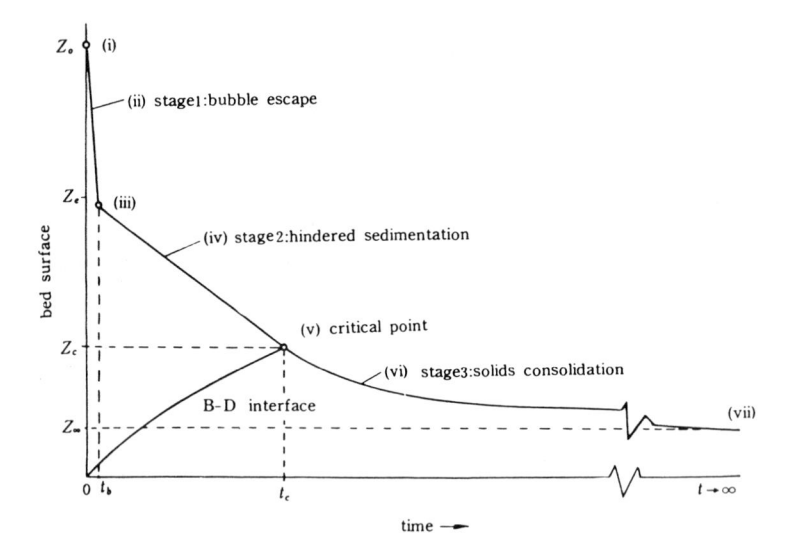

FIG. 3. Modeling the three-stage bed collapsing process (Yang et al., 1985).

$t = t_c$, shown as (v), layer B has just disappeared, all solids being now in layer D. The time which marks the end of this second stage will be denoted as the critical time t_c, and the corresponding voidage, the critical voidage ε_c. Layer D possesses a loose structure, with fluid in the interstitial spaces being slowly expelled by virtue of the weight of the accumulated solids. The rate of solids settling decreases with time, until at $t = \infty$, the ultimate bed height Z_∞ is reached as shown in (vii). The third stage is called *solids consolidation*.

Mathematical modeling of the bed collapsing process yielded the following results.

Bubble escape stage, $0 < t < t_b$:
Change of bed surface is linear with time:

$$Z_1 = Z_0 - u_1 t, \tag{1}$$

where

$$u_1 = \frac{f_B(1 + f_w \varepsilon_e) u_B + \varepsilon_e (1 - f_B - f_w f_B) u_e}{(1 - f_B)(1 - \varepsilon_e)}, \tag{2}$$

$$f_B = \frac{Z_0 - Z_e}{Z_0}. \tag{3}$$

Hindered sedimentation stage, $t_b < t < t_c$:
While layer B settles with a constant velocity $u_2 = dZ_2/dt$, layer D acquires solids from layer B, consolidating itself as a first-order process (Roberts, 1949):

$$-\frac{dZ}{dt} = K(Z_D - Z_\infty), \tag{4}$$

where K is a rate constant. The surfaces of layer B and layer D are, respectively,

$$Z_D = Z_B = Z_e - u_2 t, \tag{5}$$

$$Z_D = \left(\frac{Z_\infty}{Z_e - Z_\infty}\right) u_2 t + \left\{1 - \exp\left[-K\left(\frac{Z_e - Z_\infty}{Z_e}\right) t\right]\right\} \frac{u_2 Z_e}{(Z_e - Z_\infty) K}$$

$$\times \left[\frac{1 - \varepsilon_e}{\varepsilon_e - \varepsilon_c} - \frac{Z_e}{Z_e - Z_\infty}\right]. \tag{6}$$

Solids consolidation stage, $t_c < t < \infty$:
According to Eq. (4), the rate of bed surface descent for layer D is

$$\frac{dZ_3}{dt} = \frac{dZ_D}{dt} = -K(Z_D - Z_\infty), \tag{7}$$

which integrates to

$$Z_3 = Z_D = (Z_c - Z_\infty) \exp[-K(t - t_c)] + Z_\infty. \tag{8}$$

IV. Instrument for Automatic Surface Tracking and Data Processing

Figure 4 shows the instrument (Qin and Liu, 1986) used for automatic tracking of the surface of the collapsing bed, inclusive of a computer for data acquisition and on-line analysis. The fluidized bed is 5 cm in diameter and

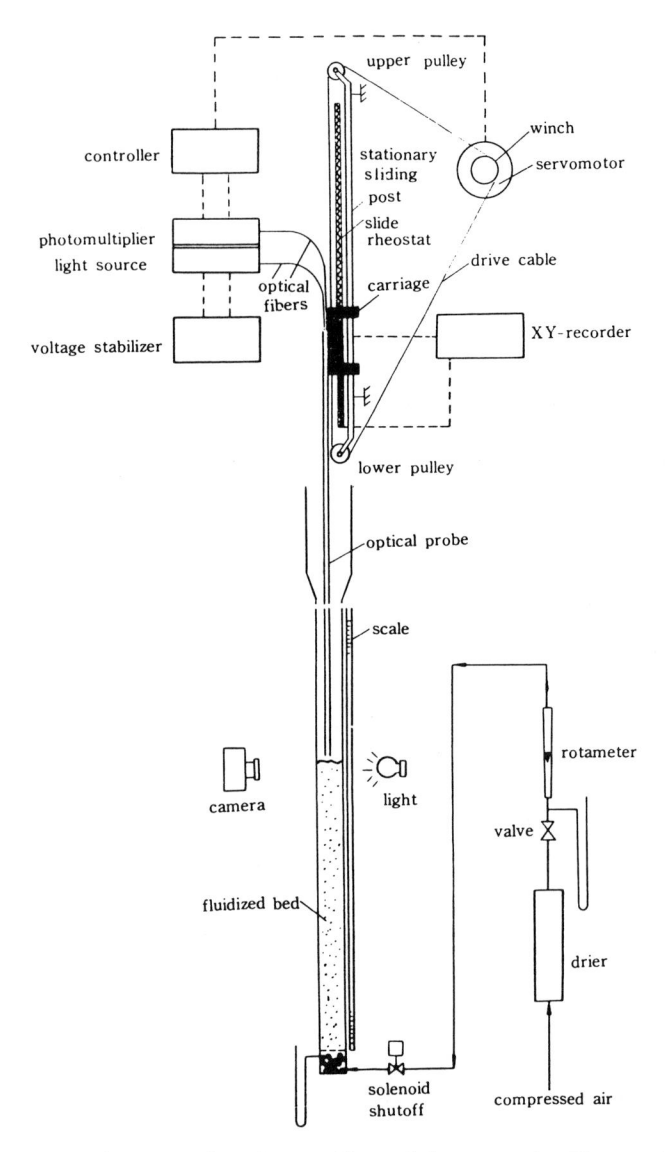

FIG. 4. Instrument for automatic surface tracking and data processing (Yang et al., 1985).

120 cm high, provided with a high pressure-drop gas distributor to insure uniform gas flow. Below the distributor is a specially designed knife valve operated by a solenoid for quick gas shutoff.

An optical-fiber probe, consisting of two separate sets of projector and receiver fibers, is used for rapid tracking and recording of the subsiding bed surface. The optical-fiber probe is mounted on a carriage which can be moved in a vertical direction on a stationary sliding post. The illuminating set of optical fibers at the lower tip of the probe projects light of constant luminosity into the space above the fluidized solids. When the probe tip is next to the solids bed surface, the reflected light is sent through the receiver fibers to a photomultiplier tube, the output of which, through proper electronic circuitry, drives a servomotor to position the tip of the probe automatically at an appropriate distance above the surface of the solids bed.

The carriage is provided with a slide rheostat which sends out a voltage proportional to the position of the carriage, to an xy-recorder which traces the subsidence curved simultaneously with the bed collapsing process. This same voltage is also fed to a microcomputer for data acquisition and on-line analysis.

Figure 5 shows a typical bed collapsing curve traced by the instrument just described.

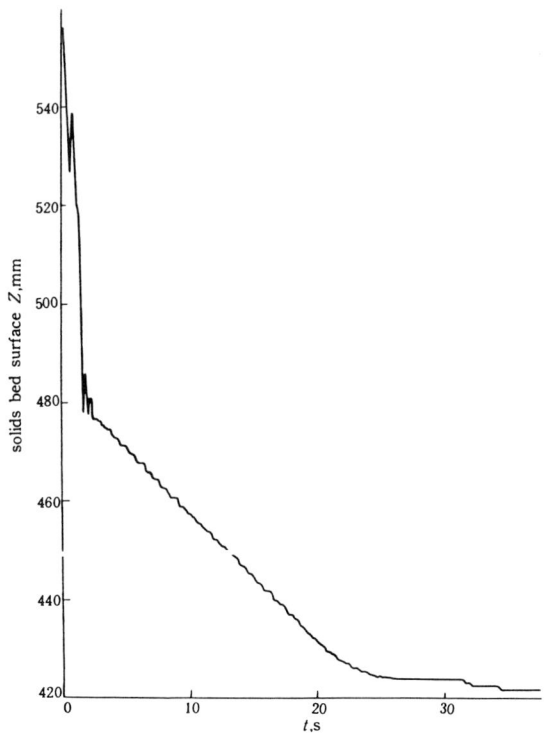

FIG. 5. Typical bed collapsing curve traced by optical-fiber probe tracing instrument (solids A66, alumina, 140–280 microns).

250   MOOSON KWAUK

The entire determination, from gas shutoff to printout from the computer, rarely exceeds three to four minutes for normal solids. Also, since all measurements are taken by the instrument, they are not subject to personal error of observation. The size of a solids test sample is of the order of a kilogram or less.

V. Qualitative Designation for Bed Collapsing

Not all powders exhibit all three of the stages described in Section III for the bed collapsing process. For G/S systems, the full three-stage curves obtain only for Geldart Group A powders and their like, as shown in Fig. 6. For Geldart Group B and D powders, which fluidize aggregatively at all gas velocities above incipient fluidization, the bed collapsing curve consists only of the first stage showing that the solids reach their final static bed height as soon as the bubbles are expelled. For L/S systems, on the other hand, the first stage is generally absent. For Geldart B and D powders, neither does the third stage appear.

These powders can be classified qualitatively by assigning appropriate three-digit designations, or descriptors, in accordance with the presence or absence of particular states in the bed collapsing test, as shown in Table II.

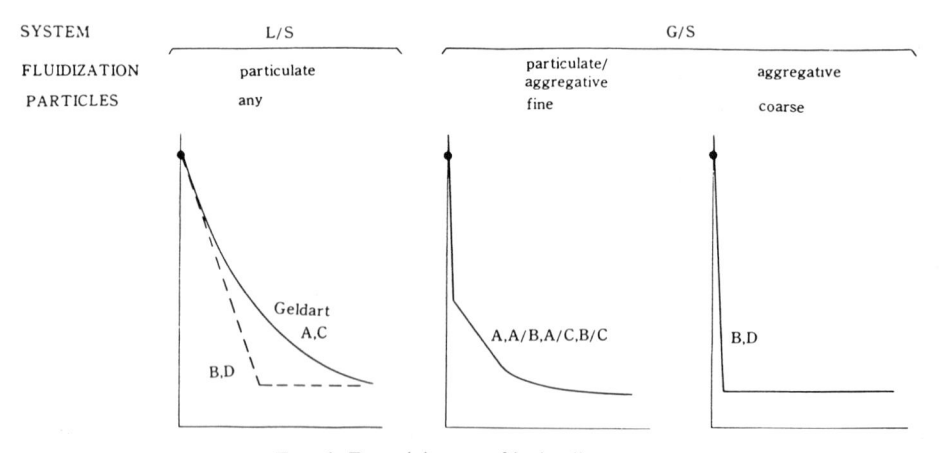

SYSTEM	L/S	G/S	
FLUIDIZATION	particulate	particulate/aggregative	aggregative
PARTICLES	any	fine	coarse

FIG. 6. Essential types of bed collapse curves.

VI. Quantifying Fluidizing Characteristics of Powders

The three-digit designation affords a convenient descriptive classification of powders in terms of their bed collapsing behavior, but some additional criterion is desirable to show the entire spectrum of continuous transition of fluidized states from particulate to aggregative.

Examination of the multitude of bed collapsing curves obtained experimentally led to the recognition of the significance of a number of variables, *viz.*, the difference between the height of the dense phase and the bed height at the critical point $(Z_e - Z_c)$, which defines the extent of particulate contraction after bubbles have escaped; the rate of hindered sedimentation u_2; and certain physical properties of the solids and the fluid involved. These factors are organized into a dimensionless number

$$\theta = \frac{d_p g}{u_t u_2}\left(\frac{Z_e - Z_c}{Z_\infty}\right). \qquad (9)$$

If the particles are small, and their terminal velocity can be expressed by Stokes' Law, $u_t = d_p^2(\rho_s - \rho_f)g/\mu$, then the dimensionless number is simplified to

$$\theta = \frac{t_c}{d_p(\rho_s - \rho_f)Z_\infty}. \qquad (10)$$

This dimensionless number, designated as the dimensionless subsidence time of a powder, portrays its dynamic behavior in relation to the overall bed collapsing process, inclusive of the significant variables Z_∞ and t_c.

To test the viability of θ in quantifying fluidizing characteristics, it is plotted against the ratio of incipient bubbling velocity to incipient fluidization velocity, u_{mb}/u_{mf}, the latter being calculated after Geldart. Figure 7 shows that a linear relation exists between $\ln(u_{mf}/u_{mf})$ and $\theta^{1/4}$ as represented by the following empirical relation:

$$\ln(u_{mb}/u_{mf}) = 4\theta^{1/4}. \qquad (11)$$

The straight line starts from $\theta = 0$ and $u_{mf}/u_{mf} = 1$, and extends without limit towards the upper right-hand corner. The value of $u_{mb}/u_{mf} = 1$ obviously signifies aggregative fluidization. The corresponding value of θ is zero, that is, t_c approaches zero, indicating that stages 1 and 2 of the bed collapsing process take place almost instantaneously. As the fluidizing characteristics improve, the value of u_{mb}/u_{mf} becomes progressively greater than unity, signifying particulate expansion. The corresponding value of θ also increases, showing a slower bed collapsing process accompanied by a large value for t_c. As the curve tends toward even larger values of θ, the characteristics of particulate fluidization become more predominant.

TABLE II
Three-Digit Designation of Bed Collapsing Curves

System		G/S					L/S		
Geldart's group	D,B	B/A	A	C		D	B	A,C
ICM designation	100	120	123	123		120	020	023
		rare					ideal	
Stage in bed collapse									
1. Bubble escape	×		×	×		×		×
2. Hindered sedimentation		×	×	×		×	×	×
3. Solids consolidation			×	×				
Fluidization characteristics								
Particulate		×	×	×		×	×	
Bubbling	×	×	×	×				
Channeling				×				×
Value of θ								
High								
Medium			×				×	
Low		×		×				
Near zero	×	×		×				

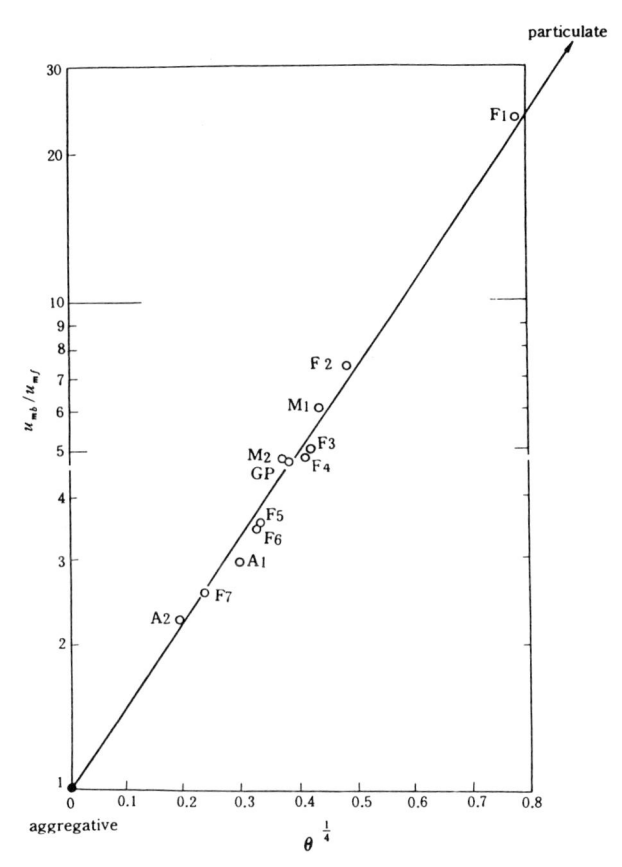

FIG. 7. Plot of u_{mb}/u_{mf} vs. $\theta^{1/4}$ (Yang et al., 1985).

VII. Improving Fluidization by Particle Size Adjustment

It was mentioned earlier in the chapter that catalyst powders with carefully specified particle size distribution have been prepared to ensure good fluidization characteristics. It is well known that even addition of fine particles to a powder of coarser particles tends to improve its fluidization characteristics. Experiments were thus conducted on binary particle mixtures, each consisting of a fairly close particle size distribution (Yang, Z., 1982).

Figure 8a shows a set of bed collapsing curves for a Geldart Group A-A binary solids mixture, two closely sized alumina powders of average particle diameter 104 and 66 microns, respectively. The curve on the extreme left

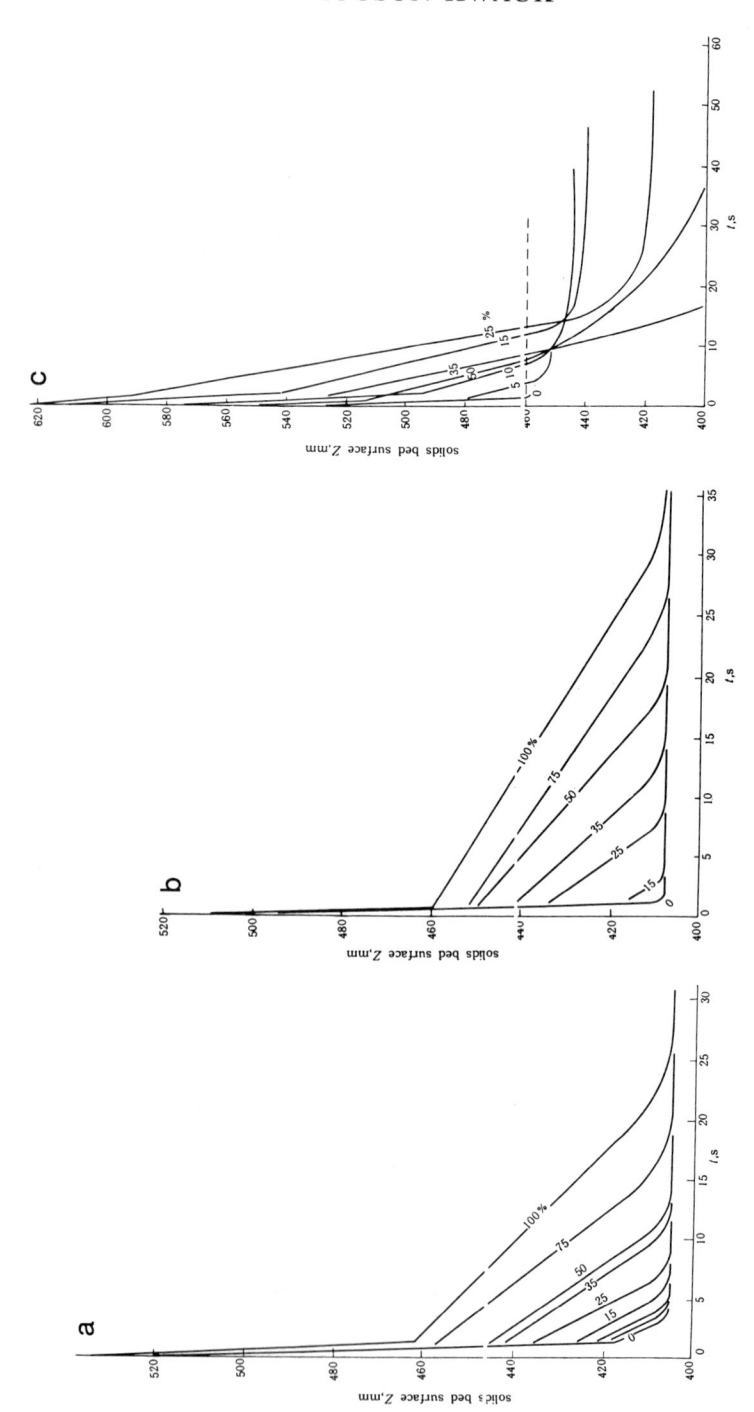

FIG. 8. Bed collapse curves for solid pairs. (a) Geldart A-A; solids A104/A66. (b) Geldart B-A; solids PE201/F27. (c) Geldart A-C; solids SA165/17 (Yang et al., 1985).

with 0% fines represents the pure coarse component, which is barely Group A in fluidizing characteristics, as can be seen from its very brief stage 2. The curve on the extreme right, representing the 100% fine component, demonstrates pronounced Group A, or "123," fluidizing characteristics with a long stage 2. Curves with intermediate compositions are shown in their ordered locations between the purely coarse and purely fine components.

Figure 8b illustrates the fluidizing characteristics of a set of Group B-A mixtures of relatively coarse (212 microns) polyethylene spheres mixed with a relatively fine (27 microns) FCC catalyst. The pure polyethylene spheres, shown as the 0% curve, demonstrate the "100" fluidizing characteristics typical of a Group B powder. As the fine FCC catalyst is added, there is a progressive predominance of the Group A fluidizing characteristics of the finer component.

Figure 8c shows the effect of adding talc (7 microns), Group C, to a coarse sea sand (165 microns), Group B. The beneficial effect of the fines on the fluidizing characteristics is shown to increase to a fines composition of around 25%, and then it diminishes with further fines addition.

Figure 9a plots the dimensionless subsidence time θ for six sets of Group A-A and Group B-A binary mixtures for different compositions x_f, showing that the improvement of fluidizing characteristics by addition of fine particles increases monotonically with increasing percentage of the fines.

Figure 9b is a similar plot, but for 10 sets of Group B-C mixtures. Most curves show that the values of θ for Group B-C binaries give maxima at certain intermediate compositions. The presence of these maxima suggests that not only may fine particles belonging to Group C improve the fluidizing characteristics of such coarse solids as Group B, but the coarse particles may also improve the fluidizing characteristics of the fine particles, which are known to possess the notorious tendency towards channelling before fluidization sets in.

This synergism (Yang, Z., 1982; Kwauk, 1984, 1986) of Group B-C mixtures testifies to the significance of particle size selection and particle size distribution design, in order to tailor a solid particulate material to certain desired fluidizing characteristics. The presence and absence of synergism have been found for ternary solid particle mixtures, too, as shown in Fig. 10 (Zheng, 1987).

As for additives, Zheng found not only that the fines needed to approach submicron size, but that their shape was also critical: fibrous, if possible. Early mention of fluidization aid seemed to favor ultrafine materials of pyrogenic origin (Kuhn, 1963), without reference to shape, however. Recent studies by Brooks and Fitzgerald (1985, 1986) showed near particulate fluidization with a tendrillar carbonaceous material. A new fibrous catalyst carrier having a porous structure, Aerogel (Chaouki *et al.*, 1985) seemed to show promise as another candidate for improving fluidization.

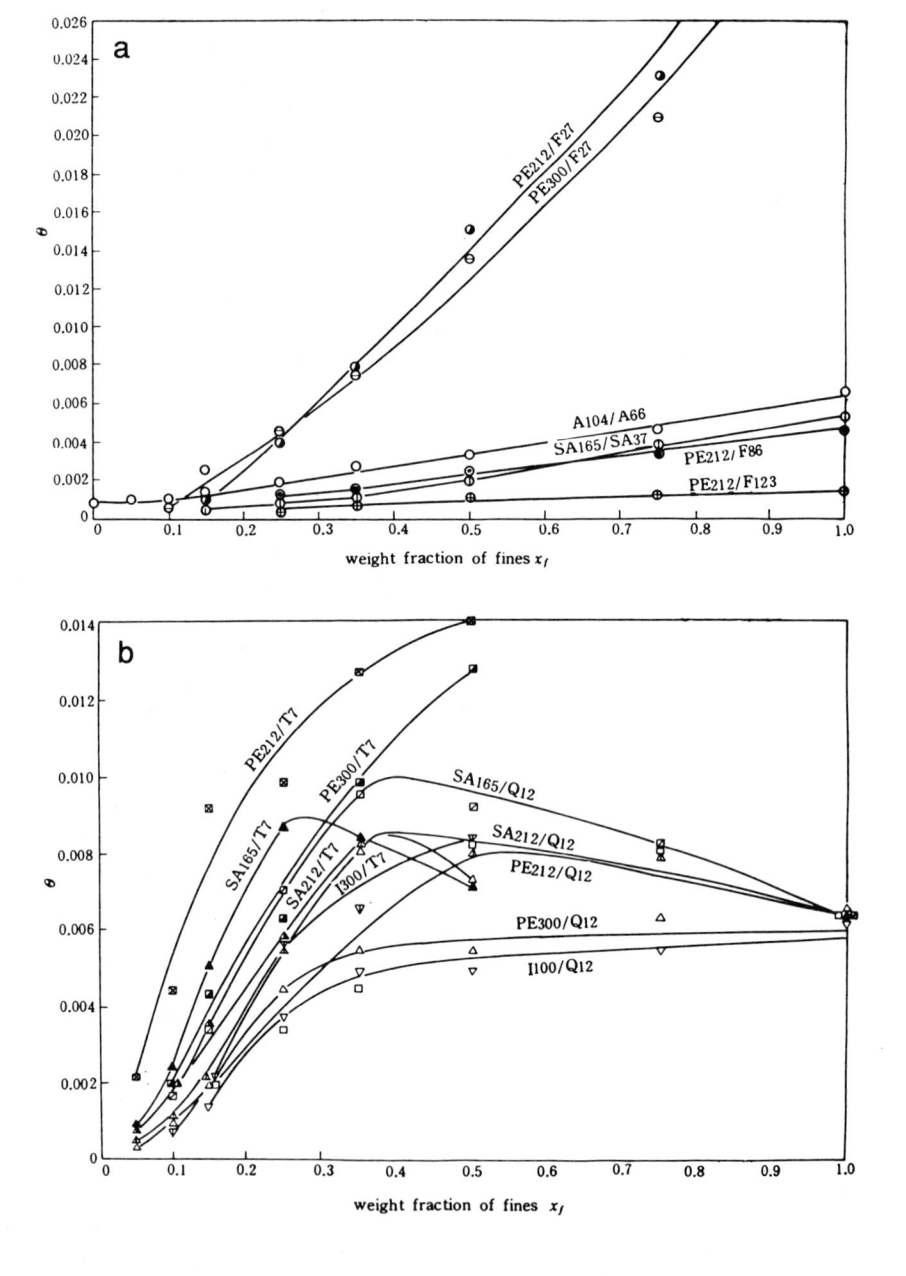

FIG. 9. Improving fluidizing characteristics by fines addition. (a) Geldart A-A and B-A; (b) Geldart A-C and B-C (Yang et al., 1985).

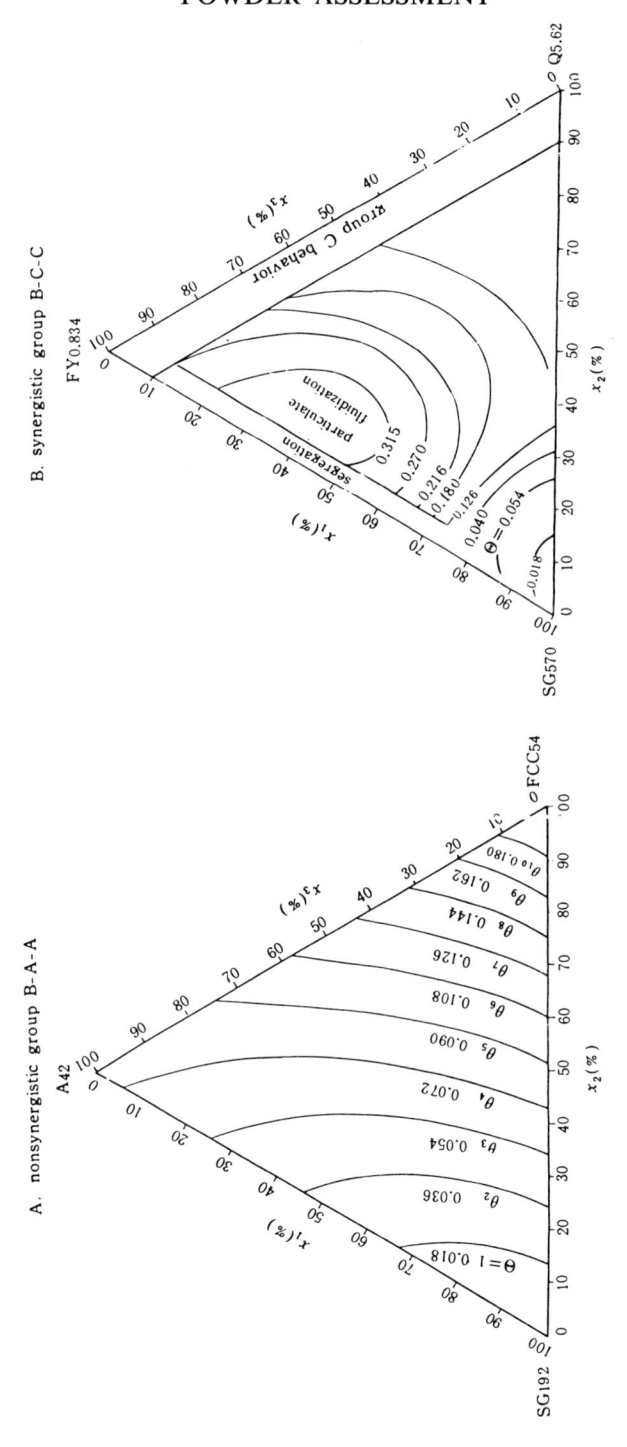

FIG. 10. Constant-θ diagrams for ternary particle mixtures (Zheng, 1987).

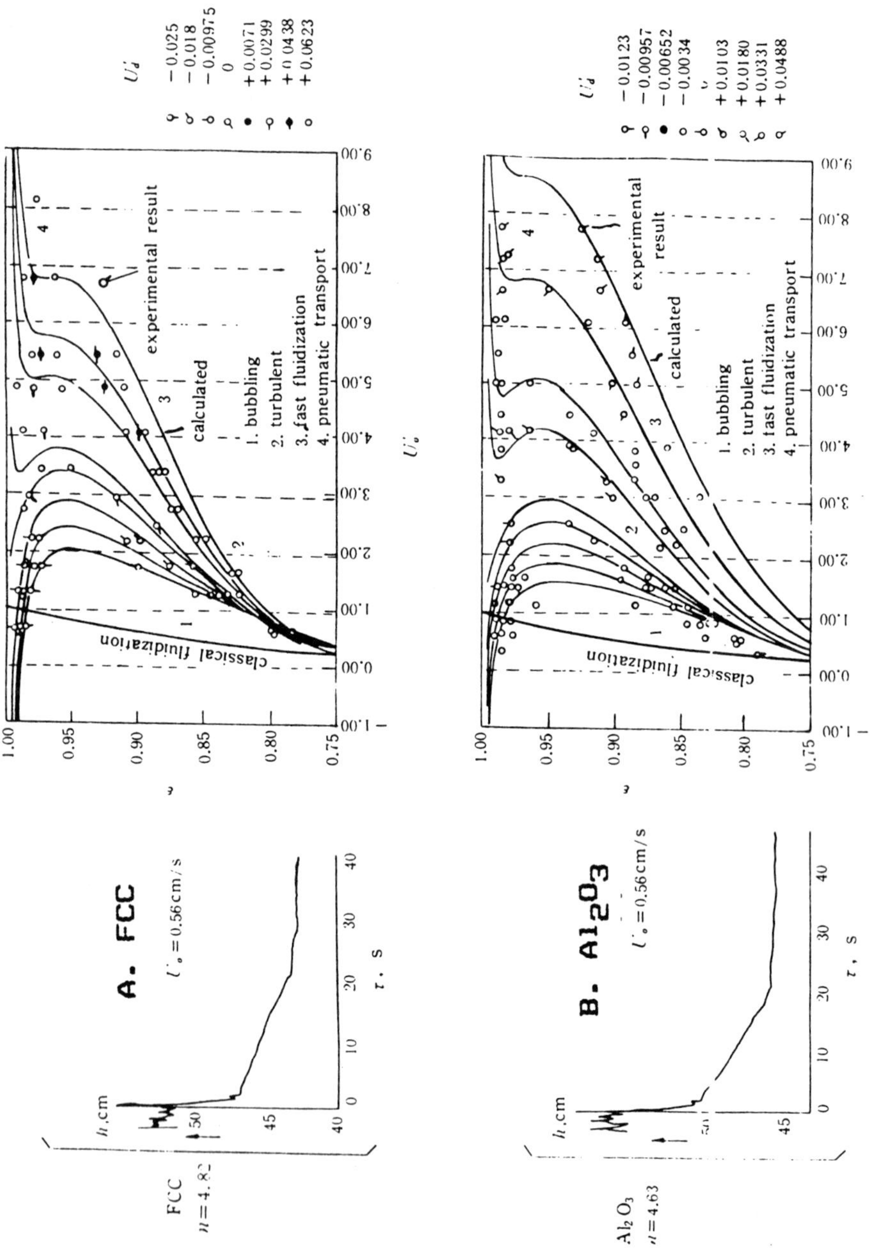
A. FCC
U_o = 0.56 cm/s
h, cm
τ, s
FCC
n = 4.82
ε
U_o'
U_d'
−0.025
−0.018
−0.00975
0
+0.0071
+0.0299
+0.0438
+0.0623
experimental result
calculated
1. bubbling
2. turbulent
3. fast fluidization
4. pneumatic transport
classical fluidization
B. Al2O3
U_o = 0.56 cm/s
Al2O3
n = 4.63
−0.0123
−0.00957
−0.00652
−0.0034
+0.0103
+0.0180
+0.0331
+0.0488

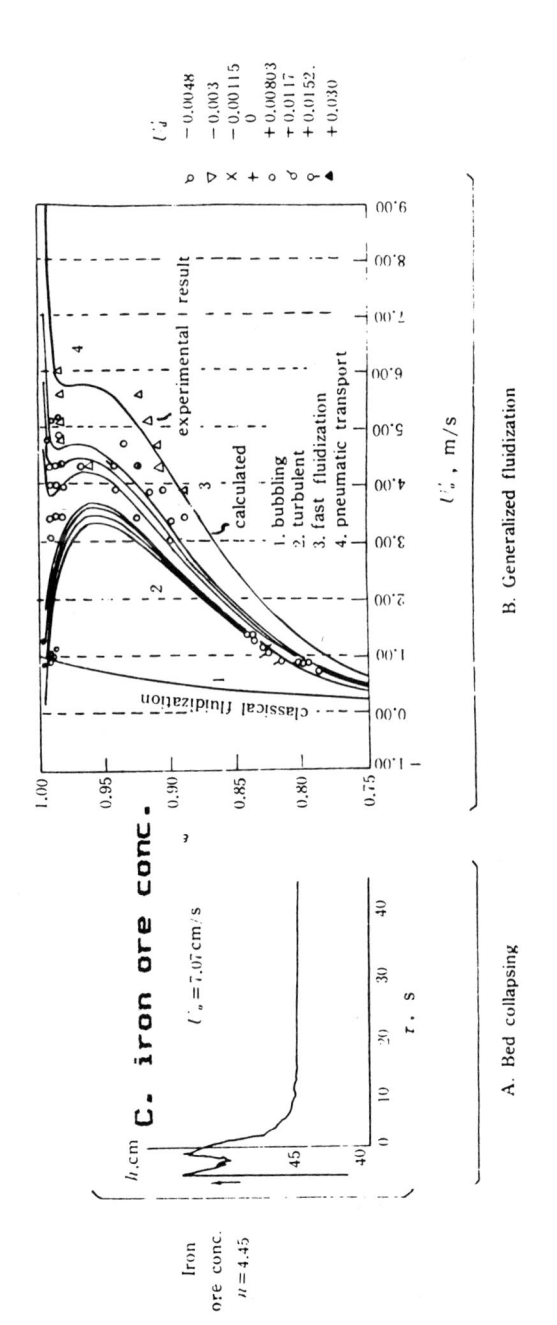

FIG. 11. Charting G/S generalized fluidization against experimental data (Kwauk et al., 1985).

Synergism caused by the addition of Group C particles was found from both SEM microscopy and video macroscopy (Xia and Kwauk, 1985) to be associated with adhesion of these fine particles on the coarse particle. Nevertheless, more appropriate analytical, or even theoretical, ground needs to be found to account for the initial rise and later fall in the value of the dimensionless time θ as the mass fraction of fines increases from zero to unity.

Figure 11 compares the generalized fluidization charts on the right-hand side of experimental data for both co-up and counter-down fast fluidization for three solids—FCC catalyst, alumina, and iron ore concentrate—with the bed collapsing curves of the solids on the left. The fluidizing behaviors of the three powders along the left-hand side are well reflected in the shapes of the families of ε–u curves for generalized fluidization along the right-hand column. Both FCC catalyst and alumina, which are Geldart A or "123" in nature, exhibit extended ranges for fast fluidization, while for iron ore concentrate, which is weakly Geldart B, or essentially "100," bubbling is followed fairly closely by pneumatic transport in the dilute phase, with a rather short intermediate range of fast fluidization.

VIII. Measure of Synergism for Binary Particle Mixtures

A mathematical model has been proposed to account for the mutual synergistic action of either particle component on the other in increasing the value of the dimensionless time θ as shown in Fig. 9b. Thus, the dimensionless time θ_1 for the coarse particles could be assumed to exert a mass-fraction-based influence on the fine particles, proportional to $\theta_1(1 - x_2)$, which is affected by certain interaction by the fines, inclusive of their ability to adhere to the surface of the coarse and form clusters among themselves, lumped in certain appropriate form, for instance, $[1 + f(x_2)]$, where the function $f(x_2)$ may again be assumed to possess certain appropriate form, for instance, exponential, $x_2^{n_1}$, where n_1 may be called the interactive exponent. This results in an overall contribution by the coarse particles, suitably corrected for the interaction of the fine particles, $\theta_1(1 - x_2)(1 + x_2^{n_1})$. This function has the property of accommodating the following boundary conditions:

$$\text{equals } \theta_1 \quad \text{at } x_2 = 0,$$
$$0 \quad \text{at } x_2 = 1.$$

Similarly, the mass-fraction-based contribution of the fine particles would be proportional to $\theta_2 x_2$, and the interaction of the coarse particles, for instance, in breaking up lumps and opening up channels of the otherwise hard-to-fluidize fine particles could be assumed to possess a similar form $[1 + (1 - z_2)^{n_2}]$. Thus, the overall contribution by the fine particles, $\theta_2 x^2[1 + (1 - x_2)^{n_2}]$, will be noted to satisfy the following boundary conditions:

$$\text{equals } \theta_2 \quad \text{at } x_2 = 1,$$
$$0 \quad \text{at } x_2 = 0.$$

Also, the overall dimensionless time θ for the binary system will consist of the algebraic sum of the contributions from the component particles:

$$\theta = \theta_1(1 - x_2)(1 + x_2^{n_1}) + \theta_2 x_2[1 + (1 - x_2)^{n_2}]. \tag{14}$$

At the maximum value of $\theta = \theta_m$, $x_2 = x_m$, and the first derivative of θ with respect to x_2, $d\theta/dx_2 = 0$, giving

$$\frac{\theta_1}{\theta_2} = \frac{n_2 x_2(1 - x_2)^{n_2-1} - (1 - x_2)^{n_2} - 1}{-(n_1 + 1)x_2^{n_1} + n_1 x_2^{n_1-1} - 1}. \tag{15}$$

Mutual synergism of binary mixtures containing fine particles can be quantified in terms of the departure of the θ–x_2 curve from a linear tie line, which signifies absence of synergism, joining θ_1 and θ_2, as shown in Fig. 12. A convenient measure of this departure is the area lying between the θ–x_2 curve and the linear tie line. This can be derived analytically from Eq. (13) in terms of what will be called the synergism number Sy, normalized with respect to $\theta_1 + \theta_2$ of both the coarse and the fine particles:

$$\text{Sy} = \frac{1}{\theta_1 + \theta_2}\left[\int_0^1 \theta \, dx - \frac{1}{2}(\theta_1 + \theta_2)\right]$$
$$= \frac{1}{(n_1 + 1)(n_1 + 2)} + \frac{\theta_2/\theta_1}{(n_2 + 1)(n_2 + 2)}. \tag{16}$$

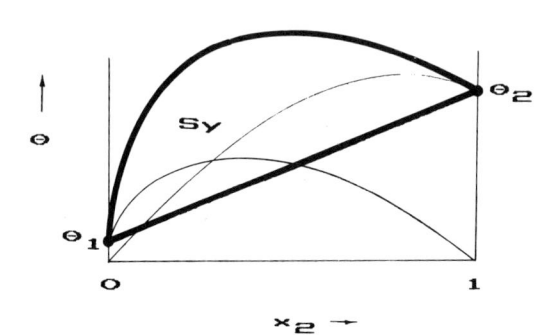

FIG. 12. Measure of synergism for binary particle mixtures (Kwauk, 1984, 1986).

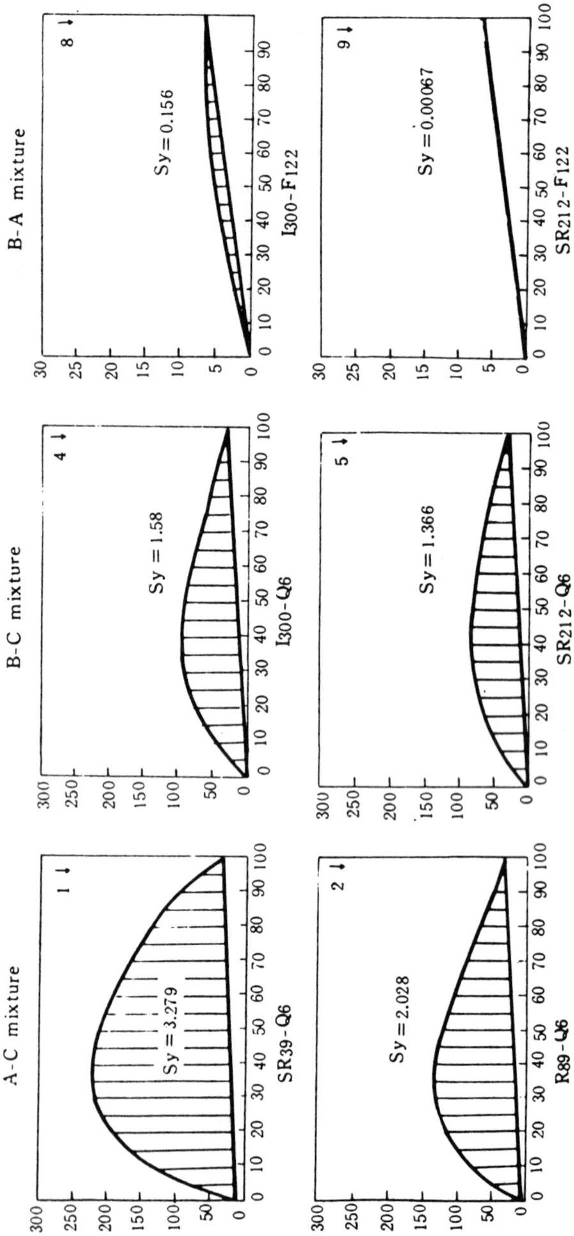
A-C mixture
B-C mixture
B-A mixture
1
2
4
5
8
9
Sy = 3.279
Sy = 2.028
Sy = 1.58
Sy = 1.366
Sy = 0.156
Sy = 0.00067
SR39-Q6
R89-Q6
I300-Q6
SR212-Q6
I300-F122
SR212-F122

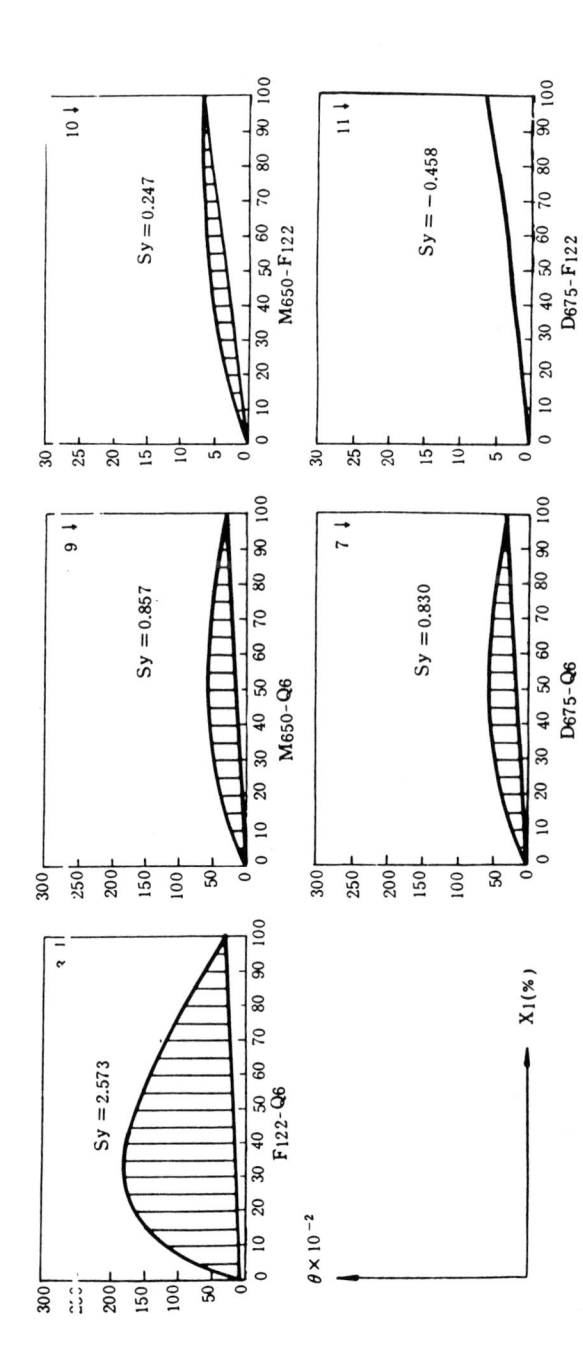

FIG. 13. Computer-generated curves showing synergism for different binary particle mixtures (Qian and Kwauk, 1986).

The larger is the value of Sy, the stronger is the mutual synergistic interaction between the coarse and fine particles. Also, an effective fine particle additive to improve the fluidizing characteristics of coarse particles calls for a large value of θ_m produced with minimal amount of the fine material, that is, a small value of x_2.

Figure 13 shows more computer-generated curves (Qian and Kwauk, 1986) showing synergism for different binary particle mixtures.

For synergism in ternary particle mixtures, the corresponding subsidence time has been expressed (Zheng, 1987) as

$$\theta = \theta_1 x_1^a [1 + x_1]^{n_{11}} [1 + x_2]^{n_{12}} [1 + x_3]^{n_{13}}$$
$$+ \theta_2 x_2^a [1 + x_1]^{n_{12}} [1 + x_2]^{n_{22}} [1 + x_3]^{n_{23}}$$
$$+ \theta_3 x_3^a [1 + x_1]^{n_{13}} [1 + x_2]^{n_{23}} [1 + x_3]^{n_{33}}.$$

It can be appreciated that regression of experimental data to obtain the constants n_{ij} would be a formidable problem when the number of components increases. Therefore, improved modeling is indicated.

Notation

Ar	$= d_p^3 \rho_f g \rho / \mu^2$, Archimedes number, dimensionless	u_B	superficial fluid velocity in bubble, cm/s
d_p	particle diameter, cm	u_e	superficial fluid velocity in emulsion, cm/s
d_{sv}	surface–volume particle diameter, cm	u_{mb}	superficial fluid velocity at incipient bubbling, cm/s
g	$= 980$ cm/s^2, acceleration of gravity	u_{mf}	superficial fluid velocity at incipient fluidization, cm/s
n_1	exponent for influence of coarse particles, dimensionless	u_t	terminal particle velocity, cm/s
n_2	exponent for influence of fine particles, dimensionless	x_f	weight fraction of fines in binary particle mixture, dimensionless
t	time, s	Z_1	height of bed surface, cm
t_b	time when all bubbles have escaped in bed collapsing, s	Z_2	bed height for layer B, cm
		Z_3	bed height for layer D, cm
t_c	critical time, or time at end of second stage in bed collapsing, s	Z_D	bed height for layer D, cm
		Z_0	initial bed height, cm
		Z_∞	ultimate bed height, cm

ε	voidage or void fraction, dimensionless		time in bed collapsing test
		μ	viscosity, g/cm-s
ε_c	voidage at end of second stage in bed collapsing, dimensionless	ρ_s	density of solid, g/cm^3
		ρ_f	density of fluid, g/cm^3
θ	dimensionless subsidence	ρ	$= \rho_s - \rho_f$, effective density, g/cm^3

References

Brooks, E. F., and Fitzgerald, T. J. "Aggregation and fluidization characteristics of a fibrous carbon," 7th Annual Meeting, A.I.Ch.E., Chicago, 1985.

Brooks, E. G., and Fitzgerald, T. J. "Fluidization of novel tendriller carbonaceous materials," in "Fluidization V" (Østergaard, K., ed.), pp. 217–224. Elsinor, 1986.

Chaouki, J., Chavarie, C., Klvane, D., and Pajonk, G. "Effect of interparticle forces on the hydrodynamic behavior of fluidized aerogels," *Powd. Technol.* **43**, 117 (1985).

Chen, M., and Song, T. "Fluidized graphite-particle furnace," in "Second National Fluidization Conference, Beijing, Dec. 1980, Proceedings," pp. 119–122 (in Chinese) (1980).

Foscolo, P. U., and Gibilaro, L. G. "A fully predictive criterion for the transition between particulate and aggregative fluidization," *Chem. Eng. Sci.* **39**, 1667–1675 (1985).

Geldart, D. "The effect of particle size and size distribution on the behavior of gas-fluidized beds," *Powder Technol.* **6**, 201–205 (1972).

Geldart, D. "Types of fluidization," *Powder Technol.* **7**, 285–290 (1973).

Geldart, D., and Wong, A. C. Y. "Fluidization of powders showing degrees of cohesiveness—I. Bed expansion. II. Experiments on rates of de-aeration," *Chem. Eng. Sci.* **39**, 1481–1488 (1984); **40**, 653–661 (1985).

Grace, J. R. "Contacting modes and behavior classification of gas–solid and other two-phase suspensions," *Can. J. Chem. Eng.* **64**, 353–363 (1986).

Harrison, D., Davidson, J. F., and deKock, J. W. "On the nature of aggregative and particulate fluidization," *Trans. Instn. Chem. Engrs.* **39**, 202 (1961).

Jin, H. "Particles and distributor for fluid-bed heat treatment furnace," 2nd National Fluidization Conference, Beijing, Dec. 1980, p. 109.

Jin, H., and Zhao, H. "Fluidized electric furnace for heat treatment," 3rd National Fluidization Conference, Taiyuan, April 1984, p. 336.

Kuhn, W. E. (ed.) "Ultrafine Particles," pp. 538–541. Electrochemical Society, New York, 1963.

Kwauk, M. "Behavior of Binary Particulates, Note 1, 84-11-1; Note 2, 86-10-20." Unpublished (1984, 1986).

Kwauk, M., Wang, N., Li, Y., Chen, B., and Shen, Z. "Fast Fluidization at ICM," *Proc. 1st Intern. Conf. Circulating Fluidized Bed, Halifax, Canada*, pp. 33–62 (1985).

Morooka, S., Nishinaka, M., and Kato, Y. "Sedimentation velocity and expansion ratio of the emulsion phase in a gas–solid fluidized bed" (in Japanese), *Kagaku Kogaku* **37**, 485 (1973).

Qian, Z., and Kwauk, M. "Computer application in characterizing fluidization by the bed collapsing method," Tenth Intern. CODATA Conf., Ottawa, July 1986.

Qin, S., and Liu, G. "Automatic surface tracking for collapsing fluidized bed," in "Sec. China–Japan Fluidization Symp., Kunming, 1985," p. 468. Elsevier, 1985.

Rietema, K. "The effect of interparticle forces on the expansion of a homogeneous gas-fluidized bed," *Proc. Intern. Symp. Fluidization, Toulouse*, pp. 28–40 (1967).

Roberts, E. J. "Thickening — art or science?" *Min. Eng. Min. Trans.* **1**, 61 (1949).
Romero, J. B., and Johanson, L. N. "Factors affecting fluidized bed quality," *Chem. Eng. Prog. Symp. Ser.* **58**(38), 28–37 (1962).
Rowe, P. N. "A rational explanation for the behavior of Geldart Type A and B powders when fluidized," *Dept. Chem. & Biochem. Eng., Univ. Coll. London*, April 1986.
Tung, Y. "Dynamics of bed collapse for fluidization systems," M.S. thesis, Inst. Chem. Metall. (1981).
Tung, Y., and Kwauk, M. "Dynamics of collapsing fluidized beds," *in* "China–Japan Fluidization Symp., Hangzhou, April 4–9, 1982," pp. 155–166. Science Press, Beijing, Gordon and Breach, 1982.
Tung, Y., Yang, Z., Xia, Y., Zheng, W., Yang, Y., and Kwauk, M. "Assessing fluidizing characteristics of powders," Sixth International Fluidization Conference, Banff, Canada, May 7–12, 1989.
Wilhelm, R. H., and Kwauk, M. "Fluidization of solid particles," *Chem. Eng. Prog.* **44**, 201–218 (1948).
Xia, Y., and Kwauk, M. "Micro-visualization of fluidizing behavior of binary particle mixtures," *in* "Intern. Symp. Heat Transfer, Beijing, 1985," Hemisphere, 1985.
Yang, Z. "Effect of particle size distribution on fluidization quality," M.S. thesis, Inst. Chem. Metall. (1982).
Yang, Z., Tung, Y., and Kwauk, M. "Characterizing Fluidization by the bed collapsing method," *Chem. Eng. Commun.* **39**, 217–232 (1985).
Zheng, W. "Influence of primary particle properties on fluidization quality," M.S. thesis, Institute of Chemical Metallurgy (1987).

7. HARDWARE DEVELOPMENT

LI HONGZHONG

INSTITUTE OF CHEMICAL METALLURGY
ACADEMIA SINICA,
ZHONGGUAN VILLAGE, BEIJING, PEOPLE'S REPUBLIC OF CHINA

I. The V-Valve	267
A. Mechanics	268
B. The Fluidized Spout	269
C. The Nonfluidized Spout	274
D. The Fluidized Dipleg	279
E. The Pneumatically Controlled Downcomer	281
II. Nonfluidized Diplegs	291
A. A Generalized Theory for the Dynamics of Nonfluidized Gas–Particle Flow	292
B. Sealing Ability for Moving Beds	302
C. Fluid–Particle Flow Phase Diagram	302
D. The Ideal Sealing State and Its Application	305
E. Bridging	307
III. The Pneumatically Actuated Pulse Feeder	314
IV. Internals	315
V. Multi-layer Fast Fluidized Beds	320
VI. Integral CFB	322
A. Characteristics	322
B. The Circular V-Valve	323
C. Multi-inlet Cyclone	325
Notation	327
References	329

I. The V-Valve

The downcomer (or dipleg, or standpipe) through which solids move from bed to bed is a key component of a circulating or a multi-stage fluidized bed. It is often connected at the lower end to a mechanical or non-mechanical valve for controllable transport. The so-called V-valve is a conical (tapered) spout installed at the lower end of a fully fluidized dipleg for controlling flow of particulate solids. It belongs to the class of non-mechanical valves which include the L-valve and the J-valve. The fully fluidized dipleg–V-valve system is a new kind of pneumatically controlled downcomer, which has been extensively studied by Kwauk (1973), Liu *et al.* (1980), Li *et al.* (1981), Li (1981), Li *et al.* (1982), and Li and Kwauk (1991).

A. Mechanics

Figure 1 presents an isometric view of the pneumatically controlled downcomer. Solids drop by gravity into a fully fluidized dipleg and pass via

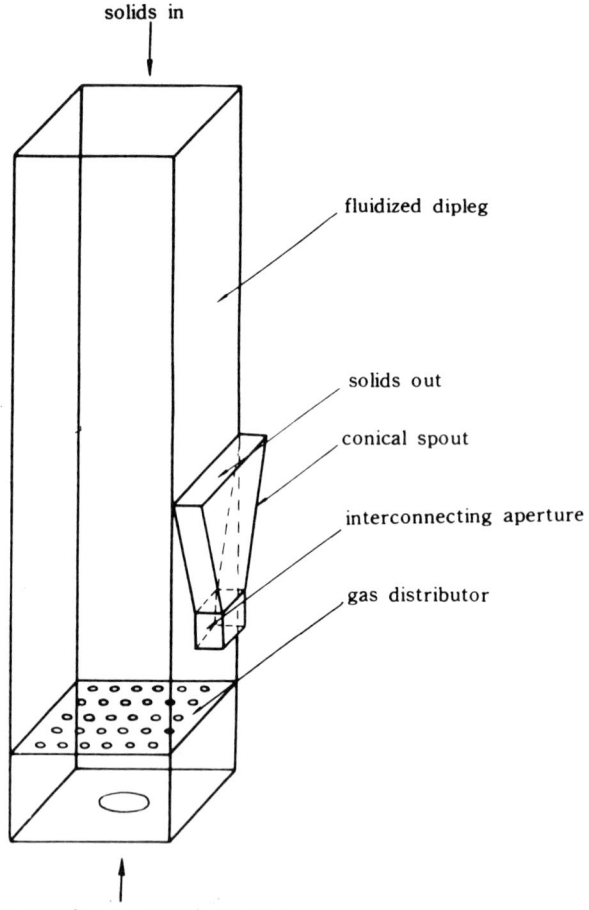

FIG. 1. Pneumatically controlled downcomer (Liu *et al.*, 1980).

an interconnecting aperture into a tapered (trapezoidal or conical) spout for upward transport in dense phase. Fluidization of the dipleg minimizes bridging of solids which often takes place for moving-bed solids descent, especially when the material exhibits poor flow properties. In view of its peculiar pressure drop characteristics, the tapered spout affords a positive seal against geysering of solids through gas bypassing. In operation, the downcomer is provided with a set gas rate which is only sufficient to insure mild fluidization. With no solids feed, a fixed bed is maintained in the tapered spout. As solids are added to the dipleg, the level in it rises until enough pressure is built up near the gas distributor to initiate upward solids transport in the tapered spout. The solids level in the dipleg is self-adjusting to accommodate the pressure differential between the solids inlet and outlet, so that the downcomer can enable solids to flow either from a high-pressure region to a low-pressure region, or vice versa. Solids flow rate can be varied from zero to the permissible maximum of the dipleg, the interconnecting aperture, or the tapered spout, whichever represents the main resistance. The fluidizing gas for pneumatic control, once set, divides itself between the dipleg and the tapered spout in accordance with the solids flow rate: For zero solids flow rate most fluidizing gas enters the dipleg, and this portion diminishes as the solids rate increases. The pattern of solids motion in the spout depends strongly on the position of the gas distributor. When the gas distributor is above the interconnecting aperture, a bubbleless region is formed between the gas distributor and the interconnecting aperture, and moving-bed up-transport would take place in the spout, while if the distributor is located below the interconnecting aperture, fluidization would take place instead (Li, 1981; Li and Kwauk, 1991).

B. THE FLUIDIZED SPOUT

1. Modes of Operation

In experimenting with the apparatus shown in Fig. 2, it was found that when solids were added slowly to the top of the dipleg, the solids level in it rose, until at some height L_{start}, the conical spout began to discharge solids, while the solids level in the dipleg descended, until at some lower height L_{stop}, solids discharge stopped. Then the cycle repeated itself. This mode of operation is recapitulated in Fig. 3 in terms of the pressure drop of the conical spout, Δp_c, height of solids level in the dipleg, L, and the solids discharge rate, S.

At sufficiently high solids rates, the frequency of the intermittent solids flow increases to a point of continual discharge. The minimal solids level in

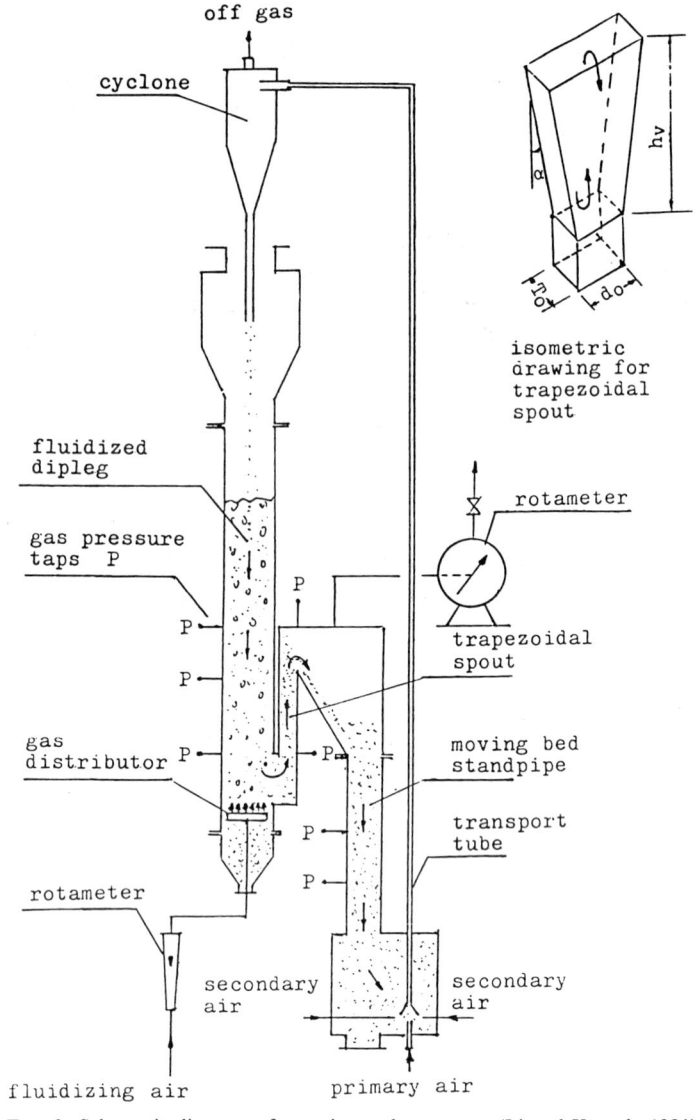

FIG. 2. Schematic diagram of experimental apparatus (Li and Kwauk, 1991).

the dipleg for continuous discharge is L_{stop}, and the corresponding solids discharge rate, S_{stop}. With further increase in solids rate, the continuous solids flow is accompanied by a rise of solids level in the risen to beyond the total height of the dipleg, L_{leg}, and the dipleg overflows. Thus, the maximum solids rate, S_{leg}, is determined by the height of the dipleg, L_{leg}.

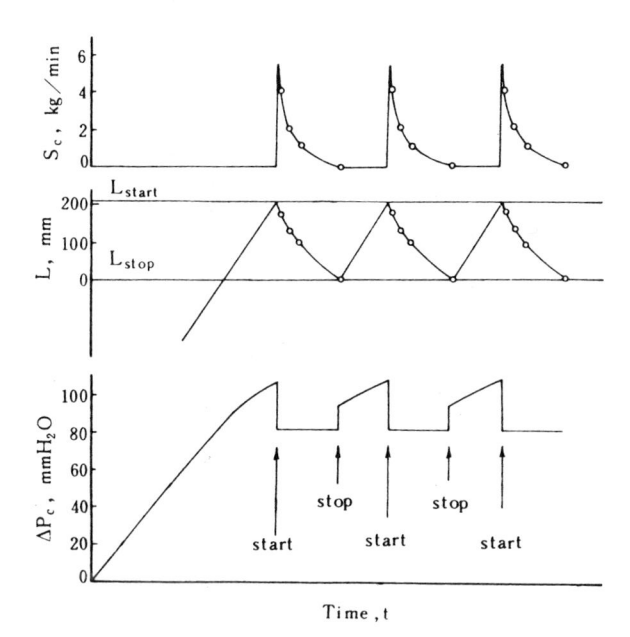

FIG. 3. Variation of solids level in dipleg L, pressure drop through conical spout ΔP_c and solids discharge rate S in intermittent solids flow (Li et al., 1981).

Figure 4 summarizes the preceding findings with respect to the various possible modes of operation. For any initial solids level L_i in the dipleg, depending on the solids rate, it assumes different final positions as shown in the five curves of Fig. 4 (see table).

Curve	Solids Rate	Solids Level in Dipleg
1	$S < S_{stop}$	fluctuates between L_{start} and L_{stop}
2	$S = S_{stop}$	$L = L_{stop}$
3	$S > S_{stop}$	$L > L_{stop}$
4	$S \gg S_{stop}$	$L \gg L_{stop}$
5	$S > S_{leg}$	$L > L_{leg}$, solids overflow

2. Capacity for Solids Discharge

For solids discharge of the fluidized spout, the interconnecting aperture represents the main resistance to solids flow. By treating the solids emulsion as an inviscid fluid, as proposed by Jones and Davidson (1965) and recommended by Leung (1980), the solids discharge is likened to the efflux

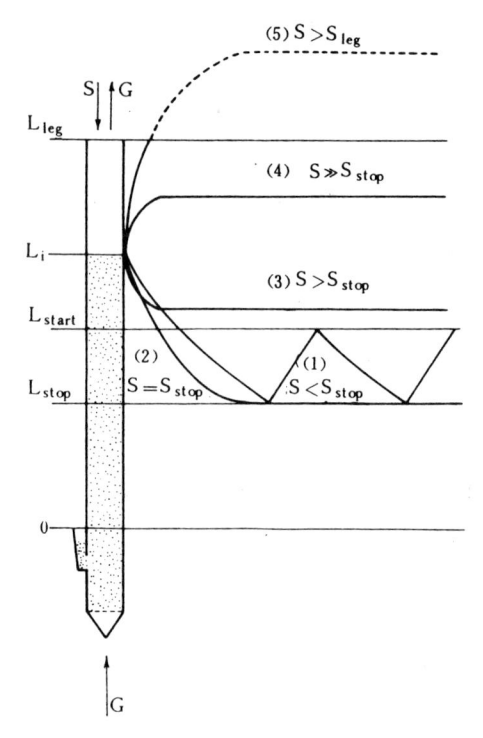

FIG. 4. Possible modes of operation for a pneumatically controlled downcomer (Li et al., 1981).

of a column of liquid through an orifice placed at the bottom:

$$G_P = C_d A_0 \rho_P (1 - \varepsilon_{mf})(2gL_{mf})^{0.5}, \tag{1}$$

in which A_0 is the cross-sectional area of the interconnecting aperture and L_{mf} is the equivalent column height, prorated to a voidage at minimal fluidization, ε_{mf}. Figure 5 shows a generalized correlation of the discharge coefficient C_d with solids velocity through the interconnecting aperture, u_{d0}, for two test materials (active carbon, $\rho_P = 1.02$ g/cm^3, $\bar{d}_P = 0.97$ mm; semi-coke, $\rho_P = 1.44$ g/cm^3, $\bar{d}_P = 0.37$ mm) used and three aperture sizes (10 × 40 mm, 15 × 40 mm, 20 × 40 mm). It should be noted that the magnitude of C_d is sensitive to solids velocity, and for high solids velocity, that is, at $u_{d0} > 1$ m/s, C_d remains essentially constant in the range of 0.6 to 0.7.

Afterward, Huang (1990) improved Eq. (1) in order to consider the effect of the pressure at the top of the dipleg, P_a, the pressure at the outlet of the spout, P_b, and the height of the spout, h_v, on the capacity of solids discharge.

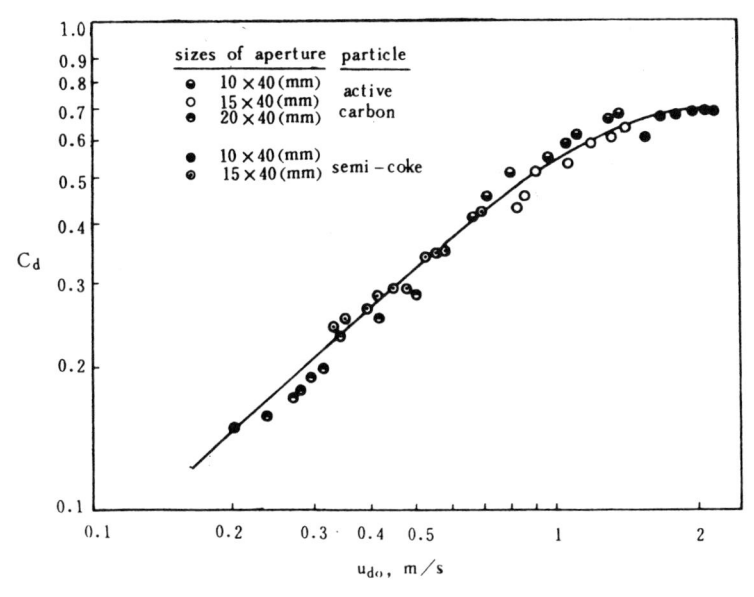

FIG. 5. Variation of discharge coefficient C_d with solids velocity u_{do} (Li et al., 1981).

That is,

$$G_P = C_d A_0 \rho_P (1 - \varepsilon_{mf}) \left[2g \left(L_{mf} - h_v - \frac{P_b - P_a}{\rho_P (1 - \varepsilon_{mf}) g} \right) \right]^{0.5}. \quad (1a)$$

Furthermore, on the basis of regression analysis for 130 experimental data, an empirical equation for evaluating the value of C_d was given as follows:

$$C_d = 0.714 u_{d0}^{0.575}. \quad (1b)$$

By comparison with the measured value of C_d, Eq. (1b) gave an average error of $\pm 7.3\%$.

3. Dimensions of the Fluidized Spout

Determination of the dimensions of the fluidized spout is based on operation experience. According to Li et al. (1982), the cone angle, θ, for smooth fluidized flow in the conical section may be taken as $\theta = 10°$. The minimum thickness of the spout is set by the requirement that the spout base area should not be less than the cross-sectional area of the interconnecting aperture. The diameter of the dipleg should be greater than four times the equivalent diameter of the aperture. These will ensure an aperture restriction capable of controlling solids flow through the valve.

4. Duplex Spout

The so-called duplex spout, which consists of a dipleg connected with two conical spouts, one at the bottom and the other near the middle of the dipleg, was developed by Huang (1990). Experiments with the duplex spout showed that when the solids level in the dipleg rose to some height, the lower spout began to discharge solids, while the upper spout did not discharge solids, and when the solids level in the dipleg reached some higher level, both the upper and lower spouts discharged solids simultaneously. Compared with the single spout, the duplex spout possesses the advantage of a higher discharge rate without flooding, in which state particles in the fluidized dipleg overflow from its top. In addition, the duplex conical spout also insures continuous and steady operation even for a very low discharge rate of solids.

C. THE NONFLUIDIZED SPOUT

In general, when the drag force of an up-flowing gas on particles equals the weight of the solids, the solids will move upward. For a cone-shaped bed, Kwauk (1973) pointed out that there is a boundary level in the bed where the drag force of a flowing gas on the solids equals the weight of the solids in the bed. At this boundary level the relative velocity between the gas and solids equals the critical (incipient) fluidization velocity. The region above this boundary level is "sub-critical" in the sense that the drag force of a flowing gas on the solids is less than the weight of the solids in this region. The region below this boundary level is "super-critical" where the drag force of a flowing gas on the solids is larger than the weight of the solids in that region. The surplus weight of the sub-critical region acts on the region below to limit the expansion and fluidization of the super-critical region. As a result, moving-bed uptransport occurs in the cone-shaped bed where the excess drag in the super-critical region at the bottom balances the deficient drag in the sub-critical region at the top, and pushes up all the solids in the cone-shaped bed. Figure 6 shows these mechanics.

By comparison with the fluidized spout, the nonfluidized spout has a number of advantages such as continual and stable operation, reduced gas bypassing, improved sealing ability, and high discharge rate.

1. Equations for Force Balance

Because the motion of particles in the tapered spout is moving-bed uptransport, there exist contact stresses among particles and between particles and the pipe wall. Following Walker's (1966, 1967) and Walters's (1973)

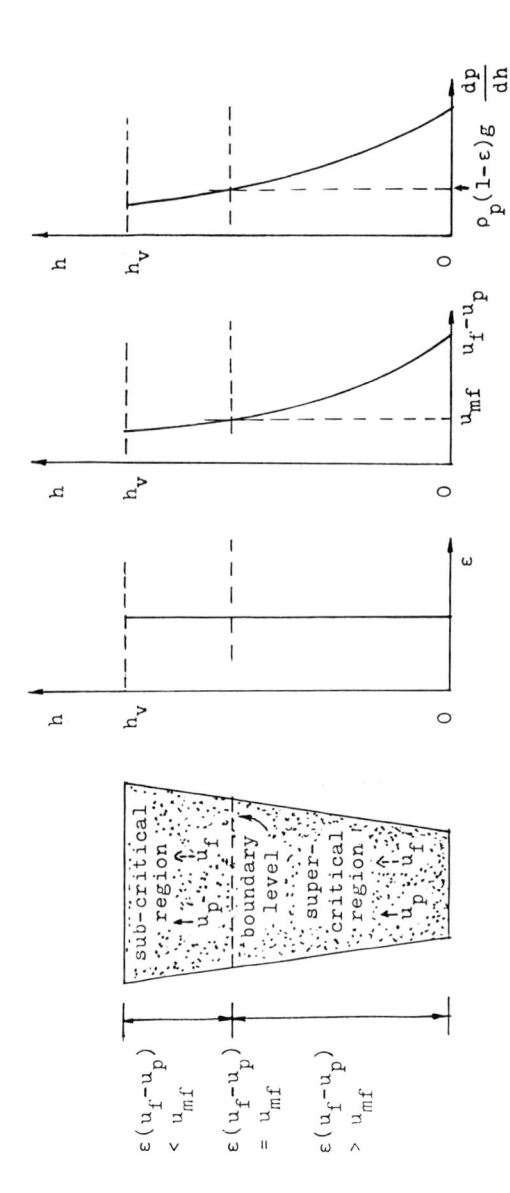

FIG. 6. Mechanics for conical moving bed transport (Li and Kwauk, 1991).

analysis of stresses for hoppers and silos, Li (1981) and Li and Kwauk (1989, 1991) analyzed the stresses in a nonfluidized tapered spout in vertical pneumatic moving-bed uptransport, and gave a series of stress ratios, such as E_v, H_v, E_s, H_s, D_t, D_c, D_s, D_v, to express the relationships among the normal stress, the shear stress, and the mean vertical stress on a horizontal cross-section in any horizontal elemental slice and adjacent to the vertical or slanting wall. The definitions of these stress ratios are given in Section II.A.

For stable moving-bed uptransport of particles in a tapered spout, the drag force of the flowing gas on the particles is used to overcome gravity, acceleration, and wall friction. Vertical force balance on a horizontal element of solid particles in a trapezoidal pneumatic moving bed, shown in Fig. 7, gives

$$\frac{d\bar{\sigma}_h}{dh} = -\left(\frac{f}{1+fh}\left(1 + \frac{E_s D_s}{\tan \alpha} - H_s D_s\right) + \frac{2}{T_0}E_v D_v\right)\bar{\sigma}_h - \rho_p(1-\varepsilon)g - \frac{dp}{dh}$$
$$+ \frac{fG_p^2}{\rho_p(1-\varepsilon)A_0^2(1+fh)^3}, \qquad (2)$$

where

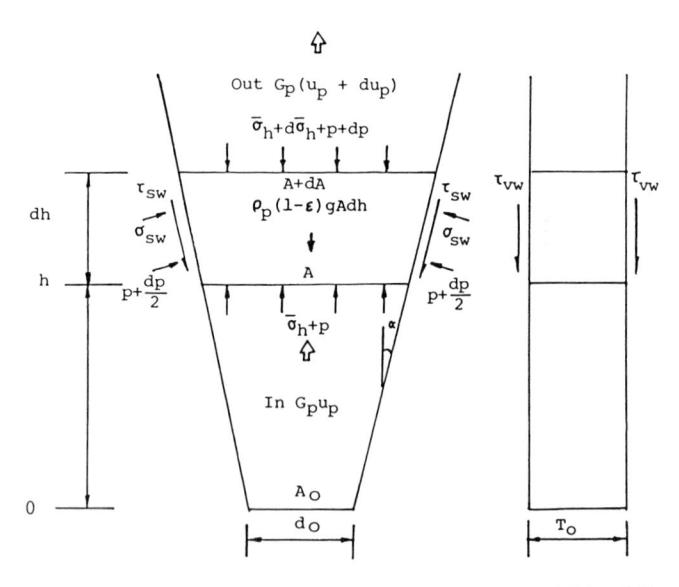

$$f = \frac{2\tan \alpha}{d_0}. \qquad (3)$$

FIG. 7. Force balance for elemental slice in trapezoidal moving bed (Li and Kwauk, 1991).

2. Equations for Gas Pressure Drop

The pressure gradient set up by fluid–particle friction is described by Ergun (1952):

$$-\frac{dp}{dh} = 150\left(\frac{1-\varepsilon}{\varepsilon}\right)^2 \frac{\mu_f(u_f - u_p)}{(\phi_s d_p)^2} + 1.75\left(\frac{1-\varepsilon}{\varepsilon}\right)\frac{\rho_f(u_f - u_p)^2}{\phi_s d_p}. \quad (4)$$

For the case of the trapezoidal spout, the Ergun equation can be modified as follows:

$$-\frac{dp}{dh} = 150\left(\frac{1-\varepsilon}{\varepsilon}\right)^2 \frac{\mu_f}{(\phi_s d_p)^2}\left(\frac{u_{f0} - u_{p0}}{1 + fh}\right) + 1.75\left(\frac{1-\varepsilon}{\varepsilon}\right)\frac{\rho_f}{\phi_s d_p}\left(\frac{u_{f0} - u_{p0}}{1 + fh}\right)^2. \quad (5)$$

Integration of Eqn (5) from $h = 0$ to $h = h_v$ gives the pressure drop equation of the trapezoidal spout:

$$-\Delta p_s = 150\left(\frac{1-\varepsilon}{\varepsilon}\right)^2 \frac{\mu_f(u_{f0} - u_{p0})}{(\phi_s d_p)^2 f}\ln(1 + fh_v) + 1.75\left(\frac{1-\varepsilon}{\varepsilon}\right)\frac{\rho_f(u_{f0} - u_{p0})^2}{\phi_s d_p f}$$

$$\times \left(1 - \frac{1}{1 + fh_v}\right). \quad (6)$$

3. Gas Flow Rate

When the properties of the solids and the fluid, the size of the spout, the solids flow rate, G_p, and the gas–solids relative velocity at the bottom of the spout, $(u_{f0} - u_{p0})$, are given, the gas flow rate through the spout G_{gv} can be calculated:

$$G_{gv} = (u_{f0} - u_{p0})\rho_f \varepsilon A_0 + G_p\left(\frac{\rho_f}{\rho_p}\right)\left(\frac{\varepsilon}{1 - \varepsilon}\right). \quad (7)$$

4. Gas–Solids Relative Velocity

Gas pressure drop across the spout, ΔP_v, and gas flow rate through the spout, G_{gv}, are the more important variables for the design and operation of the V-valve. From Eqs. (6) and (7), we can see that in order to calculate ΔP_v and G_{gv}, the gas–solids relative velocity $(u_{f0} - u_{p0})$ must be known. The value of $(u_{f0} - u_{p0})$ can be calculated by solving the simultaneous differential Eqs. (2) to (5) for the trapezoidal spout, with the boundary conditions of $\bar{\sigma}_h = 0$ at $h = 0$ and $\bar{\sigma}_h = 0$ at $h = h_v$. Solution of the simultaneous equations consists of a trial-and-error process for the numerical method.

278   LI HONGZHONG

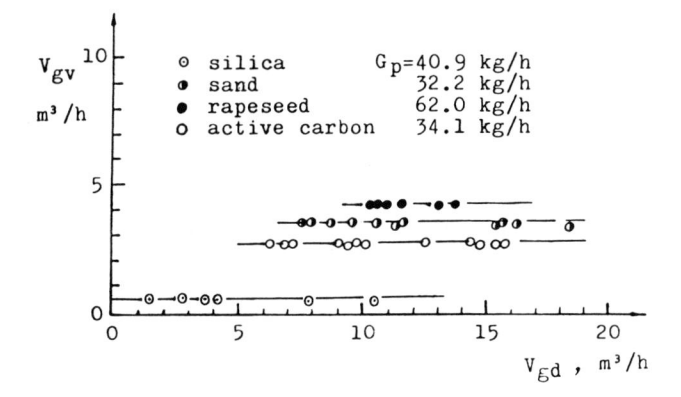

FIG. 8. Relation between V_{gv} and V_{gd} (Li and Kwauk, 1991).

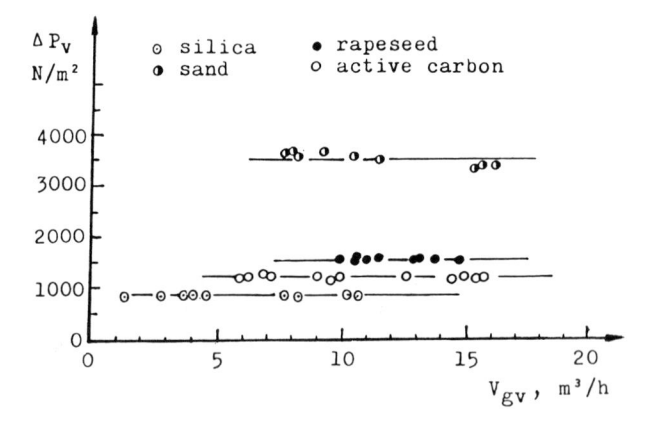

FIG. 9. Relation between Δp_v and V_{gv} (Li and Kwauk, 1991).

For a given solids flow rate, the gas flow rate through the spout V_{gv} was found to be independent of gas flow rate through the distributor V_{gd}, as shown in Fig. 8. These experimental results agree with computation by Eq. (7).

Gas pressure drop across the spout was found to be independent of gas and solids flow rates, V_{gv} and G_p, respectively, as shown in Figs. 9 and 10. These results are also in conformity with the preceding theory. It can be seen from Eq. (6) that the gas–solids relative velocity ($u_{f0} - u_{p0}$) is also constant for a gas–solids–spout system.

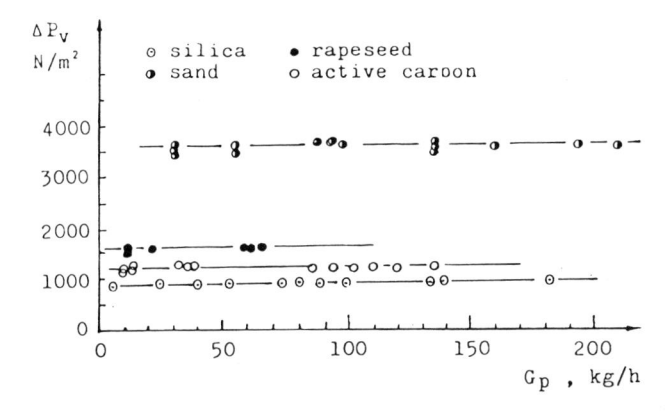

FIG. 10. Relation between Δp_v and G_p (Li and Kwauk, 1991).

5. Geometry

Solids motion was observed in the spout for different solids and gas flow rates, different positions of the distributor and different angles and sizes of the spout.

The spout with less than 10° of cone angle could not ensure against spouting of solids, but when the angle was greater than 20°, dead solids would appear near the wall. Experiments indicated that the most suitable angle lay between 10° and 20°.

Neither could a spout with a height of less than 100 mm ensure against spouting of solids. Pressure drop of the spout was found to increase with its height, thus resulting in greater stability. However, there is no need to use a height of more than 200 mm to incur unnecessarily high pressure drop. The minimum thickness of the V-valve is set by the requirement that the V-valve base area should not be less than the cross-sectional area of the interconnecting aperture.

D. THE FLUIDIZED DIPLEG

1. Pressure Drop, Δp_l

Gas pressure drop in the dipleg depends on pressure balance of the overall system. Let the pressure at the outlet of the tapered spout be p_b, and at the top of the dipleg be p_a. Then the following relation is obtained:

$$\Delta p_l = \Delta p_v + p_b - p_a. \tag{8}$$

2. Voidage, ε_1

For generalized fluidization, Kwauk (1963) put forward the following relation among voidage, particles velocity, and fluid velocity for simultaneous particle–fluid motion:

$$(\varepsilon_1)^n = \frac{u_0}{u_t} - \frac{u_d}{u_t}\left(\frac{\varepsilon_1}{1-\varepsilon_1}\right). \tag{9}$$

3. Gas Flow Rate, G_{gl}

Total gas flow rate through the distributor at the bottom of the dipleg, G_{gd}, splits into two paths: through the dipleg, G_{gl}, and through the tapered spout, G_{gv}. Therefore,

$$G_{gl} = G_{gd} - G_{gv}. \tag{10}$$

A fully fluidized dipleg to insure against bridging of solids in descent calls for a relative velocity between the gas and the solids not less than that for incipient fluidization:

$$u_{mf} = \frac{(G_{gl})_{min}}{\rho_f A_1} + \left(\frac{\varepsilon_{mf}}{1-\varepsilon_{mf}}\right)\frac{G_p}{\rho_p A_1}.$$

Therefore, minimal gas flow rate through the dipleg, $(G_{gl})_{min}$, is given by

$$(G_{gl})_{min} = \rho_f A_1 u_{mf} - \left(\frac{\varepsilon_{mf}}{1-\varepsilon_{mf}}\right)\frac{\rho_f}{\rho_p}G_p. \tag{11}$$

4. Solids Height, h_1

As a close approximation for most design, pressure drop due to friction between particles and the pipe wall can be neglected, for which the following equation holds between solids height and pressure drop:

$$h_1 = \frac{\Delta p_1}{\rho_p(1-\varepsilon_1)}. \tag{12}$$

5. Instability of Operation and Limiting Discharge Capacity, $(W_p)_{max}$

More attention was given to instability of operation. Generally, standpipes are devices used to convey particulate solids against an adverse gas pressure gradient with the aid of gravity. The standpipe must also function as a seal to prevent backflow of gas against the direction of motion of the particles. In certain circumstances, standpipes may malfunction, apparently undergoing a transition to a state in which there is much smaller pressure buildup within the pipe, and backflow of gas may occur. Because of its potentially serious consequences this instability of standpipe operation has attracted much attention. Matsen (1973) described flow instability in a standpipe as a result of rapid loss of pressure at the base of the standpipe. He attributed the cause of such instability to bubbles in the flowing fluidized solid, postulating that when the solid downflow velocity is equal to u_b, the bubble velocity, instability occurs. Concurrently, Leung and Wilson (1973) offered an alternative explanation based on a simple one-dimensional picture of flow in the standpipe. They believe that the instability is caused by a sudden change in flow pattern from fluidized bed flow to packed bed flow in the lower section of the standpipe as a result of change in operating conditions, such as circulation rate, pressure levels, and movement of slide in the slide valve.

Kwauk (1963) predicted a state of flooding, in which state the voidage of the fluidized bed increase suddenly, and as a result, the particles in the fluidized dipleg overflow from the top of the dipleg. According to the definition of flooding, it occurs in a particle–fluid system at the following condition:

$$\left[\frac{\partial u_0}{\partial \varepsilon_1}\right]_{u_d} = 0$$

or

$$\left[\frac{\partial u_d}{\partial \varepsilon_1}\right]_{u_0} = 0.$$

Therefore, by differentiating Eq. (9), the value of u_d can be derived for the flooding state:

$$u_d = -n(1 - \varepsilon_1)^2 \varepsilon_1^{n-1} u_t. \tag{13}$$

Substitution of Eq. (13) into Eq. (9) gives the corresponding value of u_0:

$$u_0 = -[n(1 - \varepsilon_1) - 1]\varepsilon_1^n u_t. \tag{14}$$

When the values of u_d or u_0 in the fluidized dipleg reach the calculated values by Eq. (13) or (14), flooding occurs. Hence, the limiting discharge capacity becomes

$$(W_p)_{max} = -n(1 - \varepsilon_1)^2 (\varepsilon_1)^{n-1} u_t \rho_p. \tag{15}$$

E. THE PNEUMATICALLY CONTROLLED DOWNCOMER

In the light of the preceding separate studies on the spout and the dipleg, the overall operation of the pneumatically controlled downcomer will now be analyzed.

1. Pressure Drop through Spout

The uptransport spout is often designed with a trapezoidal cross-section as shown in Fig. 1. A simplified analysis (Kwauk, 1992) will be given on the basis that the dipleg operates under the condition of incipient fluidization and the spout in moving bed uptransport.

The pressure gradient in the spout is

$$dP/dz = \Delta\rho(1 - \varepsilon_0)(u_s/u_f)^m, \qquad (16)$$

where the slip velocity u_s can be written in terms of the condition at the inlet (subscript i) of the spout, and of the relative cross-sectional area of the spout $(1 + fz)$, f being the rate at which the area varies with height:

$$u_s = \frac{u_{o2i}}{1 + fz} - \frac{u_{di}}{1 + fz}\left(\frac{\varepsilon_0}{1 - \varepsilon_0}\right),$$

where subscript 2 is used for the spout and 1, for the dipleg. According to the definition of the rate-of-flow ratio $r = u_{di}/u_{o2i}$,

$$u_s = \frac{u_{o2i}}{1 + fz}\left(1 + \frac{r\varepsilon_0}{1 - \varepsilon_0}\right).$$

Since u_f at incipient fluidization is equal to $\varepsilon_{o2}^n u_t$,

$$\frac{u_s}{u_f} = \frac{u_{o2i}[1 - r\varepsilon_0/(1 - \varepsilon_0)]}{(1 + fz)\varepsilon_{o2}^n u_t} = \frac{Q_i}{1 + fz},$$

where

$$Q_i = \frac{u_{o2i}[1 - r\varepsilon_0/(1 - \varepsilon_0)]}{\varepsilon_{o2}^n u_t}.$$

Therefore

$$dP/dz = \Delta\rho(1 - \varepsilon_0)[Q_i/(1 + fz)]^m. \qquad (17)$$

Force balance on a thin horizontal slice of thickness dz gives

$$dF_{\text{hydrodynamic}} \equiv a(dP/dz)\,dz = dF_{\text{gravity}} = a\Delta\rho(1 - \varepsilon_0)\,dz,$$

where a is the cross-sectional area at height z and is related to inlet area a_{2i} as $a = a_{2i}(1 + fz)$. Thus, integration from bottom to top Z of the spout has the form

$$\int_0^{Z_2} a_{2i}(1 + fz)\Delta\rho(1 - \varepsilon_0)[Q_i/(1 + fz)]^m\,dz = \int_0^{Z_2} a_{2i}(1 + fz)\Delta\rho(1 - \varepsilon_0)\,dz,$$

or

$$Q_i^m \int_0^{Z_2} (1+fz)^{1-m} dz = \int_0^{Z_2} (1+fz) dz. \tag{18}$$

Integration gives

$$\frac{Q_i^m}{(2-m)f}[(1-fZ_2)^{2-m} - 1] = Z_2 + \frac{f}{2}Z_2^2.$$

At the top Z_2 of the spout the top-to-bottom area ratio is

$$1 - fZ_2 = a_2/a_{2i} = \alpha.$$

Substitution into the integrated result gives

$$Q_i^m = \frac{(2-m)(\alpha^2-1)}{2(\alpha^{2-m}-1)}. \tag{19}$$

It can be shown that

for viscous flow, $m = 1$, $Q_i = \dfrac{\alpha+1}{2}$;

for turbulent flow, $m = 2$, $Q_i = \left[\dfrac{\alpha^2-1}{2\ln\alpha}\right]^{1/2}$.

Pressure drop through the spout can be computed by integrating Eq (17):

$$\int_{P_0}^{P_2} dP = \int_0^{Z_2} \Delta\rho(1-\varepsilon_0)[Q_i/(1+fz)]^m dz. \tag{17a}$$

Actual integration and substitution of Eq. (19) for Q_i gives the ratio of the actual pressure drop in the spout with flaring cross-section to that due to gravity of the solids:

$$N_2 = \frac{P_2 - P_0}{\Delta\rho(1-\varepsilon_0)Z_2} = \frac{2-m}{2(1-m)} \frac{(\alpha+1)(\alpha^{1-m}-1)}{\alpha^{2-m}-1}. \tag{20}$$

Computed values of Q_i from Eq. (19) and N_2 from Eq. (20) are shown in Fig. 11 as functions of α and m. It should be noted from the curves that both Q_i and N_2 increase with the area ratio α. The effect of m on Q_i is less pronounced than on N_2.

2. Gas Distribution

The expression of Q_i used for Eq. (17) was based on the inlet area of the spout. In order to analyze the distribution of gas flow between the dipleg

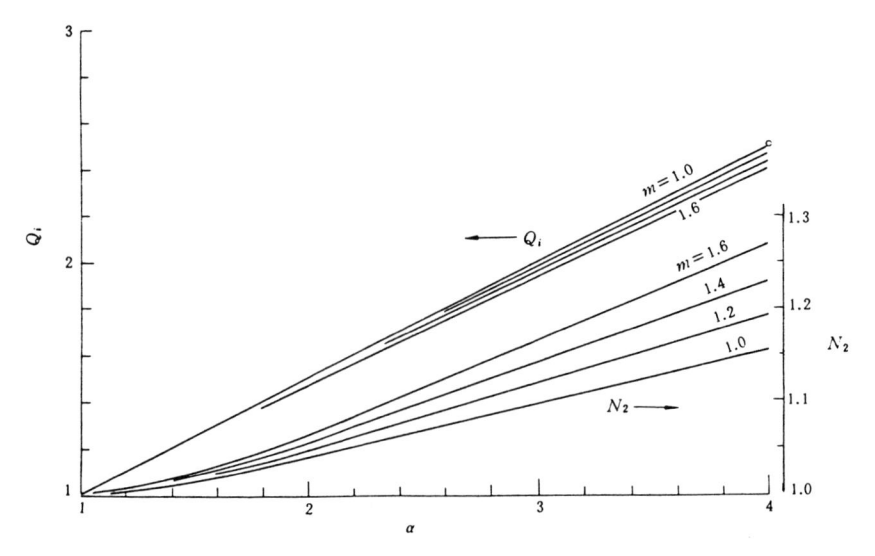

FIG. 11. Computed values for Q_i and N_2 for trapezoidal uptransport tube (Kwauk, 1992).

and the spout, Q_i will be expressed in terms of the cross-sectional area of the dipleg through the following transformations. Define

$$u'_{o2} = G_2/\rho_f a_1 u_t,$$
$$u'_{o1} = G_1/\rho_f a_1 u_t,$$
$$u'_o = G/\rho_f a_1 u_t,$$
$$u'_{d1} = S/\rho_s a_1 u_t.$$

Therefore,

$$u_{o2t} = G_2/\rho_f a_{2t} u_t = u'_{o2}(\alpha/\beta)u_t,$$
$$u_{di} = S/\rho_s a_{2i} u_t = u'_{d1}(\alpha/\beta)u_t,$$

where

$$\alpha = \alpha_2/a_{2t} \quad \text{and} \quad \beta = a_2/a_1.$$

Substitution of the last two expressions gives

$$Q_i = \frac{u_{o2i} - u_{di}\varepsilon_0/(1-\varepsilon_0)}{\varepsilon_0^n u_t} = \frac{\alpha}{\beta}\left[\frac{u'_{o2} - u'_{d1}\varepsilon_0/(1-\varepsilon_0)}{\varepsilon_0^n}\right]$$
$$= \frac{\alpha}{\beta}u'_{o2}\left[\frac{1 - r\varepsilon_0/(1-\varepsilon_0)}{\varepsilon_0^n}\right]. \tag{21}$$

Written another way, Eq. (21) becomes

$$u'_{o2} = (\beta/\alpha)Q_i\varepsilon_0^n + u'_{d1}\varepsilon_0/(1 - \varepsilon_0). \tag{22}$$

For the dipleg which is required to operate at incipient fluidization,

$$u'_{o1} = \varepsilon_0^n - u'_{d1}\varepsilon_0/(1 - \varepsilon_0). \tag{23}$$

The total gas is split into u'_{o1} for the dipleg and u'_{o2} for the spout, that is,

$$u'_o = u'_{o1} + u'_{o2}.$$

Substitution of Eqs. (22) and (23) gives

$$u'_o = [1 + (\beta/\alpha)Q_i]\varepsilon_0^n.$$

Substituting Eq. (19) for Q_i gives

$$u'_o = \left\{1 + \beta\left[\frac{(2-m)(\alpha^2-1)}{2(\alpha^2-\alpha^m)}\right]^{1-m}\right\}\varepsilon_0^n. \tag{24}$$

Equation (24) shows that the total gas to the downcomer assembly remains constant, irrespective of solids rate. However, Eq. (23) shows that the gas to the dipleg decreases linearly with solids rate, with $u'_{o1} = \varepsilon_0^n$ at $u'_{d1} = 0$. Also, at $u'_{o1} = 0$, $u'_d = (1 - \varepsilon_0)\varepsilon_0^{n-1}$. Thus, a gas distribution diagram for the downcomer assembly can be constructed as given in Fig. 12, showing the

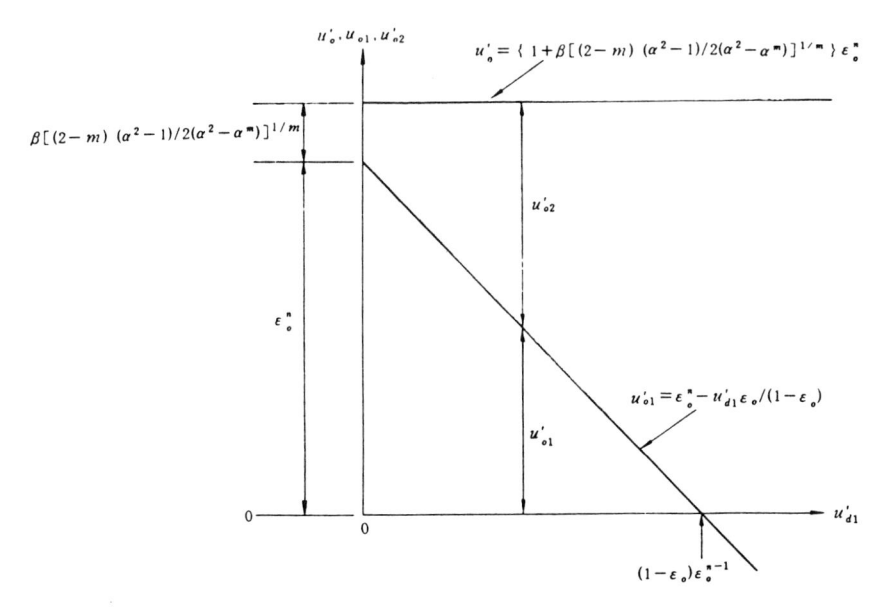

FIG. 12. Gas distribution for the pneumatically controlled downcomer (Kwauk, 1992).

total constant gas rate and its split between the dipleg and the downcomer, with expressions for the related intercepts on the abscissa and ordinate.

3. Self-Flow

For fine Geldart Group C powders which defluidize slowly, it has been shown experimentally that solids continued to be transported through the downcomer even when the fluidizing gas to the dipleg was turned off. The particulate solids move in dense phase. This operation will be designated "self-flow" (Kwauk, 1992). Under this condition, the total gas rate is zero:

$$u'_o = u'_{o1} + u'_{o2} = 0.$$

The following solid-to-gas flow ratio r for self-flow has been derived (Kwauk, 1992):

$$r = \frac{1 + (\beta/\alpha)Q_i\varepsilon^m}{\dfrac{\varepsilon'\varepsilon_{o1}}{1 - \varepsilon'\varepsilon_{o1}} + \dfrac{\beta}{\alpha}Q_i\varepsilon'^n\left(\dfrac{\varepsilon_{o1}}{1 - \varepsilon_{o1}}\right)}, \quad (25)$$

where $\varepsilon' = \varepsilon_{o2}/\varepsilon_{o1}$ represents the ratio of voidages between the spout and the dipleg. Figure 13 shows a plot of ε' against r for various values of $(\beta/\alpha)Q_i$. The plot is based on two values of $\varepsilon_{o1} = 0.4$ and 0.5.

In practice, when self-flow took place, it was noticed that the voidages in the spout and the dipleg were quite similar, thus suggesting that the value

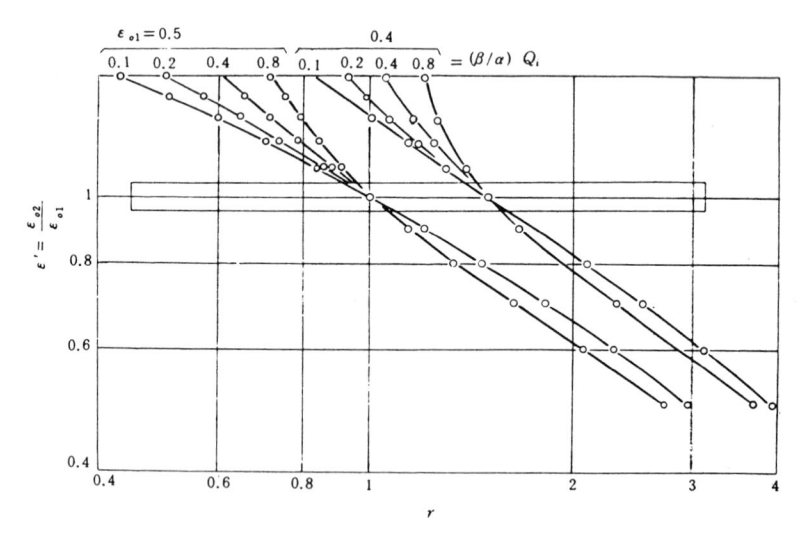

FIG. 13. Relation between parameters for "self-flow" (Kwauk, 1992).

of ε' would be close to unity. For the limiting case of $\varepsilon' = 1$, it can be shown from Eq. (25) that

$$r = \frac{1 - \varepsilon_{o1}}{\varepsilon_{o1}}. \tag{26}$$

Under this condition, the relative velocity between the solids and the gas, $1 - r\varepsilon_{o1}/(1 - \varepsilon_{o1})$, approaches zero. Now, as shown by Eq. (23), the value of reduced gas velocity u'_o needs to be inordinately high:

$$u'_{o1}[1 - r\varepsilon_{o1}/(1 - \varepsilon_{o1})] = \varepsilon_{o1}^n.$$

For practical gas and solids rates, this is possible for very fine powders, for which the terminal velocity u_t is small. What happens in self-flow is that the dropping solids fed to the dipleg carry and pump down the surrounding gas, and this compressed gas carries, in turn, the solids up the spout. This is also possible for fine powders, which do not have a sharp incipient fluidization velocity, thus preventing quick defluidization. Self-flow therefore occurs in the band marked in Fig. 13 in the vicinity of $\varepsilon' = 1$.

4. Pneumatically Started Seal

The pneumatically controlled downcomer shown in Fig. 1 stops delivering solids only when none are fed to the dipleg. In certain circumstances, it would be desirable to stop and start solids discharge from the spout at will even when the dipleg is amply supplied with solids. This is exemplified by the two solids-processing batch reactors in tandem, shown in Fig. 14, operating countercurrently against a common reacting gas (Kwauk, 1992). When the solids in Reactor 1 have reacted to some predetermined extent, Reactor 2 is first discharged and the contents of Reactor 1 need to be transferred as quickly as possible to Reactor 2, without stopping the gas flow.

The dipleg of the downcomer is connected to the gas distributor of Reactor 1 and is followed at its lower end by a spout, which is long enough to seal the solids from flowing to Reactor 2 under the normally operating pressure differential between the two reactors. At selected locations along the height of the spout are control gas inlets, which serve to unlock the fixed-bed solids inventory, according to the mechanism explained in Fig. 15. Under the normal mode of sealed operation, the solids in the spout are locked by virtue of its conical geometry. Under this condition, the pressure drop from point A to B, shown in Fig. 14, through the dipleg ΔP_1 needs to be less than the sum of pressure drops through the spout ΔP_c, the cyclone ΔP_{cy} and the distributor plate P_D of Reactor 1, that is,

$$\Delta P_1 < (\Delta P_c + \Delta P_{cy} + \Delta P_D).$$

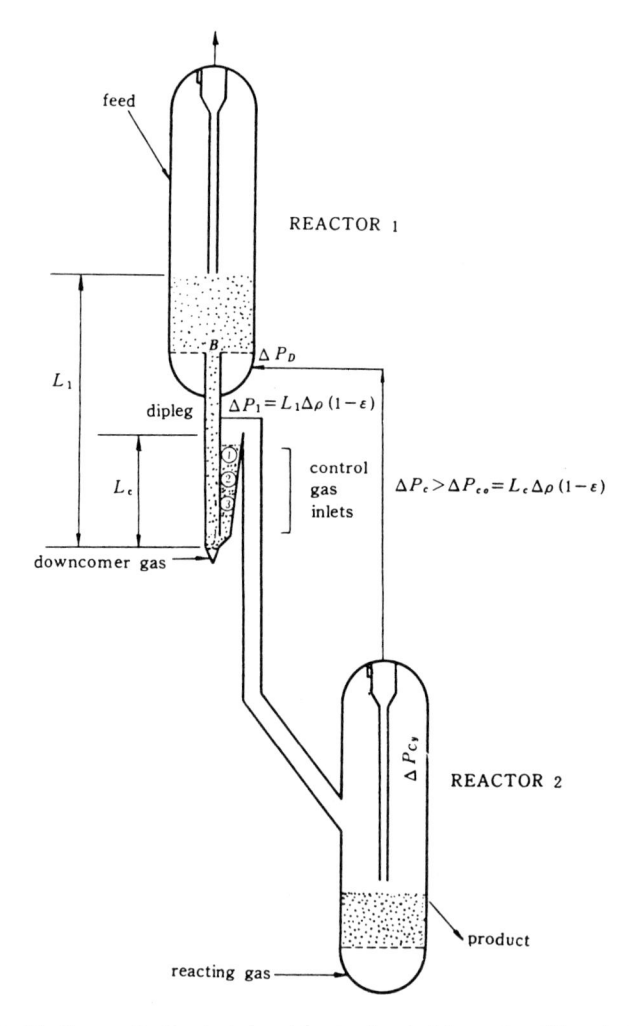

FIG. 14. Pneumatically started seal for tandem batch reactors (Kwauk, 1992).

Or, alternatively, the pressure drop through the locked solids in the spout has to satisfy the condition

$$\Delta P_c > (\Delta P_1 - \Delta P_{cy} - \Delta P_D). \tag{27}$$

If control gas at point 1 in Fig. 14 is admitted to unlock the solids above this point, the ΔP_c peak, shown in Fig. 15, will first be lowered to ΔP_{c1}, which is, however, still above the value $(\Delta P_1 - \Delta P_{cy} - \Delta P_D)$. If the unlocking gas streams at point 2 and point 3 are successively turned on, the new peak

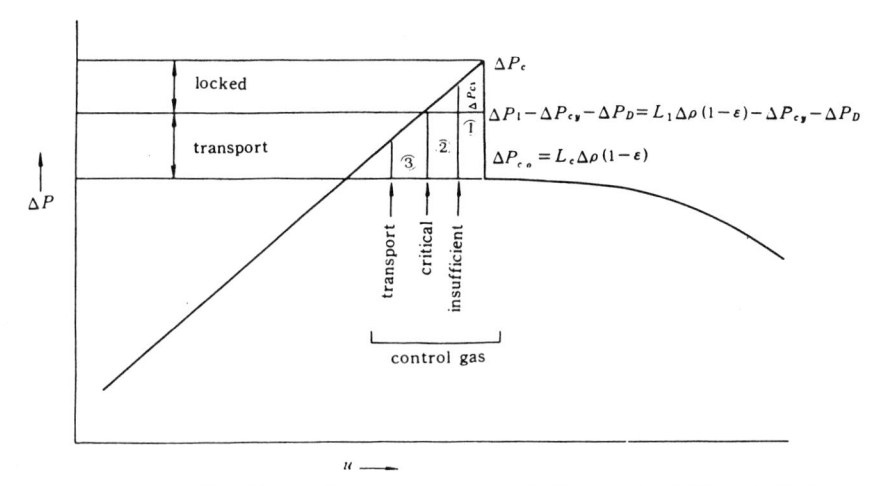

FIG. 15. Unlocking mechanism for a pneumatically started seal (Kwauk, 1992).

ΔP_{co} is eventually reached, which is below $(\Delta P_1 - \Delta P_{cy} - \Delta P_D)$. The unlocked spout now drops to its normal operating pressure drop ΔP_{co} for moving-bed uptransport, and the solids of Reactor 1 will now flow rather rapidly to Reactor 2 under the driving force $(\Delta P_1 - \Delta P_{cy} - \Delta P_D) - \Delta P_{co}$.

A method of computation has been developed for designing the pneumatically started seal.

Figure 16 shows the design variables for the pneumatically started seal. The length of the dipleg is L with a pressure drop of ΔP_1. The combined pressure drop for the cyclone and the distributor plate, $(\Delta P_{cy} + \Delta P_D)$, is represented by reducing the dipleg by an equivalent height of qL. The height of the spout is fL, with a top cross-sectional area of a_t and bottom area a_i, the ratio between these areas being $\alpha = a_t/a_i$. Thus, the area a at height z is given by

$$\frac{a}{a_i} = \frac{z}{fL}(\alpha - 1) + 1.$$

If the critically locked spout is defined by an incipient fluidization velocity u_f at the top, then velocity at height z is

$$u = u_f\left(\frac{a_i}{a}\right) = u_f\left(\frac{\alpha a_i}{a}\right).$$

The pressure gradient in the spout is

$$dP/dz = \Delta\rho(1 - \varepsilon_0)(u/u_f)^m = \Delta\rho(1 - \varepsilon_0)\left[\frac{\alpha}{(z/fL)(\alpha - 1) + 1}\right]^m.$$

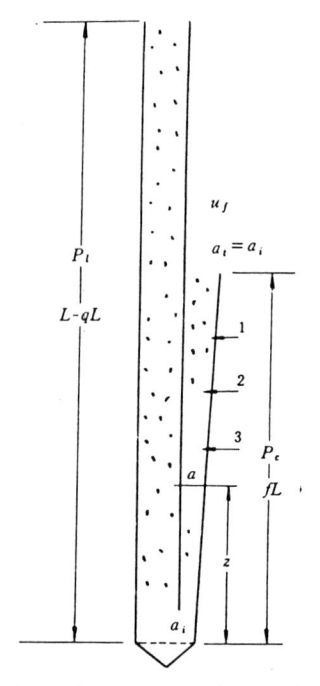

FIG. 16. Design variables for the pneumatically started seal (Kwauk, 1992).

Integrating the preceding equation,

$$\int_0^{\Delta P_c} dP = \int_0^{fL} \Delta\rho(1 - \varepsilon_0)[(z/fL)(1 - 1/\alpha) + 1/\alpha]^m \, dz,$$

gives

$$\Delta P_c = \Delta\rho(1 - \varepsilon_0)\left(\frac{fL\alpha}{1 - \alpha}\right)\left(\frac{1}{1 - m}\right)(1 - \alpha^{m-1}). \tag{28}$$

When unlocked, the spout operates in moving bed uptransport with a pressure drop given by Eq. (20):

$$\Delta P_{co} = \Delta\rho(1 - \varepsilon_0)(fL)\frac{2 - m}{2(1 - m)}\frac{(\alpha + 1)(\alpha^{1-m} - 1)}{\alpha^{2-m} - 1}. \tag{20a}$$

The maximal pressure ratio available for solids sealing is therefore

$$\frac{\Delta P_c}{\Delta P_{co}} = \left(\frac{2}{2 - m}\right)\frac{\alpha^m(\alpha^{2-m} - 1)}{\alpha^2 - 1}. \tag{29}$$

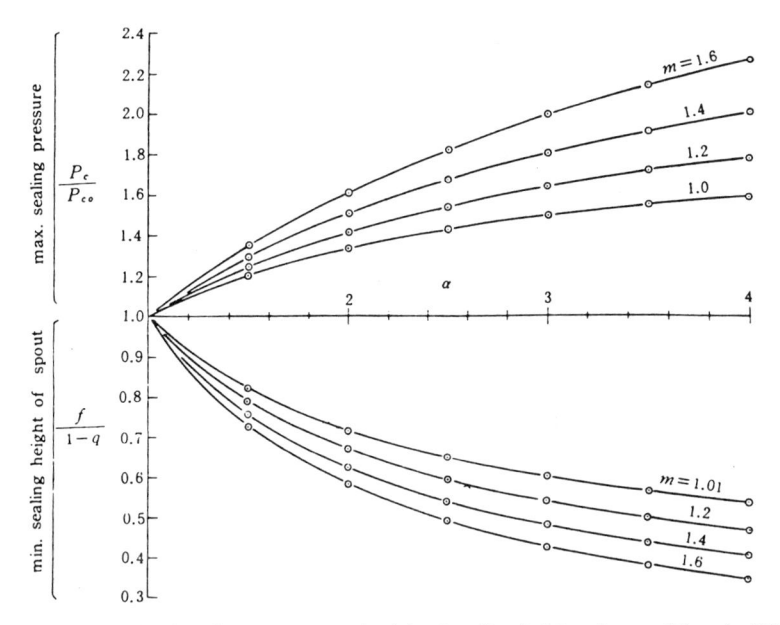

FIG. 17. Maximal sealing pressure and minimal sealing height of spout (Kwauk, 1992).

This pressure ratio is plotted in the upper part of Fig. 17. It is obvious that the greater is the spout expansion ratio α or the pressure exponent m, the greater is the sealing power of the spout.

The marginal condition for solids sealing is equality for Eq. (27),

$$\Delta P_c = (\Delta P_1 - \Delta P_{cy} - \Delta P_D) = (L - qL)\Delta\rho(1 - \varepsilon_0).$$

Substitution of Eq. (28) for ΔP_c gives the minimal sealing height required of the spout,

$$\frac{f}{1-q} = \frac{(m-1)(\alpha-1)}{\alpha^m(1-\alpha^{1-m})}. \tag{30}$$

This relation is shown graphically in the lower part of Fig. 17, indicating that a short spout is sufficient for large expansion ratios α and higher values of the exponent m.

II. Nonfluidized Diplegs

The vertical moving bed has been used in many chemical processes as a nonfluidized standpipe or dipleg. The main function of the nonfluidized

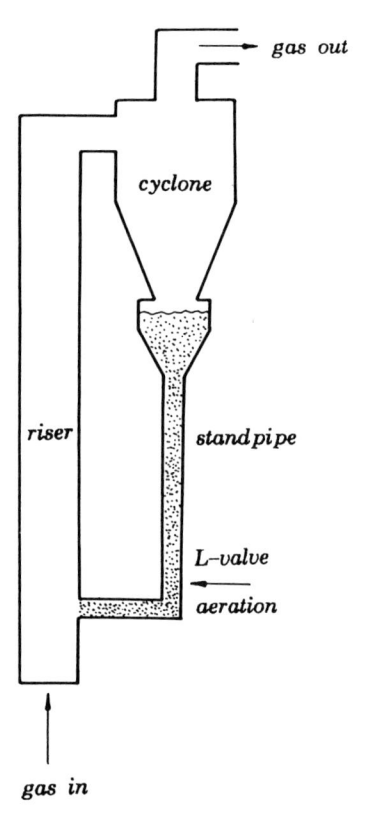

FIG. 18. Schematic of circulating fluidized bed apparatus with standpipe–L valve return system.

standpipe is transport of solid particles while minimizing fluid leakage. The nonfluidized dipleg is often connected at the lower end to a mechanical valve, orifice, or non-mechanical J-valve or L-valve for controlling transport of particulate solids, as shown in Fig. 18. In general, a moving-bed standpipe involves two phases, gas and solid or liquid and solid. Both phases are in motion, either upwards or downwards.

A. A Generalized Theory for the Dynamics of Nonfluidized Gas–Particle Flow

Investigation of nonfluidized gas–particle flow began in the 1950s, although relatively little has been published on its flow patterns, dynamics, pressure drop, or interparticle contact stress (Berg, 1953; Kwauk, 1963; Sandy et al., 1970; Yoon and Kunii, 1970; Zhang et al., 1980; Leung and Jones, 1978; Chen et al., 1984; Li and Kwauk, 1989; Li, 1981, 1990, 1991, 1992). Among the little that has been published on the dynamics of nonfluidized gas–particle flow, Li and Kwauk (1989) studied the dynamics of vertical pneumatic transport by using the theories of multiphase flow and particulate media mechanics. Afterwards, Li (1990, 1992) proposed a dynamic theory on nonfluidized gas–particle flow, which is suitable not only to vertical and cylindrical systems but also to inclined and conical systems. The frictional forces between particles and pipe walls, the compressibility of gas and particulate solids, and the interparticle contact stresses are considered in these theory. Unfortunately, little attention has been paid to the arching in nonfluidized gas–particle flow.

1. Ratios of Stresses

The basic character of non-fluidized gas–particle flow is the existence of contact pressure (or stress) among particles and between particles and the pipe wall. Both theoretical analyses and experimental results (Terzaghi, 1954; Johanson and Jenike, 1972; Li and Kwauk, 1989) showed that the pressure of the interstitial fluid in particulate material neither compresses nor increases the shear resistance of the particulate material. After Walker (1966) and Walters (1973), Li and Kwauk (1989) analyzed the stresses in a vertical pneumatic moving-bed transport tube by using the stress theory of particulate media mechanics and Mohr circles, shown in Figs. 19 and 20, and gave the following stress ratios at any point in the flow field:

$$\frac{\tau_{vw}}{(\sigma_h)_{vw}} = \frac{\sin 2\beta \sin \delta}{1 - \cos 2\beta \sin \delta} = E_v, \tag{31}$$

$$\frac{\sigma_{vw}}{(\sigma_h)_{vw}} = \frac{1 + \cos 2\beta \sin \delta}{1 - \cos 2\beta \sin \delta} = H_v, \tag{32}$$

$$\frac{\sigma_1}{(\sigma_h)_{vw}} = \frac{1 + \sin \delta}{1 - \sin \delta \cos 2\beta} = H_{1v}, \tag{33}$$

$$\frac{\tau_{sw}}{(\sigma_h)_{sw}} = \frac{\sin 2\beta \sin \delta}{1 - \cos(2\beta \mp 2\alpha) \sin \delta} = E_s, \tag{34}$$

$$\frac{\sigma_{sw}}{(\sigma_h)_{sw}} = \frac{1 + \cos 2\beta \sin \delta}{1 - \cos(2\beta \mp 2\alpha) \sin \delta} = H_s, \tag{35}$$

$$\frac{\sigma_1}{(\sigma_h)_{sw}} = \frac{1 + \sin \delta}{1 - \sin \delta \cos(2\beta \mp 2\alpha)} = H_{1s}, \tag{36}$$

294   LI HONGZHONG

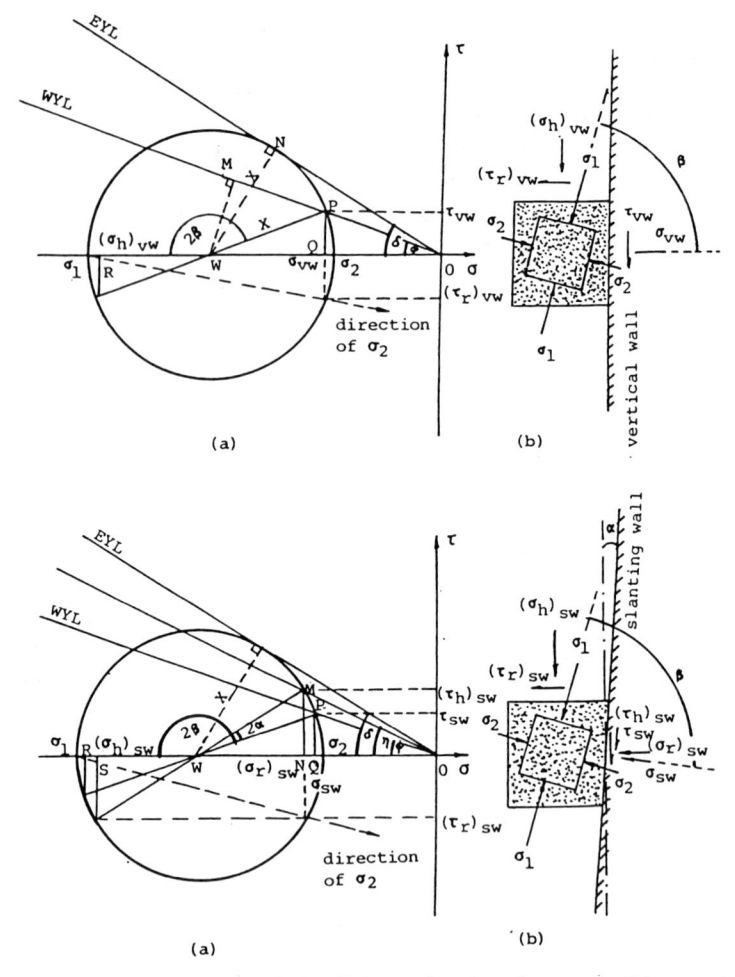

FIG. 19. Stress conditions at vertical wall (upper figure), and stress conditions at slanting wall (lower figure) (Li and Kwauk, 1989).

$$\frac{\tau_{vw}}{\sigma_{vw}} = \frac{\tau_{sw}}{\sigma_{sw}} = \frac{\sin 2\beta \sin \delta}{1 + \cos 2\beta \sin \delta} = \tan \phi = \mu, \tag{37}$$

$$\frac{(\sigma_h)_{sw}}{\bar{\sigma}_h} = \frac{\cos^2 \delta [1 - \cos(2\beta \mp 2\alpha) \sin \delta]}{[1 + \cos(2\beta \mp 2\alpha) \sin \delta][1 + \sin^2 \delta \pm y_1 \sin \delta]} = D_t, \tag{38}$$

$$\frac{(\sigma_h)_{vw}}{\bar{\sigma}_h} = \frac{\cos^2 \delta (1 - \cos 2\beta \sin \delta)}{(1 + \cos 2\beta \sin \delta)(1 + \sin^2 \delta \pm y_2 \sin \delta)} = D_c, \tag{39}$$

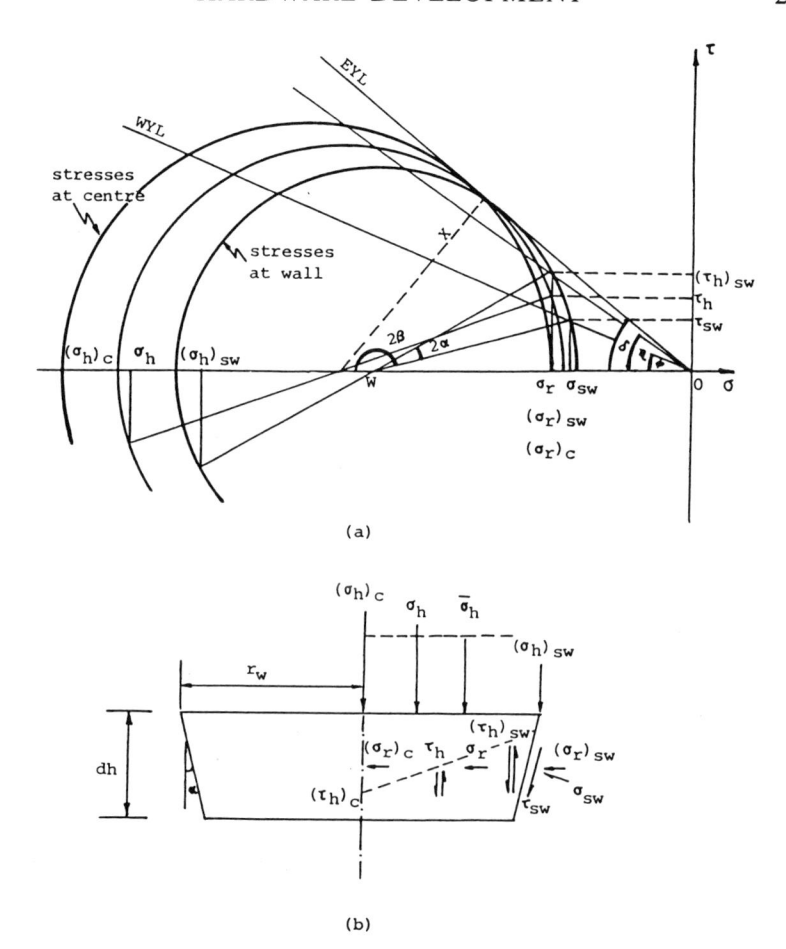

FIG. 20. Stress conditions for a horizontal elemental slice in a conical moving bed (Li and Kwauk, 1989).

$$\frac{(\sigma_h)_{sw}}{\bar{\sigma}_h} = \frac{\cos^2\delta[1 - \cos(2\beta \mp 2\alpha)\sin\delta]}{[1 + \cos(2\beta \mp 2\alpha)\sin\delta][1 + \sin^2\delta \pm y_3\sin\delta]} = D_s, \quad (40)$$

$$\frac{(\sigma_h)_{vw}}{\bar{\sigma}_h} = \frac{\cos^2\delta(1 - \cos 2\beta\sin\delta)}{(1 + \cos 2\beta\sin\delta)(1 + \sin^2\delta \pm y_4\sin\delta)} = D_v, \quad (41)$$

where

$$2\beta = \frac{\pi}{2} + \phi \pm \cos^{-1}\frac{\sin\phi}{\sin\delta}, \quad (42)$$

$$y_1 = \frac{4}{3c_1}[1 - (1 - c_1)^{3/2}], \tag{43}$$

$$y_2 = \frac{4}{3c_2}[1 - (1 - c_2)^{3/2}], \tag{44}$$

$$y_3 = \sqrt{1 - c_1} + \frac{1}{\sqrt{c_1}} \sin^{-1} \sqrt{c_1}, \tag{45}$$

$$y_4 = \sqrt{1 - c_2} + \frac{1}{\sqrt{c_2}} \sin^{-1} \sqrt{c_2}, \tag{46}$$

$$c_1 = \tan^2 \eta / \tan^2 \delta, \tag{47}$$

$$c_2 = \tan^2 \phi / \tan^2 \delta, \tag{48}$$

and

$$\eta = \tan^{-1} \frac{\sin(2\beta \mp 2\alpha) \sin \delta}{1 + \cos(2\beta \mp 2\alpha) \sin \delta}. \tag{49}$$

In the preceding equations, $\bar{\sigma}_h$ is the average axial normal stress over radial cross-section determined by the following equation:

$$\bar{\sigma}_h = \frac{2}{r_w^2} \int_0^{r_w} \sigma_h r \, dr. \tag{50}$$

Note that in all these equations we have used the convention that the upper sign of the $\pm$ and $\mp$ symbols refers to upward particle movement (active state) and the lower sign to downward particle movement (passive state). The symbols D_t, D_c, D_s, and D_v are called distribution factors, corresponding to conical, cylindrical, trapezoidal, and rectangular tube sections, respectively.

2. Generalized Force Balance Differential Equations

Many solid circulation loops require that the standpipe or at least a portion of it be inclined to the vertical because of layout constraints. Therefore, an inclined conical moving bed is regarded as a generalized case for analysis. Axial force balance on an elemental slice of material shown in Fig. 21a gives

$$(\bar{\sigma}_z + d\bar{\sigma}_z + p + dp)(A + dA) - (\bar{\sigma}_z + p)A + W_z - G_p u_s + G_p(u_s + du_s)$$
$$\pm \frac{Q \, dz}{\cos \alpha} r_{sw} \cos \alpha - \left(\sigma_{sw} + p + \frac{dp}{2}\right) \frac{Q \, dz}{\cos \alpha} \sin \alpha$$
$$\pm F_s \cos \alpha = 0. \tag{51}$$

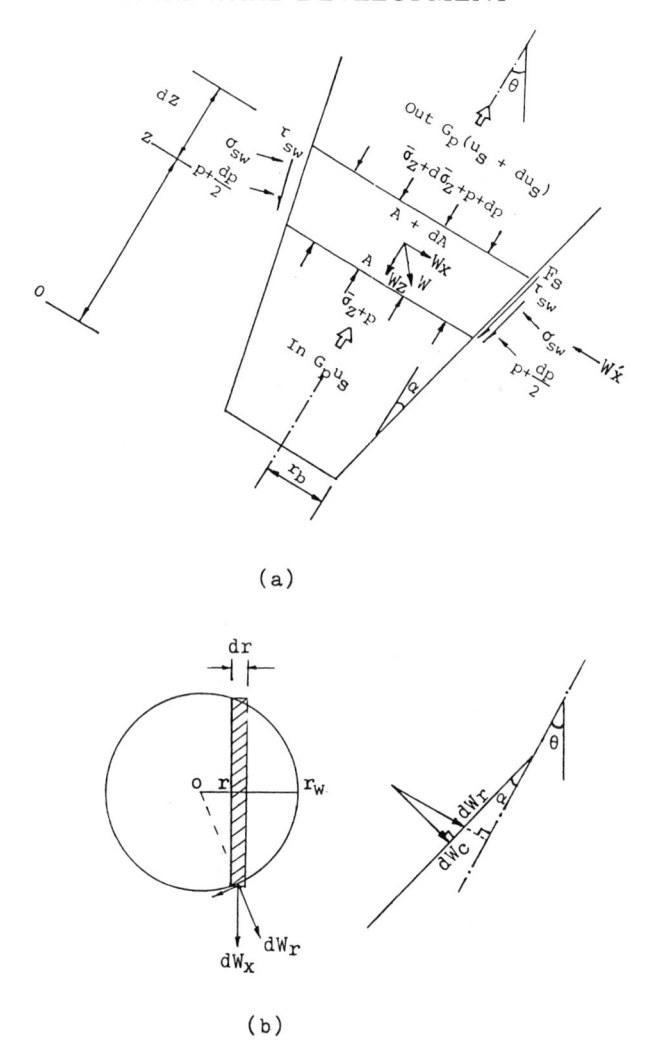

FIG. 21. Axial force balance for an elemental slice of inclined conical moving bed: (a) axial force balance; (b) lateral component of gravity of the wall (Li, 1992).

Here Q represents the perimeter of cross-sectional area A at axial position z. In Fig. 21a W_z is the axial component and W_x the lateral component of gravity W, W'_x represents the lateral supporting force of the wall which is equal to W_x but in the opposite direction, F_s is the wall friction force caused by W_x, and σ_{sw} is the normal contact stress between the particle and the inclined wall caused by the axial interparticle contact stress σ_z. They are

expressed by the following equations:
$$W = \rho_p(1 - \varepsilon)gA\,dz,$$
$$W_z = W\cos\theta,$$

and
$$W_x = W\sin\theta,$$

where θ is the angle between the axial and the vertical direction. Figure 21b shows that W_x acts on the lower semicircular zone of the inclined conical tube, with W_r the radial component of W_x, while the normal component W_c of W_r is the only effective component to produce wall friction. From the geometry of Fig. 21b,

$$W_c = 2\int_0^{r_w} 2\sqrt{r_w^2 - r^2}\,\rho_p(1-\varepsilon)g\frac{\sqrt{r_w^2 - r^2}}{r_w}\sin\theta\cos\alpha\,dz\,dr$$
$$= \tfrac{8}{3}\rho_p(1-\varepsilon)gr_w^2\sin\theta\cos\alpha\,dz.$$

In addition, note that
$$A = \pi r_w^2 = \pi(r_b + z\tan\alpha)^2,$$
$$Q = 2\pi r_w = 2\pi(r_b + z\tan\alpha),$$
$$\frac{dA}{dz} = 2\pi(r_b + z\tan\alpha)\tan\alpha,$$
$$F_s = \mu W_c = \frac{8\mu}{3\pi}\rho_p(1-\varepsilon)gA\sin\theta\cos\alpha\,dz,$$

and
$$\frac{du_s}{dz} = \frac{d}{dz}\left(\frac{G_p}{\rho_p(1-\varepsilon)A}\right) = \frac{G_p}{\rho_p}\frac{d}{dz}\left(\frac{1}{(1-\varepsilon)A}\right)$$
$$= \frac{G_p}{\rho_p A(1-\varepsilon)^2}\frac{d\varepsilon}{dz} - \frac{G_p}{\rho_p(1-\varepsilon)A^2}\frac{dA}{dz}.$$

Substituting Eqs. (34), (35), and (38) into Eq. (51), neglecting second-order differentials, and rearranging yields

$$-\frac{d\bar\sigma_z}{dz} = \frac{2}{r_b + z\tan\alpha}(\tan\alpha - H_sD_t\tan\alpha \pm E_sD_t)\bar\sigma_z + \rho_p(1-\varepsilon)g\cos\theta$$
$$+ \frac{dp}{dz} \pm \frac{8\mu}{3\pi}\rho_p(1-\varepsilon)g\sin\theta\cos^2\alpha - \frac{2G_p^2\tan\alpha}{\pi^2\rho_p(1-\varepsilon)(r_b + z\tan\alpha)^5}$$
$$+ \frac{G_p^2}{\rho_p(1-\varepsilon)^2\pi^2(r_b + z\tan\alpha)^4}\frac{d\varepsilon}{dz}. \tag{52}$$

Equation (52) is the generalized force balance differential equation for non-fluidized gas–particle flow. In this equation, for an inclined cylindrical moving bed, $\alpha = 0$; for a vertical conical moving bed, $\theta = 0$; for vertical cylindrical moving bed, $\alpha = 0$ and $\theta = 0$; and for a vertical cylindrical moving bed without interstitial gas flow, $\alpha = 0$, $\theta = 0$, and $dp/dz = 0$ simultaneously.

If non-fluidized gas–particle flow in a vertical trapezoidal tube (rectangular in cross-section, enclosed by two parallel walls and two non-parallel walls with a half angle of cone α, and the bottom of the tube rectangular with width d_0 and thickness T_0), a similar force balance yields

$$-\frac{d\bar{\sigma}_s}{dz} = \left(\frac{2}{d_0 + 2z \tan \alpha}(\tan \alpha - H_s D_s \tan \alpha \pm E_s D_s) \pm \frac{2}{T_0} E_v D_v\right)\bar{\sigma}_z$$
$$+ \rho_p(1 - \varepsilon)g + \frac{dp}{dz} - \frac{2G_p^2 \tan \alpha}{\rho_p(1 - \varepsilon)T_0^2(d_0 + 2z \tan \alpha)^3}$$
$$+ \frac{G_p^2}{\rho_p(1 - \varepsilon)^2 T_0^2(d_0 + 2z \tan \alpha)^2} \frac{d\varepsilon}{dz}. \tag{53}$$

3. Differential Equation for Gas Pressure Gradient

The gas pressure gradient set up by gas–particle friction can be described by the modified Ergun equation (Ergun, 1952) and Kwauk equation (Kwauk, 1963), shown respectively as follows:

$$\frac{dp}{dz} = 150\left(\frac{1-\varepsilon}{\varepsilon}\right)^2 \frac{\mu_f(u_s - u_f)}{(\phi_s d_p)^2} + 1.75\left(\frac{1-\varepsilon}{\varepsilon}\right)\frac{\rho_f(u_s - u_f)|u_s - u_f|}{\phi_s d_p} \tag{4a}$$

and

$$\frac{dp}{dz} = \rho_p(1 - \varepsilon)g \frac{(u_s - u_f)|u_s - u_f|^{(m-1)}}{(u_{mfp})^m}. \tag{54}$$

It is convenient to express u_f and u_s in terms of mass flow rate G_f and G_p. Thus, for a conical bed,

$$u_f = \frac{G_f}{\pi(r_b + z \tan \alpha)^2 \rho_f \varepsilon} \tag{55}$$

and

$$u_s = \frac{G_p}{\pi(r_b + z \tan \alpha)^2 \rho_p(1 - \varepsilon)}. \tag{56}$$

When the fluid is a compressible gas, the fluid density ρ_f and the actual gas velocity at incipient fluidization u_{mfp} (Kwauk, 1963) are related to gas pressure p by

$$\rho_f = \frac{Mp}{R_g T},\tag{57}$$

$$u_{mfp} = u_{mfo}\left(\frac{p}{p_o}\right)\left(\frac{1}{m} - 1\right),\tag{58}$$

where u_{mfo} is the incipient fluidization velocity at fluid pressure p_o.

4. Differential Equation for Voidage Gradient

The experiments of Li and Kwauk (1989) showed that in non-fluidized gas–particle flow, voidage varies linearly with contact pressure between particles, i.e., when $0 < \sigma < \sigma_c$,

$$\varepsilon = \varepsilon_{mf} - (\varepsilon_{mf} - \varepsilon_c)\sigma/\sigma_c,\tag{59}$$

and when $\sigma \geq \sigma_c$,

$$\varepsilon = \varepsilon_c.\tag{60}$$

Voidage gradient can be obtained by differentiation of Eqs. (59) and (60) with respect to distance z:

$$\frac{d\varepsilon}{dz} = -\frac{(\varepsilon_{mf} - \varepsilon_c)}{(\bar{\sigma}_z)_c}\frac{d\bar{\sigma}_z}{dz} \qquad (0 < \bar{\sigma}_z \leq (\bar{\sigma}_z)_c),\tag{59a}$$

$$\frac{d\varepsilon}{dz} = 0 \qquad (\bar{\sigma}_z > (\bar{\sigma}_z)_c).\tag{60a}$$

5. Longitudinal Distributions of Interparticle Contact Pressure $\bar{\sigma}_z$, Gas Pressure p, and Voidage ε

In principle, the longitudinal distribution curves, $\bar{\sigma}_z$–z, p–z, and ε–z, can be obtained by simultaneous numerical solution of Eqs. (52), (4a) or (54), (59a), and (60a), with the initial values of $\bar{\sigma}_z = 0$, $p = p_2$, and $\varepsilon = \varepsilon_{mf}$ at $z = z_2$. In fact, voidage varies mainly near the top surface and the bottom orifice within a narrow range of about 0.04 (Li and Kwauk, 1989). It is therefore advantageous to simplify the preceding set of differential equations by ignoring this slight variation and assuming voidage ε to be constant. In addition, the gas pressure gradient dp/dz approximates the average pressure gradient $\Delta p/\Delta z$ when the bed is not too long, so that Eq. (52) has the following analytical solutions with the initial value of $\bar{\sigma}_z = 0$ at $z = z_2$.

For an inclined cylindrical moving bed with average voidage ε_s,

$$\bar{\sigma}_z = \pm \frac{-r_b}{2H_v D_c \mu}\left(\frac{\Delta p}{\Delta z} + \rho_p(1-\varepsilon_s)g\cos\theta \pm \frac{8\mu}{3\pi}\rho_p(1-\varepsilon_s)g\sin\theta\right)$$
$$\times (1 - \exp(\pm 2H_v D_c \mu(z_2 - z)/r_b)); \quad (61)$$

for a vertical cylindrical moving bed with average voidage ε_v,

$$\bar{\sigma}_z = \pm \frac{-r_b}{2H_v D_c \mu}\left(\frac{\Delta p}{\Delta z} + \rho_p(1-\varepsilon_v)g\right) \times (1 - \exp(\pm 2H_v D_c \mu(z_2 - z)/r_b)); \quad (62)$$

and for a vertical cylindrical downward moving bed without interstitial gas flow with average voidage ε_0,

$$\bar{\sigma}_z = \frac{r_b}{2H_v D_c \mu}\rho_p(1-\varepsilon_0)g \times (1 - \exp(-2H_v D_c \mu(z_2 - z)/r_b)). \quad (63)$$

The last equation can be recognized as the well-known Janssen's equation (Janssen, 1895).

Voidage ε_0 can be measured in a vertical cylindrical downward moving bed with the condition of $\Delta p/\Delta z = 0$ by weighing the amount of particles W_0 received in a container connected closely to the bottom of the moving bed, and meantime measuring the gas volume V_0 flowing out of the container under the same pressure within the container, i.e.,

$$\varepsilon_0 = 1 - \frac{W_0}{\rho_p V_0}. \quad (64)$$

6. Relations among ε_s, ε_v, ε_0, and ε_{mf}

Substituting Eqs. (61) and (63) into Eq. (59) yields

$$\varepsilon_s = \varepsilon_{mf} - (\varepsilon_{mf} - \varepsilon_0)\frac{\rho_p(1-\varepsilon_s)g\left(\cos\theta - \frac{8\mu}{3\pi}\sin\theta\right) + \frac{\Delta p}{\Delta z}}{\rho_p(1-\varepsilon_0)g}.$$

Rearranging this equation gives

$$\varepsilon_s = \frac{\varepsilon_{mf} - \frac{\varepsilon_{mf} - \varepsilon_0}{1 - \varepsilon_0}\left(\cos\theta - \frac{8\mu}{3\pi}\sin\theta + \frac{\Delta p}{\rho_p g \Delta z}\right)}{1 - \frac{\varepsilon_{mf} - \varepsilon_0}{1 - \varepsilon_0}\left(\cos\theta - \frac{8\mu}{3\pi}\sin\theta\right)}. \quad (65)$$

Similarly, substituting Eqs. (62) and (63) into Eq. (59) and rearranging yields

$$\varepsilon_v = \frac{\left(\varepsilon_{mf} - \dfrac{(\varepsilon_{mf} - \varepsilon_0)}{(1 - \varepsilon_0)}\left(1 + \dfrac{\Delta p}{\rho_p g \Delta z}\right)\right)}{\left(1 - \dfrac{(\varepsilon_{mf} - \varepsilon_0)}{(1 - \varepsilon_0)}\right)}. \tag{66}$$

The voidages ε_{mf} and ε_0 are easy to measure. Therefore, ε_s and ε_v can be calculated by Eqs. (65) and (66). If the calculated values of ε_s and ε_v are less than that of ε_c, then ε_s and ε_v should be replaced by ε_c.

B. Sealing Ability for Moving Beds

The conventional sealing ability denotes the maximal negative pressure gradient with which the moving bed can be loaded. It is obvious that the theoretical sealing ability of a moving bed is equal to the bulk density of the solid particles, i.e.,

$$-\left(\frac{dp}{dL}\right)_t = \rho_p(1 - \varepsilon)g. \tag{67}$$

According to Kuo and Soon (1960), there exists a critical value of negative pressure gradient for a countercurrent–cogravity moving bed, which is less than the theoretical sealing ability. When the negative pressure gradient exceeds this critical value, the moving bed becomes unstable. The critical negative pressure gradient can be calculated by Kuo's equation:

$$-\left(\frac{dp}{dL}\right)_c = 0.65[\rho_p(1 - \varepsilon)g - 1334]. \tag{68}$$

C. Fluid–Particle Flow Phase Diagram

Fluid–particle flow patterns are often clarified by application of phase diagrams. Kwauk (1963) and Matsen (1983) proposed such phase diagrams for moving beds.

Following Kwauk (1963), Li (1991) proposed a modified fluid–particle flow phase diagram for vertical moving beds. This correlates quantitatively 13 modes of moving-bed operation by using a modification of Kwauk's equation for moving-bed pressure drop (Kwauk, 1963). The ideal sealing state, where no fluid passes through the moving bed, useful in the operation of industrial diplegs, was also put forward by Li (1973, 1991).

It is convenient to convert Eq. (54) into dimensionless form using reduced quantities, $u'_f = u_f/u_{mf}$, $u'_s = u_s/u_{mf}$, and $(dp/dL)' = (dp/dL)/(\rho_p(1-\varepsilon)g)$, while taking $(-1)^m = -1$ to accord with physical truth, such that

$$\left(\frac{dp}{dL}\right)'^{1/m} = u'_s - u'_f. \tag{69}$$

Equation (69) contains the fundamental quantitative relationship for moving beds. The four variables of Eq. (69), $(dp/dL)'$, u'_f, u'_s and m, are all dimensionless, and with the exception of m, which is a function of Re_t and ε (Kwauk, 1963), the remaining three variables are determined by operation of the system. Equation (69) can be used to provide a graphical representation of all the modes of the operation, as shown in Fig. 22. The origin, O, represents the original static bed in which both particles and fluid are at a standstill. From Eq. (69) and Fig. 22, it is obvious that u'_s–u'_f curves have a constant slope of unity, while its intercept on the u'_s-axis is $(dp/dL)'^{1/m}$. Figure 22 will be referred to as the fluid–particle flow phase diagram for vertical moving beds. The physical significance of the various points, lines and regions in Fig. 22 may be clarified from a quantitative point of view by application of Eq. (69).

With upward velocities defined as positive, and downward velocities as negative, the pressure difference between top and bottom of a moving bed is positive when the pressure at the top is higher than at the bottom. Otherwise, the pressure difference is negative.

Lines AOB, COD, and EOF divide the u'_s–u'_f plane into six regions, representing six modes of operation as follows:

Region BOC: Negative pressure difference, countercurrent–cogravity operation (abbreviated designation: $-\Delta p$ counter-down)

Region COE: Negative pressure difference, cocurrent–countergravity operation ($-\Delta p$ co-up)

Region BOF: Negative pressure difference, cocurrent–cogravity operation ($-\Delta p$ co-down)

Region DOF: Positive pressure difference, cocurrent–cogravity operation ($+\Delta p$ co-down)

Region AOD: Positive pressure difference, countercurrent–countergravity operation ($+\Delta p$ counter-up)

Region AOE: Positive pressure difference, cocurrent–countergravity operation ($+\Delta p$ co-up)

The six boundary lines represent six modes of operation as follows:

Line OD: Positive pressure difference, fixed-bed operation

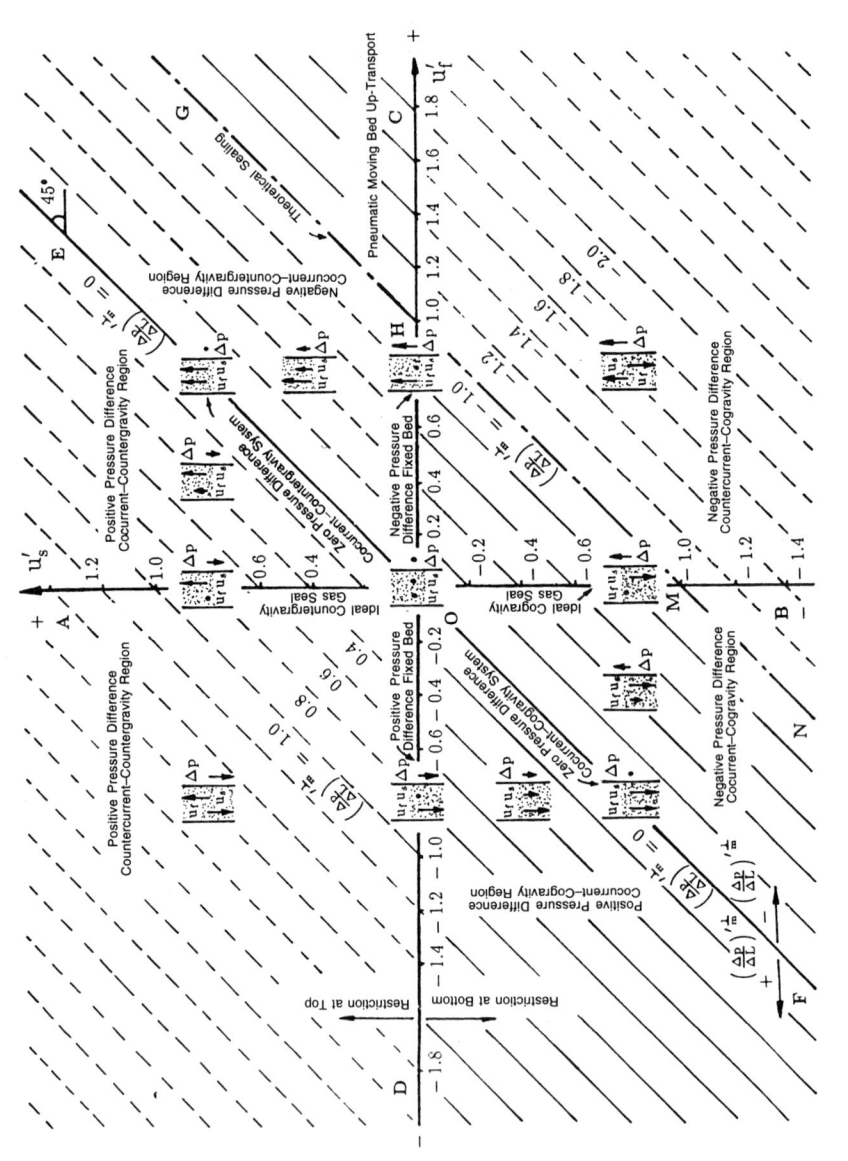

FIG. 22. Fluid-particle flow phase diagram for vertical moding bed (Li, 1991).

Line OC: Negative pressure difference, fixed-bed operation
Line OE: Zero pressure difference, cocurrent–countergravity operation
Line OF: Zero pressure difference, cocurrent–cogravity operation
Line OA: Ideal countergravity gas seal operation
Line OB: Ideal cogravity gas seal operation

The regions with fine dotted lines are inoperable when solids motion is initiated only by gravity and drag forces. However, these regions may become operable when the particles are motivated by other forces such as magnetic, mechanical, or electrical.

When particles are motivated only by gravity and drag forces, only the following regions DOF, FOMN, MOH, and CHG are operable:

DOF: $+\Delta p$ co-down, particles forced down by both gravity and drag;
FOMN: $-\Delta p$ co-down, particles fall by gravity with drag acting upward;
MOH: $-\Delta p$ counter-down, particles fall by gravity with drag acting upward (shaft furnace);
CHG: $-\Delta p$ co-up, particles moved up by drag of upflowing gas (moving-bed uptransport).

D. The Ideal Sealing State and Its Application

The ideal gas sealing condition is represented by line AOB for $u'_f = 0$, on which particles move while the fluid is at a standstill. When particles move downwards, the ideal sealing state $u_f = 0$ can be realized under a negative pressure gradient, called the value of the ideal gas seal, which can be calculated from Eq. (4a) or Eq. (54) with $u_f = 0$, i.e.,

$$\left|\frac{dp}{dL}\right|_i = 150\left(\frac{1-\varepsilon}{\varepsilon}\right)^2 \frac{\mu_f |u_s|}{(\phi_s \bar{d}_p)^2} + 1.75\left(\frac{1-\varepsilon}{\varepsilon}\right)\frac{\rho_f u_s^2}{\phi_s \bar{d}_p}, \quad (4b)$$

or

$$\left|\frac{dp}{dL}\right|_i = \rho_p(1-\varepsilon)g\left(\frac{|u_s|}{u_{mf}}\right)^m. \quad (54a)$$

From Eqs. (4b) and (54a), we can also see that the value of ideal gas seal depends on the properties of the fluid staying in the moving bed. The fluid above the top of the moving bed is usually different from that below the bottom of the moving bed, and the fluid within the moving bed can be chosen as desired by controlling the value of the ideal gas seal.

The ideal sealing state can not be obtained for all particle velocities. We can see from Eqs. (4b) and (54a) that the value of the ideal gas seal increases with the particle velocity. When the particle velocity increases to a certain value, the value of the ideal gas seal exceeds the critical negative pressure gradient, i.e., $|dp/dL|_i > |dp/dL|_c$, and may go so far as to be more than the theoretical seal ability, i.e., $|dp/dL|_i > |dp/dL|_t$. In these cases, stable operation will be dstroyed. Therefore, there exists a maximum particle velocity for maintaining the ideal sealing state. This is called the theoretical critical particle velocity u_{sc}. It can be calculated from the following equations:

$$\left|\frac{dp}{dL}\right|_t = 150\left(\frac{1-\varepsilon}{\varepsilon}\right)^2 \frac{\mu_f |u_{sc}|}{(\phi_s \bar{d}_p)^2} + 1.75\left(\frac{1-\varepsilon}{\varepsilon}\right)\frac{\rho_f (u_{sc})^2}{\phi_s \bar{d}_p}, \quad (4c)$$

or

$$\left|\frac{dp}{dL}\right|_t = \rho_p (1-\varepsilon)g \left|\frac{u_{sc}}{u_{mf}}\right|^m. \quad (54b)$$

The point $M(u'_s = -1)$ in the phase diagram, Fig. 22, represents the theoretical critical particle velocity. It is clear that

$$u_{sc} = -u_{mf}. \quad (70)$$

The application of the preceding analysis will now be illustrated by the following example taken from process engineering.

Design a return dipleg for a circulating fluidized bed. The dipleg is desired to operate in the ideal sealing state of moving bed, and the gas above the top of dipleg is desired to stay in the dipleg.
Data:

Average diameter of particles $\bar{d}_p = 4.16 \times 10^{-4}$ m
Shape factor of particles $\phi_s = 1.0$
Density of particles $\rho_p = 1{,}603$ kg/m^3
Voidage $\varepsilon = 0.477$
Circulation rate of particles $G_p = 2.133$ kg/s
Height of moving bed in dipleg $L = 3.5$ m
Pressure difference between the top and the bottom of the dipleg, $\Delta p = p_t - p_b = -14{,}813$ N/m^2
Average temperature in the dipleg $T = 763$ K
Average pressure in the dipleg $p = 47{,}096$ N/m^2
Viscosity of gas in the dipleg from top $\mu_{ft} = 2.93 \times 10^{-5}$ kg/(m·s)
Viscosity of gas in the dipleg from bottom $\mu_{fb} = 3.05 \times 10^{-5}$ kg/(m·s)
Average density of gas in the dipleg from top $\rho_{ft} = 0.565$ kg/m^3
Average density of gas in the dipleg from bottom $\rho_{fb} = 0.61$ kg/m^3

Solution:
(i) Critical pressure gradient:

$$-\left(\frac{dp}{dL}\right)_c = 0.65[\rho_p(1-\varepsilon)g - 1334] = 0.65[1{,}603(1-0.477)9.81 - 1{,}334]$$

$$= 4{,}478.8 \text{ N/m}^3.$$

(ii) Real pressure gradient:

$$\frac{dp}{dL} = \frac{-14{,}831}{3.5} = -4{,}232.3 \text{ N/m}^3,$$

$$\left|\frac{dp}{dL}\right| < -\left(\frac{dp}{dL}\right)_c.$$

Therefore the operation is stable.
(iii) Particle velocity for ideal sealing operation, u_s:
u_s can be determined from Eq. (4b) or Eq. (54a). In this case,

$$\left|\frac{dp}{dL}\right|_i = \left|\frac{dp}{dL}\right| = 4{,}232.3 \text{ N/m}^3,$$

$$\mu_f = \mu_{ft} = 2.93 \times 10^{-5} \text{ kg/(m·s)}.$$

Then,

$$u_s = -0.128 \text{ m/s}$$

can be obtained by solving Eq. (4b).
(iv) Diameter of the dipleg, D
The diameter of dipleg should be determined from

$$G_p = \frac{\pi}{4}D^2 u_s \rho_p (1-\varepsilon).$$

Hence,

$$D = \left[\frac{4G_p}{\pi u_s \rho_p(1-\varepsilon)}\right]^{0.5} = \left[\frac{4 \times 2.133}{\pi \times 0.128 \times 1{,}603 \times (1-0.477)}\right]^{0.5} = 0.159 \text{ m}.$$

E. BRIDGING

A particular hazard in the operation of non-fluidized diplegs handling cohesive powders is their tendency to bridge (or to arch) unexpectedly, even though the dipleg has been correctly designed for mass flow. Powder material below the arch empties out completely, then the circulation of powder material is stopped, leaving a huge cavity below an unstable arch, which, on collapse, can bring down perhaps several tons of bulk solid with disastrous consequences. The pipe may burst under the impact, and the supporting structure may be severely damaged by shock overloading.

The mechanism of arching refers to particulate media mechanics. Conventional studies of arching were focused on the arching in mass flow hoppers, where particulate solids move downwards with the aid of gravity (Jenike, 1964; Walker and Blanchard, 1967; Enstad, 1975). However, there is a difference between the mass flow hopper and the non-fluidized dipleg. In the non-fluidized dipleg, there exist interstitial gas flow and the pressure drop due to gas–particle friction. Based on the dynamics of non-fluidized gas–particle flow proposed by Li (1990), and an approximate theory for arching in hoppers proposed by Enstad (1975), the following assumptions are put forward (Li, 1994).

Consider the formation of a free static arch for a material which has previously been moving in a dipleg. The present concept treats the blocked mass as made up of a stack of self-supporting element arches of similar basic shape. The basic element arch can be taken as of uniform vertical thickness, because this closely approximates the requirement that the element shape, after appropriate scaling to bridge the pipe wall, should be capable of forming a stack of element arches one upon another, which then accounts for the entire volume of material over the lower part of the dipleg. When a powder arches in a dipleg, the stresses will be of a quite different kind, for the powder is not in a fully plastic state of stress. In order to determine the critical arching radius, which marks the change from arching to flow, a situation where all the powder is in a critical passive state of stress has therefore to be considered. That means that the major and minor principal stresses $\bar{\sigma}_1$ and $\bar{\sigma}_2$ belong to a Mohr circle touching the actual failure locus, line PYL, as shown in Fig. 23. The powder in the dipleg is assumed to be composed of arched layers, the cross-sections of which are shown in Fig. 24. Each cross-section is considered to be limited by eccentric circles with centers on the center line of the vertical dipleg cross-section. The major principal stress is assumed to be tangential to the circles, whereas the minor principal stress is normal to them. The stresses are assumed to vary only in direction along a layer, but not in magnitude. The angle, $\bar{\beta}$, between the tangent to the circle at the point where it intersects the pipe wall, and the direction perpendicular to the wall, is assumed to be such that the abutments get the maximum lift from the wall at the same time as the powder does not slide along the wall. In addition, the voidage of the bed is assumed to be constant, and the gas velocity, uniformly distributed over the radial cross-section.

Consider an elemental arched layer, dh, shown in Fig. 24. Four forces act

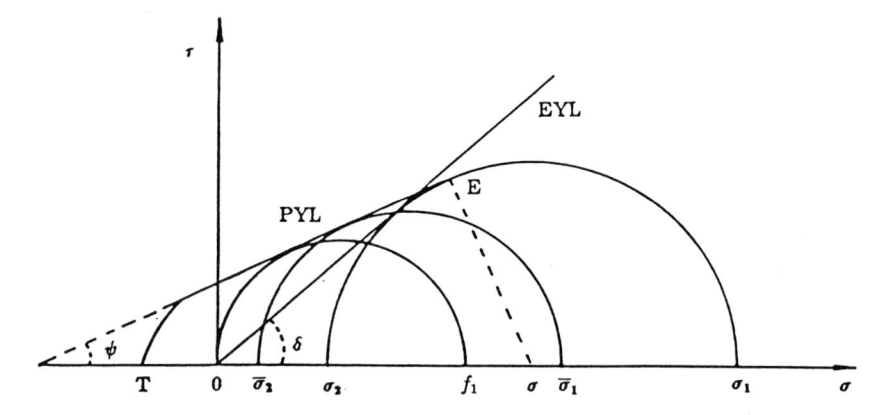

FIG. 23. Critical stresses $\bar{\sigma}_1$ and $\bar{\sigma}_2$ when the failure locus is considered linear and powder has been consolidated by the major principal stress σ_1 (Enstad, 1975).

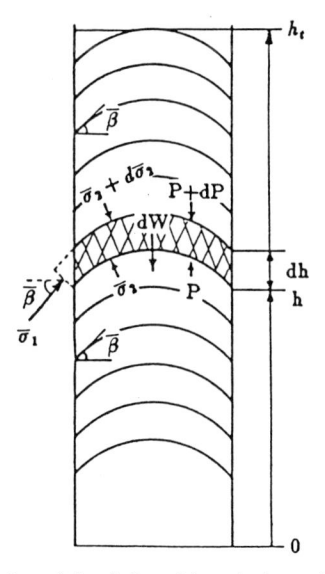

FIG. 24. Cross-section of the dipleg with arched powder layers (Li, 1994).

on the layer: the weight, the interaction with the powder above and below, the reaction from the walls, and the drag force from the gas flow.
The volume of the layer, dv, is

$$dv = \pi R^2 \, dh.$$

The weight of the layer, dw, will then be

$$dw = \rho_b g \pi R^2 \, dh.$$

The powder on top of the layer will exert a downward stress on the layer, resulting in a downward force:

$$F_a = \pi R^2 (\bar{\sigma}_2 + d\bar{\sigma}_2).$$

From the powder below, the layer will experience an upward force:

$$F_b = \pi R^2 \bar{\sigma}_2.$$

The net lift from the walls is

$$dF_f = \bar{\sigma}_1 2\pi R \cos \bar{\beta} \sin \bar{\beta} \, dh - \bar{\sigma}_2 2\pi R \sin \bar{\beta} \cos \bar{\beta} \, dh$$
$$= (\bar{\sigma}_1 - \bar{\sigma}_2) \pi R \sin 2\bar{\beta} \, dh. \tag{71}$$

It can be seen from Eq. (71) that the lifting force dF_f from the wall will be related to the angle $\bar{\beta}$. The formation of arch is considered to be more or less random. Hence, the value of $\bar{\beta}$ which gives maximum lift has to be considered. Because $\bar{\beta}$ is limited by the wall friction angle ϕ, a larger value of $\bar{\beta}$ would make the powder slide along the wall until an arch with a smaller $\bar{\beta}$ were formed. However, dF_f has its maximum when $\sin 2\bar{\beta}$ has its maximum, which occurs when $2\bar{\beta} = \frac{\pi}{2}$. Thus, $\bar{\beta}$ will be given by

$$\bar{\beta} = \begin{cases} \phi, & \text{if } \phi \leq \dfrac{\pi}{4} \\ \dfrac{\pi}{4}, & \text{if } \phi > \dfrac{\pi}{4} \end{cases}. \tag{72}$$

The drag force on the arched layer from gas flow, dF_d, is

$$dF_d = \frac{\Delta P}{\Delta h} \pi R^2 \, dh,$$

where $\Delta P/\Delta h$ represents the average gas pressure gradient in the dipleg. Vertical force balance for the elemental arched layer in the dipleg gives

$$dw + dF_d + F_a = F_b + dF_f,$$

that is,

$$\rho_b g \pi R^2 \, dh + \frac{\Delta P}{\Delta h} \pi R^2 \, dh + \pi R^2 (\bar{\sigma}_2 + d\bar{\sigma}_2)$$
$$= \pi R^2 \bar{\sigma}_2 + (\bar{\sigma}_1 - \bar{\sigma}_2) \pi R \sin 2\bar{\beta} \, dh.$$

Rearranging this equation yields

$$\frac{d\bar{\sigma}_2}{dh} - \frac{\sin 2\bar{\beta}}{R}(\bar{\sigma}_1 - \bar{\sigma}_2) = -\rho_b g - \frac{\Delta P}{\Delta h}. \quad (73)$$

From Fig. 23, the following expression can be derived:

$$\bar{\sigma}_1 = \frac{1 + \sin \psi}{1 - \sin \psi}\bar{\sigma}_2 + f_1. \quad (74)$$

Substituting Eq. (74) into Eq. (73) and rearranging give the differential equation of force balance:

$$\frac{d\bar{\sigma}_2}{dh} - \frac{\sin 2\bar{\beta}}{R}\left(\frac{1 + \sin \psi}{1 - \sin \psi} - 1\right)\bar{\sigma}_2 = \frac{\sin 2\bar{\beta}}{R}f_1 - \left(\rho_b g + \frac{\Delta p}{\Delta h}\right). \quad (75)$$

The unconfined yield strength of the powder, f_1, is a powder characteristic which decides whether it can resist flow under gravity from a certain dipleg configuration, and is a function of the consolidating stress (σ_1) present during its preparation as shown in Fig. 25.

Furthermore, the failure function $f_1(\sigma_1)$ can also be assumed to be a straight line as given by

$$f_1 = k\sigma_1 + f_c. \quad (76)$$

The major principal or consolidating stress (σ_1) depends on the stress condition just before the arching. The flowing powder in the dipleg is considered to be in passive state of stress. The axial profile of the average vertical stress over horizontal cross-section, $\bar{\sigma}_h$, has been given by Li (1990)

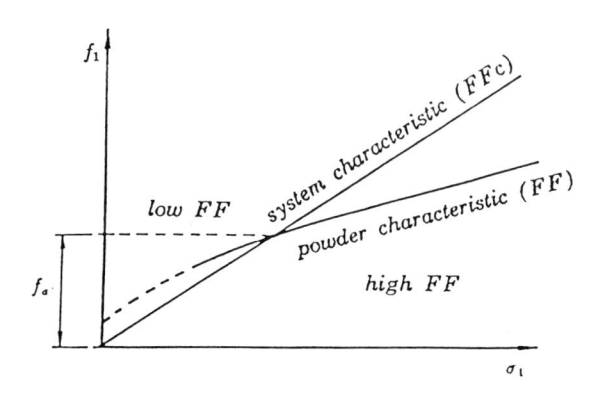

FIG. 25. Flow factor characteristics (Li, 1994).

with the initial value of $\bar{\sigma}_h = 0$ at $h = h_t$ as follows:

$$\bar{\sigma}_h = \frac{R}{2H_vD_c \tan \phi}\left[\rho_b g + \frac{\Delta p}{\Delta h}\right]\left[1 - \exp\left(-2H_vD_c \tan \phi \frac{h_t - h}{R}\right)\right]. \quad (62a)$$

For a deep dipleg, $(h_t - h) \gg R$, Eq. (62a) can therefore be simplified as follows:

$$\bar{\sigma}_h = \frac{R}{2H_vD_c \tan \phi}\left[\rho_b g + \frac{\Delta p}{\Delta h}\right]. \quad (62b)$$

From Eq. (33), Eq. (39) and Eq. (62b) it is readily found that

$$\sigma_1 = \left(\frac{1 + \sin \delta}{1 - \sin \delta \cos 2\beta}\right)\frac{R}{2H_v \tan \phi}\left[\rho_b g + \frac{\Delta p}{\Delta h}\right]. \quad (77)$$

Equation (77) shows that in a deep dipleg, the value of σ_1 is a constant independent of the height of the dipleg. Therefore, the value of f_1 is also a constant.

Solving the differential equation (75) with the initial value of $\bar{\sigma}_2 = 0$ at $h = h_t$ yields

$$\bar{\sigma}_2 = \frac{1 - \sin \psi}{2 \sin \psi}\left[\frac{\left(\rho_b g + \frac{\Delta p}{\Delta h}\right)R}{\sin 2\bar{\beta}} - f_1\right]$$

$$\times \left[1 - \exp\left(-\sin 2\bar{\beta}\left(\frac{2 \sin \psi}{1 - \sin \psi}\right)\frac{h_t - h}{R}\right)\right]. \quad (78)$$

When $(h_t - h) \gg R$, one has

$$\bar{\sigma}_2 = \frac{1 - \sin \psi}{2 \sin \psi}\left[\frac{\left(\rho_b g + \frac{\Delta p}{\Delta h}\right)R}{\sin 2\bar{\beta}} - f_1\right]. \quad (79)$$

Because the stress normal to the free bottom surface of the arch would have to be zero, the value of the unconfined yield strength f_1, which is necessary for arching, can be obtained from Eq. (79) with the condition of $\bar{\sigma}_2 = 0$, i.e.,

$$f_1 = \frac{\left(\rho_b g + \frac{\Delta p}{\Delta h}\right)R}{\sin 2\bar{\beta}}. \quad (80)$$

Here, a concept of critical flow factor of equipment (FFC), proposed by Walker (1966), is used to evaluate the flowability of the powder–equipment system:

$$\mathrm{FFc} = \frac{\sigma_1}{f_1}. \tag{81}$$

Substituting Eq. (77) and Eq. (80) into Eq. (81) gives the critical flow factor for a cylindrical dipleg,

$$\mathrm{FFc} = \frac{\sigma_1}{f_1} = \left[\frac{1 + \sin\delta}{1 + \cos 2\bar{\beta} \sin\delta}\right] \frac{\sin 2\bar{\beta}}{2\tan\phi}. \tag{82}$$

It can be seen from Eq. (82) that the critical flow factor is a constant and depends on the properties of the powder and the pipe wall. Another concept of flow factor (FF) proposed by Jenike (1964) is then used to evaluate the flowability of the powder:

$$\mathrm{FF} = \frac{\sigma_1}{f_1}. \tag{83}$$

In practice, however, the value of FF is not a constant but increases with consolidating stress. It is determined by the experimental f_1–σ_1 curve as shown in Fig. 25. The $f_1 - \sigma_1$ curve can approximately be expressed by Eq. (76). In this case a critical radius R_c of the dipleg can be calculated.

Figure 25 illustrates a case where the powder flow factor locus (FF) lies in part above and part below a critical flow factor locus (FFc). When the powder flow factor locus lies below the (FFc) locus in the graph, then the powder flow factor (FF) is desirably high, but when the powder flow factor locus is above the (FFc) line, its flow factor is less than (FFc) and arching can occur. If we let the critical value of f_1 at the crossover point be f_a, then to ensure no arching the radius R_c should be given by Eq. (84) such that

$$R_c \geq \frac{f_a \sin 2\bar{\beta}}{\left(\rho_b g + \dfrac{\Delta p}{\Delta h}\right)}. \tag{84}$$

If the powder f_1–σ_1 curve can be represented by Eq. (76), R_c should be given directly by Eq. (80) and Eq. (82), such that

$$R_c \geq \frac{f_c \sin 2\bar{\beta}}{[1 - k(\mathrm{FFc})]\left[\rho_b g + \dfrac{\Delta p}{\Delta h}\right]}. \tag{85}$$

III. The Pneumatically Actuated Pulse Feeder

Nonmechanical, pneumatic operation is the preferred guideline for designing fluid-bed devices. Figure 26 shows schematically the pneumatically actuated pulse feeder used in controlling solids circulation rate (Kwauk et al., 1985). It consists of a downcomer connected to a solids charge hopper at the top and terminating at a distance above a stop plate at the bottom. The bottom plate is adjustable to vary the clearance between it and the downcomer. Into the middle of the downcomer and above the stop plate are fed downwardly directed puffs of air from a centrally located tube connected through a solenoid valve to an air supply. The on-and-off frequency and duration of the open period of the solenoid valve are controlled by a pulse current generator. The air flow rate through the central tube is adjusted to some desired value while the solenoid valve is turned on manually.

The stop plate is adjusted to a height so that the angle its rim makes with that of the downcomer is always less than the angle of repose of the solids used. In this manner, the solids in this peripheral zone do not flow while no air issues from the central tube, and are splashed radially outward by air puffs during the on-period of the solenoid valve.

It has been found that the solids feed rate is generally linear with the frequency of the solenoid valve, as shown by a typical calibration curve shown in Fig. 27.

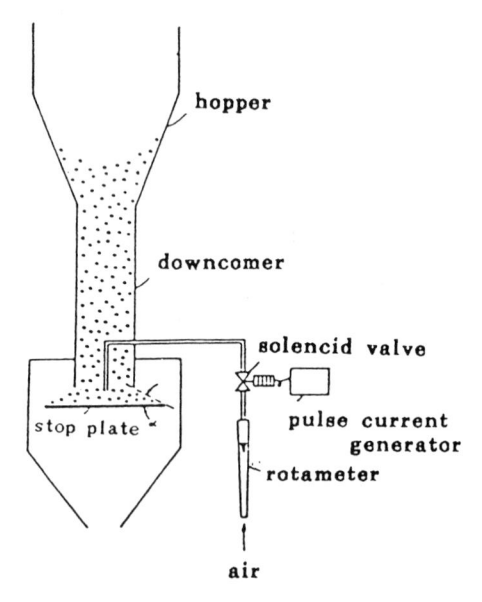

FIG. 26. Pneumatically actuated pulse feeder (Kwauk et al., 1985).

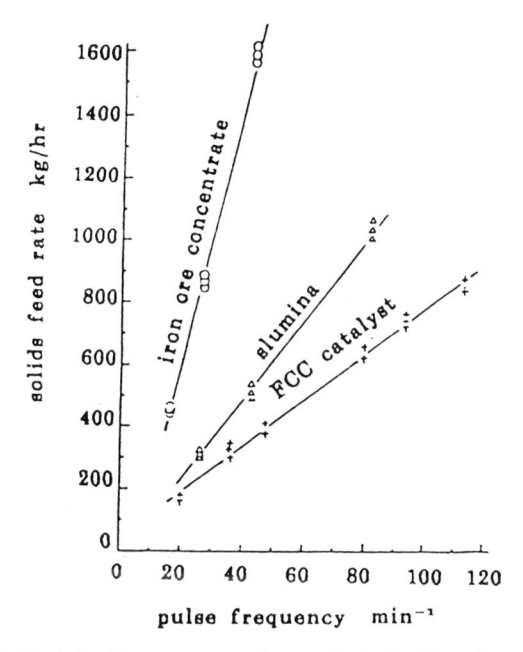

FIG. 27. Typical calibration curves for a pulse feeder (Kwauk et al., 1985).

Normally the pneumatically controlled pulse feeder operates with limited pressure difference between the hopper and the stop plate. When the pressure difference is in the direction opposite to that of the downward solids flow, the downcomer needs to be lengthened in order to seal the solids, or else other modes of downcomer design must be resorted to.

IV. Internals

Heterogeneity of radial and axial solids distributions is a basic feature of fast fluidized beds. Axially, two regions in the S-shaped voidage profile have been noted: a dilute-phase region at the top and a dense-phase region at the bottom. Between these two regions, there is a transition zone centered around the inflection point of the voidage profile. Radially, solids concentration is relatively low in the center of the bed and increases toward the wall, that is, solids tend to cluster and migrate to the tube wall where fluid moves slowly. This often results in particles downflow at the wall, and consequently leads to solids backmixing, therefore seriously affecting the selectivity and yield of some chemical reactions.

Efforts on improving radial voidage distribution were focused on appropriate design and installation of internals to change the radial profiles of gas velocities. Zheng and Tung (1990) found that the near parabolic radial voidage profiles could be flattened by perforated plates and ring internals, and among these internals, the perforated plate with an annular gap near the tube wall was considered the best among those they tested. Unfortunately, the cross-section-averaged solids concentration was thereby diluted under the influence of internals. This dilution will lead to reduction of solids inventory in the fast bed, thereby imparing the efficiency for chemical reactions.

Afterwards, Zheng *et al.* (1992) developed a new kind of ring internals, which can minimize solids backmixing in both the dilute-phase and dense-phase regions while preserving the overall solids inventory to insure adequate chemical reaction. Experiments with the new ring internals were carried out in a 10-m high by 90-mm i.d. FCC–air fast fluidized bed. Figure 28 shows a schematic diagram of the experimental apparatus. Axial voidage profiles were measured automatically by using a step-scanning transducer-valving mechanism for connecting the pressure-drop taps along the fast bed, pair by pair in succession, to an IBM-PC/XT computer, and radial voidage was measured by an optic fiber probe.

Solids backmixing was determined from residence time distribution curves by using FCC tracer particles impregnated with NaCl. Figure 29 shows that the content of tracer in particles C is related linearly to the corresponding conductivity e of their aqueous solutions. The solid tracer was pulse-injected into the bed by high-pressure air, and solids were withdrawn by a special sampler consisting of a cyclone separator and a slender storage tube, capable of collecting a series of samples in succession. The locations for tracer injection and sampling, both upstream and downstream with respect to injection, in a fast fluidized bed are shown in Fig. 30. Experimental results were compared for the fast bed both with and without internals.

When two units of the ring internals were installed at $H = 2.25$ m and $H = 4.75$ m, the dense-phase region was cut into three layers with a dilute influence zones in between and was thereby itself extended upward, while the solids inventory remained the same. Voidages ε^* for the dilute-phase region and ε_a for the dense-phase region remain essentially the same, as shown in Fig. 31. In the influence zone of the internals, the radial voidage distribution is considerably flattened, as shown in Fig. 32. Reduced solid concentration in an influence zone is instrumental in suppression of solids mixing between adjacent dense-phase layers.

Solids backmixing studies were carried out with the use of a single set of rings installed at $H = 4.00$ m. Figure 33 shows the tracer concentrations at sampling location #1, which is 0.46 m upstream of tracer injection, for both

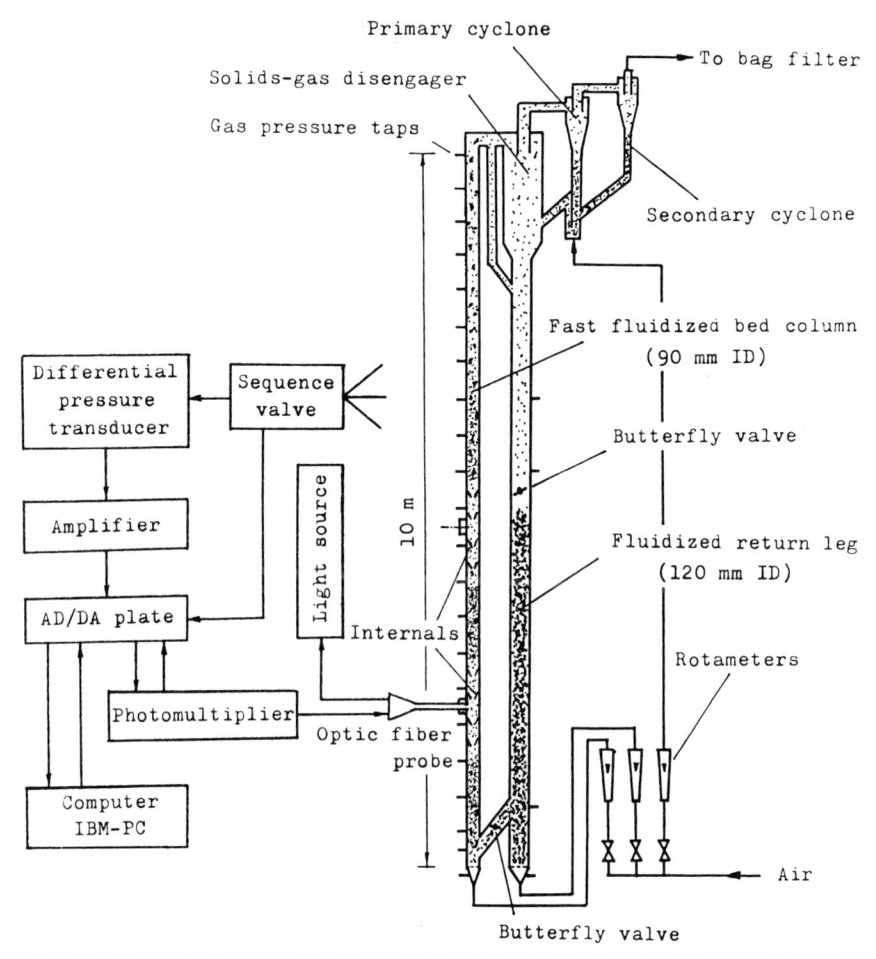

FIG. 28. Schematic diagram of the 90 mm i.d. circulation fluidized system (Zheng et al., 1992).

the empty column and for the case of a ring set. Without the ring set, Fig. 33 shows evident solids backmixing in a tracer concentration of as high as 0.1%, while the use of the ring set, tracer concentration drops down to a negligible value.

Figure 34 compares the residence time distribution in the influence zone of a single set of ring internals at location #3, which is 0.55 m downstream of tracer injection, with and without the ring set. The decrease of solids backmixing is shown by an increase of the Peclet number from 1.06 to 2.46, and from 1.97 to 3.61, when the ring set is used, for gas velocities of 1.75 and 2.62 m/s, respectively.

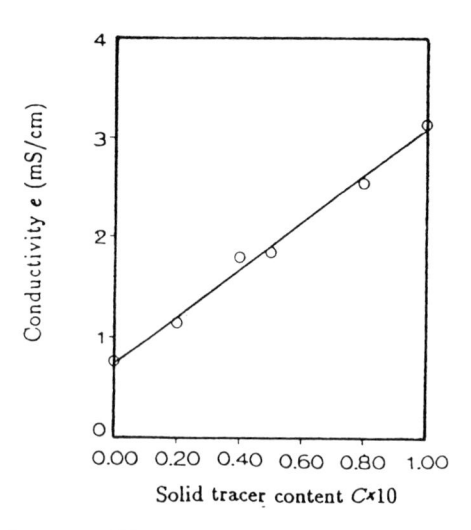

FIG. 29. Relation between solid tracer content and conductivity of its aqueous solution (Zheng et al., 1992).

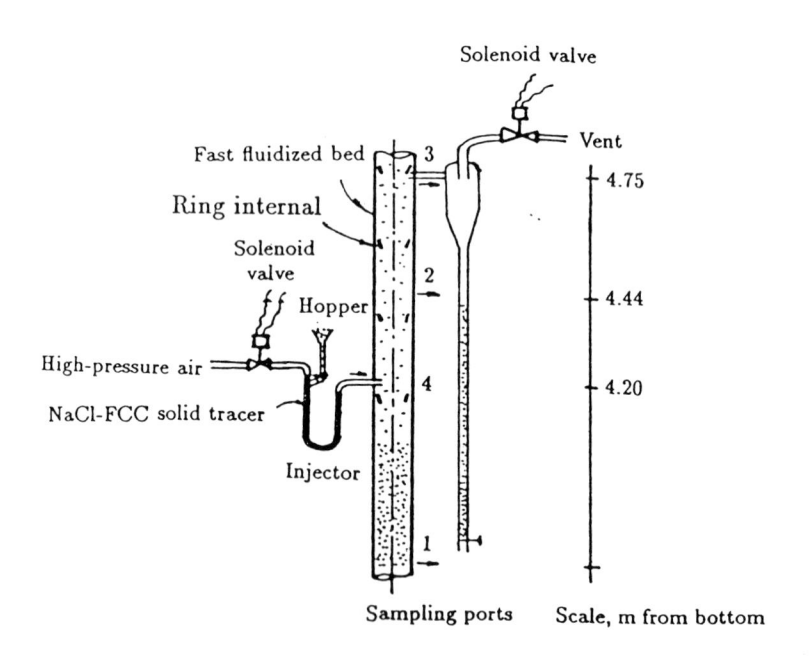

FIG. 30. Injection and sampling system (Zheng et al., 1992).

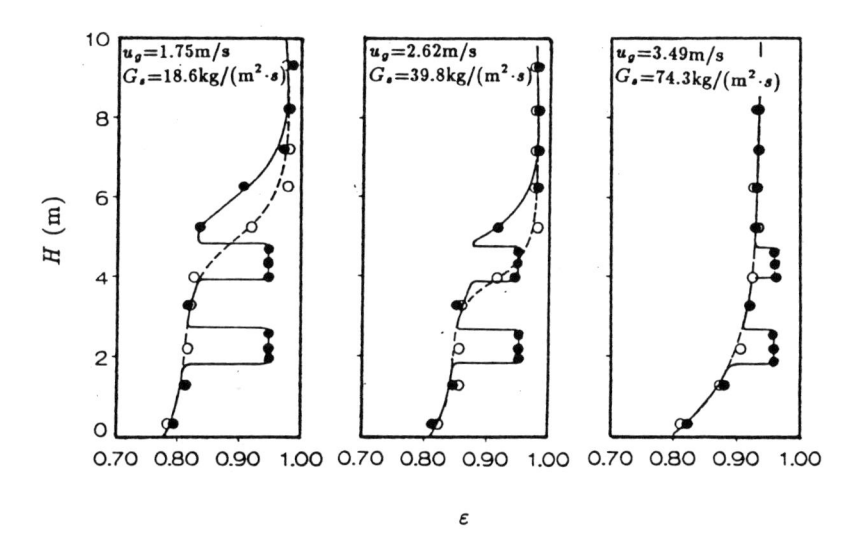

FIG. 31. Axial voidage profiles with and without ring internals (○, without ring internals; ●, with ring internals) (Zheng et al., 1992).

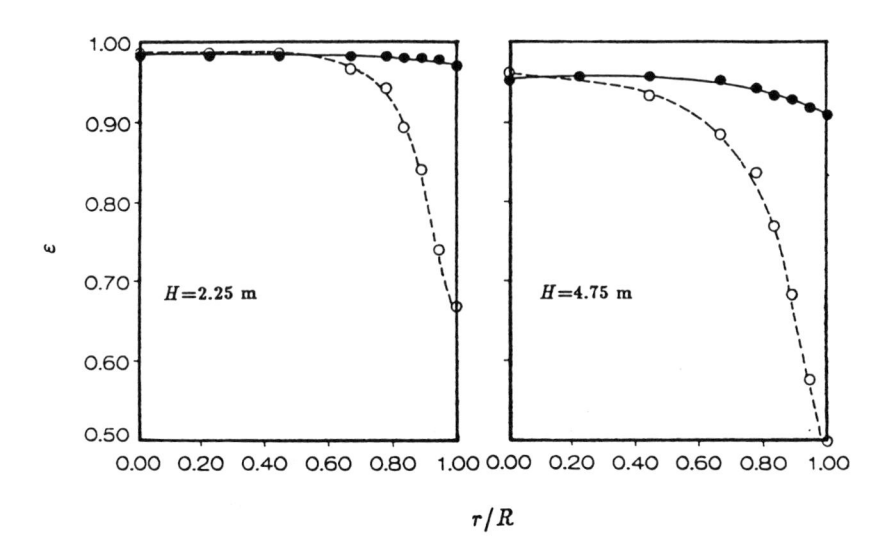

FIG. 32. Radial voidage profiles with and without ring internals. $u_g = 1.75$ m/s, $G_s = 18.6$ kg/(m^{2}s) (○, without ring internals; ●, with ring internals) (Zheng et al., 1992).

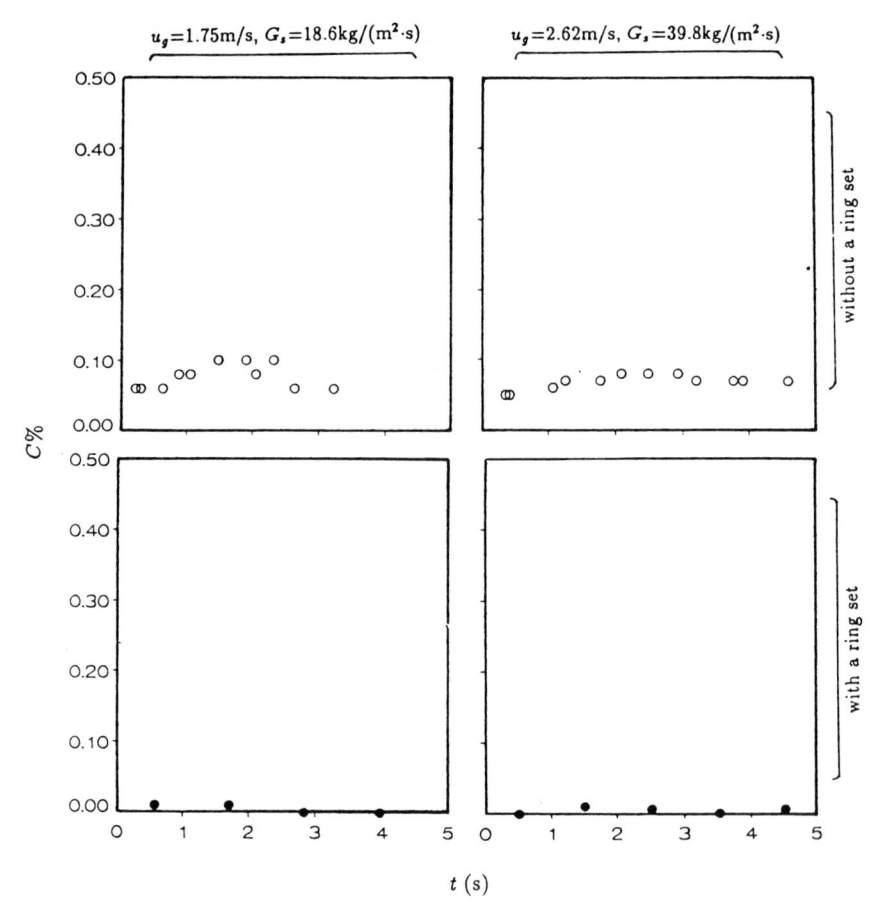

FIG. 33. Solid tracer content at sampling location #1 upstream of injection (Zheng et al., 1992).

V. Multi-layer Fast Fluidized Beds

On the basis of the experiments of ring materials, it can be seen that the fundamental characteristic behavior of a ring set is the formation of a zone of influence having low solids concentration. Such a zone of influence is capable of dissecting the dense-phase region of a fast fluidized bed into layers having limited solids intermixing between layers. Ring internals do not affect the overall solids inventory in the fast fluidized bed, which is mainly determined by the imposed pressure drop. The dense-phase region could be

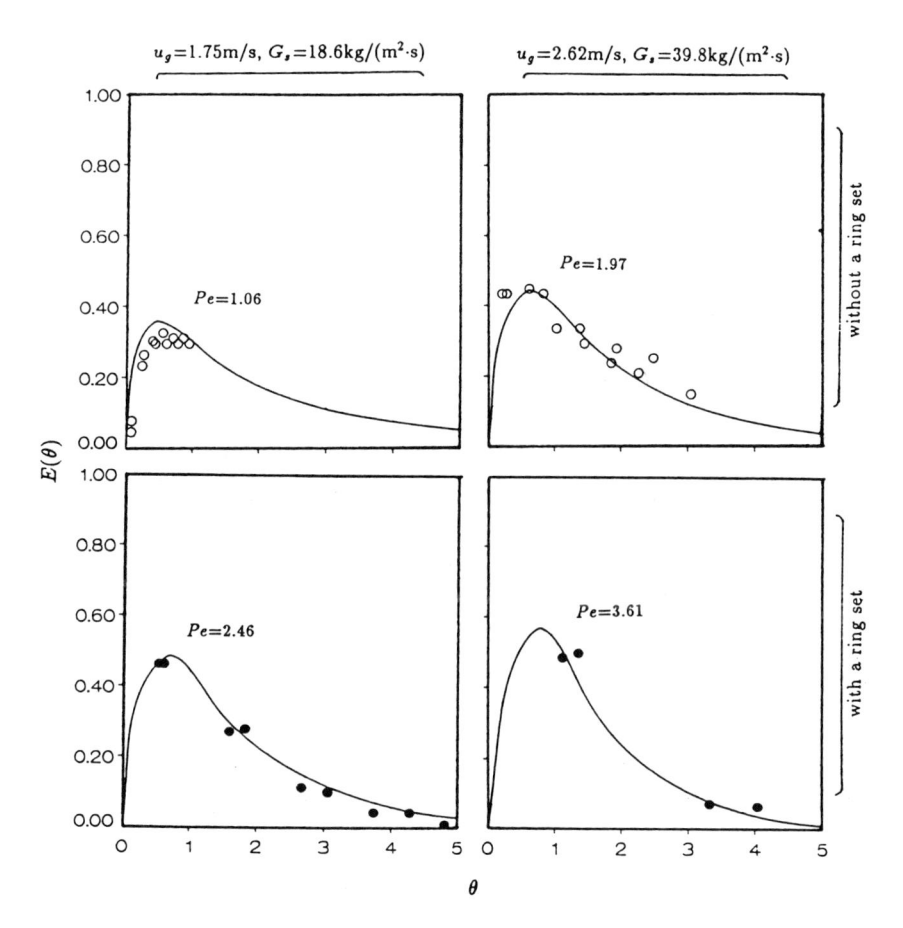

FIG. 34. Residence time distribution of solids at sampling location #3 downstream of injection (Zheng et al., 1992).

chopped into several layers by a series of ring sets, while the dilute-phase region at the top is thereby apportioned to these influence zones between the dense-phase layers, thus forming a multi-layer fluidized bed with alternating dense and dilute zones. However, the original characteristics of fast fluidization composed of a two-region structure is still preserved with the same solids inventory in the unit. Dissection of the dense-phase region with the interposition of intermediate zones of influence leads to an elevation of the point of inflection in the axial voidage profile of the fast fluidized bed, thus reducing the height of the dilute-phase region and increasing the height of the dense-phase region. Segmentation of the dense-phase region has the effect not only of suppressing overall solids backmixing but also of reducing the scale of mixing from macro toward micro, thereby intensifying gas–solids contacting.

VI. Integral CFB

Conventional CFB is generally characterized by two or more parallel columns. Recent work focused on the effect of CFB structure (including solids inlet and product exit) (Jin *et al.*, 1988; Li and Kwauk, 1980; Xia *et al.*, 1988) on flow, mass and heat transfer revealed certain inherent disadvantages of the conventional CFBs, e.g.,

1. asymmettic inlet/exit—leading to heterogeneous solids distribution in the bed, thus making scale-up, design, and control difficult;
2. large energy consumption in separating and circulating solids from an outside loop;
3. stress due to differential thermal expansion amongst components—making mechanical design complex.

An integral CFB (ICFB) was thus proposed to combine the four major component parts in a single concentric-tube structure: the fast bed, a circular V-valve (CV-valve), an internal multi-inlet cyclone and its dipleg (Liu, 1990; Kwauk *et al.*, 1991).

A. Characteristics

Among the multitude of possible configurations based on the preceding concept, Fig. 35 shows two designs: (a) a fast bed in the annular space, and the cyclone dipleg located in the center; (b) a fast bed in the center with the annular space serving as the dipleg.

For the annular-fast-bed design shown in Fig. 35a, solid particles elutriated from the fast bed are separated in a multi-inlet cyclone (MIC); they then descend through a mildly fluidized dipleg and are recirculated to the fast bed via a circular V-valve (CV-valve). Fluidizing air is supplied separately to the fast bed and the dipleg.

By incorporating the cyclone and the dipleg inside the fast bed, the ICFB is endowed with the following advantages:

1. compact structure;
2. heat can be exchanged between the solids in the dipleg and those in the surrounding fast bed, with maximal heat conservation (Hirama *et al.*, 1988);

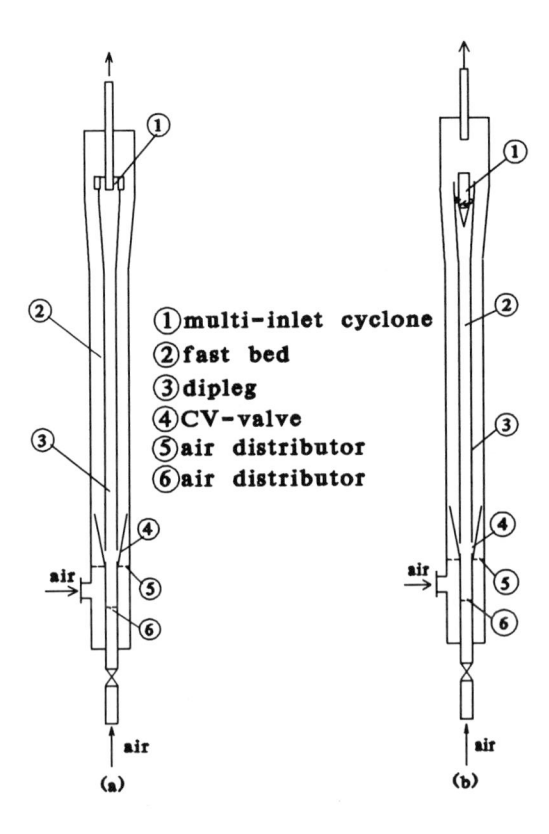

FIG. 35. Integral CFB (Liu, 1990; Kwauk et al., 1991).

① multi-inlet cyclone
② fast bed
③ dipleg
④ CV-valve
⑤ air distributor
⑥ air distributor

3. the integral internals, consisting of the cyclone, its dipleg, the CV-valve and the fast bed air distributor welded below the centrally located CV-valve, can slide up and down in the reactor shell and permits easy withdrawal for inspection and servicing;
4. the radial symmetry of the structure, from top to bottom, enhances homogeneity in flow, as shown in Fig. 36, where side A and side B are obtained along two radii in opposite directions at both the top and the bottom regions.

B. THE CIRCULAR V-VALVE

The CV-valve is designed with a short truncated conical spout, inclined at 8° to 12° from the vertical. Solids from the dipleg at the center are circulated

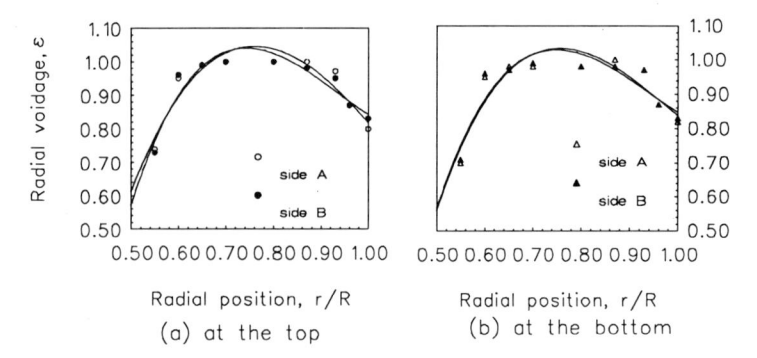

FIG. 36. Radial symmetry of the fast bed (Liu, 1990). (a) top; (b) bottom.

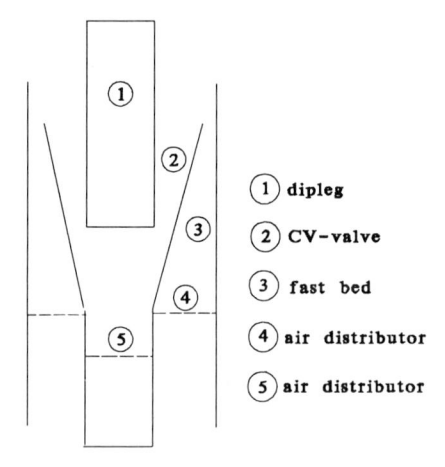

FIG. 37. Circular V-valve. (Liu et al., 1990a)

via the conical space and then spread fanwise to the annular fast bed, as shown somewhat in detail in Fig. 37.

Compared to the conventional V-valve, the CV-valve is structurally simple, possesses larger discharge capacity and is radially symmetrical. The distinct feature of radial symmetry results in uniform peripheral solids discharge into the surrounding fast bed. At low solids feed rate to the dipleg, that is, below certain critical rate, cyclic on-and-off solids discharge results as already discussed in Section I.B.1. (Li et al., 1982), with solids level fluctuating periodically between two more or less fixed levels, Z_{stop} and Z_{start}. For the CV-valve, these two levels are, however, rather closed to each other, thus yielding a high-frequency low-amplitude fluctuating solids flow, which is beneficial to the stability of the circulating system. The rate of solids flow from the dipleg can be controlled by manipulating the fluidizing air entering the CV-valve.

The discharge capacity of the CV-valve can also be expressed by Eq. (1).

C. MULTI-INLET CYCLONE

Figure 38 (Liu et al., 1990b) shows a four-inlet cyclone. Solids tend to be densified around the inlet region as the vertically entering solids-laden streams are suddenly diverted toward tangential flow in a horizontal plane. At inlet velocities exceeding 8.5 m/s, the overall separation efficiency for a typical FCC catalyst was found to be consistently above 99%, as shown in Fig. 39. This design, however, resulted in relatively high pressure drop unless the inlets were appropriately enlarged. The efficiency and pressure drop of this type of multi-inlet cyclones could be approximated by modifying the single-inlet cyclone expressions (Kou, 1987; Shi et al., 1987):

$$\eta_x = 1 - \exp\left(Bd_x \frac{2}{1.336 D^{0.14} + 2} \right), \tag{86}$$

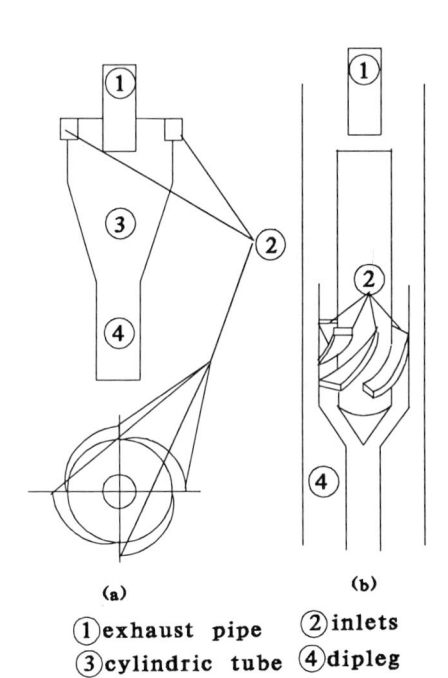

①exhaust pipe ②inlets
③cylindric tube ④dipleg

FIG. 38. Multi-inlet cyclone (MIC). (Liu et al., 1990b)

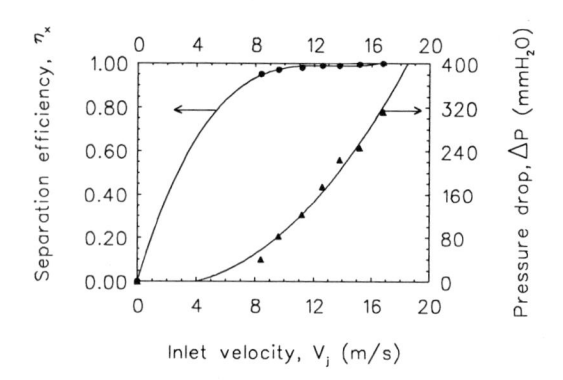

FIG. 39. Performance of a typical MIC. (Liu, 1990)

where

a = inlet width, mm

$$B = -0.365\sqrt{\frac{V_j N}{124}}$$

b = inlet height, mm
D = cylinder diameter, mm

$$d_x = d_p\sqrt{\frac{2.72}{\gamma_s}}$$

h = height of cylinder, mm

$$N = \frac{nh(R-r)}{2Rab}$$

n = inlet number
R = radius of the cylinder, mm
r = radius of the exhaust pipe, mm
V_j = inlet velocity, m/s
γ = specific gravity of gas, g/cm^3
γ_s = specific gravity of solid, g/cm^3
η_x = separation efficiency

$$\Delta P = \xi\frac{V_j^2 \gamma}{2g} \qquad (87)$$

ξ = resistance coefficient, for four-inlet cyclone, $\xi = 28.5$
ΔP = pressure drop, mm H$_2$O

A second type of multi-inlet cyclone, shown in Fig. 38b, consists of radial vanes to direct solids issuing from a central fast bed in axial upflow (see Fig. 35b) peripherally into an annular dipleg. A single stage of such radial vanes often gives an efficiency of 95% for an FCC catalyst with an average particle size of 54 μm, and the pressure drop is low. For higher separation efficiency, a series of vanes could be employed.

Notation

A	cross-sectional area, m²	f_1	unconfined yield strength of material, Nm⁻²
A_1	cross-sectional area of dipleg, m²	G_f	mass flow rate of fluid, kg s⁻¹
A_0	cross-sectional area of interconnecting aperture, m²	G_{gd}	gas mass flow rate through distributor at bottom of dipleg, kg s⁻¹
C_d	solids discharge coefficient	G_{gl}	gas mass flow rate in dipleg, kg s⁻¹
c_1	coefficient defined by Eq. (47)		
c_2	coefficient defined by Eq. (48)	$(G_{gl})_{min}$	minimum value of G_{gl}, kg m⁻¹
D	diameter of pipe, m	G_{gv}	gas mass flow rate through spout, kg s⁻¹
D_c	distribution factor for cylindrical moving bed defined by Eq. (39)	G_p	mass flow rate of particles, kg s⁻¹
D_s	distribution factor for trapezoidal moving bed defined by Eq. (40)	g	acceleration due to gravity = 9.81 ms⁻²
D_t	distribution factor for conical moving bed defined by Eq. (38)	H_s	ratio of stresses near slanting wall defined by Eq. (35)
D_v	distribution factor for rectangular moving bed defined by Eq. (41)	H_v	ratio of stresses near vertical wall defined by Eq. (32)
d_0	width of trapezoidal tube at bottom, m	H_{1s}	ratio of stresses near slanting wall defined by Eq. (36)
d_p	diameter of particle, m	H_{1v}	ratio of stresses near vertical wall defined by Eq. (33)
$\bar{d}_p$	mean diameter of particle, m	h	height from bottom, m
E_s	ratio of stresses near slanting wall defined by Eq. (34)	h_1	height of fluidized bed in dipleg, m
		h_t	height of packed bed in dipleg, m
E_v	ratio of stresses near vertical wall defined by Eq. (31)	h_v	height of spout, m
		k	slope of the failure function in Eq. (76)
F_s	wall friction force caused by the lateral component of gravity, N	L	height of solids level in dipleg, m
		L_{leg}	height of dipleg, m
f	characteristic constant of spout defined by Eq. (3), m⁻¹	L_{mf}	equivalent column height prorated to a voidage at minimal fluidization, m
f_a	critical value of f_1 at the crossover point in Fig. 25, Nm⁻²	M	molecular weight of gas, kg kgmol⁻¹
f_c	constant part of the failure function in Eq. (76), Nm⁻²	m	pressure-drop exponent

n	voidage exponent	V_{gv}	gas flow rate through spout, $m^3 s^{-1}$
p	gas pressure, Nm^{-2}	V_o	gas volume in Eq. (64), m^3
p_a	gas pressure at top of dipleg, Nm^{-2}	W	weight of particles in element slice, N
p_b	gas pressure at outlet of spout, Nm^{-2}	W_c	normal component of W_r, N
p_0	atmospheric pressure, Nm^{-2}	W_o	weight of particles in Eq. (64), N
$\Delta p_c, \Delta p_v$	pressure drop of conical spout, Nm^{-2}	$(W_p)_{max}$	maximum solids flow rate per unit cross-section in dipleg, $kg m^{-2} s^{-1}$
Δp_l	pressure drop in dipleg, Nm^{-2}		
Q	perimeter of cross-sectional area, m	W_r	radial component of W_x, N
		W_x	lateral component of gravity W, N
R	radius of tube, m	W'_x	lateral supporting force of wall, N
R_c	critical radius for arching, m	W_z	axial component of gravity W, N
Re_t	terminal Reynolds number of particle	y_1-y_4	coefficients defined by Eqs (43)–(46)
R_g	gas constant = 8,319 N m $kgmol^{-1} K^{-1}$	z	axial distance of bed from bottom, m
r	tadius, m, or rate-of-flow ratio in Section I.E	z_2	total length of bed, m
r_b	radius of conical tube at bottom, m		

GREEK LETTERS

r_w	radius of tube, m		
S	solids discharge rate, $kg s^{-1}$		
T	absolute temperature, K	α	half angle of cone, °
T_0	thickness of trapezoidal tube, m	β	angle between major principal stress and normal to the wall, °
u'	$= u/u_t$, reduced velocity		
u_d	superficial particle velocity, ms^{-1}	$\bar{\beta}$	angle between abutment of layer and the normal to the slanting wall for the powder in critical state of stress, °
u_{do}	solids velocity through interconnecting aperture, ms^{-1}		
u_f	fluid velocity or superficial fluid velocity at incipient fluidization in Section I.E, ms^{-1}	$\bar{\beta}'$	angle between abutment of layer and the normal to the vertical wall for the powder in critical state of stress, °
u'_f	reduced fluid velocity = u_f/u_{mf}		
u_{fo}	fluid velocity at bottom of spout, ms^{-1}	δ	effective angle of internal friction of material, °
u_{mf}	incipient fluid velocity, ms^{-1}		
u_{mfo}	incipient fluid velocity at p_a, ms^{-1}	ε	voidage
u_{mfp}	incipient fluid velocity at p, ms^{-1}	ε_c	voidage of closely compacted bed
u_o	superficial gas velocity, ms^{-1}	ε_l	voidage in dipleg
u_p	actual particle velocity, ms^{-1}	ε_{mf}	voidage for incipient fluidization
u_{po}	particle velocity at bottom of spout, ms^{-1}	ε_0	voidage for vertical gravity moving bed without interstitial fluid flow
u_s	particle velocity or relative velocity between particle and fluid in Section I.E, ms^{-1}	ε_s	voidage for inclined cylindrical moving bed
u'_s	reduced particle velocity = u_s/u_{mf}	ε_v	voidage for vertical cylindrical moving bed
u_{sc}	critical particle velocity, ms^{-1}		
u_t	terminal velocity of particle, ms^{-1}	η	angle defined by Eq. (49), °
V_{gd}	gas flow rate through distributor, $m^3 s^{-1}$	θ	angle between axial and vertical directions or cone angle, °

μ	wall friction coefficient = $\tan \phi$	σ_z	axial normal stress on radial cross-section, Nm^{-2}
μ_f	fluid viscosity, $kgm^{-1}s^{-1}$		
ρ_f	density of fluid, $kg\,m^{-3}$	$\bar{\sigma}_z$	mean value of σ_z over radial cross-section, Nm^{-2}
ρ_p	density of particle, $kg\,m^{-3}$		
ρ_s	density of solids, $kg\,m^{-3}$	$(\sigma_z)_c$	critical value of σ_z, corresponding to ε_c, Nm^{-2}
$\Delta\rho$	$= \rho_s - \rho_f$ effective density, $kg\,m^{-3}$		
		σ_1	major principal stress in limiting state, Nm^{-2}
σ	interparticle contact stress, Nm^{-3}		
σ_c	critical value of σ, corresponding to ε_c, Nm^{-2}	$\bar{\sigma}_1$	major principal stress in critical state, Nm^{-2}
σ_h	vertical stress on a horizontal cross-section, Nm^{-2}	σ_2	minor principal stress in limiting state, Nm^{-2}
$\bar{\sigma}_h$	mean value of σ_h over horizontal cross-section, Nm^{-2}	$\bar{\sigma}_2$	minor principal stress in critical state, Nm^{-2}
$(\sigma_h)_{sw}$	value of σ_h near slanting wall, Nm^{-2}	τ_{sw}	shear stress on slanting wall, Nm^{-2}
$(\sigma_h)_{vw}$	value of σ_h near vertical wall, Nm^{-2}	τ_{vw}	shear stress on vertical wall, Nm^{-2}
σ_{sw}	normal stress on slanting wall, Nm^{-2}	ϕ	angle of wall friction, °
		ϕ_s	shape factor of particle
σ_{vw}	normal stress on vertical wall, Nm^{-2}	ψ	angle of internal friction of powder, °

References

Berg, C. *Chem. Eng.* **60**(5), 138 (1953).
Chen, Y. M., Rangachari, S., and Jackson, R. *Ind. Eng. Chem. Fundam.* **23**, 354 (1984).
Enstad, G. *Chem. Eng. Sci.* **30**, 1273 (1975).
Ergun, S. *Chem. Eng. Progr.* **48**, 89 (1952).
Hirama, T., Takeuchi, H., Chiba, S., and Chiba, T. *In* "Circulating Fluidized Bed Technology" (P. Basu, and J. F. Large, eds.), pp. 511–518. Pergamon Press, Oxford, 1988.
Huang, S. "Standpipe with duplex conical spouts," M.S. thesis, Institute of Chemical Metallurgy, Beijing, China (1990).
Janssen, H. A. *Z. Ver. Dtsch. Ing.* **39**, 1045 (1895).
Jenike, A. W. Bulletin 123, University of Utah (1964).
Jin, Y., Yu, Z., Qi, C., and Bai, D. *In* "Fluidization '88—Science and Technology" (M. Kwauk and D. Kunii, eds.), pp. 165–173. Science Press, Beijing, 1988.
Johanson, J. R., and Jenike, A. W. *Powder Technology* **5**, 133 (1972).
Jones, D. R., and Davidson, J. F. *Rheol. Acta.* **4**, 180 (1965).
Kou, L. *Design of Chemical Equipments (China)* **5**, 23–26 (1987).
Kuo, T., and Soon, Y. *J. Chem. Ind. Eng. (China)* **2**, 97 (1960).
Kwauk, M. *Scientia Sinica* **12**(4), 587 (1963).
Kwauk, M. *Scientia Sinica* **16**(3), 407 (1973).
Kwauk, M. *In* "Fluidization—Idealized and Bubbleless, with Applications," pp. 113–124. Science Press and Ellis Horwood, New York, 1992.
Kwauk, M., Wang, N., Li, Y., Chen, B., and Shen, Z. *In* "Circulating Fluidized Bed Technology" (P. Basu, ed.), pp. 33–62. Pergamon Press, Oxford, 1985.

Kwauk, M., Yao, J. Z., Liu, S. J., Shen, T. L., and Liu, D. J. To Institute of Chemical Metallurgy, P. R. China Pat. CN91204327.X (1991).
Leung, L. S. *In* "Fluidization" (J. R. Grace and J. M. Matsen, eds.), p. 25. Plenum Press, New York, 1980.
Leung, L. S., and Jones, P. J. *Powder Technology* **20**, 145 (1978).
Leung, L. S., and Wilson, L. A. *Powder Technology* **7**, 343 (1973).
Li, H. "Gas–Solid Flow and Gas Seal in a Vertical Moving Bed under Negative Pressure Difference" (in Chinese), Institute of Coal Chemistry, Chinese Academy of Sciences, Taiyuan (1973).
Li, H. "Hydrodynamics of moving-bed conical overflow pipe," M.S. Thesis, Institute of Chemical Metallurgy, Beijing, China (1981).
Li, H. *Journal of Chemical Engineering of Chinese Universities* **4**(4), 321 (1990).
Li, H. *Powder Technology* **67**, 37 (1991).
Li, H. *Powder Technology* **73**(2), 147 (1992).
Li, H. *Powder Technology* **78**, 179 (1994).
Li, H., and Kwauk, M. *Chem. Eng. Sci.* **44**, 249 (1989).
Li, H., and Kwauk, M. *Trans. Instn. Chem. Engrs.* **69**, Part A, 355–360 (1991).
Li, X., Liu, D., and Kwauk, M. *Proc. Joint Meeting of Chem. Eng. CIESC and AIChE*, 382–391. Chem. Ind. Press, Beijing, 1981.
Li, X., Liu, D., Liu, J., and Sun, K. *Chemical Engineering (China)* **4**, 31–37 (1982).
Li, Y., and Kwauk, M. *In* "Fluidization" (J. R. Grace and J. M. Matsen, eds.), p. 537. Plenum Press, New York, 1980.
Liu, D., Li, X., and Kwauk, M. *In* "Fluidization" (J. R. Grace and J. M. Matsen, eds.), pp. 485–492. Plenum Press, New York, 1980.
Liu, D. J. "Preliminary study on an integral circulating fluidized bed (ICFB)," M. S. Thesis, Institute of Chemical Metallurgy, Beijing, China (1990).
Liu, D. J., Li, H., and Kwauk, M. To Institute of Chemical Metallurgy, P. R. China Pat. CN902257293 (1990a).
Liu, D. J., Li, H., and Kwauk, M. To Institute of Chemical Metallurgy, P. R. China Pat. CN902257005 (1990b).
Matsen, J. M. *Powder Technology* **7**, 93 (1973).
Matsen, J. M. *In* "Fluidization IV" (D. Kunii and R. Toei, eds.), pp. 225–232. Engineering Foundation, New York, 1983.
Sandy, C. W., Daubert, T. E., and Jones, J. H. *Chem. Eng. Prog. Symp. Ser.* **66**(105), 133 (1970).
Shi, M. X., Wang, Y. Y., Liu, J. R., Ying, X. T., Shi, C. N., and Lao, J. R. *In* "Handbook of Chemical Engineering (China)," (Board of Handbook of Chemical Engineering, eds.), Vol. 5, Chapter 21, pp. 101–122. Chemical Industry Press, Beijing, 1987.
Terzaghi, K. "Theoretical Soil Mechanics," Seventh Printing. P. M. Wiley, New York, 1954.
Walker, D. M. *Chem. Eng. Sci.* **21**, 975 (1966).
Walker, D. M., and Blanchard, M. H. *Chem. Eng. Sci.* **22**, 1713 (1967).
Walters, J. K. *Chem. Eng. Sci.* **28**, 779 (1973).
Xia, Y., Tung, Y., and Kwauk, M. *Engineering Chemistry & Metallurgy (China)* **9**(3), 14–24 (1988).
Yoon, S. M., and Kunii, D. *Ind. Eng. Chem., Proc. Des. Dev.* **9**, 559 (1970).
Zhang, G., Yang, G., Yang, S., Shu, S., and Li, H. *J. Chem. Ind. Eng. (China)* **3**, 229 (1980).
Zheng, C., and Tung, Y. *Engineering Chemistry & Metallurgy (China)* **11**(4), 296 (1990).
Zheng, C., Tung, Y., Li, H., and Kwauk, M. *In* "Fluidization VII" (O. E. Potter and D. J. Nicklin, eds.), pp. 275–283. Engineering Foundation, New York, 1992.

8. CIRCULATING FLUIDIZED BED COMBUSTION

YOUCHU LI
INSTITUTE OF CHEMICAL METALLURGY
ACADEMIA SINICA,
ZHONGGUAN VILLAGE, BEIJING, PEOPLE'S REPUBLIC OF CHINA

XUYI ZHANG
TSING HUA UNIVERSITY,
BEIJING, PEOPLE'S REPUBLIC OF CHINA

I. Experimental Studies on Coal Combustion	333
A. Devolatilization of Coal	333
B. Kinetic Characteristics of Combustion	335
C. Intraparticle Gas Diffusion	340
D. Model for SO_2 Retention	341
II. Behavior of CFB Combustor	350
A. Axial Voidage Profile in Combustor	351
B. Axial Temperature Profile	353
C. Axial Residual Carbon Content Profile	354
D. Axial Particle Size Profile	354
E. Reduction of SO_2 and NO_x Emissions	355
F. Gas–Solid Separation	356
III. Modelling of CFB Combustor	359
A. Physico–chemical Hydrodynamic Basis	359
B. Governing Conservation Equations for CFB Combustor	362
C. Comparison of Model with Experiments	364
D. Effect of Various Operating Parameters	365
IV. Process Design for CFBC Boilers	371
A. CFBC Boilers with Low Steam Capacities	371
B. CFBC Boilers for Co-generation Power Plants	373
C. CFBC Boilers with Co-generation of Gas and Steam	374
D. Heat-Carrier Gasification Combined Cycle (CGCC) for Power Generation	377
V. Improvement of Desulphurization	377
A. Influence of Pore Characteristic of SO_2 Sorbent	377
B. High-Activity Artificial Sorbent	378
C. Temperature Characteristics of Desulphurization	381
Notation	384
References	385

According to forecasts on world fuel use, coal use in the next 70 years may increase by a factor of three or more (Daman, 1991; Reh, 1991). In China, about 75% of total energy consumption comes from coal, 80% of which is directly burned in various combustors. This situation will continue for a long period of time in the future. In some districts of China, low-grade coal containing high ash and high sulphur, or even coal refuse has to be used because of the non-uniform distribution of energy resources and the shortage of energy supplies, leading to low combustion efficiency and high emission of harmful pollutants such as dust, SO_2, NO_x and others. Statistical data show that the major pollutants in the atmosphere come from the utilization of coal, as listed in Table I. Advanced energy conversion technology of coal is urgently required to improve the utilization of coal and to suppress the emission of pollutants.

In the past 10 years, the circulating fluidized bed (CFB) boiler as an efficient and clean combustion technology has rapidly developed and is now widely accepted throughout the world (Kullendorff and Anderson, 1986; Yerushalmi, 1986; Engstrom and Lee, 1991; Xu and Zhang, 1991). Its pronounced characteristics are

- wide fuel flexibility
- high combustion efficiency
- small excess air ratio
- low emission of harmful pollutants
- large turndown ratio and good load following capacity
- relatively less heated surface area

As a result, to enhance the overall efficiency of coal utilization and to reduce specific investment and cost, CFB is expected to be of considerable significance, especially for China, where coal is used as the main energy source (Li and Kwauk, 1981). Now in China CFB boilers of 220 t/h (steam) are being developed.

Intelligent design of circulating fluidized bed boilers depends on sufficient understanding of the physico-chemical hydrodynamics occurring in the combustors, such as chemical kinetics of coal combustion and pollutant formation, hydrodynamics of gas–solid two phase flow, mixing of gas and solids, distribution of heat released, and heat transfer between immersed

TABLE I
PERCENTAGE OF POLLUTANTS FROM COAL UTILIZATION IN TOTAL POLLUTANTS OF ATMOSPHERE
(XU AND ZHANG, 1991)

Air pollutants	SO_2	CO	NO_x	particulates
Percent from coal combustion	87	71	67	60

surfaces and the fluidized bed. In this chapter the issue to be discussed is principally associated with the macro kinetics of the combustion process and the design of the CFB combustors, since other features are covered in the corresponding chapters.

I. Experimental Studies on Coal Combustion

A. DEVOLATILIZATION OF COAL

Knowledge of coal devolatilization is important for design of coal combustors and gasifiers. There have been many models for devolatilization of coal particles in fluidized beds (Robert, 1982; Agarwal et al., 1984; Borghi et al., 1985), but the kinetic parameters of these are mostly dependent on coal type. Thus these models are inconvenient for practical design. Based on universal kinetic characteristics for different types of coal, as shown in Fig. 1 and 2, Fu et al. (1987) proposed a general model in which apparent kinetic parameters for pyrolysis, i.e., activation energy E and kinetic constant k, are independent of coal type but dependent on heating rate and final reaction temperature. The overall rate of devolatilization can be expressed by the

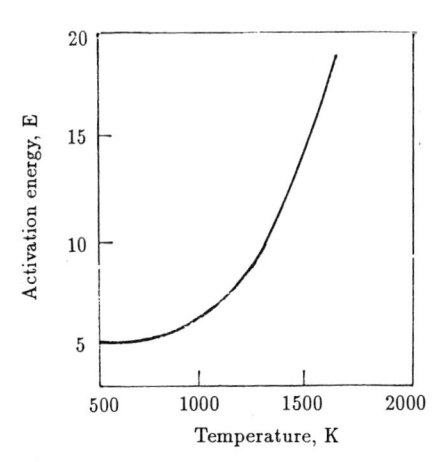

FIG. 1. Universal curve of activation energy for pyrolysis (after Fu et al., 1987).

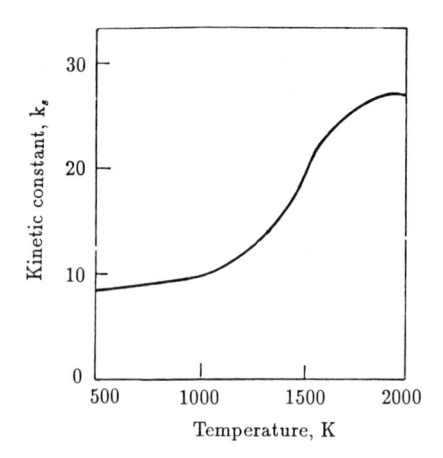

FIG. 2. Universal curve of kinetic constant (after Fu et al., 1987).

Arrhenius formula as follows:

$$\frac{d(\rho_c/\rho_{co})}{dt} = (V_\infty - V)k \exp(-E/R_g T), \quad (1)$$

where the final volatile yield V_∞ depends on coal type, coal particle size and heating conditions. The heating of coal particles is described by the following equation:

$$\rho_c C_{ps} \frac{\partial T}{\partial t} = \lambda \left(\frac{\partial^2 T}{\partial r^2} + \frac{2}{r} \frac{\partial T}{\partial r} \right), \quad (2)$$

with initial and boundary conditions:

$$\begin{aligned} t = 0, & \quad T(t, r) = T_0; \\ r = 0, & \quad \frac{\partial T}{\partial r} = 0; \\ r = R, & \quad \lambda \frac{\partial T}{\partial r} = h(T - T_\infty); \end{aligned} \quad (3)$$

in which

$$h = h_g + h_p + h_r \quad (4)$$

and h_g, h_p, h_r can be calculated from Martin's correlations (1982). The universal kinetic parameters of this model are listed in Table II.

Kinetic curves of pyrolysis in Fig. 3 for different types of coal under

TABLE II
KINETIC PARAMETERS FOR COAL PYROLYSIS

T K	λ W/m K	E kJ/mol	k, 1/s Medial heating	Fast heating	Very fast
1,000	0.0293	61.0	1.246×10^4	1.90×10^4	3.80×10^4
1,050	0.0315	64.0	1.083×10^4	2.35×10^4	5.55×10^4
1,100	0.0337	68.0	2.226×10^4	3.08×10^4	8.78×10^4
1,150	0.0360	73.0	4.136×10^4	4.21×10^4	1.54×10^5
1,200	0.0385	79.4	8.087×10^4	6.10×10^4	3.00×10^5
1,250	0.0414	86.7	1.799×10^5	9.20×10^4	6.68×10^5
1,300	0.0442	95.0	4.424×10^5	1.55×10^5	
1,350	0.0470	105.0	1.329×10^6	2.85×10^5	
1,400	0.0494	117.0	5.127×10^6	6.00×10^5	
1,450	0.0517	130.0	2.186×10^7	1.45×10^6	
1,500	0.0536	144.5	1.461×10^8	3.78×10^6	
1,550	0.0551	160.0	1.319×10^9	1.20×10^7	
1,600	0.0563	175.0	1.190×10^{10}	5.70×10^7	
1,650	0.0570	194.0	1.600×10^{11}	3.70×10^8	
1,700	0.0575	215.0	2.270×10^{12}	1.85×10^9	
1,750	0.0579	237.0	3.550×10^{13}	5.70×10^9	
1,800	0.0580	263.0	1.586×10^{15}		

different conditions demonstrate that the prediction is in good agreement with experiments, in the range 1,173 to 1,733 K.

B. KINETIC CHARACTERISTICS OF COMBUSTION

Combustion of coal in fluidized beds is generally related to temperature, particle size and transfer rate of oxygen to the surface of the particles. In bubbling fluidized beds at about 850°C, most experiments showed that the reaction rate is determined by diffusion of oxygen (Avedesian and Davidson, 1973; Basu, 1977; Beer, 1977; LaNauze; 1985 and others), while in fast fluidized beds, it is likely to be kinetics controlled because of the greatly increased slip velocity between the gas and the solids (Basu et al., 1986). However, further investigation on the kinetics of combustion in fast fluidized beds is needed when the origin of the coal varies. Recently, Luo (1990) and Mi et al. (1993) tested a bituminous coal from Datong in Shanxi Province, with properties listed in Table III. The bed materials consisted of sand No. 1, with mean size of 294 μm and density of 1,280 kg/m^3, and No. 2, 650 μm and 1,360 kg/m^3. Combustion rate was determined by monitoring the carbon dioxide in the flue gas using an infrared analyzer connected to a data acquisition

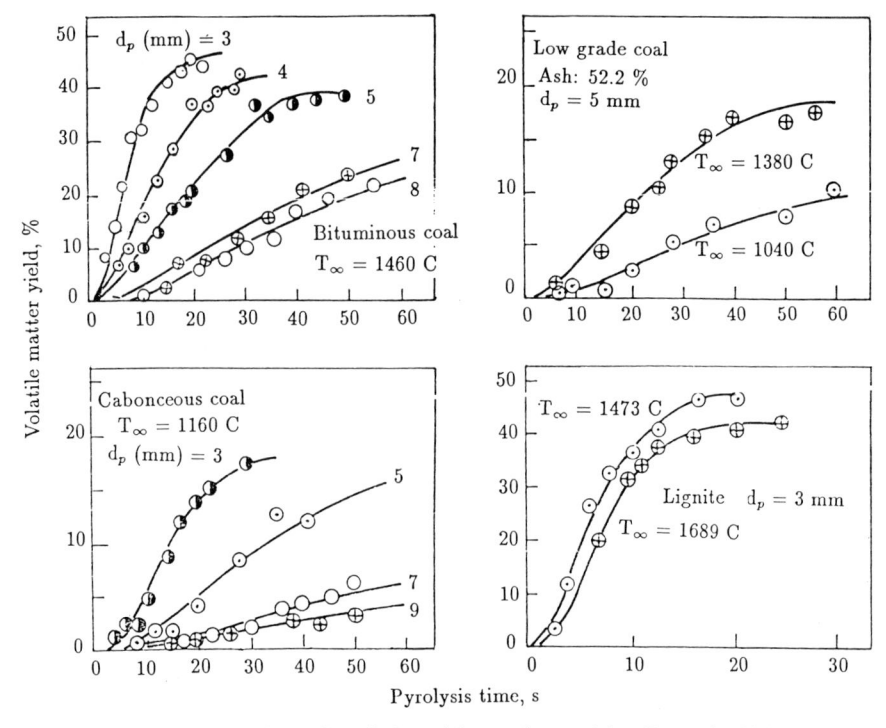

FIG. 3. Comparison of prediction with experiments (after Fu et al., 1987).

TABLE III
SOME PROPERTIES OF DATONG COAL

Particle Size d_p, μm	Carbon Content kgmol/m^3	Fixed Carbon %
160	29.6	57.05
206	30.8	59.47
294	33.0	60.91
650	35.9	62.45
900	36.7	62.71
1,125	38.0	63.22
2,500	39.9	64.42
3,500	42.2	64.77

computer system. Influence of particle size, combustion temperature and gas velocity on the combustion rate was studied.

Figure 4a compares combustion in the bubbling and fast fluidized bed operating at 0.86 m/s and 2.02 m/s, respectively, indicating much faster

a. Comparison between CFB and bubbling fluidized bed

$d_p = 0.65$ mm
$T = 1123$ K
1: u=2.02 m/s
2: u=0.86 m/s

b. Effect of gas velocity in fast fluidized bed

$d_p = 0.65$ mm
$T = 1123$ K
1: u=2.81 m/s
2: u=4.06 m/s

c. Effect of coal particle size

$T = 1123$ K   u = 4.06 m/s
1: $d_p = 0.65$ mm
2: $d_p = 1.125$ mm
3: $d_p = 2.5$ mm
4: $d_p = 3.5$ mm

d. Effect of temperature for coarse coal

$d_p = 2.5$ mm
u = 2.02 m/s
1: T= 1173 K
2: T= 1073 K
3: T= 937 K

e. Effect of temperature for fine coal

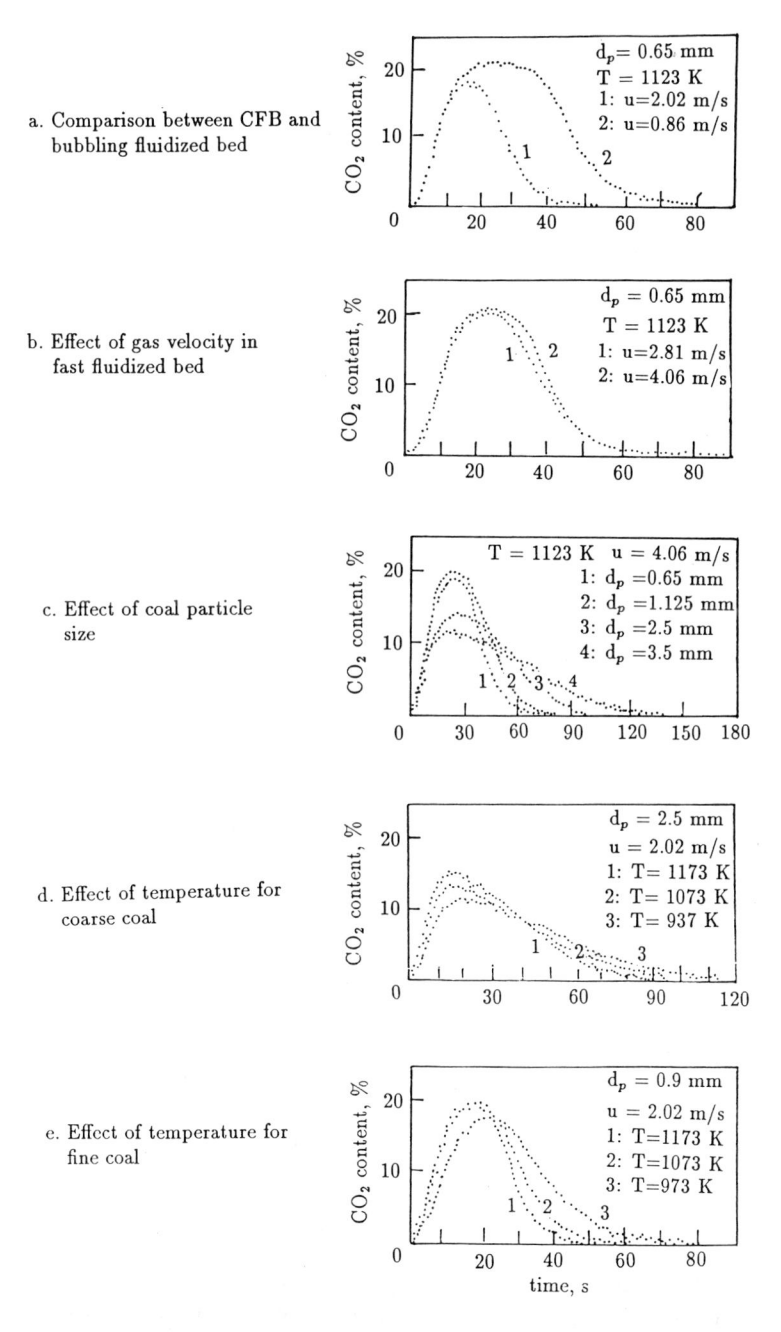

$d_p = 0.9$ mm
u = 2.02 m/s
1: T=1173 K
2: T=1073 K
3: T=973 K

FIG. 4. Burnout time of Datong coal in sand fluidized bed (after Mi et al., 1993).

burnout in the FFB. Figure 4b shows that in FFB, the burnout time is little affected by raising the gas velocity from 2.81 to 4.06 m/s, indicating that combustion is essentially kinetics controlled. Figure 4c shows that the burnout process is increased from 70 to 130 seconds as the coal particle size is increased from 0.65 to 3.5 mm. Figures 4d and 4e show that burning is enhanced by increased temperature, even more pronounced in the case of fine particles.

The overall process of coal combustion involves the following three steps in sequence: transport oxygen from the bulk stream of gas to the outer surface of the particles, diffusion of oxygen through the ash layer to the reaction interface, and finally reaction with the combustible matter in the unreacted core. The overall flux of oxygen per unit surface area can be expressed as follows:

$$q_{O_2} = \frac{C_o - C_o^*}{\frac{1}{k_g} + \frac{\delta}{\varepsilon D_e} + \frac{1}{k_s}}, \tag{5}$$

in which

C_o = oxygen concentration of the bulk stream,
C_o^* = oxygen concentration at the reaction surface,
k_g = convective mass transfer coefficient,
k_s = intrinsic reaction rate constant,
D_e = molecular diffusion coefficient,
δ = thickness of ash layer,
ε_e = effective diffusion porosity.

The kinetic constant k_s has been obtained through experimental measurements (Mi et al., 1993):

$$k_s = 41.93 \exp(-113,000/RT). \tag{6}$$

And k_g can be calculated from the modified Sherwood Number correlation:

$$Sh = 2\varepsilon + 0.64 Re^{0.6} Sc^{0.35}, \tag{7}$$

in which

$$Sh = \frac{d_p k_g}{D_g}. \tag{8}$$

Fu and Zhang (1991) developed a general method for calculating burning rate of coal and char particles, and a chart was made through numerical solution for the model equations, as shown in Fig. 5. The dimensionless

CIRCULATING FLUIDIZED BED COMBUSTION

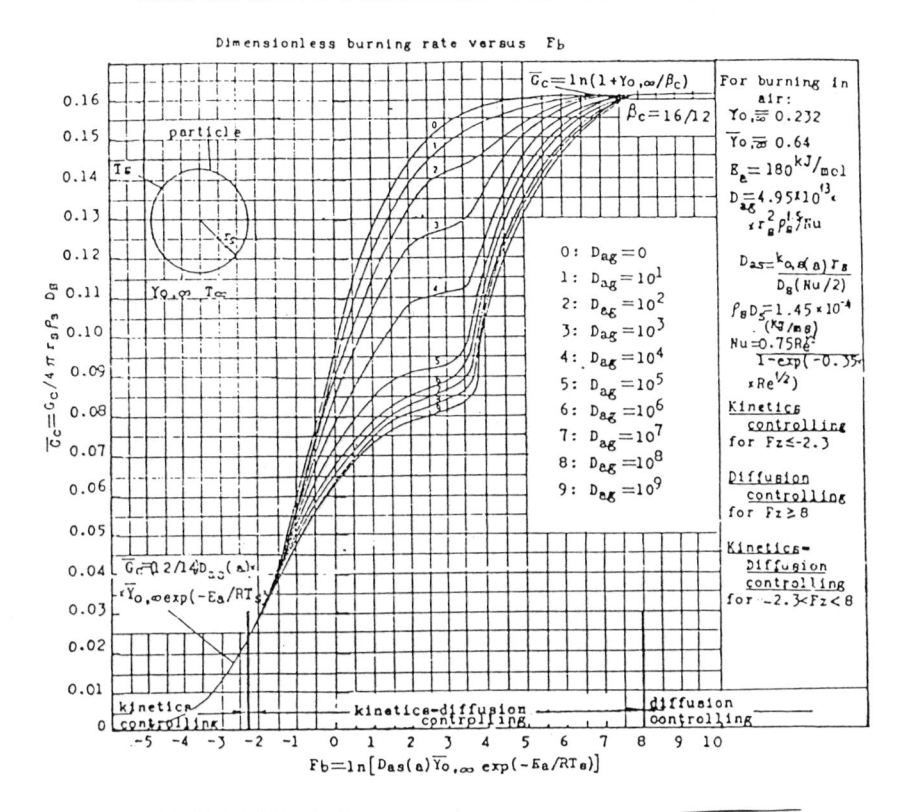

FIG. 5. Universal curves for calculating $\bar{G}_c$ (after Fu and Zhang, 1991).

burning rate $\bar{G}_c$ can be found from this diagram:

$$\bar{G}_c = \frac{G_c}{4\pi r_s \rho_s D_s}, \tag{9}$$

when the following two dimensionless numbers are known:

$$D_{ag} = 4.95 \times 10^{13} \frac{r_s^2 \rho_s^{1.5}}{Nu}, \tag{10}$$

$$F_b = \ln\left[D_{as}(b) Y_{O_2} \exp\left(-\frac{E_b}{RT_s}\right)\right], \tag{11}$$

in which

$$D_{as} = r_s K_{o,s}(b)/D_s. \tag{12}$$

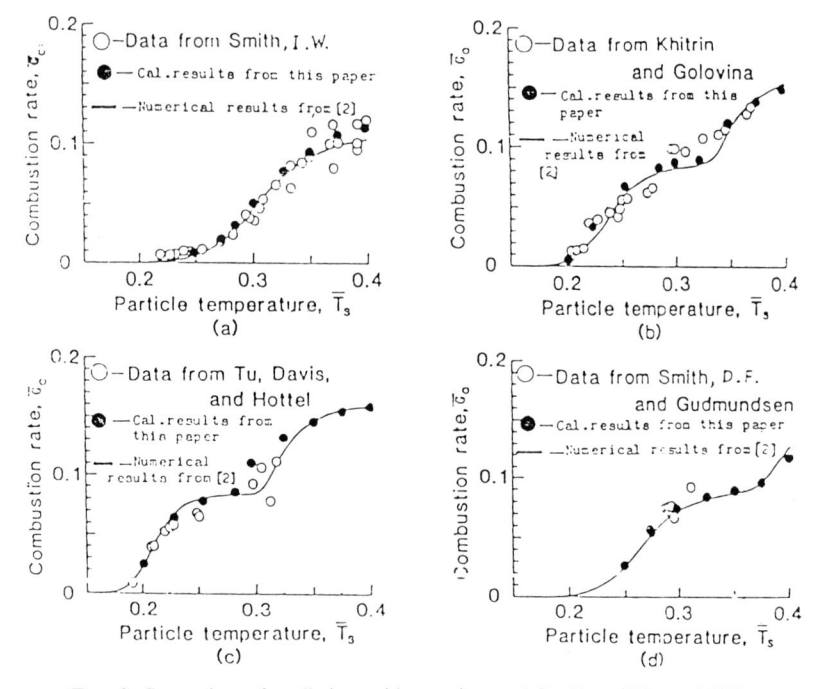

FIG. 6. Comparison of predictions with experiments (after Fu and Zhang, 1991).

Y_{O_2} is a dimensionless oxygen concentration, equal to 0.223; $K_{o,s}(b)$ is a frequency factor for reaction of carbon with carbon dioxide, equal to 3.37×10^{10}; and $E_b = 270$ kJ/mol.

Figure 6 compares prediction with experiments, demonstrating good agreement. From Fig. 5, the reaction state can be determined by the F_b value:

$F_b < -2.3$, kinetics controlled;

$-2.3 < F_b < 8$, diffusion-kinetics controlled;

$F_b > 8$, diffusion controlled.

C. Intraparticle Gas Diffusion

For coarse coal particles with high ash content, diffusion of oxygen through the ash layer often determines the overall combustion rate. The inner pores in the ash layer can be divided into micropores of 0.004–0.0012 μm, transition pores of 0.0012–0.03 μm and macropores of 0.03–1 μm diameter. Experiments showed that the macropores constitute the main channels for mass transfer.

TABLE IV
Oxygen Diffusion Parameters through Ash Layer

Coal Kinds	ε_0	$B \times 10^{-3}$, 1/m
Wangfeng waste coal	0.061	1.59
Xishan waste coal	0.108	0.95
Yangshan anthracite coal	0.108	0.30
Schuicheng coal	0.124	1.54
Jinxi anthracite coal	0.199	1.39
Datong coal	0.236	0.60
Wangfeng coal	0.254	0.58
Songzao coal	0.282	0.92
Xinglong coal	0.347	1.00
Zhangcum coal	0.460	2.31

If ε_e represents the effective porosity for gas diffusion during char burning, and ε_0, the effective diffusion porosity for the initial char surface when δ reaches zero, suppose that there exists the following relation between ε_0 and ε_e (Yan et al., 1991):

$$\varepsilon_e = \varepsilon_0 \exp[-B\delta], \tag{13}$$

where B is a characteristic parameter associated with ash structure for a given kind of coal. The practical diffusion parameters, ε_e and B, for 10 Chinese coals are listed in Table IV. It is obvious that the greater the initialy porosity ε_0 is, the smaller the diffusion resistance of oxygen through the ash layer is. The parameter B stands for the ratio of effective diffusion channels in the ash layer, which decreases with burning time. The greater the value of B is, the more rapidly the char combustion rate decreases during the later period of burning.

D. Model for SO_2 Retention

There has been significant interest in the reaction of calcium oxide with sulphur dioxide because of its application in the retention of sulphur oxide resulting from fluidized bed coal combustion. Different models have been proposed to describe this reaction, for example, the shrinking core model (Szekely and Evans, 1971) and the pore model (Petersen, 1957). The pore model has been extended from the single pore model (Ramchandram and Smith, 1977), through the random pore model (Bhatia and Perlmutter, 1981) and the pore size distribution model (Christman, 1983), to the generalized pore model (Yu and Sotirchos, 1987; Prasannan, 1985). The reaction of granular calcium oxide with sulphur dioxide is a complex process not only dealing with product-layer pore diffusion, depending on pore structure (pore diameter and its length), but also associated with interfacial chemical reaction. This reaction is accompanied by a solid volume expansion resulting from conversion of CaO to CaSO$_4$, leading to plugging of the pores and changing the pore structure. Prediction for the latter period of reaction becomes even more difficult. Recently, Zhang's (1992) experiments showed that while silicon dioxide and aluminium oxide in the sorbent are inert toward sulphur dioxide, ferric oxide acts as a catalyst. In order to study the effect of these materials, sorbents were prepared of calcium oxide, silica dioxide, aluminium oxide and ferric oxide in different weight ratios, and then calcined under high temperature. These authors showed that the sulfuration process belongs characteristically to gas–solid surface chemical reaction, the reaction rate of which depends on temperature and sorbent composition, and can be expressed in the range of 800–1,080°C by the following correlation:

$$r = k_s P_{SO_2}, \tag{14}$$

$$k_s = \frac{2.12}{1 + 1.26 X_i} \exp\left(-\frac{E}{RT} - 24 X_c\right), \tag{15}$$

in which

$$E = 37.7 \exp(-11.07 X_c) \tag{16}$$

and X_c is the weight fraction of ferric oxide in the sorbent, and X_i, the weight fraction of inert matter in the sorbent.

Figure 7 shows the relation of the kinetic constant k_s with reaction temperature T. Figure 8 shows how ferric oxide can accelerate the initial sulfuration reaction.

The diffusion coefficient of SO$_2$ through the product layer has also been measured, the value of which is about 10^{-12} m^2/s. Figure 9 shows comparison of intraparticle gas diffusivity between the sulfurated sorbent and its calcined sample, indicating calcination can enhance gas diffusivity. The inert matter in the sorbent, however, is beneficial for improving gas diffusion. Measurements (Zhang, 1992) indicated that pores greater than 700 Å in diameter for different limestones possess the same distribution function $f_1(r)$, as can be expressed by the following correlation:

$$f_1(r) = 0.0119 r^{-3.6}, \tag{17}$$

while for pores less than 700 Å, there is notable difference in pore distribution, as shown in Fig. 10, leading to different reactivity and calcium utilization rate.

Zhang (1992) then proceeded to formulate a distribution pore size model

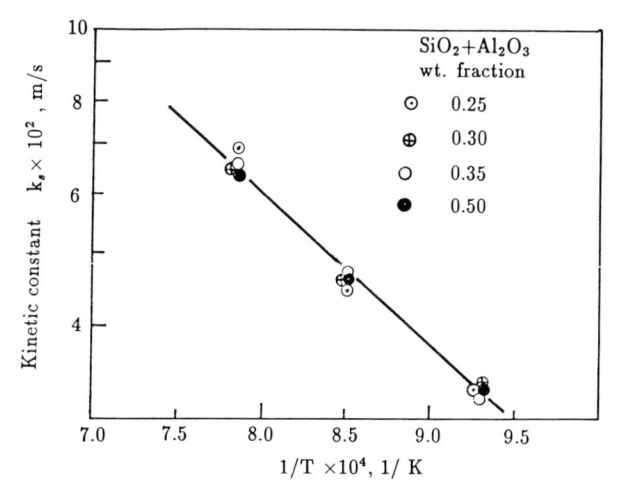

FIG. 7. Relation of kinetic constant k_s with reaction temperature T (after Zhang, 1992).

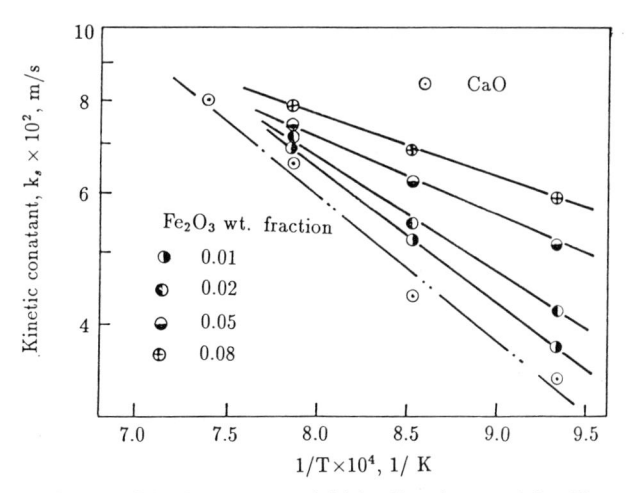

FIG. 8. Influence of Fe_2O_3 content on initial sulfuration rate (after Zhang, 1992).

by first summarizing his findings as follows:
(1) Pore dimension change:

$$r_1^2 = (1-\alpha)r_2^2 + \alpha r_0^2, \qquad (18)$$

in which

$$\alpha = \frac{\rho_{CaO}}{\rho_{CaSO_4}}. \qquad (19)$$

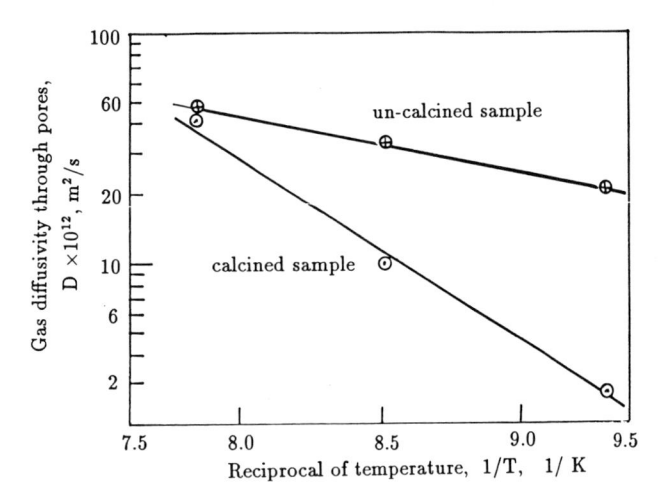

FIG. 9. Comparison of gas diffusivity between calcined and uncalcined samples for sulfurated sorbent (after Zhang, 1992).

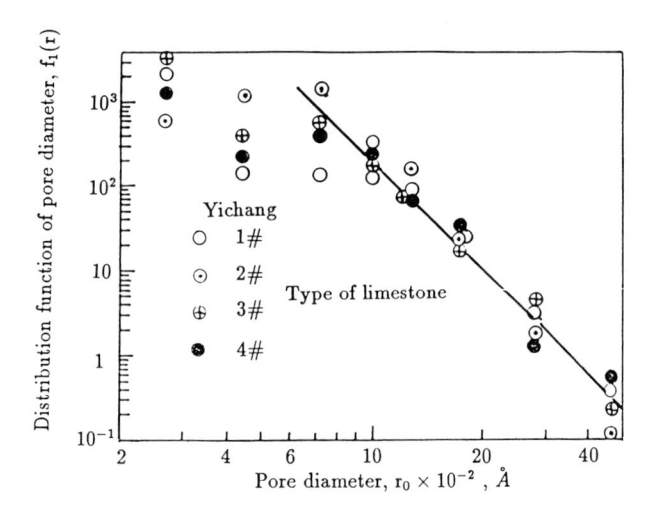

FIG. 10. Distribution of pore diameter for different limestones (after Zhang, 1992).

(2) Gas diffusion via the pore length:

$$\frac{\partial}{\partial z}\left(D(r_1)r_1^2\frac{\partial c_z}{\partial z}\right) = \frac{2k_g r_2 c_z}{(1 + r_2 k_s/D_s \ln(r_2/r_1))}; \qquad (20)$$

boundary conditions:

$$z = 0, \quad \frac{\partial c_z}{\partial z} = 0; \tag{21}$$

$$z = l_g, \quad c_z = c_0$$

(3) Gas diffusion via the product layer:

$$\frac{\partial^2 c_r}{\partial r^2} + \frac{1}{r}\frac{\partial c_r}{\partial r} = 0; \tag{22}$$

boundary conditions:

$$r = r_1, \quad c_r = c_z;$$

$$r = r_2, \quad k_s c_r = -D_s \frac{\partial c_r}{\partial r}. \tag{23}$$

(4) Interface reaction:

$$\frac{\partial r_2}{\partial t} = \frac{k_s c_x}{\rho_{CaO}(1 + r_2 k_s/D_s \ln(r_2/r_1))}; \tag{24}$$

initial conditions:

$$t = 0, \quad r_1 = r_2 = r_0. \tag{25}$$

Therefore, the local conversion of CaO for a pore is

$$x_i = \frac{\varepsilon_0}{1 - \varepsilon_0}(r_2^2 - 1). \tag{26}$$

The longitudinal average conversion of CaO for a pore is

$$y_i = \frac{1}{l_g}\int x_i \, dz. \tag{27}$$

The total conversion of CaO for a grain of sorbent is

$$W_t = \frac{\int_0^\infty l_g y_i f_1(r) \, dr}{\int_0^\infty l_g f_1(r) \, dr}. \tag{28}$$

For a spherical sorbent of radius R_0,

$$l_g = \frac{R_0 r_0}{3}\left(\frac{\rho_{CaO}S}{2\varepsilon}\right)^{1/2}\tau, \tag{29}$$

$$\tau = \frac{tk_s C_o}{r_0 \rho_{CaO}}. \tag{30}$$

S is surface area per unit solid weight, and ε_0 is the initial voidage inside the particle.

By defining the following dimensionless parameters,

$$Z = \frac{z}{l_g}; \quad C_z = \frac{c_z}{c_0}; \quad R_1 = \frac{r_1}{r_0}; \quad R_2 = \frac{r_2}{r_0}; \quad D_g = \frac{D}{D_9};$$

$$\tau = \frac{t}{\frac{r_0 \rho_{CaO}}{k_s C_0}} \tag{31}$$

the foregoing differential equations can be transformed into the following dimensionless forms:

$$R_1^2 = \alpha + (1 - \alpha) R_2^2, \tag{32}$$

$$\frac{\partial R_2}{\partial \tau} = \frac{C_x}{1 + \mathrm{Bi}\, R_2 \ln(R_2/R_1)}, \tag{33}$$

$$\frac{\partial}{\partial Z}\left(D_g R_1^2 \frac{\partial C_z}{\partial z}\right) = H_0^2 \frac{R_2 C_z}{1 + \mathrm{Bi}\, R_2 \ln(R_2/R_1)}, \tag{34}$$

Initial and boundary conditions are

$$\begin{aligned}\tau = 0, & \quad R_1 = R_2 = 1; \\ Z = 0, & \quad C_z = 1; \\ Z = 1, & \quad \frac{\partial C_z}{\partial Z} = 0,\end{aligned} \tag{35}$$

where

$$\mathrm{Bi} = \frac{k_s r_0}{D_s}, \tag{36}$$

$$H_0 = l_g \left(\frac{2k_g}{r_0 D_0}\right), \tag{37}$$

$$D_g = \frac{R_1 + a}{R_1 + b}(1 + b), \tag{38}$$

$$a = \frac{D_s}{D_{k0}}, \tag{39}$$

$$b = \frac{D_m}{D_{k0}}, \tag{40}$$

$$D_{k0} = 9,700 r_0 (T/M)^{0.5}, \tag{41}$$

$$D_m = \frac{0.001 T^{1.75}\{(M_1 + M_2)/(M_1 M_2)\}^{0.5}}{P(\Sigma V_1^{1/3} + \Sigma V_2^{1/3})}. \tag{42}$$

Computed results by this model are in fairly good agreement with experimental measurements, as shown in Fig. 11. The figure shows that high SO_2 concentration would accelerate plugging of the pores, leading to reduced conversion of calcium oxide, i.e., lower utilization rate. The influence of sorbent particle size on CaO conversion is shown in Fig. 12, demonstrating that conversion can be raised by using fine particles, especially those less than 0.5 mm. Temperature of reaction can change the structure of the pores but is beneficial for enhancing gas diffusion in the pores. Therefore, there is an optimum reaction temperature, depending on the extent of SO_2 retention, as can be seen in Fig. 13, about 950°C for 10% conversion of CaO, and about 800°C for 20% conversion.

Xu et al. (1991) developed an improved shrinking core model for SO_2 retention, in which the effect of both increased porosity of a particle due to CO_2 generation and reduced porosity due to SO_2 absorption during reaction were considered. The parameters were defined as follows:

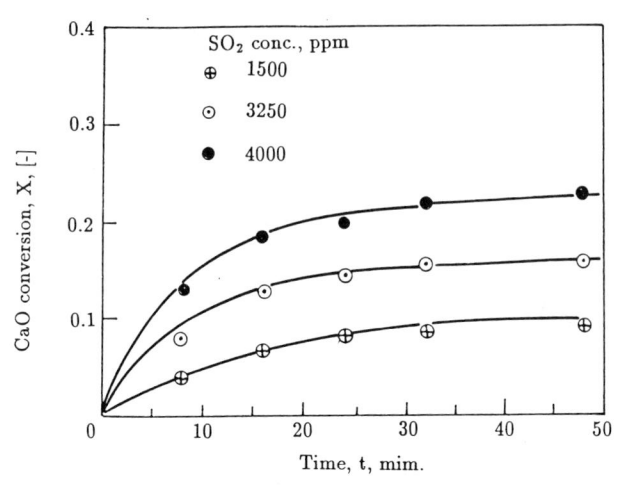

$$X_R = 1 - (r_c/r_o), \tag{43}$$

FIG. 11. Influence of SO_2 concentration on CaO conversion (after Zhang, 1992).

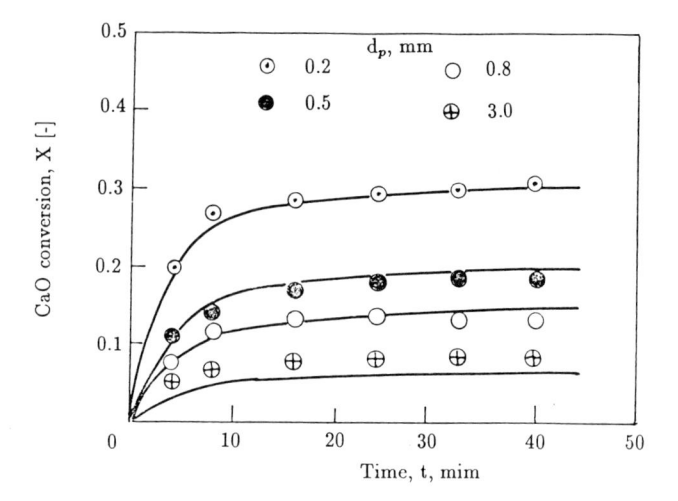

FIG. 12. Influence of sorbent particle size on CaO conversion (after Zhang, 1992).

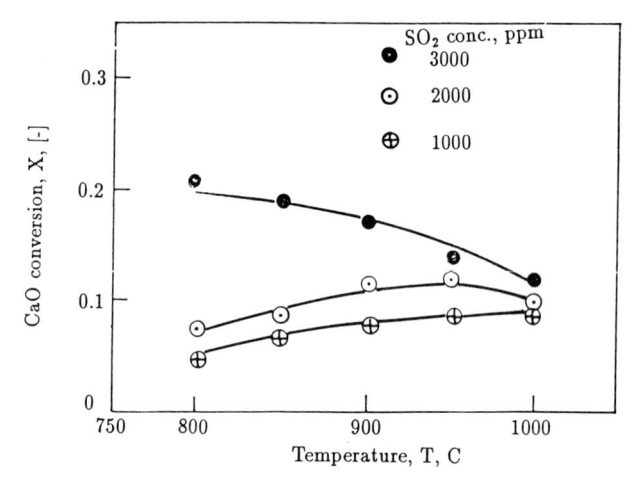

FIG. 13. Influence of reaction temperature on CaO conversion for different SO_2 concentrations (after Zhang, 1992).

Overall conversion of CaO to $CaSO_4$:

$$X = \frac{3}{R} \int X_R r^2 \, dr. \tag{44}$$

Actual radius of a swollen grain:

$$\frac{V_{CaSO_4}}{V_{CaO}} = \frac{r_s^3 - r_c^3}{r_o^3 - r_c^3}. \tag{45}$$

The calcined rate of a limestone particle can be expressed by the following equation:

$$\frac{\partial r_c}{\partial t} = \frac{D_e \varepsilon_R R[P_{CO_2}^* - P_{CO_2}]}{\rho_{CaO} R_g r_c^2 [(R/r_c) - 1 + (D_e/k_c) \cdot R]}, \tag{46}$$

$$\varepsilon_R = 1 - (1 - \varepsilon_c)[Y \cdot X_R(V_{CaSO_4}/V_{CaO} - 1) + 1], \tag{47}$$

with initial condition

$$t = 0, \quad r_c = R. \tag{48}$$

The unreacted radius of a particle during sulfuration is

$$\frac{\partial r_c}{\partial t} = -\frac{kM_{CaO}c_s c_{oi}^{0.5}}{\rho_t \left[1 + (k/D_s)\left(1 - \frac{r_c}{r_g}\right) r_c c_{oi}^{0.5}\right]}. \tag{49}$$

with initial condition

$$t = 0, \quad r_c = R.$$

The radial SO_2 concentration profile is

$$\frac{\partial^2 c_s}{\partial r^2} + \frac{2 \partial c_s}{r \partial r} = \frac{3(1 - \varepsilon_c) k r_c^2 c_{oi}^{0.5} c_s}{D_e r_c \left[1 + (k/D_s)\left(1 - \frac{r_c}{r_s}\right) r_c c_{oi}^{0.5}\right]} \tag{50}$$

with the boundary conditions

$$r = R, \quad c_s = c_{so};$$
$$r = R_s, \quad \frac{\partial c_s}{\partial r} = 0, \tag{51}$$

in which

$$\varepsilon_c = 1 - (1 - \varepsilon_{ls})[Y(C_{CaO}/V_{CaSO_4} - 1) + 1]. \tag{52}$$

Similarly, the radial O_2 concentration profile is

$$\frac{\partial^2 c_o}{\partial r^2} + \frac{2 \partial c_o}{r \partial r} = \frac{3(1 - \varepsilon_c) R_s r_c D_s (c_o - c_{oi})}{D_e r_o^3 (R_s - r_c)}. \tag{53}$$

Fig. 14 shows comparison of the predicted values with experimental results for different conditions, indicating that they are in agreement.

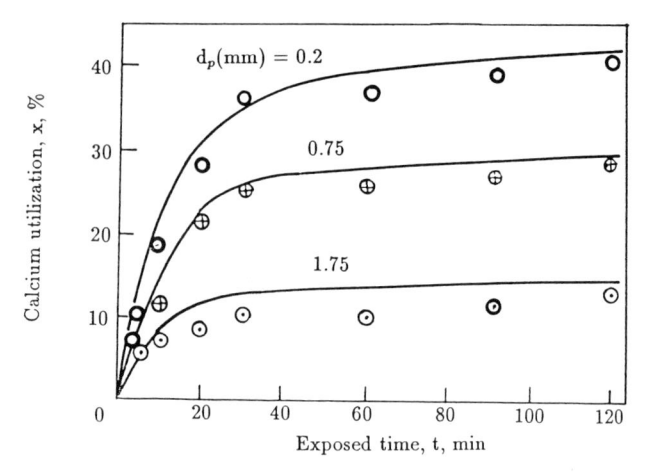

FIG. 14. Effect of exposed time on calcium utilization (after Xu et al., 1991).

II. Behavior of CFB Combustor

The circulating fluidized bed combustor is generally composed of a fast fluidized column, primary cyclone for solids recycling and a downcomer with a seal or a solids rate controlling device. However, variations in design do exist to accommodate special requirements and to produce special effects (Darling et al., 1986; Engstrom et al., 1986; Kobro and Brereton, 1986; Boyd and Friedman, 1991; Gierse, 1991; Ishida et al., 1991; Lu et al., 1991; Ohtake, 1991; Xu and Zhang, 1991; Yan et al., 1991; Tawara et al., 1991). The uniqueness of an improved pilot plant combustor designed by ICM, Academia Sinica (Li et al., 1991), as shown in Fig. 15, consists of the following features:

- an ultralow pressure drop air distributor for the fast fluidized column, with an aperture ratio far greater than what is normally employed for bubbling fluidized bed, thus leading to a reduction of some 20–40% in the total pressure drop of the combustor;
- a special downcomer for solids recycling incorporating a heat exchanger; the downcomer is capable of maintaining a solid bed height automatically in response to the total pressure drop of the fast fluidized column, that is, self-adjusting to thermal load;
- special design for heated tubes arrangement in the downcomer to take advantage of the high heat transfer coefficient in the dense solids bed and low erosion of low velocity fluidization in the solids downcomer;
- location of coal feed at the exit of the V-valve at the bottom of the downcomer for effective premixing, thus minimizing caking of the fresh coal feed;

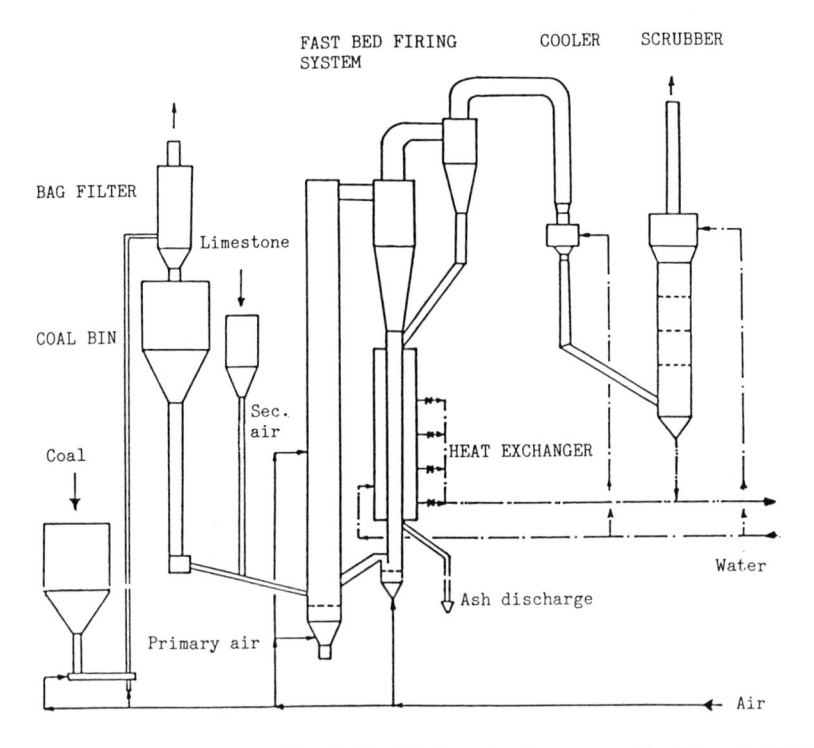

FIG. 15. Schematic diagram of fast fluidized bed combustion system (after Li et al., 1991).

- operation with high solids circulating rates in order to effect uniform axial temperature profile for improved fuel combustion, SO_2 retention and bed-to-heated-surface heat transfer.

Experiments of burning have been conducted for different kinds of coal, including bituminous coal (No. 1), sub-bituminous coal (No. 2), caking coal (No. 3) and low-grade coal (No. 4). Proximate and ultimate coal analyses are listed in Table V. The operating behavior of this combustor will be briefly summarized below.

A. Axial Voidage Profile in Combustor

The operation of this pilot plant combustor is similar to that of the Type B fast fluidized bed in Chapter 3, Section II. Figure 16 (Li et al., 1991) shows the axial profiles of bed voidage for different gas velocities and total pressure drops across the fast fluidized combustor, demonstrating the

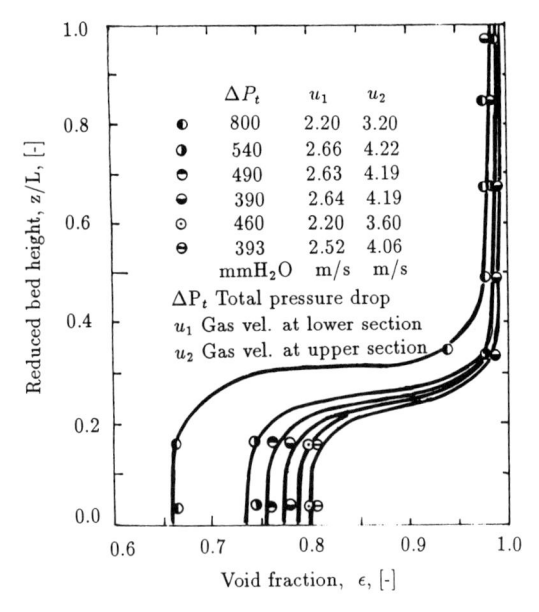

FIG. 16. Void fraction vs. reduced bed height under various operating conditions (after Li et al., 1991).

TABLE V
PROXIMATE AND ULTIMATE ANALYSES OF COAL TESTED

	Units	No. 1	No. 2	No. 3	No. 4
Moisture	%	2.64	3.14	1.05	1.00
Ash	%	25.53	35.32	33.13	82.70
Fixed carbon	%	30.02	21.05	16.68	2.94
H	%	47.66	49.87	56.68	12.20
N	%	0.67	0.76	1.03	—
S	%	2.30	0.48	1.94	1.15
%	%	7.17	7.32	2.84	—
Heat. value	MJ/kg	22.27	19.02	22.80	4.02
Range of size	mm	0–6	0–3	0–6	0–1.5
Mean size	mm	0.66	0.64	0.41	0.27

characteristic segregation of the fast fluidized bed into an upper dilute region and a lower dense region. However, owing to the additional provision of a secondary air at the middle of the combustor, the position of inflection point remains essentially near to the port of the secondary air. For practical CFB boilers, as gas velocity increases, the voidage in the lower dense phase region becomes greater, implying that a part of the solids is transferred to the upper dilute region, by which the heat released in combustion is carried up to the membrane wall arranged in the upper section. The upward mass flux determines the corresponding heat flux and the load of the boiler.

B. AXIAL TEMPERATURE PROFILE

Experimental results (Li et al., 1991) indicate that the axial temperature distribution is highly dependent on the secondary-to-total air ratio. When this ratio is less than about 0.3, the axial temperature profile is essentially uniform except for the region approaching the exit of the combustor, as shown in Fig. 17. When this ratio exceeds about 0.4, however, the temperature in the middle of the combustor becomes lower than those at the two ends. On the other hand, experimental results also indicate that increasing the secondary air ratio is favorable for supressing NO_x emission. If this ratio is greater than 0.5, NO_x emission can be controlled to less than 65 ppm. Therefore, it is necessary to locate the secondary air inlet properly and choose the secondary air ratio in order to optimize between efficient combustion and low NO_x emission.

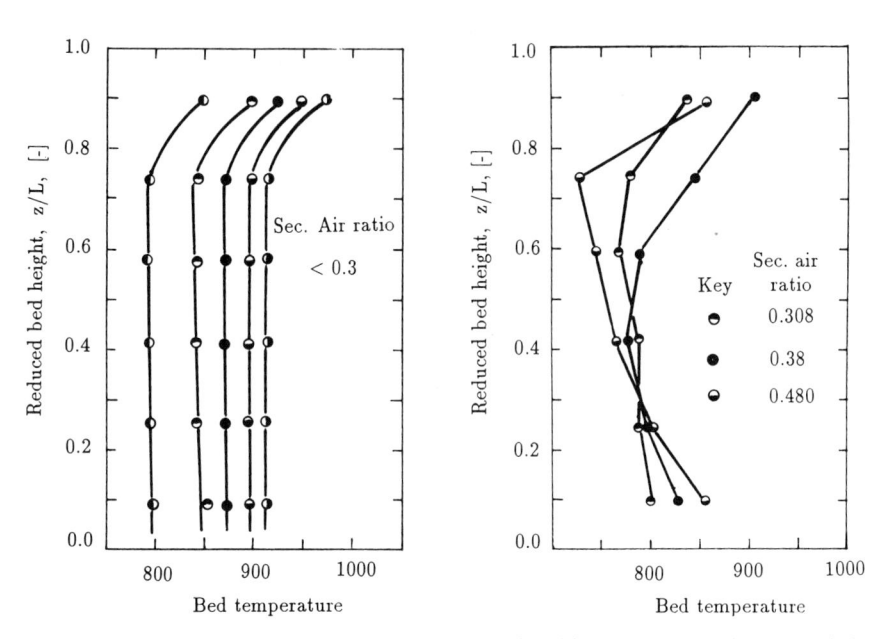

FIG. 17. Distributions of bed temperature along reduced bed height for various second air ratios (after Li et al., 1991).

C. AXIAL RESIDUAL CARBON CONTENT PROFILE

The residual carbon contents at different axial locations of the combustor were measured in the pilot plant tests (Li et al., 1991), as shown in Fig. 18. These data show that axial variations in carbon content with temperature (from 810 °C-923 °C) are as a whole rather slight, but mean carbon content increases with decreasing excess air ratio. Besides, for excess air ratios greater than 1.2, the carbon content at the top of the combustor is somewhat less than that at the bottom, while for excess air ratio less than 1.2, the opposite tendency is evident. In conclusion, for this improved combustor, an excess air ratio of 1.2 is considered enough for carbon burn-out, leading to reduced flue gas and increased heat efficiency as compared to bubbling fluidized bed combustion. That is probably attributable to bubbleless gas–solid contacting for increased mass transfer between gas and solids in the fast fluidized bed, as explained by combustion kinetics.

D. AXIAL PARTICLE SIZE PROFILE

Figure 19 shows the size distribution of solids sampled at the top, middle, and bottom of the fast fluidized bed combustor and the downcomer (Li et al., 1991). It can be seen that all particle size distribution curves are rather

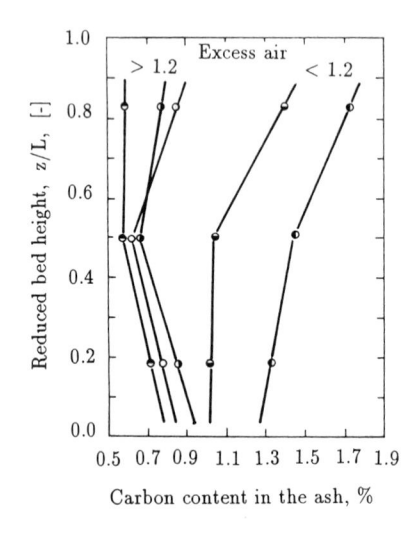

FIG. 18. Residual carbon content profile along reduced bed height (after Li et al., 1991).

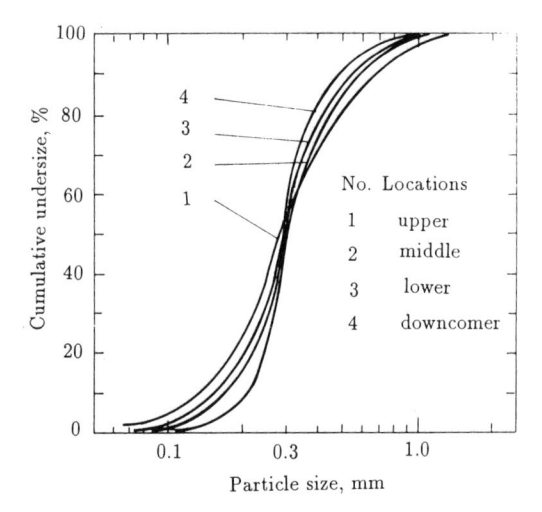

FIG. 19. Distributions of particle size for solids samplings at different locations of the bed (after Li et al., 1991).

similar—less than 2 mm, averaging between 0.3 and 0.5 mm, and spanning the major range of between 0.1 and 1 mm. Regional size segregation is thus not evident. Coupled with the axial carbon content profile, it can be seen that mixing of solids in the fast fluidized bed combustor is nearly complete.

E. REDUCTION OF SO_2 AND NO_x EMISSIONS

In this pilot plant test, natural limestone containing 34% CaO, crushed to below 2 mm, was used as the SO_2 sorbent. Below a bed temperature of about 850°C, the SO_2 retention increases with Ca/S molar ratio. When this ratio approximates to 2, the SO_2 content in the flue gas is reduced to about 100 ppm, corresponding to an SO_2 retention efficiency of about 83%, as can be seen in Fig. 20.

When a part of the air is supplied as secondary air entering at the middle of the combustor, resulting in the combination of a reducing zone in the bottom and an oxidizing zone at the top, NO_x content in the flue gas would decrease. Emission of NO_x can thus be controlled by changing the secondary-to-total air ratio. Figure 21 indicates that when this ratio is raised from 0 to 0.6, NO_x emission drops from about 300 ppm to 50 ppm for Datong coal. If a secondary-to-total air ratio of 0.5 is used, the expected NO_x emission is about 65 ppm.

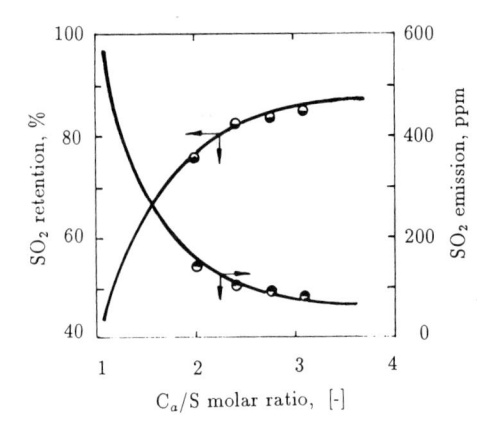

FIG. 20. SO$_2$ retention and emission as a function of Ca/S ratio for fine powder of limestone (after Li et al., 1991).

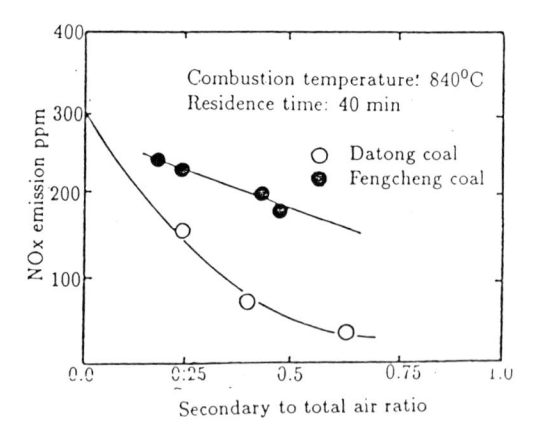

FIG. 21. Effect of the secondary-to-total-air ratio on NO$_x$ emission (after Li et al., 1991).

F. Gas–Solid Separation

The load characteristic of circulating fluidized bed boilers is closely related to the gas–solid flux through the combustor. Thus, choice of coal size and the proper gas–solid separator deserves attention. Generally, the fraction of particles below 0.1 mm is smaller than 10%, which could be removed only by a cyclone separator, while the major fraction of coarser particles could be easily separated by inertia or gravity (Li and Kwauk, 1981). The diagram of ash balance for a circulating fluidized bed boiler, shown in Fig. 22, can

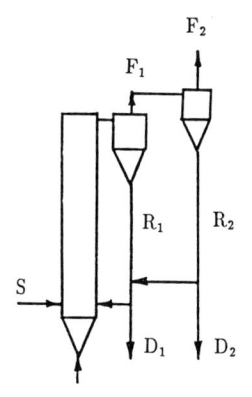

FIG. 22. Diagram of ash balance for CFB boilers (after Li and Wang, 1992).

be used for an analysis of gas–solid separation (Li and Wang, 1992). Total gas–solid separation efficiency of this system is given by

$$\eta_t = \frac{D_1 + D_2}{S} = 1 - \frac{F_2}{S}. \tag{54}$$

Thus, efficiency of the first separator is

$$\eta_1 = \frac{R_1}{S + R_1 + R_2 - D_1 - D_2}, \tag{55}$$

and efficiency of the secondary separator is

$$\eta_2 = \frac{R_2}{F_1} = 1 - \frac{F_2}{F_1}. \tag{56}$$

Solids recirculation ratios are defined as:
  total:

$$N_t = \frac{R_1 + R_2 - D_1 - D_2}{S}; \tag{57}$$

  first stage:

$$N_1 = \frac{R_1 - D_1}{S}; \tag{58}$$

  second stage:

$$N_2 = \frac{R_2 - D_2}{S}. \tag{59}$$

Then,

$$\eta_1 = 1 - \frac{1-\eta_t}{(1-\eta_2)(1+N_1+N_2)}. \qquad (60)$$

Under ordinary operating conditions of fast fluidization,

$$N_1 \gg N_2 \gg 1.$$

Therefore,

$$\eta_1 = 1 - \frac{1-\eta_t}{(1-\eta_2)N_1}. \qquad (61)$$

Combustion efficiency depends on the total separation efficiency. For a given total efficiency, it is most important that η_1 and η_2 are properly assigned. Figure 23 indicates that when the efficiency of the secondary separator η_2 rises, especially beyond 96%, the efficiency of the first stage η_1 drops to relatively low values. The influence of solids recirculation ratio N_1 on the first stage efficiency η_1 is shown in Fig. 24, showing η_1 increases with solids circulation rate. In the case of 99% for η_t and η_2, the first stage efficiency required is about 95% to 99%. According to this pilot plant test, the fraction of ash particles below 0.1 mm is less than about 5%, while the major fraction is about 0.35 mm. These coarser particles could easily be collected by an inertia or gravity separator instead of a cyclone. The gas–solid separation system for a CFB boiler could consist of a combination of a first-stage inertia or gravity separator and a second-stage cyclone, resulting in a simplified configuration.

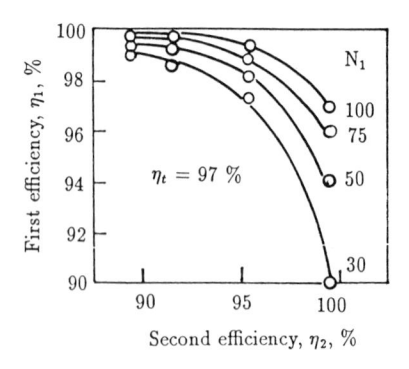

FIG. 23. First-stage separation efficiency as function of second-stage separation efficiency (after Li and Wang, 1992).

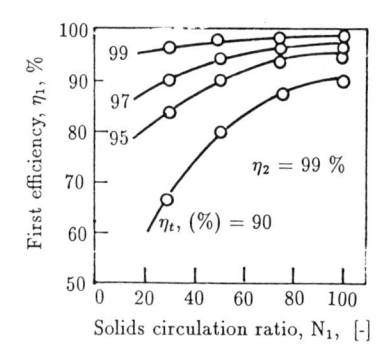

FIG. 24. Influence of solids circulation ratio on separation efficiency for first-stage separation (after Li and Wang, 1992).

III. Modelling of CFB Combustor

The modelling of CFB combustors for scale-up has been given attention. Recently, Weiss and Fett (1988), Lee and Hyppanen (1989), Hyppanen et al. (1991), Lou et al. (1991), and Wang (1992) have proposed models to predict CFB combustion.

Among the factors affecting CFB combustion, hydrodynamic behavior is the dominant and decisive factor. For the bubbling fluidized bed, coal burns mostly in the dense-phase region at the lower section of the combustor where the evaporator is installed, while for the fast fluidized bed, the combustion zone is surrounded by a water-cooled membrane as an evaporator, most of which is placed in the upper section. The heat generated in the lower combustion region is carried up to the upper dilute region through solids recirculation, and is then transferred to the water-cooled wall. A two-phase model of CFB combustor for axial gas–solid distribution has been developed (Lin, 1991; Li et al., 1992) to predict combustion efficiency and load turndown ratio as well as pollutant emissions, and to rationally select operating parameters for optimal design.

A. Physico-chemical Hydrodynamic Basis

Figure 25 shows a physical model of CFB combustion. According to experimental findings (see Chapter 3), the following assumptions are proposed.

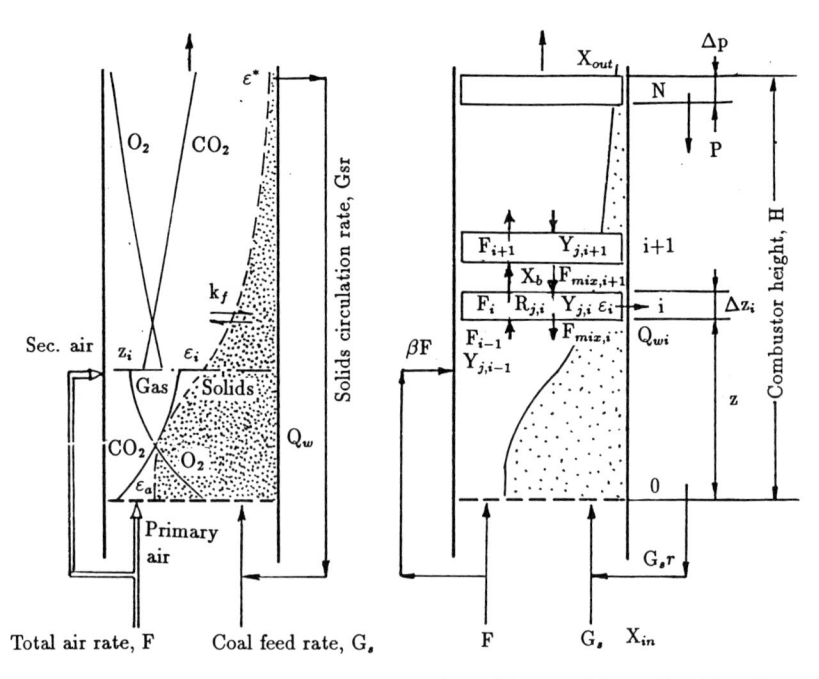

FIG. 25. Physical model of CFB combustion for axial gas–solids profiles (after Li et al., 1992).

1. Hydrodynamics

- Whether axially or radially, gas–solid flow is heterogeneous in structure. Axial voidage profile characterizes a dilute region at the top and a dense region at the bottom, i.e., an S-shaped distribution curve.
- Radial voidage profile displays the structure of dilute at the center and dense near the wall.
- There exists both axial and radial gas diffusion in the CFB combustor, the radial gas dispersion coefficient being much smaller than the axial.
- Solids mixing is nearly complete, with essentially the same carbon content at any point in the combustor.

Therefore, for coal burning, neglecting radial gas dispersion and radial heterogeneity of solids concentration might be acceptable. Gas–solid flow can thus be simplified to a one-dimensional model with axial distributions only.

Using the axial voidage profile (Li and Kwauk, 1980; Kwauk et al., 1986),

$$\ln\left(\frac{\varepsilon - \varepsilon_a}{\varepsilon^* - \varepsilon}\right) = -\frac{1}{Z_0}(z - z_i), \tag{62}$$

in which four parameters ε_a, ε^*, z_i, and Z_0, can be calculated from their corresponding correlations, Li and Wu (1991) were able to determine axial dispersion in a CFB combustor.

2. Transfer Process

- Contrary to bubbling fluidization, gas flow in the form of bubbles is transformed to a continuous dilute phase, while solids in the emulsion are transformed into a discontinuous phase as clusters.
- In most cases of experiments, the size of the bed materials ranges from 0.3 to 0.5 mm. The gas–solid suspension can be taken to possess a uniform temperature since heat transfer between them is very fast (Grace, 1986).
- The bed-to-wall heat transfer coefficient depends mainly on the suspension density in the bed. For specified operating parameters, what needs to be considered is therefore only the transfer between the suspension and the heat transfer surface, although the actual mechanism involves various heat transfer processes.

For mass transfer outside particles, the transfer coefficient can be calculated by using Equation (7). Inside the particles, the diffusion coefficient can be modified by means of the effective porosity (see Section I.B., or determined by another method such as that of Lou et al., 1991). For bed-to-wall heat transfer, the transfer coefficient can be computed by using relevant correlations presented in Chapter 5.

3. Chemical Reactions

When coal is heated, it first pyrolyzes into volatile matter, and the remaining char then burns. Char burning is much slower than that of the volatiles, and the rate controlling process has been described by Rajan and Wen (1980) as follows:

$$C + \theta O_2 \rightarrow 2(1 - \theta)CO + (2\theta - 1)CO_2, \tag{63}$$

$$\theta = \left[0.5\left(\frac{CO}{CO_2}\right) + 1\right] / \left(\frac{CO}{CO_2} + 1\right), \tag{64}$$

in which

$$\frac{CO}{CO_2} = 2{,}500 \exp(-6239/T). \tag{65}$$

The carbon consumption rate is expressed as

$$\frac{dm_c}{dt} = \pi d_p^2 K C_{O_2}. \tag{66}$$

The constant K involves chemical kinetics, and diffusion both outside and inside a particle.

The carbon monoxide produced from the preceding reaction continues to burn. At the same time, the nitrogen in the coal is converted into nitrogen oxides through complicated mechanisms, but the overall reaction can be expressed as

$$N_{char} + O_2 \rightarrow 2NO. \tag{67}$$

In the presence of a reducing agent such as carbon or carbon monoxide, reduction of nitrogen oxide will take place (Chan et al., 1983):

$$2NO + C \rightarrow N_2 + CO_2, \tag{68}$$

$$2NO + 2CO \rightarrow N_2 + 2CO_2. \tag{69}$$

As to sulfur dioxide, limestone as a SO_2 sorbent, when heated, is decomposed into CaO, which then reacts with SO_2:

$$CaCO_3 \rightarrow CaO + CO_2, \tag{70}$$

$$CaO + SO_2 + \tfrac{1}{2}O_2 \rightarrow CaSO_4. \tag{71}$$

The background information on these reactions has been discussed elsewhere (Oguma et al., 1971; Rajan et al., 1978; Chan et al., 1982; Borghi et al., 1985; Basu et al., 1986; Fu et al., 1987; Moritomi et al., 1991; Xu et al., 1991; Zhang, 1992; Luo, 1990; Mi et al., 1993).

B. Governing Conservation Equations for CFB Combustor

Based on the two-phase model for axial gas–solid distribution, the steady operation of the combustor can be described by using N compartments in series. Therefore, for the ith compartment with height Δz_i at combustor height z, according to mass, momentum and heat conservation, a set of operation equations can be established as follows.

1. Mass Conservation

For gas species j,

$$F_i y_{j,i} = F_{i-1} Y_{j,i-1} + F_{mix,i+1} Y_{j,i+1} - F_{it1} Y_{it1} - F_{mix,i} Y_{j,i} + A_{j,i} + \sum_0^k R_{kj,i}. \tag{72}$$

for solids,

$$(1 - \varepsilon_i)\rho_s A_t \Delta z_i = G_s \bar{t}_i, \tag{73}$$

$$E(t) = \frac{1}{\bar{t}_i} \exp\left(1 - \frac{t}{\bar{t}_i}\right), \tag{74}$$

$$1 - X_b = \Sigma(1 - t/\bar{t}_i)E(t)\Delta t, \tag{75}$$

$$\Sigma(1 - \varepsilon_i)\rho_s A_t \Delta z_i = G_s \Sigma \bar{t}_i, \tag{76}$$

$$X_{\text{out}} = X_{\text{in}}(1 - X_b). \tag{77}$$

2. Momentum Conservation

For the ith compartment,

$$\Delta P_i = (1 - \varepsilon_i)\rho_s g \Delta z_i. \tag{78}$$

For the whole combustor,

$$\Delta P = \Sigma(1 - \varepsilon_i)\rho_s g \Delta z_i, \tag{79}$$

in which voidage ε_i is determined by a set of hydrodynamic correlations (Li et al., 1982; Kwauk et al., 1986).

3. Heat Conservation

For the ith compartment,

$$\frac{(1 - \varepsilon_i)\rho_s A_t \Delta z_i}{\bar{t}_i} Q_{\text{dw}} + \sum_j^M Q_{\text{gi},j,i} + Q_{\text{si},i}$$

$$= \sum_j^M Q_{\text{go},j,i} + Q_{\text{so},i} + Q_{\text{w},i}. \tag{80}$$

For the whole,

$$G_s Q_{\text{dw}} + \sum_j^M \sum_i^N Q_{\text{gi},j,i} + \sum_i^N Q_{\text{si},i}$$

$$= \sum_j^M \sum_i^N Q_{\text{go},j,i} + \sum_i^N Q_{\text{so},i} + \sum_i^N Q_{\text{w},i}. \tag{81}$$

Boundary conditions:

$$Y_{O_2}(0) = 0.21; \qquad Y_{SO_2} = 0;$$

$$Y_{NO}(0) = 0; \quad Y_{CO_2}(0) = 0;$$
$$Y_{CO}(0) = 0; \quad Y_{H_2O}(0) = 0; \qquad (82)$$
$$C_{in} = X_c + X_R; \quad F_{mix,j,N} = 0.$$

C. COMPARISON OF MODEL WITH EXPERIMENTS

Experimental results for the pilot plant CFB combustor at ICM, Academia Sinica, have been simulated by the preceding axial gas–solids distribution two-phase model. In computation, the parameters of the model were taken from actual operations. A bituminous coal from Datong was used, the approximate and ultimate analyses of which are listed in Table VI.

Natural limestone containing 34% CaO was employed as SO_2 sorbent, and the sorbent size ranges from 0 to 2 mm. The main operating parameters are listed in Table VII. The simulated results are shown in Table VIII.

Table VIII shows that the predicted values are in good agreement with experimental results. Since the predicted pressure drop is higher than the experimental (equivalent to longer burnout time), the predicted carbon content is slightly less than the experimental value.

Figure 26 compares the computed axial voidage profiles with the experimental, indicating that they agree well. The axial gaseous compositions

TABLE VI
APPROXIMATE AND ULTIMATE ANALYSES OF DATONG COAL

Ultimate %	O_f 7.17	C_f 57.64	H_f 2.93	N_f 0.637	S_f 2.3
Approximate %	W_y 3.77	A_y 25.53	V_y 23.04	C_y 47.66	Q_{dw}, kJ/kg 22,347

TABLE VII
MAIN OPERATING PARAMETERS OF PILOT PLANT COMBUSTOR

Mean size of bed material	mm	0.35
Coal feed rate	kg/h	30
Excess air ratio	—	1.20
Primary/total air ratio	—	0.68
Ca/S molar ratio	—	2.0
Combustion temperature	K	1,123
Bed internal diameter	m	0.25
Combustion chamber height	m	6.25

TABLE VIII
COMPARISON OF PREDICTED VALUES WITH EXPERIMENTS

		Predicted	Experimental
Pressure drop across the combustor	Pa	5,270	4,800
Residual carbon content	%	0.524	0.635
At combustor exit			
O_2	%	4.04	4.20
CO_2	%	15.60	15.40
SO_2	ppm	130	160
NO_x	ppm	205	170

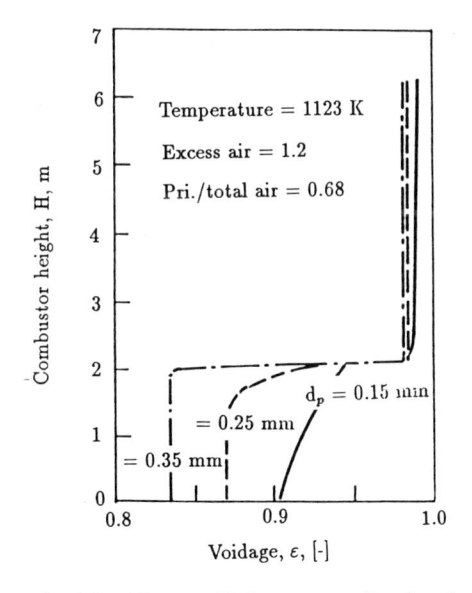

FIG. 26. Comparison of axial voidage profile between predicted and experimental values (after Lin, 1991).

are shown in Figs. 27 and 28. The axial variation of oxygen concentration is essentially a mirror image of that of carbon dioxide. At the exit, the predicted results are close to the experimental data.

D. EFFECT OF VARIOUS OPERATING PARAMETERS

Effect of the various operating parameters on combustion can be estimated by using this axial gas–solid distribution two-phase model.

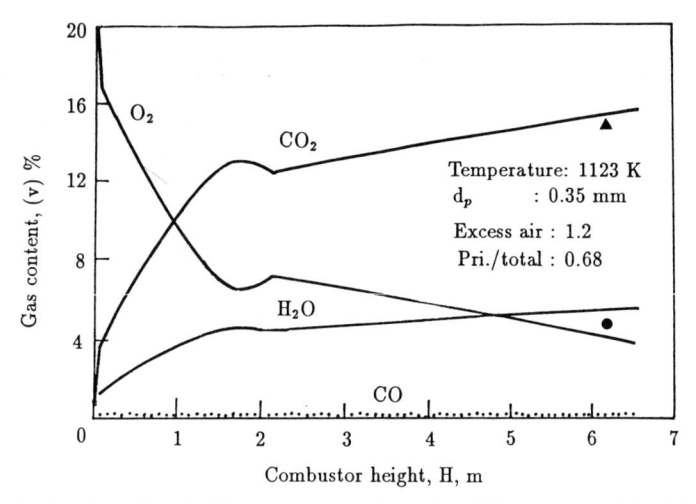

FIG. 27. Axial profiles of different gas compositions in CFB combustor (after Lin, 1991).

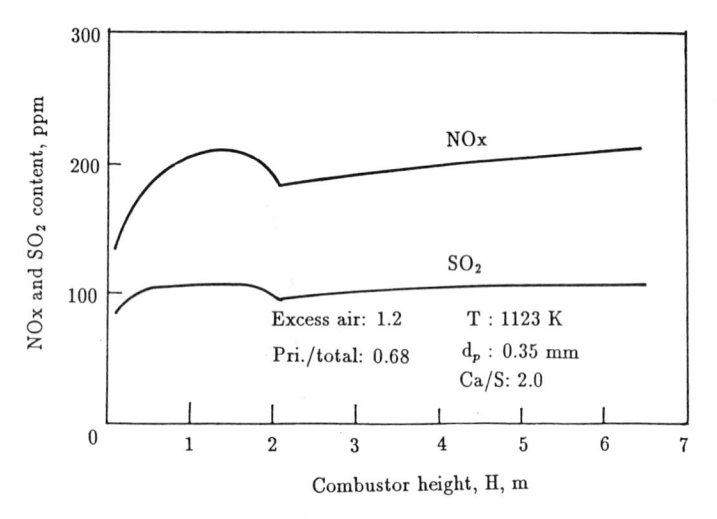

FIG. 28. Axial profiles of harmful gaseous pollutants (after Lin, 1991).

1. Particle Size of Bed Material

For any given operating velocity, the axial voidage profile changes with particle size. Figure 29 shows that the smaller the particle is, the more gradual the axial voidage changes, indicating that most solids in the lower dense

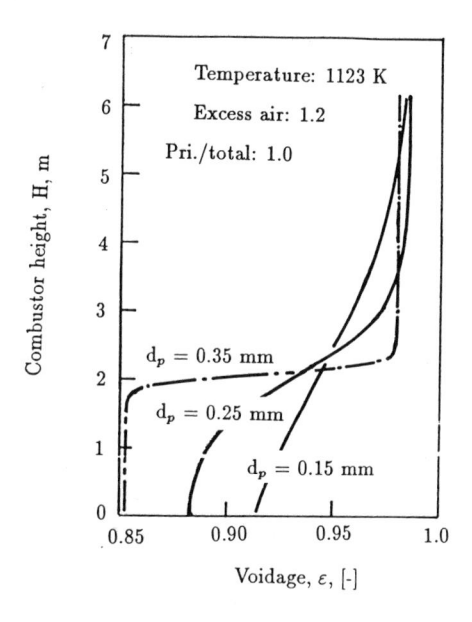

FIG. 29. Influence of particle size on axial voidage profile (after Lin, 1991).

region are transported to the upper dilute region where the heated surface is placed. The resulting upward heat flux favors meeting any specified thermal load. Secondary air has the same effect, though the voidage profile in the dilute region is slightly different, as can be seen in Fig. 16. According to combustion kinetics, burnout time diminishes notably with decreasing particle size. Therefore, selecting the appropriate particle size is of decisive significance for operating CFB boilers. A narrow range of coal size should generally be adopted for high ash content coal.

2. Optimum Operating Temperature

The reaction rate for CFB combustion is usually determined by coal reactivity, combustion temperature and oxygen diffusion. Figure 30 shows that for a given type of coal, when the combustion temperature is below 1,200 K, the combustion efficiency decreases sharply with temperature, while above 1,200 K, it is about 99%. On the other hand, SO_2 retention is also affected by temperature. Figure 31 shows the dependence of SO_2 emission on temperature, indicating that at about 1,127 K, SO_2 emission is minimal. Therefore, the optimum operating temperature ranges roughly from 1,123 to 1,173 K.

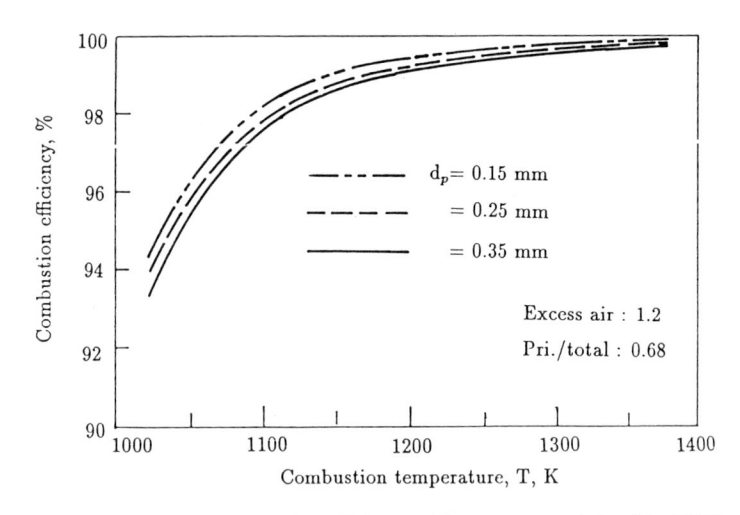

FIG. 30. Variation of combustion efficiency with temperature (after Lin, 1991).

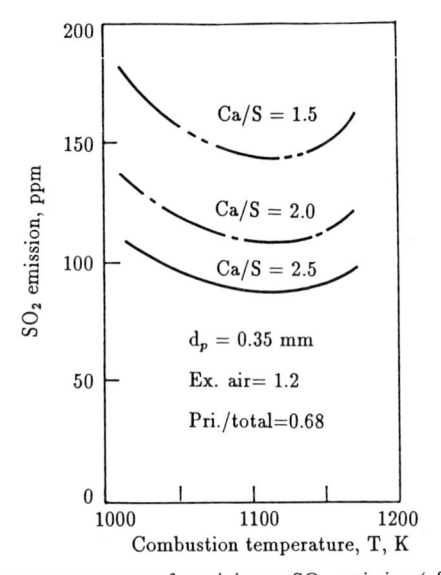

FIG. 31. Optimum temperature for minimum SO_2 emission (after Lin, 1991).

3. Rational Excess Air Ratio

High excess air ratio will result in increasing sensible heat loss in the flue gas, while insufficient air supply means low combustion efficiency. To gain high heat efficiency, a rational excess air ratio needs to be specified. Figure

32 shows that as the excess air ratio reaches 1.2 or more, a combustion efficiency of 98% to 99% can be obtained. NO_x emission is also associated with excess air ratio. Figure 33 indicates that when this air ratio is less than 1.3, NO_x emission decreases perceptibly. In conclusion, CFB combustion should be carried out in an excess air ratio of 1.2 to 1.3.

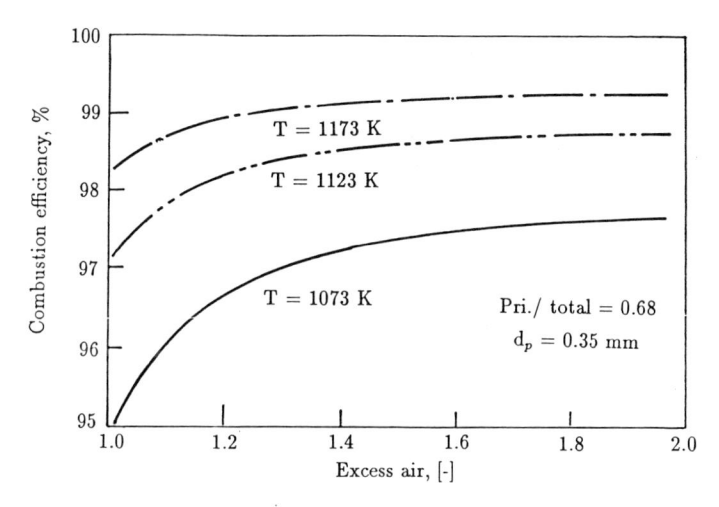

FIG. 32. Relation of combustion efficiency to excess air ratio (after Lin, 1991).

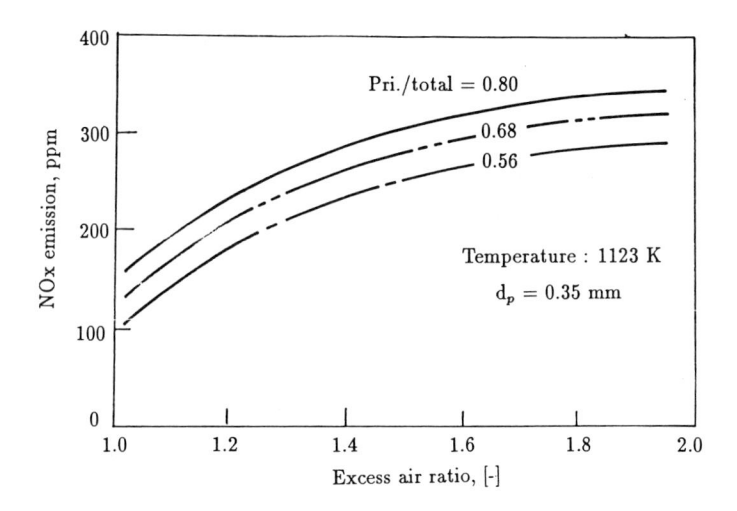

FIG. 33. Variation of NO_x emission with excess air ratio (after Lin, 1991).

4. Ca/S Molar Ratio

Figure 34 gives predicted variation of SO_2 retention efficiency with the Ca/S ratio. When this ratio exceeds 1.5, this efficiency would be beyond 80%. At low SO_2 retention, the predicted and experimental values are in good agreement, while for high SO_2 retention, the values of the prediction are higher than measurements. Volumetric expansion of the reaction product always accompanies sulfurazation, leading to plugging of the inner pores. In this case, intraparticle diffusion of SO_2 becomes difficult, resulting in insufficient utilization of the sorbent CaO.

Constitutionally speaking, the axial gas–solid distribution two-phase model gives reasonable prediction for CFB combustion, but the sub-models of apparent reaction kinetics for SO_2 sorption and NO_x reduction still calls for further improvement.

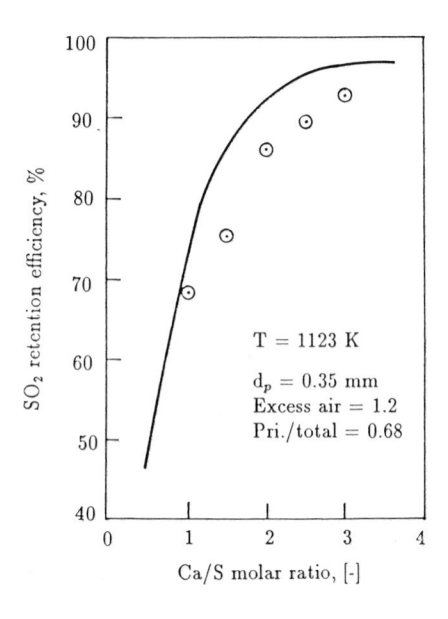

FIG. 34. Comparison of SO_2 retention efficiency between predicted and experimental values (after Lin, 1991).

IV. Process Design for CFBC Boilers

A. CFBC Boilers with Low Steam Capacities

Most boilers in China are of the stoker type, which consume about one-third of total coal production, at an average efficiency of about 60% with high pollutant emissions. To raise efficiency and to lower pollutant emissions, recent R&D efforts have been focused on CFFBs with low steam capacity, possessing the following features:

- low steam pressure
- low air velocity
- natural water-steam circulation
- low furnace height
- relatively coarse coal to simplify coal preparation
- channel gas–solid separator

Figure 35 shows a 6 t/h packaged CFBC boiler. At the bottom of the furnace is a bubbling fluidized bed operated at a gas velocity of 2 m/s. The maximum size of the fuel coal could be as much as 30 mm in spite of the low fluidizing velocity, mainly due to the well-designed air distribution system. The height of the furnace is less than 3 m. The major portion of the entrained solid particles is separated from the flue gas by a channel separator mounted at the exit of the furnace, falls into an ash silo, and is then returned to the fluidized bed.

Figure 36 is an illustration of a CFBC boiler retrofitted from a stoker-fired boiler. Also, at the bottom is a bubbling bed operated at a higher velocity of 3.5 m/s. A tube bundle is installed at an inclination of 15° in the dense bed, which extends to a height of 1.5 m from the air nozzles. A channel separator is also used, and the separated solid particles are returned through an air seal to the combustor.

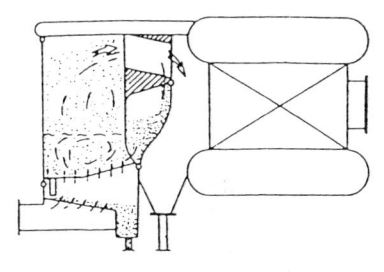

FIG. 35. 6 t/h packaged CFBC boiler.

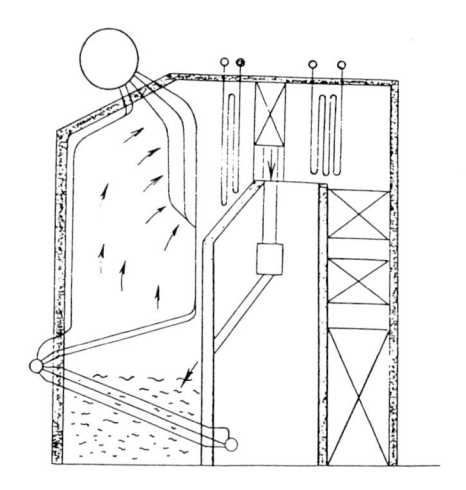

FIG. 36. A CFBC boiler retrofitted from a stoker.

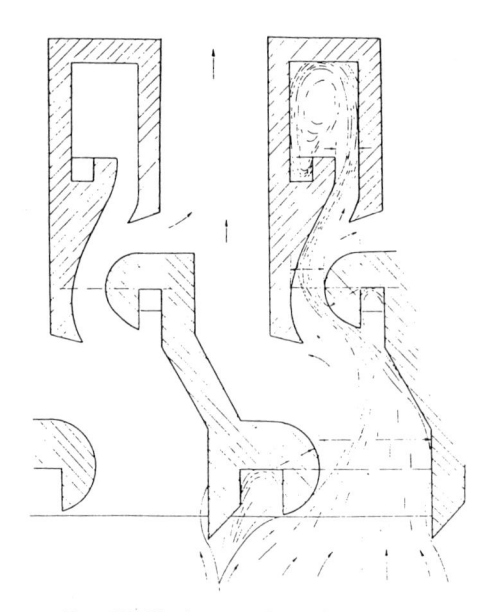

FIG. 37. Single-stage channel separator.

Figure 37 shows the schematic design of the channel separator, which is constructed of shaped bricks. When flue gas changes direction in conformity with the contour of the bricks, the entrained solids run into the ash collection grooves due to inertia, and then fall into the ash silo below. The channel separator normally extends through the entire cross-section of the exit conduit.

The combustion efficiency of these low-capacity CFB combustors varies from 93% to 96%, with 20% excess air at the furnace exit. Measurements have shown that CO concentration in the flue gas could be reduced from 1,000–2,000 ppm to 100–200 ppm through the separator, indicating continued and rather efficient combustion in the separator. With an artificial sorbent (see Section V), the SO_2 removal could reach 83–93% at a Ca/S ratio of 1.4–1.7. The NO_x in the exit gas is about 100 mg/Nm^3.

B. CFBC Boilers for Co-generation Power Plants

At present, a large number of co-generation power plants with steam capacity ranging from 35 to 220 t/h are being built in China. Figure 38 shows a 65 t/h CFBC boiler developed by Tsinghua University. Unlike the low-capacity CFBC boilers mentioned above, this boiler is characterized by the absence of any tube bundle in the dense phase. Consequently, the combustion fraction (ratio of local heat release to the total) in the dense bed should be limited by a proper choice of the amount of fine coal particles entrained by the uprising combustion gas. A simple criterion has been established: if the heat carried by the recycled materials is less than 10% of the total combustion heat, then the sum of the percentage of fines and volatile matter in the coal on an ash-water-free basis needs to be larger than 85%.

The suspended superheater below the furnace roof worked well in regulating the steam temperature. Three 75 t/h CFBC boilers of similar design have been operating in Anshan Second Thermal Power Plant during

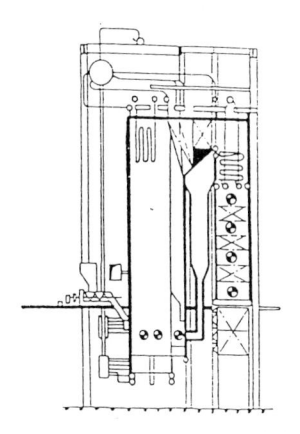

FIG. 38. 65 t/h CFBC boiler.

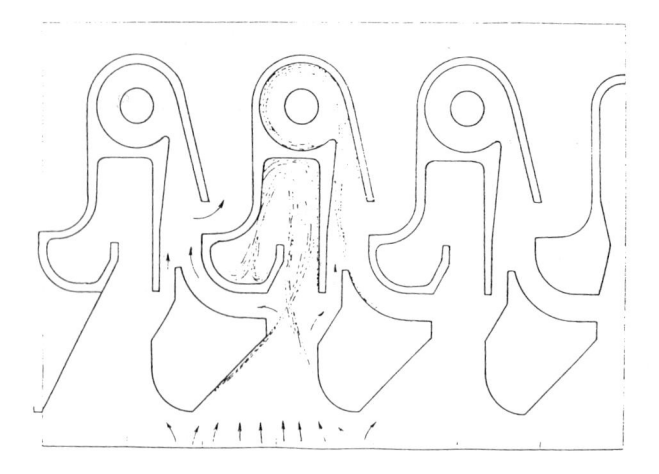

FIG. 39. Two-stage channel separator.

the last three years with no obvious erosion of the suspended superheater at a fluidizing velocity of 5.5 m/s. The steam temperature remains within the standard range for the turbine while the steam output varies from 24% to 110%. The exit flue gas temperature is 137°C, and excess air, 37%. These boilers have no fly ash reburn facilities, and the carbon content of the fly ash and bottom-bed ash are 3.5% and 1%, respectively.

Based on the experience of these 75 t/h boilers, a 220 t/h CFBC boiler has been designed and is now being fabricated. A two-stage channel separator, as shown in Fig. 39, is used on these CFBC boilers. This is followed by multi-cyclones. By using this combination of gas–solid separators, and with fly ash reburn, the combustion efficiency has reached 97.5% for a 12,000-kJ/kg low-grade coal.

C. CFBC Boilers with Co-generation of Gas and Steam

In order to provide fuel gas for both municipal and industrial use, a gas–steam co-generation CFBC, as shown in Fig. 40, has been developed, which utilizes coal more rationally by producing gas from its volatile matters and burning the remaining char for steam generation. The hot circulating char from the CFBC is used to pyrolyze the coal feed to produce a fuel gas with a heating value of 14,000 to 21,000 kJ/Nm^3. Char is recycled to the combustor where it is burned to reheat the char for further coal pyrolysis.

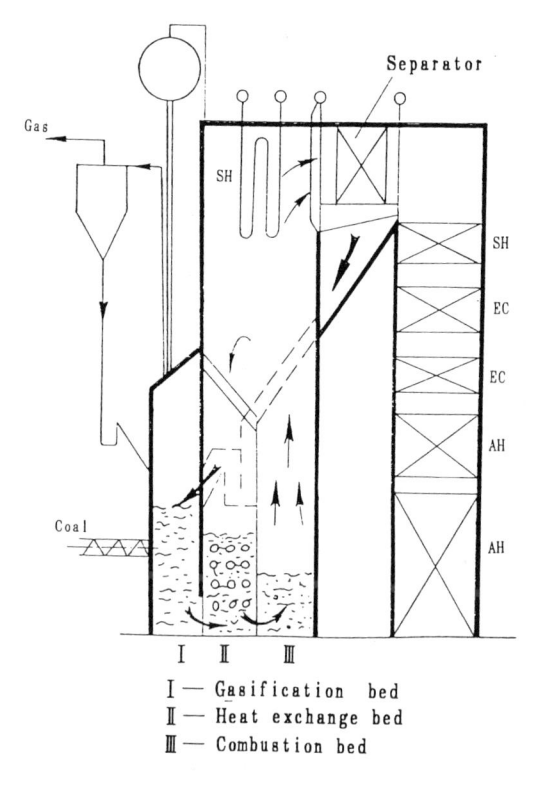

FIG. 40. Gas-steam cogeneration CFB boiler.

I — Gasification bed
II — Heat exchange bed
III — Combustion bed

A 35 t/h CFBC boiler with a co-generation capacity of 55,000 Nm^3/day of fuel gas is currently being tested.

The bottom section of this CFBC boiler is divided into three chambers. Coal is fed into a 830°C gasification chamber, fluidized by cooled and cleaned recycle fuel gas at 0.8 m/s, where it is instantly mixed with the hot ash collected by the channel separator, generating fuel gas that is separated from the entrained solids in a cyclone. The char then enters a 730°C heat exchange bed, fluidized by air at 0.8 m/s, to release heat to immersed tube bundles generating steam. It then goes to the combustion chamber, fluidized with air at 9 m/s, to heat the recycling ash. The combustion flue gas merges with the hot air from the heat exchange bed at the top and goes to the channel separator, at a combined velocity of 5.5 m/s, to precipitate the solid particles. The high-solids loading in fuel gas lessens the stringent requirement on particle size distribution needed by low-heat release in the dense bed.

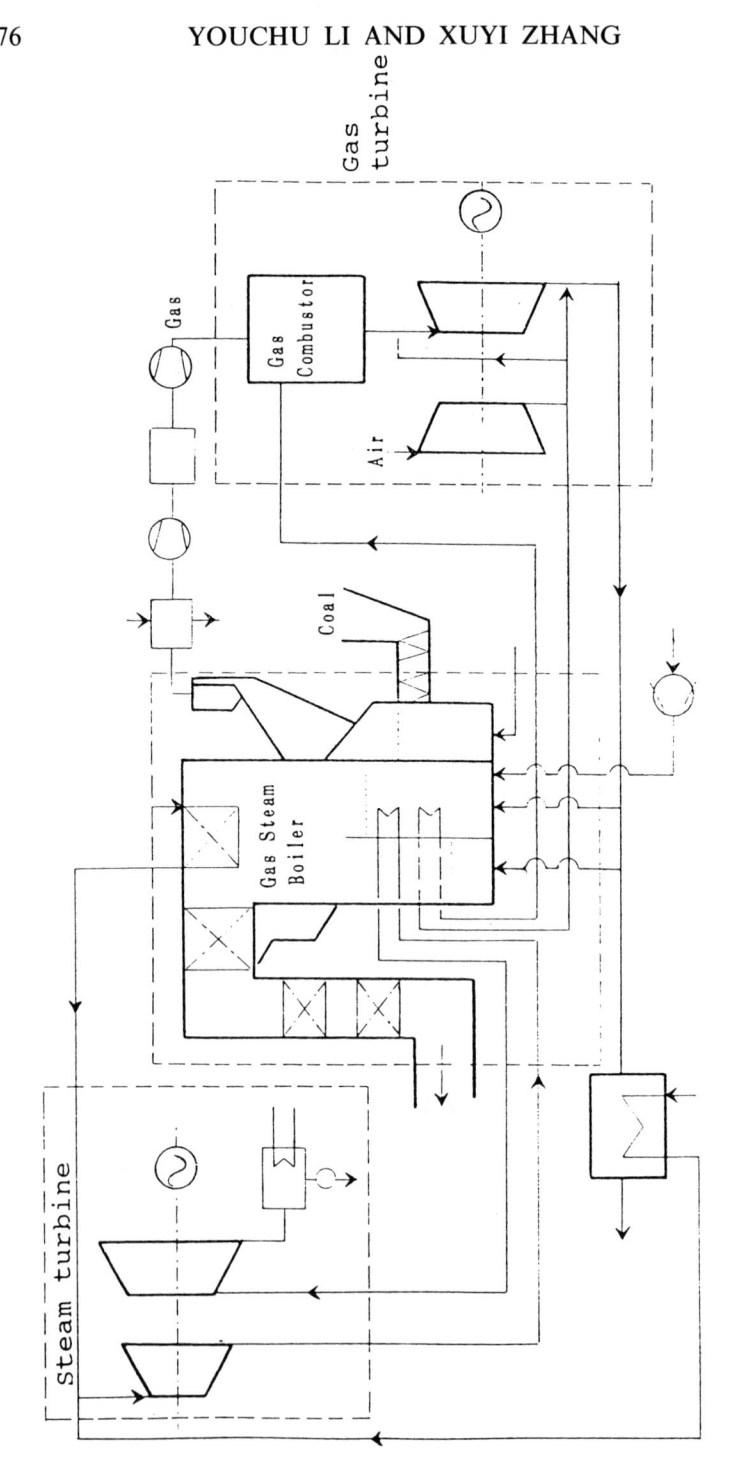

FIG. 41. A combined cycle power generation system.

D. Heat-Carrier Gasification Combined Cycle (CGCC) for Power Generation

The gas–steam co-generation CFBC boiler leads to the conceptual design of a heat-carrier gasification combined cycle (CGCC), illustrated in Fig. 41. The fuel gas is first cleaned and compressed, heated in a tube bundle in the heat transfer bed of the CFBC, and then burned in a pre-combustor to a temperature of between 1100 to 1400°C. It then enters the gas turbine. The hot turbine exhaust is used as combustion air in the CFBC. This CGCC uses no oxygen, and, compared to IGCC, it consumes less energy in gas cooling and cleaning. Compared to pressured PFBC, it is less complex in hardware.

V. Improvement of Desulphurization

A. Influence of Pore Characteristic of SO_2 Sorbent

Figure 42 shows that the availability of calcium for desulfurization in coal combustion in a bubbling fluidized bed decreases with increasing particle diameter and is less than 20% for particles larger than 1 mm. On the other hand, fly ash collected from flue gas shows a calcium availability of not more than 10%, possibly due to its short residence time in the combustor.

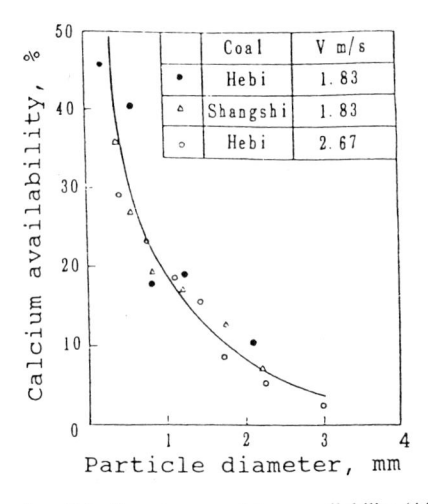

	Coal	V m/s
•	Hebi	1.83
△	Shangshi	1.83
○	Hebi	2.67

FIG. 42. Influence of particle diameter on calcium availability (After Zhang, 1981).

Experimental studies have shown that the pore size profile of a limestone sorbent is related to the combustion temperature, the origin of limestone and the partial pressure of SO_2. The pore size profile of a calcined limestone, illustrated in Fig. 43, shows that 75% of its volume is occupied by pores 500–1,600 Å in diameter. The molar volume of $CaSO_4$ (45.99 cm^3/gmol) is 1.346 times that of $CaCO_3$ (34.16 cm^3/gmol) and 2.79 times that of CaO (16.49 cm^3/mol), indicating that whatever pores are present in calcined limestone would be readily reduced in size or even plugged upon sulfation, thus reducing the calcium availability on account of the diminished diffusion of SO_2.

Smaller limestone particles possess shorter pores. Figure 44 shows that as particle diameter decreases, the pore volume increases, especially for particles measuring below 250 μm in diameter. Wu and Zhang (1988) attributed this phenomenon to increased absorption of mechanical energy as particles become smaller. An effective way to improve pore structure, and therefore, SO_2 capture, is to reduce particle size. For CFBC, addition of fine limestone is feasible without the carryover problem in the case of bubbling fluidization, often resulting in a calcium availability of as high as 50%.

B. HIGH-ACTIVITY ARTIFICIAL SORBENT

Another recourse is to develop artificial sorbents with more favorable pore structure. Figure 45 compares the desulfurization characteristics of

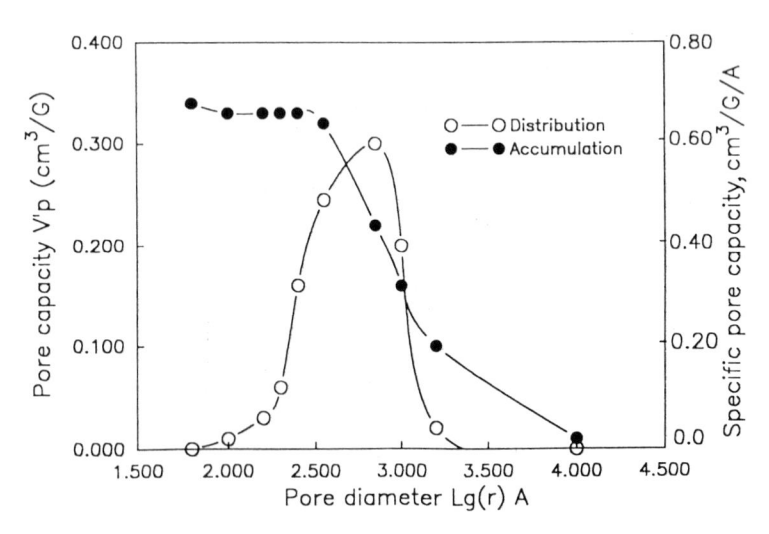

FIG. 43. Pore volume distribution for calcined limestone.

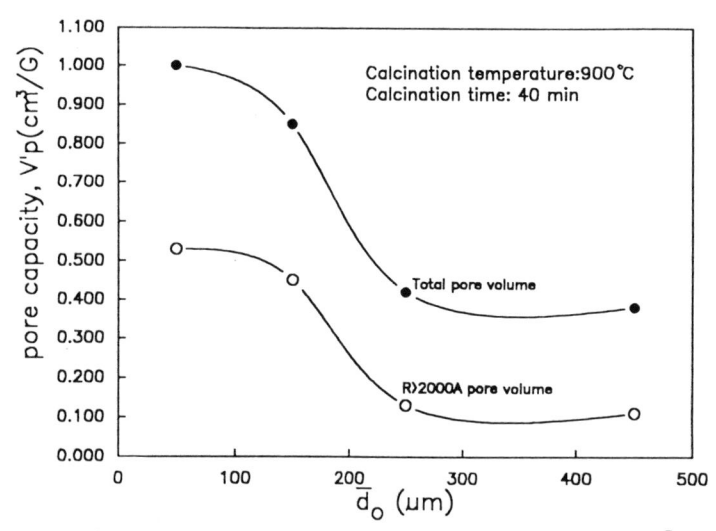

FIG. 44. Specific pore volume V'_p as a function of average particle diameter $\bar{d}_o$ (after Wu, 1988).

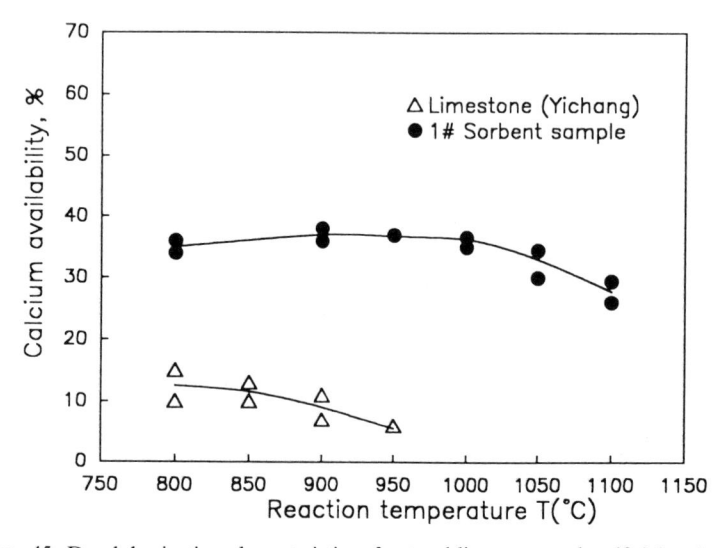

FIG. 45. Desulphurization characteristics of natural limestone and artificial sorbent.

natural limestone with that of an artificial sorbent developed by the Thermal Engineering Department of Tsinghua University (Huang et al., 1991) by agglomerating pulverized limestone with sulfur-retentive coal ash. Figure 46 compares the pore size distribution of the limestone and the artificial sorbent,

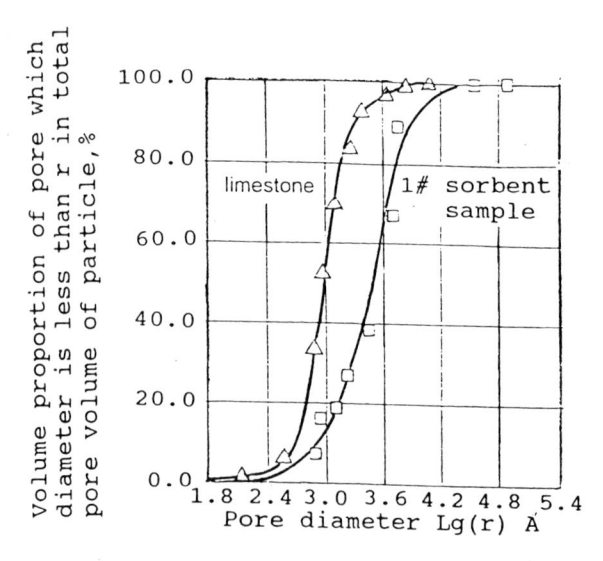

FIG. 46. Pore volume distribution of natural limestone and artificial sorbent.

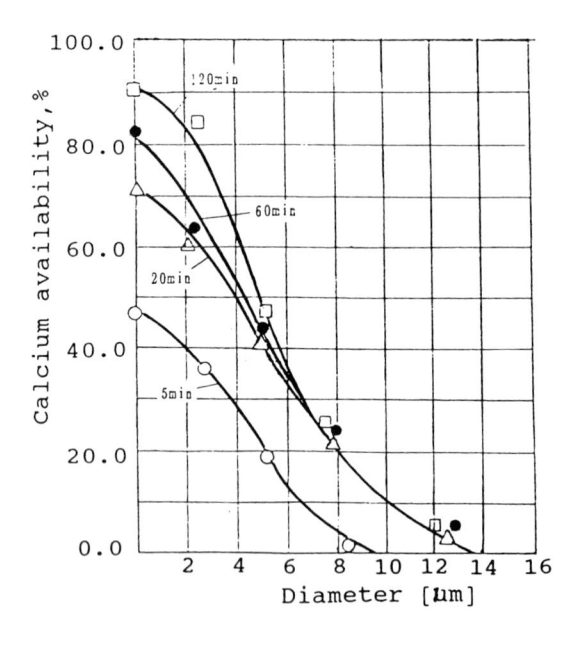

FIG. 47. Reaction characteristics along diameter of nature limestone.

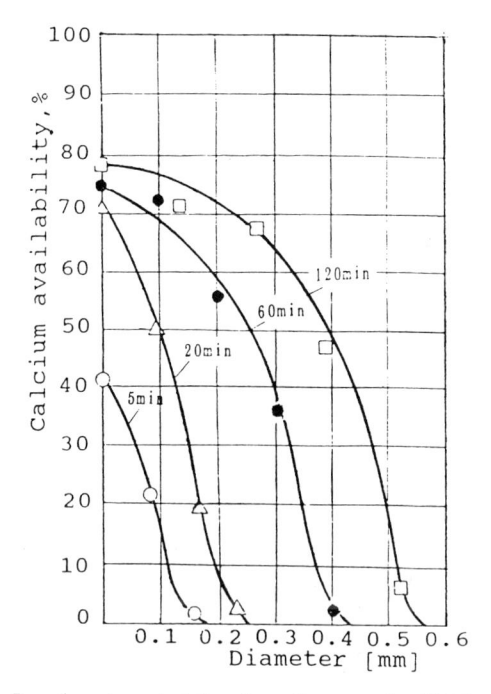

FIG. 48. Reaction characteristics along diameter of artificial sorbent.

showing that the latter possesses more pores with diameters larger than 2000Å. Figures 47 and 48 compare the variation of calcium availability along the particle diameter toward the center for natural limestone and the artificial sorbent, respectively. At the end of parallel tests lasting 120 minues, penetration of SO_2 was not more than 14 μm for the limestone, but as high as 0.55 mm for the artificial sorbent.

C. Temperature Characteristics of Desulphurization

Figure 49 shows an optimal temperature for calcium availability of the artificial sorbents with compositions listed in Table IX, indicating that the maximum calcium availability occurs in the range of 950 to 1000°C. Figure 50 illustrates the different time characteristics of SO_2 adsorption at different temperatures, demonstrating that calcium availability increases continuously with reaction time, and implying that coarser pore structure can hardly be plugged. Figure 51 and 52 show the effect of inert additives on altering, respectively, both the temperature and time characteristics of sorbents,

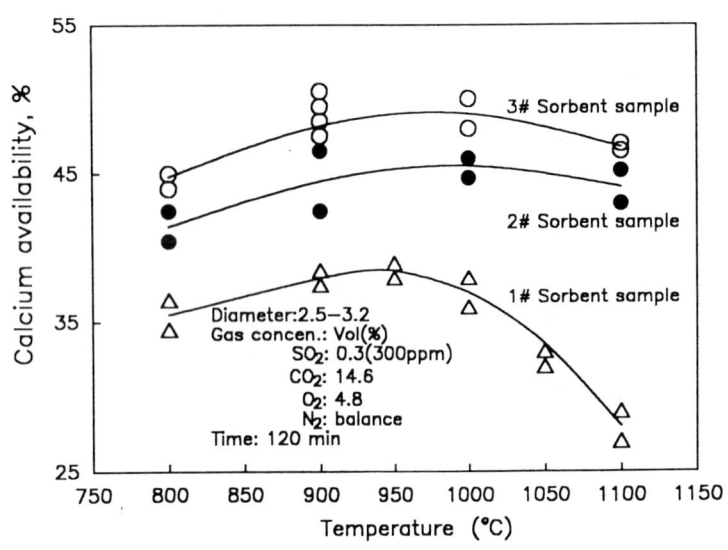

FIG. 49. Variation of calcium availability with temperature.

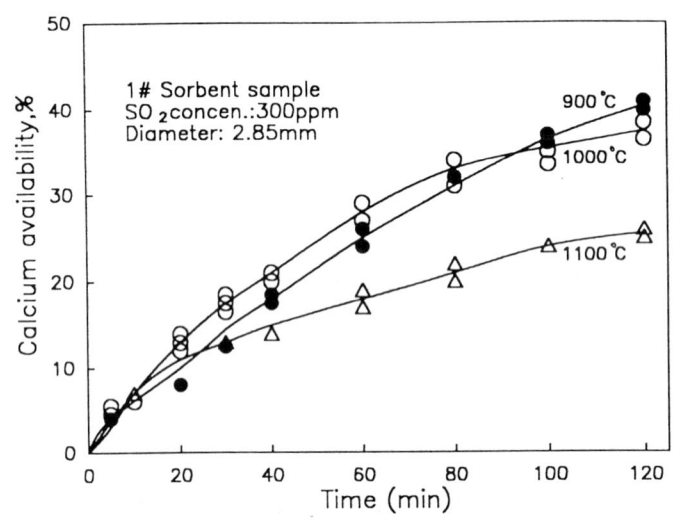

FIG. 50. Variation of calcium availability with time.

TABLE IX
COMPOSITIONS FOR DIFFERENT ARTIFICIAL SORBENTS, IN PERCENTAGES

	Limestone	Clay	Quicklime	Gypsum	FBB ash
Sample 1 #	85	15	—	—	—
Sample 2 #	70	—	8	2	20
Sample 3 #	65	—	8	2	25

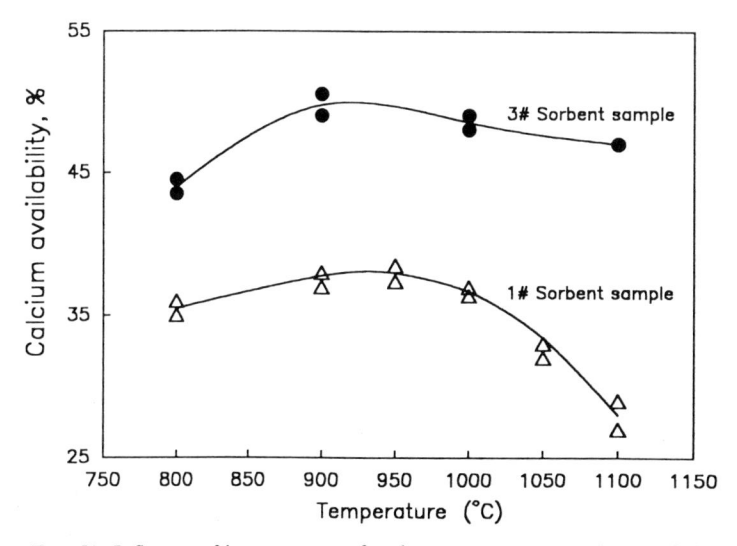

FIG. 51. Influence of inert content of sorbent on temperature characteristics.

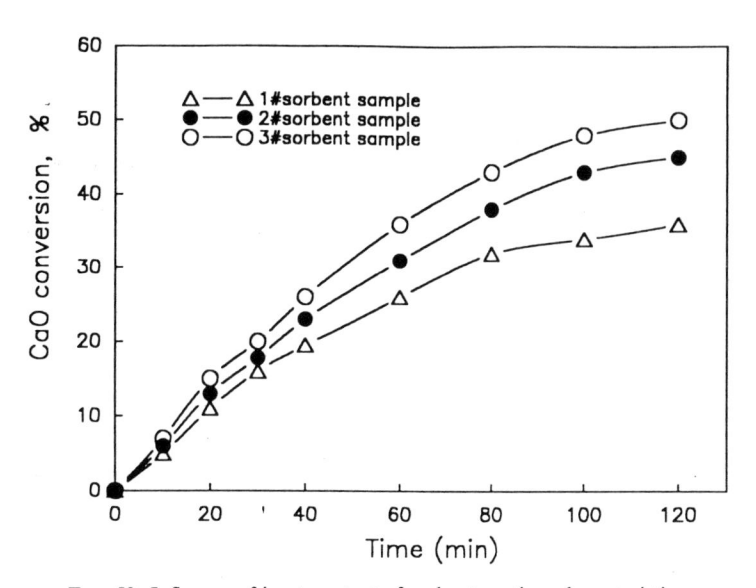

FIG. 52. Influence of inert content of sorbent on time characteristics.

indicating that the optimal desulfurization temperature of artificial sorbents are about a 100°C higher than that of natural limestone. This is beneficial for improving combustion, heat transfer and N_2O emission.

Notation

A_t	cross sectional area, m^2	k_s	intrinsic reaction rate constant, 1/s
A_j	additional air rate, M^3/s	$K_{o,s}(b)$	frequency factor for reaction of carbon with carbon dioxide, (m^3/mol)$^{1/2}$
B	experimental constant, 1/m		
Bi	dimensionless number, defined by Eq. (36)	l_g	length of pore, m
C_{ps}	specific heat of coal, kJ/kg K	M	molecular weight
c	concentration, mol./m^3	M_{CaO}	CaO molecule mass, kg
c_o	oxygen concentration of bulk stream, mol/m^3	N	solids circulation ratio, dimensionless
c_o^*	oxygen concentration at reaction surface, mol/m^3	P_{CO_2}	partial pressure of CO_2 in bulk stream, kg-wt/m^2
c_{oi}	oxygen concentration at active reaction surface, mol/m^3	$P_{CO_2}^*$	partial pressure of CO_2 on decomposing surface, kg-wt/m^2
c_s	concentration of sulphur dioxide, mol/m^3	P	pressure, kg-wt/m^2
		ΔP	pressure drop, Pa
c_{so}	concentration of sulphur dioxide in bulk stream, mol/m^3	Q	heat flux, kJ/s
		Q_{dw}	lower heating value of coal, kJ/kg
D	gas diffusivity, m^2/s	q_{O_2}	mass flux of oxygen, mol/m^{2}s
D_{ag}	dimensionless number, defined by Eq. (10)	R	radius of particle, m
		R_g	gas constant, kJ/mol K
D_c	carbon dioxide diffusivity through CaO layer, m^2/s	$R_{kj,i}$	amount of reaction for each reaction, mol/s
D_e	effective gas diffusivity, m^2/s	Re	Reynolds number, dimensionless
D_g	reduced gas diffusivity inside pore, dimensionless	Rs	radius of sulphur retention reaction in grain, m
D_k	Nusselt diffusivity, m^2/s	r	radial coordination
D_m	molecule diffusivity, m^2/s	r_0	initial radius of pore, or grain m
D_o	diffusivity through pore, m^2/s	r_c	radius of moving reaction interface, m
D_s	gas diffusivity through product layer in particle, m^2/s	r_s	radius of sulphur absorption reaction in grain, m
F	gas flow rate, nm^3/s		
F_b	dimensionless number, defined by Eq. (11)	S	specific surface area, m^2/kg
		Sc	Schmidt number, dimensionless
F_{mix}	gas backmixing rate, nm^3/s	Sh	Sherwood number, dimensionless
$f_1(r)$	pore diameter distribution function	t	time, s
E	activation energy, kJ/mol K	T	temperature, K
E_b	activation energy for reaction of carbon with carbon dioxide	T_0	initial temperature, K
		T_∞	final temperature of reaction, K
G_c	burning rate of carbon, lg/s	V	volatile yield, %
$\bar{G}_c$	reduced burning rate of carbon, dimensionless	V_∞	final volatile yield, %
		v	atom diffusion volume
G_s	coal feed rate, kg/s	W_t	total conversion of CaO, %
h	overall heat transfer coefficient, W/m^2 K	X_b	carbon conversion, %
		X_c	weight fraction of ferric oxide, [−]
h_g	heat transfer coefficient for gas, W/m^2 K	X_i	weight fraction of inert matter in sorbent, [−]
h_p	solids particle heat transfer coefficient, W/m^2 K	x_i	local conversion of CaO, %
		Y	content of gaseous compositions, %
h_r	radiative heat transfer coefficient, W/m^2 K	y_i	longitudinal average conversion of CaO, %
k	kinetic constant, 1/s	Z_0	characteristic length of voidage profile for fast bed, 1/m
k_c	transfer coefficient of CO_2 through pore, m/s		
		z	longitudinal coordination, m
k_g	convective mass transfer coefficient, m/s	z_i	position of inflection point of voidage profile curve, m

Δz	compartment height, m	ε_x	porosity of sulfated limestone, [–]
α	volume change rate, [–]	θ	proportion coefficient, [–]
δ	thickness of ash layer, m	λ	heat conductivity, W/m K
ε	bed voidage, [–]	ρ_{CaO}	CaO content in sulphur dioxide sorbent, mol/kg
ε^*	voidage at top of bed, [–]		
ε_a	voidage at bottom of bed, [–]	ρ_c	carbon content in coal particle, mol/kg
ε_e	effective diffusion porosity, [–]		
ε_0	initial effective porosity, [–]	ρ_{co}	initial carbon content, mol/kg
ε_R	local porosity of particle, [–]	ρ_s	solids density, kg/m^3
ε_c	porosity of calcined limestone, [–]	ρ_t	true density of CaO, kg/m^3
ε_{ls}	porosity of limestone, [–]	τ	reduced time, [–]

References

Agarwal, P. K., Genetti, W. E., and Lee, Y. Y. *Fuel* **63**, 1157 (1984).

Avedesian, M. M., and Davidson, J. F. "Combustion of carbon particles in a fluidized bed," *Trans. Instn. Chem. Eng.* **51**, 121 (1973).

Basu, P. "Burning rate of carbon in fluidized beds," *Fuel* **56**, 390 (1977).

Basu, P., Ritchie, C., and Harder, P. K. "Burning rates of carbon particles in circulating fluidized bed," *in* "Circulating Fluidized Bed Technology" (P. Basu, ed.), pp. 229–238. Pergamon Press, 1986.

Beer, J. M. "The fluidized combustion of coal," *in* "16th International Symposium on Combustion," pp. 439–460. The Combustion Institute, 1977.

Bhatia, S. K., and Perlmutter, D. D. The effect of pore structure on fluid–solid reaction: Application to the SO$_2$-lime reaction, *AIChE J.* **27**, p. 226 (1981).

Bi, H., Jin, Y., Yu, Z., and Bai, D. "An investigation on heat transfer in circulating fluidized bed," *in* "Circulating Fluidized Bed Technology III" (P. Basu, M. Horio and M. Hasatani, eds.), pp. 233–238. Pergamon Press, 1991.

Borghi, G., Sarofim, A. F., and Beer, J. M. "A model of coal devolatilization and combustion in fluidized beds," *Combustion and Flame* **61**, 1–16 (1985).

Boyd, T. J., and Friedman, M. A. "Operations and first program summary at the 110 MWE Nucla CFB," *in* "Circulating Fluidized Bed Technology III" (P. Basu, M. Horio, and M. Hasatani, eds.), pp. 297–311. Pergamon Press, 1991.

Chan, L. K., Sarofim, A. F., and Beer, J. M. "Kinetics of the NO–carbon reaction at fluidized bed combustor conditions," *Combustion and Flame* **52**, 37–45 (1983).

Christman, P. G. Analysis of simultaneous diffusion and chemical reaction in the calcium oxide/sulfur dioxide system, Ph. D. Dissertation (1983).

Daman, E. L. "Coal combustion in the 21st century," *Proceedings of 2nd International Symposium on Coal Combustion: Science and Technology, China Machine Press*, pp. 15–26 (1991).

Darling, S. L., Beisswenger, H., and Wechsler, A. "The Lurgi/combustion engineering circulating fluidized bed boiler — Design and operation," *in* "Circulating Fluidized Bed Technology" (P. Basu, ed.), pp. 297–308. Pergamon Press, 1986.

Engstrom, F., and Lee, Y. Y. "Future challenges of circulating fluidized bed combustion technology," *in* "Circulating Fluidized Bed Technology III" (P. Basu, M. Horio, and M. Hasatani, eds.), pp. 15–26. Pergamon Press, 1991.

Engstrom, F., Osakeyhtio, A. A., and Sahagian, S. "Operating experiences with circulating fluidized bed boilers," *in* "Circulating Fluidized Bed Technology" (P. Basu, ed.), pp. 309–316. Pergamon Press, 1986.

Fett, F. N., Burkle, K. J., Dersch, J., Scholer, J., Edelmann, H., and Heimbockel, I. "Progress in fluidized bed modelling," *Proceedings of 2nd International Symposium on Coal Combustion: Science and Technology, China Machine Press, Beijing,* pp. 555–566 (1991).

Fu, W. B., and Zhang, B. L. "A simple and general method for calculating the burning rate of coal char/carbon particle in air," *Proceedings of 2nd International Symposium on Coal Combustion: Science and Technology, China Machine Press, Beijing* (1991).

Fu, W. B., Zhang, Y. P., and Han, H. W. "A general devolatilization model for large coal particles in fluidized beds," *Proceedings of 1st International Symposium on Coal Combustion: Science and Technology, China Machine Press, Beijing* (1987).

Gierse, M. "Some aspects of the performance of three different types of industrial circulating fluidized bed boilers," *in* "Circulating Fluidized Bed Technology III" (P. Basu, M. Horio and M. Hasatani, eds.), pp. 347–352. Pergamon Press, 1991.

Grace, J. R. "Heat transfer in circulating fluidized bed," *in* "Circulating Fluidized Bed Technology" (P. Basu, ed.), p. 63. Pergamon Press, 1986.

Horio, M., and Wen, C. Y. "Simulation of fluidized bed combustors: Part 1. Combustion efficiency and temperature profile," *AIChE Symp. Ser.* **74**(176), 101–111 (1978).

Huang, X. Y., Li, Z. J., and Zhang, X. Y. "Experimental studies on the highly active water-resistant sulfur sorbent for high temperature desulfurization in fluidized bed boilers," *in* "Proceedings of 2nd International Symposium on Coal Combustion: Science and Technology (X. Xu, L. Zhou, X. Zhang, and W. Fu, eds.), pp. 436–443. China Machine Press, 1991.

Hyppanen, T., Lee, Y. Y., and Rainio, A. "A three-dimensional model for circulating fluidized bed combustion," *in* "Circulating Fluidized Bed Technology III" (P. Basu, M. Horio and M. Hasatani, eds.), pp. 563–568. Pergamon Press, 1991.

Ishida, K., Hasegawa, Y., Nishioka, K., Kono, M., Kiuchi, H., Shintani, K., Yoshinaka, K., and Vemoto, M. "Experience with 300 t/h Mitsui circulating fluidized bed boiler," *in* "Circulating Fluidized Bed Technology III" (P. Basu, M. Horio, and M. Hasatani eds,), pp. 313–319. Pergamon Press, 1991.

Kobro, H., and Brereton, C. "Control and fuel flexibility of circulating fluidized bed," *in* "Circulating Fluidized Bed Technology" (P. Basu, ed.), pp. 263–272. Pergamon Press, 1986.

Kullendorff, A., and Anderson, S. "A general review on combustion in circulating fluidized beds," *in* "Circulating Fluidized Bed Technology" (P. Basu, ed.), pp. 83–96. Pergamon Press, 1986.

Kwauk, M., Wang, N., Li, Y., Chen, B., and Shen, Z. "Fast fluidization at ICM," *in* "Circulating Fluidized Bed Technology" (P. Basu, ed.), pp. 63–82. Pergamon Press, 1986.

LaNauze, R. D. "Fundamentals of coal combustion," *in* "Fluidization," pp. 631–673. Academic Press, London, 1985.

Lee, Y., and Hyppanen, T. "A coal combustion model for circulating fluidized bed boilers," *International Conference on Fluidized Bed Combustion, San Francisco, California, April 30–May 3,* Vol. 2, pp. 753–764 (1989).

Li, Y., and Kwauk, M. "The dynamics of fast fluidization," *in* "Fluidization" (J. R. Grace and J. M. Matsen, eds.), p. 573. Engineering Foundation, 1980.

Li, Y., and Kwauk, M. Fast fluidized bed combustion of coal, *Chemical Metallurgy* (in Chinese), No. 4, pp. 87–94 (1981).

Li, Y., and Wang, F. "Analysis on existing problems for CFB boilers," Research report (in Chinese), Institute of Chemical Metallurgy, Academia Sinica (1992).

Li, Y., and Wu, P. "Axial gas mixing in circulating fluidized bed," *Preprints for 3rd International Conference on Circulating Fluidized Bed, Japan* G-7-1 (1990).

Li, Y., Chen, B., Wang, F., and Wang, Y. "Hydrodynamic correlations for fast fluidization," *in* "Fluidization: Science and Technology" (M. Kwauk and D. Kunii, eds.), pp. 124–134. Science Press, Beijing, 1982.

Li, Y., Wang, F., and Kwauk, M. "An improved circulating fluidized bed combustor," *in* "Circulating Fluidized Bed Technology III" (P. Basu, M. Horio, and M. Hasatani, eds.), pp. 359–364. Pergamon Press, 1991.

Li, Y., Lin, X., and Wang, F. "Modelling for fast fluidized bed combustion of coal," Research report (in Chinese), Institute of Chemical Metallurgy, Academia Sinica (1992).

Lin, W., and Van Den Bleek, C. M. "The SO_x/NO_x emissions in the circulating fluidized bed combustion of coal," *in* "Circulating Fluidized Bed Technology III" (P. Basu, M. Horio and M. Hasatani, eds.), pp. 545–550. Pergamon Press, 1991.

Lin, X. "A two-phase model for fast fluidized bed combustion," M. S. Thesis (in Chinese), Institute of Chemical Metallurgy, Academic Sinica. To be published (1991).

Lou, Z. Y., Cen, K. F., and Ni, M. J. "A comprehensive mathematical model for circulating fluidized bed combustion," *in* "Proceedings of 2nd International Symposium on Coal Combustion: Science and Technology" (X. Xu, L. Zhou, X. Zhang, and W. Fu eds.), pp. 488–495. China Machine Press, 1991.

Lu, H. L., Bao, Y. L., Zhang, Z. D., and Yang, L. D. "Design of CFB boiler with low circulation ratio," *in* "Proceedings of 2nd International Symposium on Coal Combustion: Science and Technology" (X. Xu, L. Zhou, X. Zhang, and W. Fu, eds.), pp. 370–374. China Machine Press, 1991.

Luo, G. "A kinetic model of coal combustion in fast fluidized bed," M. S. thesis (in Chinese), Institute of Chemical Metallurgy, Academia Sinica (1990).

Martin, H. "Heat and mass transfer in fluidized beds." *International Chemical Engineering*, **22**, No. 1 (1982).

Mi, W., Li, Y., Wang, F., and He, K. "Combustion behavior of coal in a fast fluidized bed," *Proceedings of 6th National Conference on Fluidization, WuHan, China* (1993).

Moritomi, H., Suzuki, Y., Kido, N., and Ogisu, Y. "NO_x emission and reduction from a circulating fluidized bed combustor," *in* "Circulating Fluidized Bed Technology III" (P. Basu, M. Horio, and M. Hasatani, eds.), pp. 399–404. Pergamon Press, 1991.

Neidel, W. "Computer simulation of coal combustion in circulating fluidized bed units," Proceedings of 2nd International Symposium on Coal Combustion: Science and Technology, China Machine Press, Beijing, pp. 480–482 (1991).

Ntouros, Z., Karas, E., and Papageorgiou, N. "Experimental investigation of coal combustion in the circulating fluidized bed reactor of NTUA," Proceedings of International Symposium on Coal Combustion: Science and Technology, China Machine Press, Beijing, pp. 381–388 (1991).

Oguma, A., Yamada, W., Furusawa, T., and Kunii, D. "Reduction reaction rate of char with NO," *11th Fall Meeting of Soc. of Chem. Eng., Japan*, p. 121 (1971).

Ohtake, K. "Fundamentals and application of coal combustion engineering in Japan," *Proceedings of 2nd International Symposium on Coal Combustion, China Machine Press, Beijing*, pp. 82–89 (1991).

Petersen, E. E. "Reaction of porous solids," *AIChE J.*, **3**, 433 (1957).

Prasannan, P. C. "A model for gas–solid reactions with structure changes in the presence of inert solids," *Chem. Eng. Sci.*, **40**, 1251 (1985).

Rajan, R. R., and Wen, C. Y. "A comprehensive model for fluidized bed coal combustion," *AIChE Journal* **26**(4), 642–655 (1980).

Rajan, R. R., Krishnan, R., and Wen, C. Y. "Simulation of fluidized bed combustion: Part II: Coal devolatilization and sulphur oxides retention," *AIChE Symp. Ser.* **74**(176), 112 (1978).

Ramchrandram, P. A., and Smith, J. M. "A single pore model for gas–solid noncatalystic reactions," *Chem. Eng. Sci.*, **34**, 1072 (1977).

Reh, L. "Cleaner energy from coal with circulating fluidized beds," *Proceedings of 2nd International Symposium on Coal Combustion: Science and Technology, China Machine Press, Beijing*, pp. 44–57 (1991).

Szekely, J., and Evans J. W. "A structure model for gas–solid reactions with a moving boundary—II," *Chem. Eng. Sci.*, **26**, 1901–1913 (1971).

Tawara, Y., Furuta, M., Hiura, F., Ishida, Y., Vetani, J., and Harajiri, H. "NSC CFB Boiler: Pilot plant test and industrial application in Japan," *in* "Circulating Fluidized Bed Technology III" (P. Basu, M. Horio, and M. Hasatani, eds.), pp. 353–358. Pergamon Press, 1991.

Wang, S. "Modelling of fast fluidized bed boilers," M. S. Thesis (in Chinese), Institute of Chemical Metallurgy, Academic Sinica (1992).

Weiss, V., and Fett, F. N. "Mathematical modelling of coal combustion in a circulating fluidized bed reactor," *in* "Circulating Fluidized Bed Technology" (P. Basu, ed.), p. 167. Pergamon Press, 1986.

Wu, B., and Zhang, X. "Limestone microstructure and its effect on desulfurization," *in* "Coal Combustion: Science and Technology of Industrial and Utility Applications" (J. Feng, ed.), pp. 899–908. Hemisphere Publishing Corporation, 1988.

Xu, B. S., Chen, I. J., and Yu, J. Q. "A model for sulphur absorption of limestone," *Proceedings of 2nd International Symposium on Coal Combustion: Science and Technology, China Machine Press, Beijing*, pp. 452–458 (1991).

Xu, X. C., and Zhang, X. Y. "Recent research work on coal combustion in China," *Proceedings of 2nd International Symposium on Coal Combustion: Science and Technology, China Machine Press, Beijing*, pp. 8–22 (1991).

Yan, C. M., Lin, X., Wang, F., and Li, Y. "Conceptual design of a 220 t/h improved fast fluidized bed boiler," *Proceedings of the 2nd International Symposium on Coal Combustion: Science and Technology, China Machine Press, Beijing*, pp. 375–380 (1991).

Yan, J. H., Cen, K. F., Ni, M. J., and Zhang, H. T. "Gas diffusion through the ash layer of coal particle during the combustion process," *Proceedings of the 2nd International Symposium on Coal Combustion: Science and Technology, China Machine Press, Beijing*, pp. 340–347 (1991).

Yerushalmi, J. "An overview on combustion in circulating fluidized beds," *in* "Circulating Fluidized Bed Technology" (P. Basu, ed.), pp. 97–104. Pergamon Press, 1986.

Yu, H. C., and Sotirchos, S. "A generalized pore model of noncatalystic gas–solid reaction exhibiting pore closure," *AIChE J.*, **33**, 782 (1987).

Zhang, H. "Research and development of high efficient SO_2 sorbents for fluidized bed coal combustion," Ph. D. Thesis (in Chinese), Tsing Hua University, Beijing (1992).

Zhang, L., Ti, T. D., Zheng, Q. Y., and Lu, C. D. "A general dynamic model for circulating fluidized bed combustion system with wide particle size distributions," Proceedings of 2nd International Symposium on Coal Combustion: Science and Technology, China Machine Press, Beijing, pp. 515–522, 1991.

Zhang, X. "Desulfurization with limestone in fluidized bed combustion," *J. Engineering Thermophysics* (in Chinese) **2**(1), 86–88 (1981).

Zheng, Q. Y., Wang, X., and Wang, Z. "Heat transfer in circulating fluidized beds," *in* "Circulating Fluidized Bed Technology III" (P. Basu, M. Horio and M. Hasatani, eds.), pp. 263–268. Pergamon Press, 1991.

9. CATALYST REGENERATION IN FLUID CATALYTIC CRACKING

CHEN JUNWU, CAO HANCHANG, AND LIU TAIJI

LUOYANG PETROCHEMICAL ENGINEERING CORPORATION
SINOPEC
LUOYANG, HENAN, PEOPLE'S REPUBLIC OF CHINA

I. Basic Requirements and Design Criteria for Catalyst Regeneration in FFB	389
II. Research and Development on FFB Catalyst Regeneration	395
A. Fast Fluidization	395
B. Coke Burning Kinetics	399
III. China's Commercial FFB Regenerators	401
A. Type I: Basic FFB Regenerator	401
B. Type II: FFB Regenerator as First Stage	403
C. Type III: FFB Regenerator as Second Stage	403
D. Type IV: FFB Regenerators in Series	404
E. Regenerator Heat Extraction	405
IV. Commercial Operating Data	406
A. General Survey	406
B. Typical Operating Data	407
C. Additional Data	410
V. Analysis of Commercial FFB Regenerator Operation	413
A. Fluidization Behavior	413
B. Chemical Reaction Engineering Characteristics	414
Notation	418
References	419

I. Basic Requirements and Design Criteria for Catalyst Regeneration in FFB

Catalytic cracking is one of the most important refinery processes, in which a hot heavy oil feed, such as virgin gas oil, coker gas oil, or residual oil, is brought into intimate contact with catalyst particles to convert it into more valuable light oil and olefinic gas products. As cracking proceeds, some

FIG. 1. A 1 MMTA fluid catalytic cracking unit with fast fluidized bed regenerator (courtesy Gaoqiao Petrochemical Company). 1, regenerator; 2, first-stage FFB regenerator; 3, second-stage FFB regenerator; 4, main fractionator; 5, FFB catalyst cooler; 6, catalyst hoppers; 7, cyclone separator.

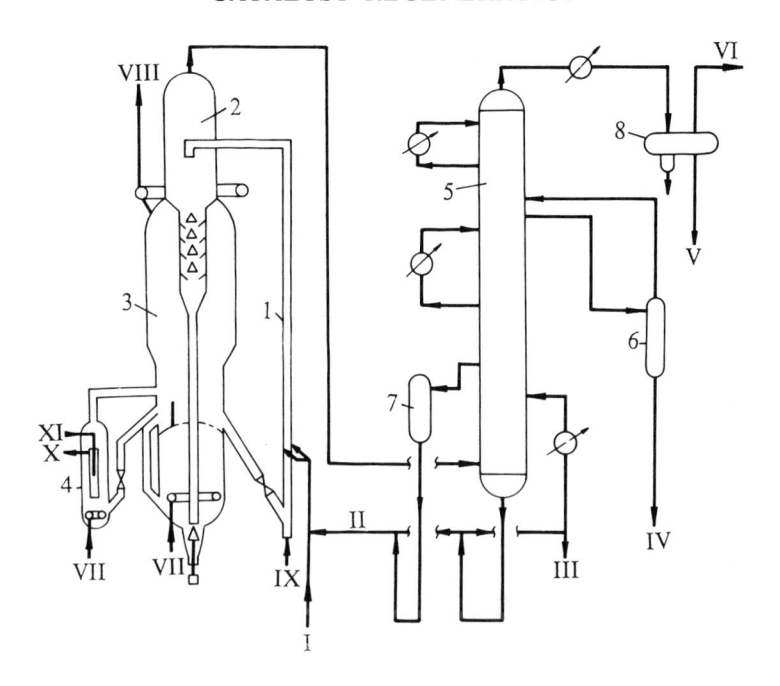

FIG. 2. Flowsheet of a typical FCCU. 1, riser; 2, disengager; 3, FFB regenerator; 4, catalyst cooler; 5, main fractionator; 6, LCO stripper; 7, HCO drum; 8, accumulator; I, fresh feed; II, recycle feed; III, slurry oil; IV, LCO; V, catalytic naphtha; VI, rich gas; VII, air; VIII, flue gas; IX, steam; X, water/steam mixture; XI, water.

components of the feed deposit in the form of coke on the catalyst and rapidly lower its activity. In order to maintain the catalyst activity at a useful level, and to supply the heat needed for vaporizing and cracking the feed, it is necessary to regenerate the catalyst by burning off this coke with air. Within 50 years of their development, fluidized catalytic cracking units (FCCU), shown in Figs. 1 and 2, have been widely installed in refineries employing catalyst in the form of very fine particles (average particle size about 60 micrometers) which behave as a fluid when aerated with vapor. The fluidized catalysts is circulated continuously between a reactor and a regenerator and acts as a medium to transfer heat from the latter to the former.

In the early days, a conventional single stage turbulent bed (TB) regenerator was used, because the amorphous silica alumina microspheric catalyst used at that time was insensitive to carbon content and its hydrothermal stability was poor. Regeneration temperature was kept at

590–610°C, and the carbon content of the regenerated catalyst (CRC) was typically 0.5–0.7 wt%.

Now zeolite catalysts have been employed by most FCCUs. Although zeolite catalysts have a much higher initial activity as compared to amorphous catalysts, coke deposit on the catalyst particles rapidly lowers their activity. As the carbon content of zeolite catalysts increases by 0.1 wt%, the activity decreases by 2–3 units. Generally the carbon content of regenerated zeolite catalysts should not be allowed to exceed 0.2 wt. %, or preferably less than 0.1 wt. % in the case of ultrastable Y zeolite (USY). Therefore, how to decrease CRC efficiently for zeolite catalysts in FCCUs has become a significant problem.

In the early 1970s, processes such as two stage regeneration, high-temperature regeneration and CO combustion-promoted regeneration were developed to burn off the coke more completely, achieving a level of CRC of about 0.05% with a CO content in the flue gas of less than 0.2 vol. %. In order to increase the carbon burning intensity (CBI), the regeneration temperature was raised.

Shortly thereafter, improved regeneration processes, such as high-efficiency regeneration and ultra cat regeneration, were announced. At the same time, research and development on fast fluidized bed (FFB) regeneration was started in China. As soon as the prototype of the basic FFB regenerator was developed, it was quickly succeeded by even newer regeneration techniques, notably in the past decade, such as (1) FFB used as the first stage; (2) FFB used as the second stage; and (3) series flow FFB. Table I gives the essential parameters of China's various regeneration processes.

It can be seen from Table I that, in order to maintain high equilibrium catalyst activity and to reduce fresh catalyst makeup, the trend is to increase CBI and to decrease CRC, so as to minimize catalyst inventory in the reactor–regenerator system. To this end, the superiority of FFB regeneration has become quite evident.

Gas transfer between the bubble phase and the emulsion phase in the fluidized catalyst bed has a major influence on CBI. In the TB, even if the gas velocity in the dense bed is as high as 1.2 m/s, the maximum O_2 transfer rate is only one-third of that required to satisfy the carbon burning rate calculated by chemical kinetics. Solid backmixing is vigorous in TB, resulting in equal carbon on catalyst in the whole bed, and this low average carbon on catalyst (which equals CRC) reduces carbon burning intensity (CBI), thus calling for high catalyst inventory and therefore a large regenerator. In order to raise regeneration efficienty, it was soon realized that both improved gas/solid contact efficiency and more favorable axial gradients could be achieved by adopting regeneration in FFB instead of TB.

As early as the 1940s, Murphree et al. (1943) described the first commercial

TABLE I
ESSENTIAL PARAMETERS OF CHINA'S REGENERATION PROCESSES

Process Type	CO_2/CO Ratio in Mixed Flue Gas v/v	CBI kg/(t·h)	CRC wt. %
Single-stage TB			
Conventional regeneration	1–1.3	80–100	0.15–0.20
CO combustion-promoted regeneration	3–200	80–120	0.10–0.20
High-temperature regeneration	200–300	80–120	0.05–0.10
Two-stage TB			
Two stages in single vessel	1.5–200	150–200	0.05–0.10
Two stages in two vessels (without catalyst cooler)	2–2.5	80–120	0.03–0.05
Two stages in two vessels (with catalyst cooler)	2–150	80–120	0.03–0.10
Two stages countercurrent	3–5	60–80	0.03–0.05
FFB			
FFB as basic regenerator	50–200	150–320	0.10–0.20
FFB used as 1st stage	50–200	150–320	0.10–0.20
FFB used as 2nd stage	3–200	90–300	0.05–0.20
Series flow FFB	50–200	200–360	0.05–0.10

FCCU, Model I, designed with gas–solid upflow. The gas velocity was merely about 0.75 m/s, and regeneration temperature about 566°C, and the resulting CBI was only 42 kg/(t·h) and the CRC as high as 0.5% (Reichle, 1992). The advantage of high gas velocity was not recognized in those days, and the upflow type was soon replaced by the downflow type.

After 20 years, the microspherical silica alumina catalyst was replaced by zeolite catalyst, and development of new regeneration processes under different fluidization regimes was also on the way.

Early in the 1970s, UOP make a comparison of catalyst residence time required at a given degree of regeneration between ideal backmixed flow and ideal plug flow regenerators operating under identical regeneration conditions (Cabrera and Mott, 1982). It was reported that the plug flow regenerator required only approximately one-tenth the residence time of that of the backmixed regenerator.

This decrease in residence time led to a smaller catalyst inventory in the reactor–regenerator system, and consequently much lower catalyst deactivation rates due to hydrothermal effects. A lower catalyst inventory has also the additional operational advantage that the FCCU can respond more quickly to changes in catalyst management, allowing the refiner to restore the desired catalyst activity level more promptly following an upset.

Based on these viewpoints, UOP developed the high-efficiency regenerator in 1974 with the following characteristics (Cabrera and Mott, 1982):

1. The spent catalyst and air move through the regenerator in plug flow to permit the unit to be designed with a low catalyst inventory and thereby reduce fresh catalyst makeup requirement.
2. The spent catalyst and air distribution are uniform to prevent oxygen and carbon monoxide from bypassing the bed.
3. To permit controlled carbon monoxide combustion without combustion promotor, the carbon monoxide and oxygen must be held in intimate contact with the catalyst at a temperature, residence time and bed density to allow adequate carbon monoxide burning to proceed, while the heat evolved is absorbed by the catalyst, and to protect regenerator internals and downstream equipment from the excessive temperature associated with afterburning.

Following these international development and guided by domestic experiences, China's Ministry of Petroleum Industry (MPI) decided to upgrade catalyst regeneration technology with the application of FFB to meet the requirements of the newer types of zeolite catalysts and the use of residue oil blended feedstocks. The philosophy of process development and design was soon evolved as summarized below:

1. The FFB vessel can be used either alone or coupled with a TB vessel in different configurations to satisfy the specific demands of expansion or revamping of existing units or the actual conditions of newly built plants.
2. The process should be able to handle various types of feedstocks by the inclusion of appropriate heat extraction facilites as dictated by the heat balance between the reactor and the regenerator.
3. Special gas/solid contacting devices with the exclusion of the ordinary air distributors should be avoided for regenerator design and construction; hence the CO-promoted combustion is preferred in order to lower regeneration temperature and to guard against afterburning and operational instability.
4. The CBI for a given CRC and a given maximal regeneration temperature should be high for the sake of lowering total catalyst inventory in the regenerator and thereby improving catalyst maintenance.
5. Minimal investment, maximal operational flexibility and less installed space should be taken into account.

Therefore, basic targets were set as follows:

1. The CRC should be below 0.1%, preferably below 0.05%.

2. Overall CBI should be higher than 200 kg/(t·h), which is almost twice that for conventional regeneration.
3. Fresh catalyst makeup rate should be reduced appreciably, depending upon the quality of feedstock.
4. Maximum temperature of regeneration should be kept below 750°C, preferably below 730°C, in order to maintain the activity of regenerated catalyst at a reasonable level.
5. Extent of CO combustion and stages of regeneration can be selected according to specific cases with emphasis on overall economy.

Research and development proceeded for several years and fruitful results were obtained. MPI and SINOPEC have since designed, constructed and operated more than 20 FCCUs with a variety of FFB regenerator configurations. Typical gas velocity was 1.0 to 1.7 m/s, exit temperature 650–730°C, resulting in CBI (in FFB) of 350–650 kg/(t·h), which far exceeded that an ordinary turbulent bed.

II. Research and Development on FFB Catalyst Regeneration

For successful design of FFB regenerators, the Institue of Chemical Metallurgy (ICM) of Academia Sinica, the Research Institute of Petroleum Processing (RIPP) and the Luoyang Petrochemical Engineering Corporation (LPEC), both formerly under MPI and now under SINOPEC, have since the 1970s made joint efforts in R & D programs on FFB catalyst regeneration, as will be regiewed in the following paragraphs.

A. Fast Fluidization

The following studies on FFB of FCC catalyst have been carried out, some under simulated plant conditions.

1. Transition from Turbulent to Fast Fluidization

Li et al. (1980) found the superficial velocity of gas and mass flow rate of solid particles at the transition point between turbulent and fast fluidization through a series of test conditions at $u_f = 0.5$–1.7 m/s and $G_s = 20$–670 t/(m²·h), in d_t 0.043, 0.081 and 0.495 m beds, and obtained the following correlation:

$$G_{str} = \frac{u_{tr}^{2.25} \rho_f^{1.627}}{0.164[gd_p(\rho_p - \rho_f)]^{0.627}}, \qquad (1)$$

where

$$u_{tr} = au_t \quad \text{and} \quad a = 3.5\text{–}4.0. \qquad (2)$$

2. Average Bed Density in FFB Regenerator

ICM studied the bed density at $u_f = 0.5\text{–}1.5$ m/s, $G_s = 6\text{–}26$ t/(m²·h) in a d_t 0.495 m cold model and deduced the following expression:

$$\bar{\rho} = \frac{G_s}{1.252 u_f L^{-1.226} - 0.651 - 0.405 \ln L}. \qquad (3)$$

Equation (3) was confirmed by RIPP in hot simulation tests in a d_t 0.081 m vessel. Further experiments were conducted by LPEC in a d_t 1.2 m vessel in a semicommercial FCCU, yielding the following conclusions:

1. Temperature shows no significant influence on bed density (Fig. 3).
2. Average bed density decreases slightly as pressure increases (Fig. 4).
3. Solid mass velocity should increase with u_f in order to keep $\bar{\rho}$ constant (Table II).
4. Evidence of the existence of an upper dilute phase and a lower dense phase is shown by the axial distribution of bed density (Table III).

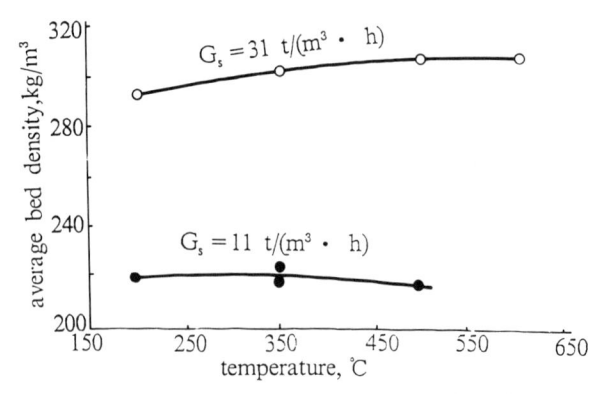

FIG. 3. Effect of temperature on average bed density on FFB. $d_t = 1.2$ m, $p = 0.14$ MPa, $u_f = 0.8$ m/s.

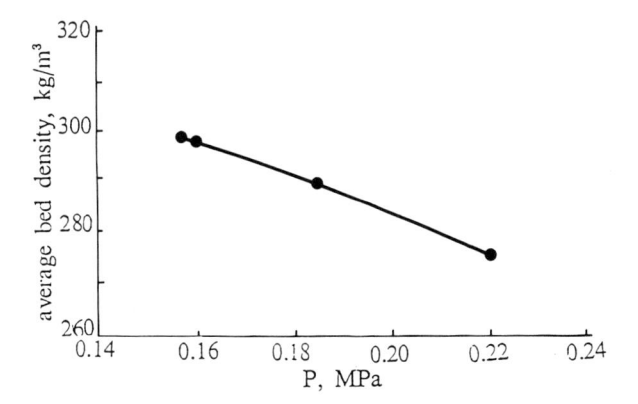

FIG. 4. Effect of pressure on average bed density in FFB. $d_t = 1.2$ m, $G_s = 2.1$ t/m²h, $T = 500$ °C, $u_f = 0.8$ m/s.

TABLE II
RELATIONSHIP BETWEEN G_s AND u_f AT CONSTANT BED DENSITY

	1	2	3	4
P, MPa (ga)	0.138	0.140	0.143	0.144
T, °C	517	490	500	600
$\bar{\rho}$, kg/m³	161	163	158	157
G_s, t/(m²·h)	6.2	21.2	36.7	44.5
u_f, m/s	0.817	1.090	1.328	1.510

TABLE III
BED DENSITY DATA FROM d_t 1.2 m SIMULATION TEST UNIT
(TEST PRESSURE 0.20 MPa(abs))

	1	2	3	4
T, °C	470	479	496	497
u_f, m/s	1.10	1.26	1.54	1.47
G_s, t/(m²·h)	73	90	168	184
$\bar{\rho}$, kg/m³	79	100	97	129
$\bar{\rho}$ at bed height				
$h = 0$–4.5 m	106	136	127	166
$h = 4.5$–6.5 m	70	70	70	90
$h = 6.5$–7.2 m	29	57	57	57

3. Bed Density in FFB Catalyst Cooler

During the R & D of the upflow type catalyst cooler, using catalyst CRC-1 ($\rho_p = 1{,}700$ kg/m), the influence of vertical heat transfer tubes on bed density was investigated by LPEC in a d_t 0.36 m cold test model, with 0.04 m o.d. tubes and 0.08–0.2 m shell side equivalent diameters. The gas velocities and solid mass velocities were 1.0–1.6 m/s and 70–180 t/(m²·h), respectively. Average bed density was correlated by the equation

$$\bar{\rho} = 2.36 u_f^{-4.859} FG_s^{1.333} d_{sc}^{0.4573}. \tag{4}$$

4. Catalyst Circulation in Two Stage FFB Regenerator

A cold model shown schematically in Fig. 5 was used by LPEC in developing the two stage FFB regenerator, with the following objectives:

1. Study the effect of superficial gas velocity on dense bed density in the second stage FFB where u_f is high.

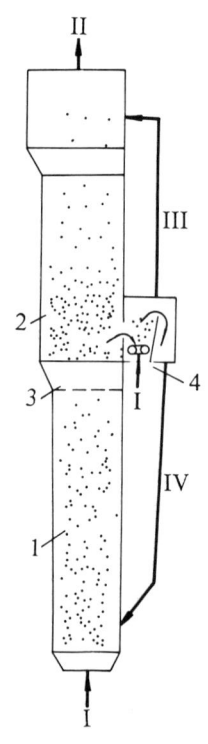

FIG. 5. Schematic diagram of two-stage FFB cold model. 1, first-stage FFB; 2, second-stage FFB; 3, distributor; 4, degassing chamber; I, air; II, flue gas; III, gas; IV, recirculating catalyst.

2. Study the performance of the second stage distributor.
3. Study the performance of the degassing chamber with submerged flow and overflow baffles.
4. Investigate the stability of dense phase catalyst circulation from the degassing chamber to the first stage FFB vessel.

The experimental results were promising: The bed voidage of the second stage can be maintained at a reasonable level and the internal recirculation rate can be controlled smoothly by the fluidization air to the degassing chamber. Thus, the design criteria are satisfied for subsequent scale-up to commercial size.

B. Coke Burning Kinetics

The main reactions in catalyst regeneration consist of combustion of carbon and hydrogen in the coke on catalyst and subsequent oxidation of CO from the primary combustion.

The burning of coke on FCC catalyst in FFB was first studied by RIPP in a d_t 0.081 m FFB regenerator at 660–680°C. The calculated CBI reached 165–230 kg/(t·h), which is 2.0 to 3.5 times that for ordinary turbulent bed regenerators. The corresponding CRC was found to be 0.18 to 0.31%.

Later studies were carried out by several investigators of the East China Petroleum College (ECPC).

Yang (1991) and Wang et al. (1986) found in a microreactor that the initial rate of formation of (CO + CO_2) was zero; it then climbed to a maximum and decreased gradually according to an exponential function, as shown in Fig. 6. They proposed the formation of an intermediate compound (C_yO_x) of carbon oxidation and suggested the following mechanism:

$$\text{C in coke} \xrightarrow[k_{1c}]{O_2} C_yO_x \xrightarrow[k_{2c}]{O_2} CO + CO_2. \tag{5}$$

They proposed the following kinetic equation:

$$\frac{d(CO + CO_2)}{dt} = \frac{k_{1c}k_{2c}C_oP_{O_2}}{k_{1c} - k_{2c}}[\exp(-k_{1c}t) - \exp(-k_{2c}t)], \tag{6}$$

where

$$k_{1c} = 1.15 \times 10^{10} \exp(-1.72 \times 10^5/RT), \tag{7}$$

$$k_{2c} = 3.8 \times 10^3 \exp(-4.49 \times 10^4/RT). \tag{8}$$

The same authors reported their experimental study on the oxidation of

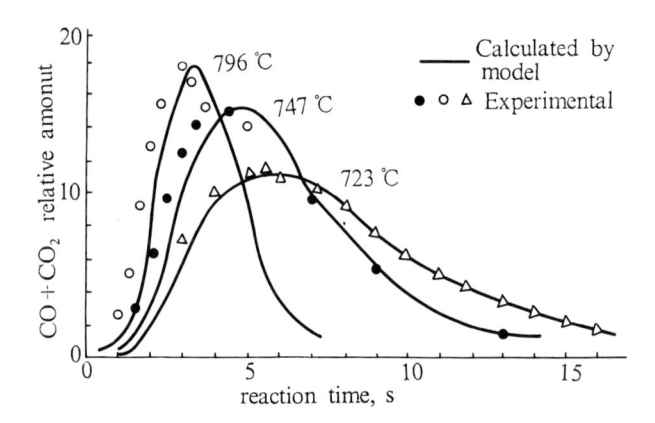

FIG. 6. Catalyst regeneration in a microreactor: effluent (CO+CO$_2$) vs. time (Wang et al., 1986).

hydrogen in the coke by both a pulsation feed method and a continuous feed method, and proposed the following rate equation:

$$-\frac{dH}{dt} = f_H k_H P_{O_2} H, \qquad (9)$$

where

$$k_H = 4.2 \times 10^9 \exp(-1.58 \times 10^5/RT), \qquad (10)$$

$$f_H = 1 - 2.67 \times 10^{30} \exp(-7.34 \times 10^4/T) \quad \text{when} \quad T \geq 973 \text{ K}, \quad (11)$$

$$f_H = 1 \quad \text{when} \quad T < 973 \text{ K}. \qquad (12)$$

Wei (1987) carried out experiments on the oxidation of CO in the presence of both clean and contaminated FCC catalyst, and gave the following kinetic equations:

$$T < 933 \text{ K}, \quad -\frac{dCO}{dt} = k_{CO} x_{CO}^{0.53} x_{O_2}^{0.35} (P/RT)^{0.87}, \qquad (13)$$

$T \geq 933$ K,
$$-\frac{dCO}{dt} = k_{CO} x_{CO}^{0.90} x_{O_2}^{0.217} [0.18 + \exp(-80 x_{H_2O})]^{-1.971} \times (P/RT)^{1.117}, \quad (14)$$

where

$$k_{CO} = 4.86 \times 10^7 \exp(-1.68 \times 10^5/RT) \quad \text{kmol/(m}^3\cdot\text{s)}, \quad (15)$$

$$k_{CO} = 5.479 \times 10^{24} \exp(-5.32 \times 10^5/RT) \quad \text{kmol/(m}^3\cdot\text{s)}. \quad (16)$$

He also calculated the initial CO_2/CO ratio from the kinetic equations of carbon and CO oxidation:

$$CO_2/CO = A \cdot \exp(E/RT), \qquad (17)$$

where A varies from 6.67 to 16.3 $\times$ 10^3 and E from 3.37 to 3.66 $\times$ 10^4 kJ/kmol depending on the type of catalyst tested.

Liu (1991) reported his experiments on the heterogeneous oxidation of CO in the presence of a FCC catalyst with Pt combustion promoters and proposed the following rate equation:

$$\gamma_{CO} = k_{CO} P_{CO} P_{O_2}^{0.5}, \qquad (18)$$

where

$$k_{CO} = 2.19 \times 10^6 \exp(-5.62 \times 10^4/RT) C_{Pt} \eta \quad \text{kg(CO)/(t·MPa}^{-1.5}\text{·s)}. \quad (19)$$

III. China's Commercial FFB Regenerators

Based on studies on gas–solids flow in cold models, coke burning of catalyst and development of FFB regenerator prototypes including plant-scale testing and operation, various types of FFB regenerators in FCC units have been put into commercial operation in the People's Republic of China within the past 15 years. All of these FFB regenerators use simple reaction vessels without internals as basic units. Air and catalyst flow upward in the FFB regenerators from bottom to top at a superficial gas velocity of 1.0 to 1.8 m/s, with solids mass flow rates of 15 to 30 kg/(m^2·s). The difference between the configurations of FFB regenerators lies in the location and function of the FFB unit in the overall regeneration system. A description of four configurations of commercial FFB regenerators will be presented in the following sections.

A. Type I: Basic FFB Regenerator

The FFB regenerator shown in Fig. 7a is used as the basic regeneration reactor. Coke-laden spent catalyst and recirculated regenerated catalyst enter the bottom of the FFB vessel and are fluidized by the combustion air in the operating regime of fast fluidization. The near regenerated catalyst flows cocurrently with the flue gas out from the top of the FFB vessel. After gas/solid separation, the catalyst falls into the bottom of a disengager situated on top of the FFB vessel and maintained in a mildly fluidized state by air

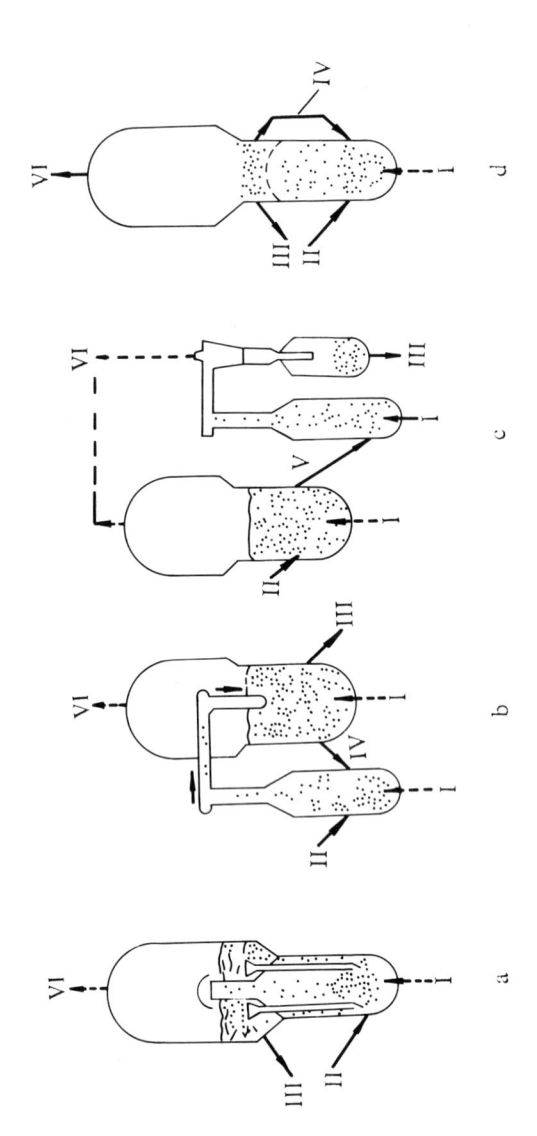

FIG. 7. Four configurations of FFB regenerator. (a) FFB as basic stage; (b) FFB as first stage; (c) FFB as second stage; (d) FFB in series flow. I, combustion air; II, spent catalyst; III, regenerated catalyst, IV, recirculating regenerated catalyst; V, semi-regenerated catalyst; VI, flue gas.

at a superficial velocity of 0.15 to 0.25 m/s. The holdup of catalyst in the disengager is based on an average retention time of about 0.5 minute per recycle in the overall circulating system. Then the regenerated catalyst is split into two streams—one for the riser reactor and the other for recirculation of the FFB vessel via either an external pipeline or internal overflow pipes. The load of coke combustion for the FFB vessel is approximately 90% of the total coke formed, and the rest is burnt in the disengager.

This type is selected mainly for grassroots units. It is relatively simple in construction and easy to operate. The principal drawback is the large catalyst inventory in the disengager, which contributes only a few percentage to carbon burning.

B. Type II: FFB Regenerator as First Stage

The FFB vessel in Fig. 7b is used as the first stage of regeneration while a conventional turbulent bed is used as the second stage. Combustion air is introduced separately into the two stages, and a part of the regenerated catalyst is circulated back to the first stage. The coke burnt in the first stage is about 40 to 60%. Since the average carbon content of the catalyst is high in the first stage, the CBI is also high. The second stage has nearly the same CBI as an ordinary turbulent bed regenerator, and therefore the overall CBI of the two stages combined is greater than that for a single stage turbulent bed regenerator. The CO oxidation rate lags behind the carbon burning rate in the first stage, some CO in the flue gas therefrom remaining unoxidized. In order to avoid afterburning in the dilute phase and also for effective utilization of excess O_2, it is necessary to introduce the semiregenerated catalyst and the flue gas from first stage into the dense bed of second stage. The vessel diameter of the second stage is almost the same as a single turbulent bed regenerator, but the catalyst inventory is reduced substantially. This type of regenerator is selected for revamping existing turbulent bed units with the purpose of further reduction of carbon on regenerated catalyst.

C. Type III: FFB Regenerator as Second Stage

As shown in Fig. 7c, the conventional turbulent bed is used prior to the FFB vessel which is used as the second stage. Combustion air is introduced separately into the two stages. The coke burnt in the FFB vessel is about 10 to 20%. Although the average carbon content of catalyst in the FFB vessel is quite low, the CBI is still higher than ordinary turbulent bed owing to more efficient O_2 transfer between phases. There are two modes of CO combustion in regeneration: (1) near complete combustion in both stages; (2) partial combustion in both stages. The latter mode is adopted for FCCU with a coke yield of 5.5 to 6.5 wt. % on fresh food and without catalyst cooling devices, but the carbon on the regenerated catalyst is somewhat higher—about 0.3%. The former mode is widely practiced in commercial units, for which the operating conditions of the first stage should be carefully controlled (regenerator temperature and excess O_2 are important parameters) to allow the rate of CO combustion to keep pace with the rate of carbon burning so that less than 1% of CO is left in the flue gas of the first stage. Thus, the distribution of coke combustion load between the two stages depends on reaction kinetics of carbon and CO combustion, and it varies with operating conditions. If the temperature of the first stage becomes higher, then the carbon on the semiregenerated catalyst will be lower, with the result that less coke enters the second stage. Since the quantity of air to the FFB vessel cannot be varied to a large extent, it is advisable to do precise engineering calculations for choosing the operating conditions so as to avoid excessive air consumption.

There are three ways to blend the flue gas streams from the two stages for the near complete combustion mode: (1) blending them outside the regeneration vessels (the CO content should be less than 0.5%); (2) directing the flue gas of the second stage into the dilute phase of the first stage, in order to make use of the "heat sink" effect of the catalyst in the dilute phase, thereby allowing a higher concentration of CO (about 2%) in the flue gas; (3) passing the flue gas of the second stage into the dense bed of the first stage with no restriction on CO content, since all CO is thus oxidized in the dense bed.

This type is specially adapted for the expansion or debottlenecking of existing FCCUs in order to raise the coke burning capacity by 20% to 30%, for which only a small diameter FFB regenerator with minimal occupied space is added. An example is the expansion of an FCCU, for which the major job can be accomplished within 6 weeks of turnaround time with very low capital expense. It is also recommended for grassroot units processing residue containing feedstocks, the FFB regenerator of which can operate under more favorable conditions to bring the CRC down to less than 0.1%.

D. Type IV: FFB Regenerators in Series

Combustion air and catalyst flow cocurrently and consecutively through two FFB vessels, superimposed one on the other (Chen et al., 1989; Wang et al., 1991), as shown in Fig. 7d. The lower one is an ordinary FFB vessel. Semiregenerated catalyst and flue gas with 4–6% excess O_2 leave the lower stage and enter the second stage on top through a specially designed low pressure drop distributor grid. The second stage regenerator operates under high gas velocity (2 m/s) in the regime of fast fluidization, but its upper section is expanded to a large diameter dilute phase region so that the gas velocity drops to 0.6–0.7 m/s and a dense phase region with a porosity of $\varepsilon = 0.85$–0.90 is formed, underneath which, however, the fluidized catalyst bed is considerably different from that existing in the disengager of Type I.

Regeneration continues in the second stage and the regenerated catalyst is split into two streams, one to the riser reactor, and the other returning to the inlet of the first stage after passing through a specially designed underflow baffle and an overflow well. This type has the following advantages: (1) The oxygen content of the flue gas leaving the first stage is higher as compared to Type I, which is favorable for coke combustion in the lower CBI bottleneck zone of the first stage regenerator; (2) although the average oxygen partial pressure in the second stage is low, it can be compensated by the higher mass transfer coefficient of O_2 due to high gas velocity, resulting in moderate CBI in the second stage; (3) if a catalyst cooler is used, the temperature at the top zone of the first stage can be maintained at a somewhat higher level than normal, in order to achieve higher CBI in that zone without appreciable catalyst deactivation; (4) the reactor disengager vessel and the two stage FFB regenerator can be arranged in a co-axial configuration to rest on a common supporting structure with minimal space. The mechanical engineering design is compact, and it also facilitates catalyst flow and plant operation.

E. Regenerator Heat Extraction

For most FCCUs processing feedstocks containing residual oil, it is necessary to extract heat from the regenerator catalyst bed as demanded by heat balance. All of the four types of FFB regenerator as well as the conventional TB regenerator can be provided with heat removal facilities. Various types of external catalyst coolers have been developed in China during the past decade. The "upflow" type cooler operates in the fast fluidization regime. Heat transfer elements are placed inside the vertical FFB vessel, and heat is transferred from the catalyst bed to the steam/water mixture inside the tubes to generate medium pressure (3.5–4.5 MPa) steam. Superficial air velocity is 1.0 to 1.5 m/s, and average catalyst bed density, 100 to 200 kg/m^3. Hairpin tubes and bayonet tubes have been used as the heat transfer elements, the latter being highly adaptable in most casts. Water from the steam drum flows down the inner tube of a bayonet element and is delivered upward through the annular space; 5 to 10% of water is evaporated by heat

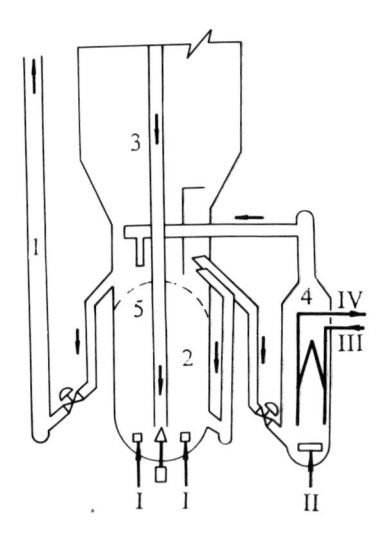

FIG. 8. Installation of external catalyst cooler. 1, riser; 2, first-stage regenerator; 3, second-stage regenerator; 4, catalyst cooler; 5, distributor; I, combustion air; II, fluidization air; III, water; IV, steam and water.

transfer and the steam/water mixture returns to the steam drum. Water flows by natural convection or forced circulation.

The bayonet tube element is fitted on the outside with a plurality of small diameter tubes or welded with longitudinal fins to enhance heat transmission from the fluidized bed (Zhang et al., 1990; Jiao et al., 1991). The heat flux is normally about 150 to 200 kW/m^2, and the heat duty is about 11–22 MW for a typical "upflow" catalyst cooler with vessel diameter 1.6 to 2.2 m.

Hot catalyst from the regenerator is introduced to the bottom of the catalyst cooler through a standpipe, while fluidization air flows upward with the catalyst, passing along the surfaces of the heat transfer elements and leaving the cooler at the top. The cooled catalyst and air are sent back to the dense bed of an ordinary regenerator or the middle zone of an FFB regenerator for continued utilization of O_2. A typical assembly of an "upflow" catalyst cooler is shown in Fig. 8.

IV. Commercial Operating Data

A. GENERAL SURVEY

Among the 30 commercial FCCUs with FFB regenerators (including semi-commercial units) now operating in China, regenerator diameter (i.d.

of lining) varies from 1.2 to 5.8 m, tangential height from 8 to 12 m, and catalyst inventory from 1 to 16 tons. The amount of coke burnt in an FFB regenerator is generally 0.4 to 9 t/h, which accounts for 15–90% of the total coke burning load. The regeneration temperature is kept at 650 to 730°C, and CO promotor is used for complete combustion.

Here CBI, defined as the carbon burnt in kilograms per hour per ton of catalyst inventory, is used as a unified index for comparison of coke burning efficiency in different units. Both CBI and the carbon content of catalyst at the outlet and the inlet of a FFB regenerator are used in combination for evaluating the performance of FFB regenerators. Owing to the limited number of sampling and measurement devices in commercial units, it is difficult to obtain exact flue gas composition, carbon content of catalyst (Type I, II, IV) at an FFB regenerator outlet and the internal circulating rate of regenerated catalyst (Type I, II, IV). The same situation exists for accurate measurements of the axial distribution and the radial distribution of temperature as well as of voidage. Although difficulties exist in procuring accurate data, from a macroscopic point of view, technical analysis and comparison are still possible for evaluating the performance of FFB regenerators.

B. Typical Operating Data

Typical operating data of the four types of commercial FFB regenerator units are presented in Tables IV, V, VI and VII. These tables show the following distinct features.

1. CRC

The CRC from FFB regeneration is generally 0.1–0.20 wt. %, some even reaching the lower ranges of 0.03–0.08 wt. %. This clearly indicates that CRC from FFB regeneration is generally much lower than that for the turbulent bed. However, the newer USY catalyst requires CRC of less than 0.1 wt. %, so the Type I and II cannot reach this goal without modifying its equipment size and operating conditions.

2. CBI

The CBI of an FFB regenerator is as high as 400–650 kg/(t·h) for Types I, II, and IV and 90–300 kg/(t·h) for Type III, depending on bed temperature,

TABLE IV
OPERATING DATA OF COMMERCIAL FCCUs WITH BASIC FFB REGENERATORS

Operating Parameter	Unit A	Unit B
Top pressure, MPa (abs)	0.26	0.32
Top temperature FFB, °C	700	710
Bottom temperature FFB, °C	692	663
Disengager dense bed temperature, °C	725	716
Superficial gas velocity of FFB, m/s	1.7	1.4
Flue gas composition, vol. %		
CO_2	15.2	15.4
CO	0	0
O_2	2.2	2.2
Carbon content of catalyst, wt. %		
Spent	0.94	1.12
Regenerated	0.13	0.11
CBI, kg/(t·h)		
FFB	631	652
Regenerator overall	295	219

TABLE V
OPERATING DATA OF A COMMERCIAL FCCU WITH 1ST STAGE FFB REGENERATOR

Top pressure (1st stg.), MPa(abs)	2.6
Top temperature FFB, °C	687
Catalyst inventory, t	
1st stg.	0.34
2nd stg.	3.18
Carbon content of catalyst, wt. %	
Spent	1.26
Semiregenerated	0.68
Regenerated	0.12
CBI, kg/(t·h)	
FFB	932
2nd stg.	97
Overall	177

average partial pressure of oxygen, gas velocity, the required CRC, etc. It can be seen that the CBI of an FFB regenerator is 2.5 to 3 times higher than that of any up-to-date TB. If all the catalyst inventory in the regenerator is taken into account, it can be seen from the tables that the overall CBI lies between 180 and 310 kg/(t·h), that is, 200–250% of that for the TB. This is equivalent to saying that the catalyst inventory for FFB regeneration is much lower than that for any other type.

TABLE VI
Operating Data of Commercial FCCUs with 2nd Stage FFB Regenerator

Operating Parameter	Unit A	Unit B	Unit C
Top pressure (1st stg.), MPa(abs)	0.26	0.25	0.23
Dense bed temperature (1st stg.), °C	640	681	650
Dilute phase temperature (1st stg.), °C	660	702	681
Top temperature FFB, °C	681	673	681
Superficial gas velocity, m/s			
TB (1st stg.)	0.84	1.34	0.82
FFB (2nd stg.)	1.24	1.96	1.65
Mode of CO combustion	Cplt.	Cplt.	Part.
Mixed flue gas composition, vol. %			
CO_2	14.1	12.6	13.8
CO	0.1	0.6	4.6
O_2	2.3	4.2	0.3
Carbon content of catalyst, wt. %			
Spent	1.10	0.85	1.13
Semiregenerated	0.40	0.11	0.43
Regenerated	0.20	0.07	0.15
CBI, kg/(t·h)			
TB	182	244	154
FFB	227	87	298
Regenerator overall	190	228	180

TABLE VII
Operating Data of a Commercial FCCU with Series Flow FFB Regenerator

Top pressure, MPa(abs)	0.32
Top temperature FFB, °C	733
Bottom temperature FFB, °C	697
Bed temperature (2nd stg.), °C	739
Superficial gas velocity, m/s	
1st stg.	1.68
2nd stg.	2.08
Carbon content of catalyst, wt. %	
Spent	1.24
Regenerated	0.05
CBI, kg/(t·h)	
1st stg.	541
2nd stg.	58
Regenerator overall	311

3. CO Promoter

When a CO promoter is used, near complete combustion of CO can be achieved within a FFB regenerator, which is confirmed by the low differential temperature (< 30°C) between the dilute phase and the dense phase of the regenerator.

4. Top Temperature

The top temperature of a FFB regenerator should not be more than 730°C unless specified for special occasions. Superficial gas velocity is 1.2–1.8 m/s. Normal operation is quite flexible, and control is easy during plant startup, shutdown and upset.

C. ADDITIONAL DATA

To facilitate detailed technical analysis, additional technical data besides normal operating parameters for FFB regeneration will be presented in the following sections.

1. Temperature Distribution

The temperature of an FFB regenerator rises gradually from the bottom to the top with an axial temperature gradient of 20–40°C, as compared to less than 10°C for the turbulent bed. The calculated temperature in the bottom zone should be in the neighborhood of 600°C by the assumption of perfect mixing of catalyst and air streams without any backmixing, yet it is far beyond the actual measured temperature. Therefore, ideal plug flow does not exist in the whole FFB, especially in the bottom zone, but the extent of mixed flow is nevertheless much less than that of TB.

Certain circumferential temperature differences exist on FFB regenerator from the bottom to the top. The temperature of two catalyst streams entering at the diagonal entrances may differ by as much as 200°C, resulting in a 20 to 50°C measured circumferential bed temperature difference at different elevations. The two curves in Fig. 9 show the axial temperature distribution on two sides of a commercial FFB unit with d_t 5 m from 3 m upward from the bottom, indicating a circumferential temperature difference of ± 20°C, which is evidence of reduced mixing.

2. Distribution of Carbon Content and Gas Composition

Table VIII presents data on gas and catalyst composition at different axial positions of a commercial FFB regenerator. It can be seen that both carbon content of the catalyst and oxygen content of the gas diminish with height, while CO_2 shows a general trend of gradual increase.

3. Bed Density and Density Distribution

Figure 10 shows a general decrease in average bed density with superficial

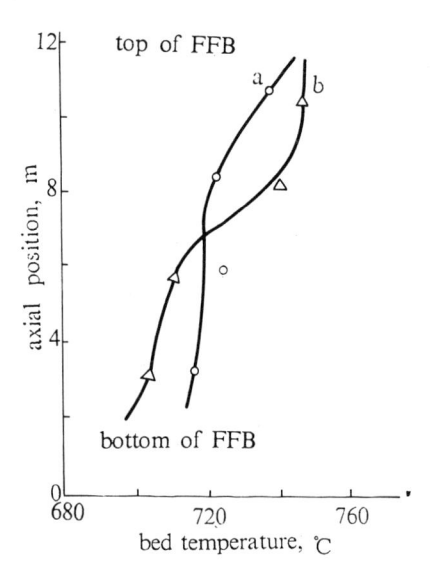

FIG. 9. Temperature profile of a commercial FFB regenerator. a, circulating catalyst entrance side; b, spend catalyst entrance side.

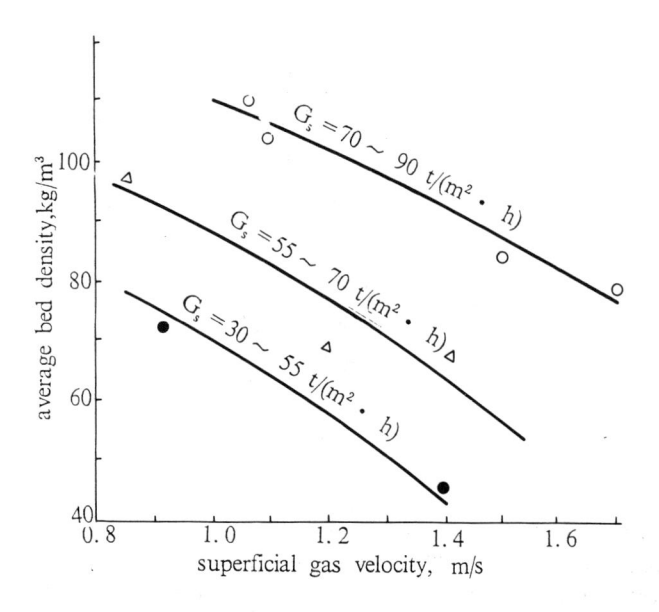

FIG. 10. Average bed density of commercial FFB vessels.

TABLE VIII
AXIAL PROFILE OF TEMPERATURE, GAS COMPOSITION, CARBON ON CATALYST OF A COMMERCIAL FFB REGENERATOR

Axial Distance from FFB Bottom m	Carbon wt. %	O_2 in Gas vol. %	CO_2 in Gas vol. %	Temperature °C
9.70	0.0430	4.10	12.21	703
6.90	0.0575	4.25	10.71	
4.93				695
4.90	0.0977	5.50	9.22	
2.90	0.1220	2.97	12.00	
1.19				682
0.9	0.1522	9.77	8.63	

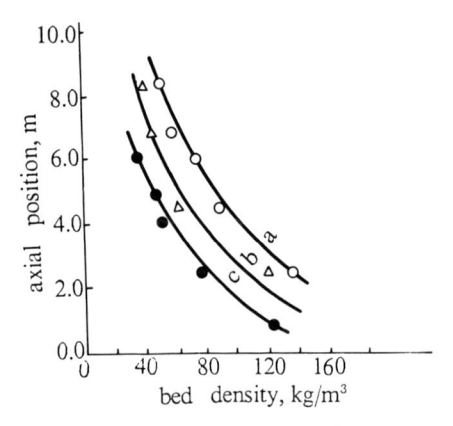

FIG. 11. Variation of axial bed density with operating parameters (Zhao, 1983).

Curve	u_f, m/s	Holdup, t
a	1.3	12.5
b	1.12	10.8
c	1.17	10.0

gas velocity and superficial mass flow rate of catalyst in some commercial units. Figure 11 (Zhao, 1983) presents the axial density distribution taken from a commercial unit, showing the distinct feature of a dilute phase at the top and a dense phase at the bottom.

TABLE IX
OPERATING DATA OF COMMERCIAL EXTERNAL FFB CATALYST COOLERS

	Unit A	Unit B
Heat duty, MW	11.63	9.3
Fluidizing air rate, Nm^3/s	2.1	1.8
Superficial gas velocity, m/s	1.0	
Catalyst inlet temperature, °C	700	680
Catalyst outlet temperature, °C	550	600
Steam pressure, MPa	3.2	3.0
Circulating water to steam ratio, w/w	20	>30
Surface area, m^2	75	46 (bare)
Heat transfer coefficient, W/(m^2·K)		
Steam regenerator bundle	494	505 (bare)
Superheater bundle	294	w/o
Type of tube bundle	bare	longitudinal finned
Type of water circulation	natural	natural
Catalyst inventory, t	1.8	

4. Heat Transfer Data

Table IX presents heat transfer data for an external catalyst cooler of the FFB type. Heat flux of 70–150 kW/m^2 and heat transfer coefficient of 350–490 W/(m^2·k) were obtained, both somewhat less than those of a turbulent bed internal cooler (horizontal coil type), but approximately equal to those of the external cooler of the bubbling bed type.

V. Analysis of Commercial FFB Regenerator Operation

A. FLUIDIZATION BEHAVIOR

The fast fluidized bed operates with high gas velocity and high solid mass flow rate as compared to the ordinary turbulent bed, thus intensifying gas/solid contacting and improving the mass transfer of oxygen from the gas phase to the solid phase, thereby resulting in a severalfold increase in CBI.

Owing to the rapid formation and dissolution of particle clusters which contribute to high slip velocities and solid backmixing but preserve a limited extent of gas backmixing, the fast fluidized bed regenerator exhibits unique axial and radial profiles for voidage, temperature and carbon concentration (see Figs. 9 and 11 and Table VIII).

The axial pressure gradient, expressed as local bed density under different gas velocities and catalyst inventories, is shown in Fig. 11. It diminishes gradually from the bottom to the top. The average bed density of the lower 3 m section is 100 to 160 kg/m^3 (corresponding to a voidage $\varepsilon = 0.87$ to 0.92), indicating the presence of a zone of vigorous gas/solid contacting and solid backmixing. In the middle section, between 3 and 4 m, the average bed density is 50 to 100 kg/m^3, while in the upper 3 m section, it is 30 to 60 kg/m^3. No inflection point in bed density can be found from the curves. The shape of the bed density curves is in general agreement with the mathematical models postulated by many authors.

The operating parameters influencing the bed density of an FFB regenerator are somewhat different from those of the turbulent bed. In the latter case, bed density changes only with gas velocity and catalyst inventory, while in the former, solid circulation rate exerts a major influence, as shown in Fig. 10.

Figure 9 shows the temperature difference between the top and the bottom of the catalyst bed for a typical FFB regenerator, which is rather low—only 20 to 40°C. The calculated mixed inlet temperature (without accounting chemical reaction) of spent catalyst (~ 500°C), recirculated regenerated catalyst (~ 700°C), plus air for combustion (~ 180°C), should be in the neighborhood of 600°C, but the actual bottom temperatures run much higher. This phenomenon is explained by the intensive solid backmixing in the lower section of the FFB regenerator, showing that a large amount of heat of combustion is efficiently transmitted to the bottom section to raise the initial reaction temperature by 60 to 80°C, which is very favorable from the viewpoint of chemical kinetics.

Wei (1990) examined the voidage profiles of some commercial FFB regenerators and fitted those data into the mathematical models proposed by Li et al. (1987, 1988) with good agreement, as shown in Fig. 12.

B. Chemical Reaction Engineering Characteristics

The process of coke combustion in bubbling or turbulent bed generally consists of the following steps:

1. transfer of oxygen from the bubble phase to the catalyst particles in the emulsion phase;
2. mass transfer of oxygen through the exterior gas film of particles to particle surface and then to the inner pores by internal diffusion;
3. chemical reactions both on the external surface and inside the particles;
4. internal diffusion of gaseous reaction products (CO_2, CO, H_2O, etc.)

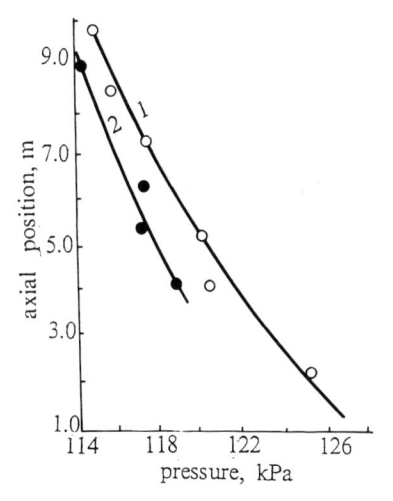

FIG. 12. Comparison of measured axial voidage distribution with mathematical model of Li et al. (1988).

outward to the particle external surface, followed by transfer to the emulsion phase;
5. transfer of reaction products from the emulsion phase to the bubble phase.

According to Weisz and Goodwin (1963), the rate of mass transfer of gaseous reactant or products through the gas film surrounding the particles is very fast as compared to the burning rate of coke, and the rate of internal diffusion of gases inside the microspherical catalyst at 650–750°C is also fast, so that the influence of bed diffusion steps can be neglected. Hence, the major rate controlling steps for catalyst regeneration are (1) and (3). The carbon burning rate for step (3) is represented by the following equations:

$$-\frac{dC}{dt} = k_\text{C} P_{\text{O}_2} C, \tag{20}$$

$$k_\text{C} = A \exp(-E/RT). \tag{21}$$

Van Deemter (1980) compared the kinetic constant k_c calculated from A and E obtained in the laboratory with the pseudo-kinetic constants k'_c from several commercial and pilot plants, and found that the values for k'_c are always lower than those corresponding to k_c, as shown in Fig. 13. It appears, therefore, that the process of coke burning is principally controlled by gas interchange and mass transfer in the bubbling or turbulent bed. Fast

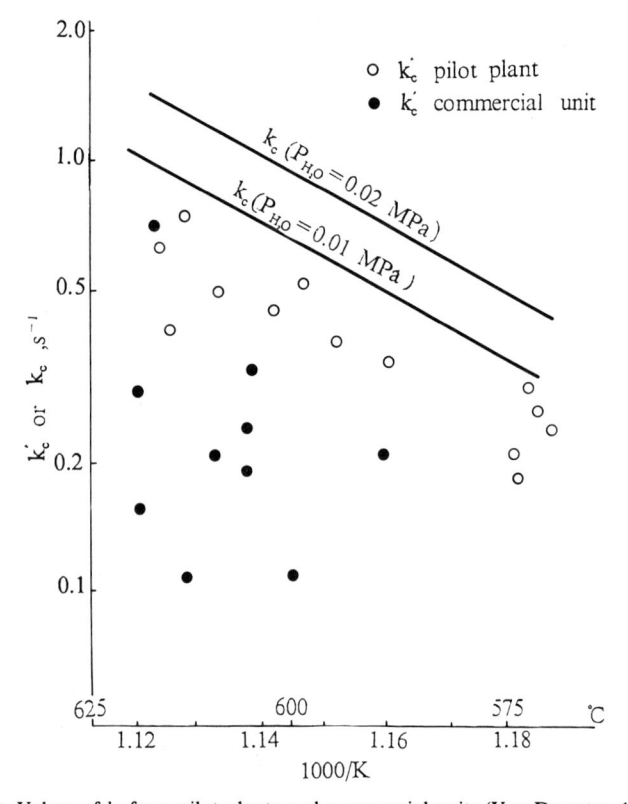

FIG. 13. Values of k_c from pilot plants and commercial units (Van Deemter, 1980).

fluidization, however, brings about a reversal of phases, with the gas phase continuous and the solids clusters dispersed, thus improving gas/solid contacting.

The data in Table X show the influence of temperature on CBI and CRC. The data in Table XI show that the calculated CBI based only on chemical kinetics is close to the actual CBI. This is evidence that the chemical reaction rate is the rate controlling step in commercial FFB regenerators.

To couple the intrinsic coke burning kinetics described in Section II.B with gas and solid flow models, the simplest approach is provided by the one-dimensional solid dispersion model based on following assumptions:

1. Gas is in plug flow without backmixing.
2. Gas component concentrations, particle concentrations, and carbon contents on catalyst are uniform radially and vary only with reactor height.

CATALYST REGENERATION

TABLE X
RELATIONSHIP BETWEEN OUTLET TEMPERATURE AND CBI IN A COMMERCIAL FFB REGENERATOR

Case	I	II	III
Pressure, MPa(ga)	0.16	0.16	0.16
Gas velocity, m/s	1.27	1.29	1.12
Outlet temperature, °C	730	720	690
Catalyst inventory, t	7.0	10.0	10.8
O_2 in flue gas, vol. %	2.0	4.0	5.0
CRC, wt. %	0.12	0.12	0.26
CBI, kg/(t·h)	658	542	474

TABLE XI
CALCULATED CBI FROM DATA IN TABLE X ($A = 2.74 \times 10^9$ (MPa·s), $E = 161{,}100$ kJ/kmol)

Sections	Bottom	Medium	Top
Pressure, MPa(ga)	0.2	0.2	0.2
O_2, vol. %	9.77	5.5	4.1
C, wt. %	0.1522	0.0977	0.0430
Temperature, °C	682	695	703
CBI, kg/(t·h)			
Calculated	705	324	129
Calculated (whole)		523	
Actual (whole)		519	

3. Axial velocities of particles and effective diffusion coefficients of solids are constant throughout the reactor.
4. Carbon combustion reaction is first order and the rate of reaction is the rate controlling step.

Then the degree of conversion of carbon versus reactor height can be expressed by the following equation:

$$\frac{1}{\text{Pe}} \frac{d^2 x}{dz^2} - \frac{dx}{dz} - k_c t C_o (1 - x) P_{O_2} = 0. \tag{22}$$

The analytical solution of Equation (22) is

$$\frac{C}{C_o} = \frac{4a \exp\left(\frac{\text{Pe}}{2}\right)}{(1 + a)^2 \exp\left(\frac{a \cdot \text{Pe}}{2}\right) - (1 - a)^2 \exp\left(-\frac{a \cdot \text{Pe}}{2}\right)}, \tag{23}$$

TABLE XII
CALCULATED VALUES OF CBI VERSUS PECLET NUMBER ($k_c = 0.8$, $C_o = 1.0$)

Pe	0.01	0.25	1	4	16	100
t	49	41	33	26	21	20
CBI	352	421	523	665	823	864

$$a = \sqrt{1 + 4k_c \cdot P_{O_2} \cdot t/\text{Pe}}. \tag{24}$$

This solution was obtained by assuming both P_{O_2} and regeneration temperature to be constant. It should be noted that both k_c and E_z are functions of temperature, while k_c is more sensitive to temperature changes. Since the actual temperature difference from top to bottom of commercial FFB regenerators is 20 to 40°C, as already mentioned, it will cause about a twofold change in k_c. Rigorous numerical integration of Eq. (22) should be applied for accurate solution. For the purpose of rough estimation one may use the logarithmic mean value of P_{O_2} and k_c, and then the CBI value can be approximated by

$$\text{CBI} = 3600(C_o - C)/t. \tag{25}$$

Table XII gives values of CBI calculated for different values of Pe by Eqs. (23) to (25), at fixed values of k_c and C_o.

Table XII shows that CBI varies directly with Pe, being low for mixed flow (low Pe) and high for plug flow (high Pe). Commercial FFB regenerators operate between the extremes of complete mixing and plug flow, while the conventional turbulent bed operates essentially with complete mixing. This explains the high efficiency of catalyst regeneration in FFB.

Notation

A	exponential factor $(\text{MPa·s})^{-1}$		d_t		bed or tube diameter, m
			d_{sc}		shell side equivalent diameter, m
C	carbon content of catalyst, wt. %		E		activation energy, kJ/kmol
C_o	initial carbon content of catalyst, wt. %		E_z		axial diffusion coefficient of solid particles, m²/s
C_{Pt}	Pt concentration calculated on catalyst plus promotor, wppm		f_H		correction factor
			G_s		mass flow rate of solid, t/(m²·h)
CO, CO_2	concentration of CO, CO_2, respectively, mol%		H		hydrogen content of coke, wt. %

k_{1c}, k_{2c}	kinetic constants, $(MPa·s)^{-1}$		$MPa/(kmol·K)]$
		t	time of reaction, s
k_c	kinetic constants of carbon combustion, $(MPa·s)^{-1}$	$\bar{t}$	time of complete conversion of single solid particle, s
k_{CO}, k'_{CO}	kinetic constants of CO combustion	T	temperature of reaction, K
		u_f	superficial gas velocity, m/s
k_H	kinetic constants of hydrogen combustion, $(MPa·s)^{-1}$	x	degree of conversion of carbon at Z
		$x_{CO}, x_{O_2}, x_{H_2O}$	mole fraction of CO, O_2, H_2O, respectively
l	longitudinal distance from entrance, m	Z	dimensionless distance, $Z = l/L$
L	bed height, m		
P	total pressure of reaction, MPa	γ_∞	rate of CO oxidation, $kg(CO)/[t(catalyst)·s]$
P_{CO}, P_{O_2}	partial pressure of CO, O_2, respectively, MPa	η	relative kinetic activity of equilibrium Pt relative to fresh Pt
Pe	Peclet number $Pe = ul/E_z$		
R	gas constant, kJ/(kmol·K) [except Eq. (3) is	$\bar{\rho}$	average bed density, kg/m³

References

Cabrera, C. A., and Mott, R. W. *Katalistiks' 3rd Annual Fluid Cat. Cracking Symposium, May 26 and 29, Amsterdam*, Paper No. 6 (1982).
Chen Junwu, Wang Zhengze, and Geng Lingyun. Chinese Patent, 89109293.55 (1989).
Jiao Fengqi, Zhang Lixin, Chen Lipei, and Dong Shizheng. Chinese Patent, 91227483.2 (1991).
Li Jinghai, Dong Yuanji, and Kwauk Mooson. *Chemical Metallurgy* (in Chinese) **2**, 1 (1987).
Li Jinghai, Dong Yuanji, and Kwauk Mooson. *Chemical Metallurgy* (in Chinese) **9**, 29 (1988).
Li Youchu, Chen Bingyun, Wang Yongsheng, and Kwauk Mooson. *Chemical Metallurgy* (in Chinese) **4**, 20 (1980).
Liu Xianfeng. M.Sc. Thesis. Zhengzhou Institute of Technology, China (1991).
Murphree, E. V., Brown, C. L., Fischer, H. G. M., Gohr, E. J., and Sweeney, W. J. *Ind. Eng. Chem.* **35**(7), 768 (1943).
Reichle, A. D. *Oil and Gas Journal* **91**(20), 41 (1992).
Van Deemter, J. J. *Chem. Eng. Sci.* **35**(1), 69 (1980).
Wang Guangxun, Lin Shixiong, and Yang Guanghua. *Ind. Eng. Chem. Proc. Des. Dev.* **25**(6), 626 (1986).
Wang Zhengze, Wang Zhuan, and Geng Linyun. Chinese Patent 91193976.X (1991).
Wei, Fei. M.Sc. Thesis, University of Petroleum, China (1987).
Wei, Fei. D.Sc. Dissertation, University of Petroleum, China (1990).
Weisz, P. B., and Goodwin, J. J. *Catalysis* **2**, 397 (1963).
Yang, Guanghua. Overview paper presented to International Symposium on Heavy Oil Upgrading and Utilization. May 5–8, International Academic Publisher, Fushun, China, 1991.
Zhang Fuyi, Li Zhanbao, Pi Yunpeng, and Shi Baozhen. Chinese Patent 90103414.4 (1990).
Zhao Weifan. *Petroleum Processing* (in Chinese) **3**, 32 (1983).

INDEX

A

Academia Sinica (China)
 butadiene synthesis, 49
 catalyst regeneration, 395
 circulating fluidized bed coal combustor, 350–351, 364
 boiler, 371, 373
 co-generation power plant, 373–374
 Institute of Chemical Metallurgy, 86, 395, 396
 Institute of Coal Chemistry, 49
Acrylonitrile synthesis, 49–50
Aerated standpipe, 22, 23, 27
Aerogel, 255
Aggregative fluidization, 240
 bed collapsing, 243
Ahlstrom Corporation, 29
Air pollution mitigation, fast fluidized bed technology, 56
Alumina powder, 24, 55, 85, 260
Ammonia synthesis, 14–15
Ammoxidation reactions, 49–50
Anglo–Iranian Oil Company, 19–20
Anshan Second Thermal Power Plant (China), 373–374
Arching, 308
Average solids concentration, fast fluidization, 97, 99
Average voidage, 181
Axial dispersion coefficients
 circulating fluidized bed combustor, 360–361
 solids, 133, 134
Axial force balance, inclined conical moving bed, 296
Axial gas dispersion, 127–130
Axial heterogeneity, 189
Axial hydrodynamics, 189–190
Axial particle size profile, circulating fluidized bed combustor, 354–355
Axial residual carbon content profile, circulating fluidized bed combustor, 354
Axial temperature profile, circulating fluidized bed combustor, 353
Axial voidage profile, 190
 circulating fluidized bed combustor, 350–353, 360–361, 366–367
 fast fluidization, 87, 107–114, 174–177

B

Backmixing
 gas, 125–127
 solids, 315–316
The Badger Company, 19
Baton Rouge plant (Louisiana), 22, 23
Bed collapsing, 243–251
Bed density, fast fluidized bed catalyst regenerator, 410–411
Bed structure, fluidized bed, 104, 116–121
Bed temperature, heat transfer, 207–208
Bed voidage, 181
 fast fluidization, 89, 128–139, 155
 heat transfer, 228
Belchetz, Arnold, 22
Binary particle mixtures, 253–260
 synergism, 255, 260–264
Blowout velocity, 23
Boiler, circulating fluidized bed combustor, 50–51, 332, 352–353, 356, 371–377
Boron ore, calcining, 32
Boudart, Marcel, 14
Boundaries, fast fluidization, 135–139, 152
Bridging, nonfluidized diplegs, 307–320
Brownsville plant (Texas), 2, 3, 15
Bubble escape stage, 245, 247
Bubbleless fluidization, 30–33
Bubbling bed design, 24–25, 30, 31, 32–33, 205

INDEX

Bubbling fluidization, 93, 94
  coal combustion, 335, 359, 361
Burning rate, 339
Butadiene synthesis, 49
Butane, oxidation to maleic anhydride, 49
Butene
  oxydehydrogenation to butadiene, 49
  synthesis, 44

C

Calcination, 24, 25, 27, 32, 53–55, 85
Calcium oxide, sulfur dioxide retention, 341–350, 370
Campbell, D. L., 22–23
Carbon burning intensity, 392, 407–408
Carrier gasification combined cycle power generator, 376–377
Carthage Hydrocol, 2, 3, 15
Catalyst regeneration, 389–419
  coke burning, kinetics, 399–401
  commercial regenerators, 401–419
  design criteria, 389–395
  fast fluidization, 395–399
  research and development, 395, 401
Catalytic cracking, 4, 11, 41–47, 389, 391
  catalyst regeneration, 389–419
    coke burning, kinetics, 399–401
    commercial regenerators, 401–419
    design criteria, 389–395
    fast fluidization, 395–399
    research and development, 395, 401
  fluid catalytic cracking, 41–43
  fluid catalytic pyrolysis, 44–45
  heavy oil, 43–44
  quick contact reactor, 45–47
Catalytic Research Associates, 19–20
CBI, see Carbon burning intensity
CFB processes, see Circulating fluidized bed processes
CGCC, see Carrier gasification combined cycle power generator
Chambers, John, 21
Changling Refinery (China), 53
Channel separator, 371–373
Char, 361, 374, 375
China
  bubbleless fluidization, 30–32
  catalyst regeneration, 392–395
    commercial plants, 401–419

circulating fluidized regenerator, 52–53
  coal combustion, 51–52
  coal use, 332
  kinetics, 335, 336
    gas diffusion, 341
  olefin synthesis, 45
  quick contact reactor, 46–47
Choking, 155
Choking velocity, 179
Circular V-valve, 322, 323–325
Circulating fluidized bed processes
  boiler, 50–51, 332, 352–353, 356–359
  coal combustion, 50–52, 352–353, 356
    boiler, 50–51, 332, 352–353, 356, 371–377
    combustor, 350–377
    hydrodynamics, 359–362
    modeling, 359–370
    rational excess air ratio, 368–369
  concept, 4–5, 7, 8
  hydrodynamics
    circulating fluidized bed combustor, 359–362
    flow structure, 177–184
    heat transfer enhancement, 219, 221, 223
    mass transfer, 235
  integral CFB, 322–326
  integral circulating fluidized bed, 322–326
  modeling, 359–370
  regenerator, 52–53
  schematic diagram, 317
  V-valve, 267–291
Cities Service Corporation, 8, 14, 15
Clustering, 103, 104
Cluster residence time, 224
Clusters, 117–119, 161
  diameter, 167–168
  force balance, 165–166
  heat transfer, 223, 224
Coal
  axial voidage profiles, 108–109
  combustion, 49–52, 331–385
    chemical reactions, 361–362
    circulating fluidized bed combustor, 350–377
    desulfurization, 355, 377–383
    devolatilization, 333–335
    gas diffusion, 340–341
    heat transfer, 361
    hydrodynamics, 360–361
    kinetics, 335–340

INDEX

mass transfer, 361
sulfur dioxide retention, 341–350
gasification, 29, 53
Co-generation power plants, 373–377
Coke, 399–401, 414–415
Combined cycle power generation system, 376–377
Combustion, 24–30
  coal, 50–52, 331–385
    chemical reactions, 361–362
    circulating fluidized bed combustor, 350–377
    desulfurization, 355, 377–383
    devolatilization, 333–335
    gas diffusion, 340–341
    heat transfer, 361
    hydrodynamics, 360–361
    kinetics, 335–340
    mass transfer, 361
    sulfur dioxide retention, 341–350
  coke, 399–401, 414–415
Combustor
  circulating fluidized bed, 350–359
    axial temperature profile, 353
    axial voidage profile, 351–353
    emissions reduction, 355–356, 377–383
    particle size, 354–355
    residual carbon content, 354
Computer software, Retrieval Fast Fluidization Application, 59
Conservation equations, circulating fluidized bed combustor, 362–364
Contacting
  fast fluidization, 32–35
  gas–solids, 184–188
  particle–fluid, 239–240
  relative contacting intensity, 187
  turbulent fluidization, 32–35
Contact pressure, 293, 300
Copper ore, 31
CO promoter, 409
Countercurrent–cogravity moving bed, 302
CRA, see Catalytic Research Associates
Cracking, see also Catalytic cracking
  catalyst regeneration, 389–419
  heavy oil, 43–44
  naphtha, 44–45
Critical flow factor, 313
Cross-sectional average voidage, 153
Cross-sectional mean voidage, fast fluidization, 97, 99

CV valve, see Circular V-valve
Cyclone, 322, 325

D

Data search, Retrieval Fast Fluidization Application, 59
Datong coal, 336, 337
Dayeh plant (China), 31
Desulfurization, 355, 377–383
Devolatilization, coal, 333–335
Diffusion, coal combustion rate, 340
Dipleg
  fluidized, 279–281
  nonfluidized, 291–313
    bridging, 307–313
    sealing, 302
  V-valve, 267
Distribution factors, 296
Distribution pore size model, 342, 344
Dorr-Oliver, 29
Double tip optical fiber, 102
Downcomer, 267, 281–291
Downflow reactor, 46–47
Drag coefficient, 111, 114, 154, 184
Drying processes, fast fluidized bed dryer, 55–56
Duplex spout, 274
Dynamics, nonfluidized gas-particle flow, 292–302

E

Eastman, DuBois, 15
E. I. du Pont de Nemours, 49
EMMS model, see Energy-minimization multi-scale model
Energy analysis, flowing particle–fluid system, 161–164
Energy minimization, ST subsystem, 169–170
Energy-minimization multi-scale model, 148, 159–171
Engstrom, Fluke, 29
Equal pressure drop, 166

F

FAB calciner, 55

INDEX

FAB gasification, 53
Failure function, 311
Fast bubbles, 241, 250
Fast fluid bed processes, 1–35; *see also* Fast fluidization; Fast fluidized bed reactors
   applications collocation chart, 59–61
   boilers, 50–51
   bubbleless fluidization, 30–32
   calcination, 24, 25, 27, 32, 53–55
   catalytic cracking, 4, 11, 41–47
      catalyst regenerators, 389–419
      coke burning kinetics, 399–401
      commercial regenerators, 401–419
      design criteria, 389–395
      fast fluidization, 395–399
      heavy oil, 43–44
      research and development, 395–401
   characteristics, 86–91, 94
   combustion, 24–30
      coal, 50–52, 331–385
   contacting, 32–35
      gas–solids, 184–188
   drying, 55–56
   future applications, 62
   gas cleaning, 56–57
   gasification, 29, 53
   hardware, 267–329
   integral circulating fluidized bed, 322–326
   multi-layer fast fluidized beds, 320–322
   prereduction, iron ore, 31, 55
   reactor characteristics, 39–41
   regenerator, 44
   research at M.I.T., 19–23
   synthesis reactions, 47–50
      acrylonitrile, 49–50
      ammonia, 14–15
      butadiene, 49
      gasoline, 1–14, 16–19, 42–43
      maleic anhydride, 49
Fast fluidization, *see also* Fast fluid bed processes; Fast fluidized bed reactors
   characteristics, 86–91, 94
   conditions for, 93
   heat transfer, 203–234
      enhancement, 219, 221, 223
      experimental studies, 204–223
      between gas and particles, 232–233
      gas convective transfer, 223–234
      particle convective transfer, 224–229
      radiative transfer, 224
      theoretical analysis and models, 223–231
   hydrodynamics, 85–141, 172–197
      average solids concentration, 97, 99
      backmixing, 125–127, 315–316
      bed structure, 116–121
      boundaries, 135–139, 152
      clustering, 117–119, 165–168
      coal combustion, 359–361
      flow behavior, 103
      flow structure, 103–107, 148–156, 177–184
      fluid velocity, 152–153
      free falling velocity, 154–155
      gas backmixing, 125–127
      gas dispersion, 127–132
      gas velocity, 123–134
      local solids concentration, 100
      mapping, 91, 93
      mass flux, 100–101
      minimum fluidization velocity, 155
      mixing, 124–135
      particle velocity, 101–103, 122–123, 152–153
      phases, 171–188
      pneumatic transport, 91
      regions, 188–197
      saturation carrying capacity, 155
      solids backmixing, 315–316
      solids distribution, 315
      solids mixing, 133–134
      terminal velocity, 154–155
      voidage, 153, 162
      voidage profile, 86–87, 89, 107–116, 158, 174–177, 196, 315–316
   mass transfer, 235
   modeling
      energy analysis, 161–164
      energy-minimization multi-scale model, 159–171
      pseudo-fluid models, 156–159
      ST subsystem, 165–168
      two-phase model, 157, 159–160
Fast fluidization velocity, 87
Fast fluidized bed reactors, 38–41
   heat transfer, 203–234
      axial distribution of heat transfer coefficient, 216
      enhancement, 219, 221, 223
      gas convective transfer, 223–224
      between gas and particles, 232–233

particle convective transfer, 224–229
radial distribution of heat transfer
coefficient, 210–216
radiative transfer, 224
variation of heat transfer along surface,
216–219
integral circulating fluidized bed, 322–326
mass transfer, 235
multi-layer beds, 320–322
solids distribution, 315–320
V-valve, 267–291
FCC process, see Fluid catalytic cracking
FCCU, see Fluid catalytic cracking
FD regime, see Fluid-dominated regime
FF, see Flow factor
FFB processes, see Fast fluid bed processes
FFB reactor, see Fast fluidized bed reactors
FFC, see Critical flow factor
Fischer-Tropsch synthesis, 1–4, 15
Flooding, 281
Flow behavior, fast fluidization, 103
Flow factor, 313
Flow rate
dipleg, fluidized, 280
spout, nonfluidized, 277
Flow regime maps, 91–93
Flow structure
circulating fluidized bed, 177–184
fast fluidization, 103–107, 148–156
Flue gas, 355–356, 373, 377–383, 404
Fluid bed
bed collapsing, 243–251
bubbleless, 30–32
bubbling, 24–25, 30, 31, 32–33, 241
contacting, 32–35, 184–188, 239–240,
243
turbulent, 25, 33
two-D, 29
Fluid-bed devices, pneumatically actuated,
314–315
Fluid catalytic cracking, 4, 11, 41–43, 239
catalyst regeneration, 389–419
coke burning, kinetics, 399–401
commercial regenerators, 401–419
design criteria, 389–395
fast fluidization, 395–399
research and development, 395–401
dynamics, 86
powder catalyst, 260
Fluid catalytic pyrolysis, 44–45
Fluid-dominated regime, 170, 179, 180, 185

Fluidization, 240–241
improving by particle size adjustment,
253–260
incipient, 299
Fluidized dipleg, 279–281
Fluidized spout, 269–274
Fluid–particle system
contacting, 239–240
energy analysis, 161–164
flow, phase diagrams, 302–305
voidage, 153
Fluid velocity, 152–153
Force balance
clusters, 165–166
differential equations, 296–299
Foster-Wheeler, 29
Free falling velocity, 154–155
Friend, Leo, 24
Fuller-Kinyon pumps, 22, 23

G

Gas backmixing, 125–127
Gas cleaning, 56–57
Gas convective transfer, 223
Gas diffusion, 340–341
Gas dispersion
axial, 127–130
radial, 130–132
Gas distribution, downcomer, pneumatically
controlled, 283–286
Gas flow rate
dipleg, fluidized, 280
nonfluidized spout, 277
Gasification, 29, 53
Gasoline synthesis, 1–14, 16–19, 42–43
Gas-particle system
flow dynamics, 292–302
heat transfer, 232–233
non-fluidized flow, 293
voidage, 300, 301
Gas pressure, 299, 300
Gas pressure drop, fluidized dipleg, 279
Gas pressure gradient, gas-particle friction,
299–300
Gas–solid system
contacting, 184–188
relative velocity, 277–278
separation, 356–359
Gas–steam co-generation power plants, 374–377

Gas transfer, 392
Gas velocity, fast fluidization, 123–124
Geldart's classification, 241–243
Gilliland, Edwin R., 21
Global average voidage, 153
Gotaverken Angteknik, 29
Group A–A binary mixture, 253, 255
Group A powders, 27, 30, 241–243
Group B–A binary mixture, 255
Group B–C binary mixture, 255
Group B powders, 27, 29, 30, 241–243, 250
Group C powders, 241–243, 286
Group D powders, 241–243, 250

H

Hairpin heat exchanger, 19
Hanford, W. E., 13
Heat-carrier gasification combined cycle, 377
Heat conservation, circulating fluidized bed combustor, 363–364
Heat extraction, catalyst regenerator, 405–406, 413
Heat probes, 208
Heat transfer
 circulating fluidized bed combustion, 361
 fast fluidized beds, 203–234
  enhancement, 219, 221, 223
  experimental studies, 204–223
  between gas and particles, 232–233
  gas convective transfer, 223–224
  particle convective transfer, 224–229
  radiative transfer, 224
  theoretical analysis and models, 223–231
 fast fluidized bed catalyst regeneration, 405–406, 413
 between gas and particles, 232–233
 hairpin heat exchanger, 19
Heat transfer coefficient
 axial distribution, 216
 bed temperature, 207
 circulating fluidized bed combustion, 361
 components, 223
 radial distribution, 210–216
 solids concentration, 205
 time averaged, 226
 variation along the surface, 216–219
Heavy oil, cracking, 43–44
Heterogeneity factor, 192, 193
High-efficiency regenerator, 44

Hill, Luther R., 4
Hindered sedimentation stage, 245, 247
Holographic imaging technique, clusters, 105, 107
Houdry, Eugene, 20
Hugoton plant (Kansas), 2, 3
Hydrocarbon Research, Inc., 2–3, 15, 23
Hydrocarbon synthesis, 2–3
Hydrocol process, 2
Hydrodynamic mean diameter, 150
Hydrodynamics
 axial, 189–190
 coal combustion, 359–362
 fast fluidization, 85–141, 97, 99, 172–197
  bed structure, 116–121
  boundaries, 135–139
  circulating fluidized bed, 177–184
  clustering, 117–119, 165–168
  cross-sectional mean voidage, 97, 99
  flow behavior, 103
  flow structure, 103–107
  fluid velocity, 152–153
  free falling velocity, 154–155
  local solids concentration, 100
  mass flux, 100–101
  measurement models and instrumentation, 97–107
  minimum fluidization velocity, 155
  mixing, 124–135
  particle velocity, 101–103, 152–153
  phases, 171–188
  regions, 188–197
  saturation carrying capacity, 155
  terminal velocity, 154–155
  test apparatus, 95–97
  test materials, 97
  voidage, 153, 162
  voidage profile, 86–87, 89, 107–116, 158, 174–177, 196, 315–316
 radial, 191–196

I

ICFB, see Integral circulating fluidized bed
I. G. Farben, 19–20
Imposed pressure drop, 152, 161
Incipient fast fluidization velocity, 90
Incipient fluidization, 299
Inclined conical moving bed, 296
Inlet design, fast fluidization, 136–139

Input energy, 167
Institute of Chemical Metallurgy (China), 86, 395, 396
Institute of Coal Chemistry (China), butadiene synthesis, 49
Institute of Petroleum Processing (China), 395, 396, 399
Integral circulating fluidized bed, 322–326
Iron ore
  copper extraction, 31
  powder catalyst for fast fluidization, 17, 360
  prereduction, 55

J

Jewell, Sr., Joseph W., 5
Jinan Refinery (China), 45
J-leg, 27, 30
Johnson, W. Benjamin, 5, 13, 14, 29

K

Keith, P. C., 2
Kellex Corporation, 2
Kellogg, M. W., 1–4, 6, 7
Kellogg Company, *see* M. W. Kellogg Company
k-epsilon turbulence model, 157–158
Kinetic constant, 338
Kinetics, coal combustion, 335–340
Kwauk, Mooson, 30–32

L

Lake Charles plant (Louisiana), 8, 15
Lanneau, K. P., 32–33
Laser doppler velocimeter, 102–103, 122
LDV, *see* Laser doppler velocimeter
Lewis, Warren K., 21, 22
Limestone, sulfur dioxide sorbent, 355, 378
Limiting discharge capacity, dipleg, fluidized, 280–281
Literature search, Retrieval Fast Fluidization Application, 59
Local average fluid velocity, 153
Local average particle velocity, 153
Local average voidage, 153
Local gas velocity, 124
Local heat transfer coefficient, 216–219, 226, 230–231
Local solids concentration, fast fluidization, 100
LPEC, *see* Luoyang Petrochemical Engineering Corporation
Luoyang Petrochemical Engineering Corporation (China), catalyst regeneration, 395, 398
Lurgi fast bed boiler, 24, 25, 27, 29
Lurgi GMBH, 53

M

Ma'anshan plant (China), 31
McGrath, H. G., 6, 11
Macro-heterogeneity, 149, 160, 161
Magnesite ore, calcining, 32
Maleic anhydride synthesis, 49
Manhattan Project, 2
Mapping, fast fluidization, 91, 93
Martin, H. Z., 22–23
Massachusetts Institute of Technology, 19–23
Mass conservation
  circulating fluidized bed combustor, 362–363
  fast fluidization, 110
  ST subsystem, 165–168
Mass flux, fast fluidization, 100–101
Mass transfer, fast fluidized beds, 235
Mass transfer coefficient, 235, 361
Mathematical modeling
  axial hydrodynamics, 189
  bed collapsing, 246–247
  binary particle mixtures, 260–264
  radial hydrodynamics, 191–192
Meniscus, 25, 27, 32
Mercury recovery, 31
Meso-heterogeneity, 149, 160, 161, 185
MIC, *see* Multi-inlet cyclone
Minimal gas flow rate, 280
Minimum energy dissipation theorem, 168
Minimum fluidization velocity, 155
Minimum pressure drop, 169
Minimum solids circulation rate, 91
Ministry of Petroleum Industry (China), catalyst regeneration, 394, 395
M.I.T., *see* Massachusetts Institute of Technology

Mixing
  fast fluidization, 124–135
    axial gas dispersion, 127–130
    gas back mixing, 125–127
    radial gas dispersion, 130–132
    solids mixing, 133–135
Mobil Oil Corporation, 34
Modeling
  bed collapsing process, 245–247
  binary particle mixtures, synergism, 260–264
  circulating fluidized bed combustor, 359–370
  fast fluidization, 156–171
    energy, analysis, 161–164
    energy-minimization multi-scale model, 159–171
    pseudo-fluid models, 156–159
    ST subsystem, 165–168
    two-phase models, 157, 159–160
Model KR, 192–193
Model LG, 189–191
Model LR, 191
Model OR, 192–196
Momentum, ST subsystem, 165–168
Momentum conservation
  circulating fluidized bed combustor, 262
  fast fluidization, 110
Moving bed
  countercurrent cogravity, 296
  inlined conical, 296
  sealing, 302
MPI, see Ministry of Petroleum Industry
Multi-inlet cyclone, 322, 325, 327
Multi-layer fast fluidized beds, 320–322
Murphree, Eger, 21, 22
M. W. Kellogg Company, 1–18, 22, 23
  Catalytic Research Associates, 19–20
  Sasolburg plant, 16–18, 24
  synthetic gasoline, 47

N

Nack, Herman, 29
Naphtha, cracking, 44–45
Natural gas, 3, 15
Nitrogen oxides, emission reduction, 355, 370
Nonfluidized diplegs, 291–313
  bridging, 307–313
  sealing, 302
Nonfluidized spout, 274–279

O

Olefin synthesis, 45
Operating temperature, coal combustion, 367–368
Optical fiber probe
  collapsing bed surface studies, 248–249
  flow behavior measurement, 103, 105
  local solids concentration, 100
  particle velocity measurement, 102, 103
Optimization, EMMS model, 171–172
Outlet design, fast fluidization, 135–136
Overall heat transfer coefficient, 231
Oxidation reactions
  butane to maleic anhydride, 49
  coke burning, 400–401
  natural gas, 15

P

Particle convective transfer, 224–229
Particle diameter, 150
Particle-dominated regime, 169
Particle–fluid-compromising regime, 170, 179–183
Particle–fluid system
  contacting, 239–240
  energy analysis, 161–164
  flow, phase diagrams, 302–305
  voidage, 153
Particle–gas system
  non-fluidized flow, 292–302
  voidage, 300, 301
Particle momentum, fast fluidization, 101
Particle properties
  fast fluidization, 239
  heat transfer, 205, 207
Particle size
  bed material, 366–367
  circulating fluidized bed combustor, 354–355
Particle suspension, 166–167
Particle velocity, fast fluidization, 101–103, 122–123, 152–153
Particulate fluidization, 240
  bed collapsing, 243
Pattern, 150
PD regime, see Particle-dominated regime
People's Republic of China, see China
PFC regime, see Particle–fluid-compromising regime

INDEX

Phase diagrams, fluid–particle flow, 302–305
Phases, 149, 160, 171–188
Pneumatically actuated pulse feeder, 314–315
Pneumatic transport, 91, 293
Pollution mitigation, fast fluidized bed technology, 56
Pore model, 341
Pore size distribution model, 341
Powder catalyst, 11, 21, 23, 239–265
 arching, 308
 bed collapsing, 243–251
  modeling, 245–247
 bridging, 307–320
 fluidizing characteristics, quantifying, 251
 Geldart's classification, 241–243
 improving fluidization behavior, 253–260
Power plants
 co-generation, 373–375
 heat-carrier gasification combined cycle, 376–377
Pressure drop
 downcomer, pneumatically controlled, 282–283
 fluidized dipleg, 279
 moving bed, 302
 nonfluidized spout, 277
Pressure fluctuation analysis, 103
Primary instability, 149
Propylene, ammoxidation to acrylonitrile, 49–50
Propylene synthesis, 45
Pseudo-fluid models, 156–159
 k-epsilon turbulence model, 157–158
 two-channel model, 158
 two-dimensional diffusion model, 158–159
Pullman Company, 2
Pulse feeder, pneumatically controlled, 314–315
Pyrolysis, fluid catalytic pyrolysis, 44–45

Q

Quick contact reactor, 45–47

R

Radial gas dispersion, 130–132
Radial heterogeneity, 191–192
Radial hydrodynamics, 191–196
Radial voidage profile, fast fluidization, 87, 89, 114–116, 158, 315, 316, 361

Radiative transfer, 224
Random pore model, 341
Rational excess air ratio, coal combustor, 368–369
Rearick, J. S., 7
Rectisol units, 17
Regenerators
 circulating fluidized, 52–53
 commercial plants, 401–419
 fast fluidized bed, 44, 392–419
 fluid catalytic cracking, 389–419
 heat extraction, 405–406
 high efficiency, 44
 turbulent bed, 391–392
Regime, 149
Regions, 150, 160, 188–197
Reh, Lothar, 24, 25, 27
Relative contacting intensity, 187
Relative velocity, gas–solids, 277–278
Research Institute of Petroleum Processing (China), 395, 396, 399
Residual carbon content, circulating fluidized bed combustor, 354
Retrieval Fast Fluidization Application, 59
RFFA, *see* Retrieval Fast Fluidization Application
Ring internals, 315–320
RIPP, *see* Research Institute of Petroleum Processing
Riser catalytic pyrolysis process, 45
Riser FCC unit, 42, 43
Riser regenerator, 53
Roberts, Jr., George, 2
Rubin, Louis C., 6, 7

S

Sasol, synthetic gasoline, 47
Sasolburg plant (South Africa), 17–19, 24
Saturation carrying capacity, 155, 171, 173
Schlesinger, M. D., 6
Sealing, 302, 305–307
 downcomer, pneumatically started, 287–291
 ideal sealing state, 305–307
Secondary instability, 149
Self-flow, 286–287
Seubert, Edward, 2
Shell Oil Company, 19–20
Sherwood number correlation, 338
Shrinking core model, 341, 347

Single pore model, 341
SINOPEC (China), 395
Slip velocity, 183, 184, 193–195
Slugging, 11, 33–34
Slyngstad, C. E., 14
Smith, Martin R., 6, 11
Snake reactor, 21, 23
Software, Retrieval Fast Fluidization Application, 59
Solid–gas system
  contacting, 184–188
  relative velocity, 277–278
Solids concentration
  fast fluidization, 100, 205
  heterogeneity, 315
Solids consolidation stage, 246, 247
Solids distribution, 315
Solids height, dipleg, fluidized, 280
Solids, mixing, 133–134, 315–316
Solids recirculation ratios, 357
Solids velocity, 183
Sorbents
  artificial, 378
  calcium oxide, 341–350
  pore characteristics, 377–383
Spout
  fluidized, 269–274
    duplex, 274
  nonfluidized, 274–279
Standard Oil Company of Indiana, 1–4, 19
Standard Oil Company of New Jersey, 19–20, 21, 22
Standpipe
  function, 280–281
  nonfluidized, 291–313
  V-valve, 267
Stanolind, 1–2, 4, 11, 14
Stationary fluid bed system, gasoline synthesis, 6–7
Steam–methane reforming, 14–15
Stress ratios, 293–296
Studsvik Energiteknik, 29
Subbarao cluster model, 224
Sulfur dioxide
  emission reduction, 377–383
    limestone sorbent, 355, 370
  retention, 341–350, 367, 370
Superficial fluid velocity, 152
Superficial particle velocity, 152
Surface orientation, heat transfer, 208
Surface-to-volume mean diameter, 150

Suspension and transport subsystem, momentum and mass conservation, 165–168
Suspension density, heat transfer, 205
Syndennis, E. J., 14
Synergism
  binary particle mixtures, 255, 260–264
  ternary particle mixtures, 255
Synthesis gas
  ammonia synthesis, 14–15
  gasoline synthesis, 7, 10
  Sasolburg plant, 17, 23
Synthesis reactions
  acrylonitrile, 49–50
  ammonia, 14–15
  butadiene, 49
  gasoline, 1–14, 16–19, 42–43
  maleic anhydride, 49
Synthol process, 1, 3, 7, 8, 13–15
Synthol reactor, 47

T

Taylor, Sir Hugh S., 14
TB regenerator, *see* Turbulent bed regenerator
Terminal velocity, 154–155
Ternary particle mixtures, synergism, 255, 264
The Texas Company, 2, 15, 19–20
Time-averaged heat transfer coefficient, 226
Titanomagnetite, gaseous reduction, 32
Tongguanshan plant (China), 31
Total energy, 161–162
Total input, 167
Transition velocity, 91
Tsing Hua University (China)
  artificial sorbent, 379, 381
  circulating fluidized bed combustor boiler, 373
  heat transfer studies, 204–223
  pilot FCC riser regenerator, 53
  research topics, 46
Tulsa plant (Oklahoma), 1, 12
Turbulent bed regenerator, 391–392, 403, 414
Turbulent fluidization, contacting, 32–35
Turkevick, John, 14
Two-channel model, 158
Two-D fluid bed, 29
Two-dimensional diffusion model, 158–159
Two-phase model, 159–160
Tyson, C. W., 22–23

U

Ultrastable Y zeolite, 392
Universal Oil Products, 19–20
University of Petroleum (China), catalytic regeneration, 393, 394
UOP, *see* University of Petroleum
USY, *see* Ultrastable Y zeolite

V

Valve
  CV-valve, 322, 323–325
  V-valve, 267–291
Vanadium ore, 32
Verinigte Aluminium Werke AG, 53
Virginia Polytechnic Institute and State University, 34–35
Voidage
  average, 162
  cross-sectional average, 153
  cross-sectional mean, 97, 99
  definition, 153
  dipleg, fluidized, 280
  gas–particle flow, 300, 301
  global average, 153
  local average, 153
  particle–fluid system, 153
Voidage gradient, 300

Voidage profile
  fast fluidization, 86–87, 89, 107–116
    axial, 87, 107–114, 174–177, 315, 351–353, 360–361, 366–367
    radial, 87, 89, 114–116, 158, 315, 316, 361
V-valve, 267–291
  dipleg, fluidized, 279–281
  downcomer, pneumatically controlled, 281–291
  spout
    fluidized, 269–274
    nonfluidized, 274–279

W

Walker, Scott, 22
Wansan plant (China), 31
Watts, George, 2
Wetterau, Stanley, 23

Z

Zeolite, 42, 392–393
  catalyst regeneration, 394
  ultrastable Y zeolite, 392

Contents of Volumes in This Serial

Volume 1

J. W. Westwater, *Boiling of Liquids*
A. B. Metzner, *Non-Newtonian Technology: Fluid Mechanics, Mixing, and Heat Transfer*
R. Byron Bird, *Theory of Diffusion*
J. B. Opfell and B. H. Sage, *Turbulence in Thermal and Material Transport*
Robert E. Treybal, *Mechanically Aided Liquid Extraction*
Robert W. Schrage, *the Automatic Computer in the Control and Planning of Manufacturing Operations*
Ernest J. Henley and Nathaniel F. Barr, *Ionizing Radiation Applied to Chemical Processes and to Food and Drug Processing*

Volume 2

J. W. Westwater, *Boiling of Liquids*
Ernest F. Johnson, *Automatic Process Control*
Bernard Manowitz, *Treatment and Disposal of Wastes in Nuclear Chemical Technology*
George A. Sofer and Harold C. Weingartner, *High Vacuum Technology*
Theodore Vermeulen, *Separation by Adsorption Methods*
Sherman S. Weidenbaum, *Mixing of Solids*

Volume 3

C. S. Grove, Jr., Robert V. Jelinek, and Herbert M. Schoen, *Crystallization from Solution*
F. Alan Ferguson and Russell C. Phillips, *High Temperature Technology*
Daniel Hyman, *Mixing and Agitation*
John Beek, *Design of Packed Catalytic Reactors*
Douglass J. Wilde, *Optimization Methods*

Volume 4

J. T. Davies, *Mass-Transfer and Interfacial Phenomena*
R. C. Kintner, *Drop Phenomena Affecting Liquid Extraction*
Octave Levenspiel and Kenneth B. Bischoff, *Patterns of Flow in Chemical Process Vessels*
Donald S. Scott, *Properties of Concurrent Gas–Liquid Flow*
D. N. Hanson and G. F. Somerville, *A General Program for Computing Multistage Vapor–Liquid Processes*

Volume 5

J. F. Wehner, *Flame Processes–Theoretical and Experimental*
J. H. Sinfelt, *Bifunctional Catalysts*
S. G. Bankoff, *Heat Conduction or Diffusion with Change of Phase*
George D. Fulford, *The Flow of Liquids in Thin Films*
K. Rietema, *Segregation in Liquid–Liquid Dispersions and Its Effect on Chemical Reactions*

Volume 6

S. G. Bankoff, *Diffusion-Controlled Bubble Growth*
John C. Berg, Andreas Acrivos, and Michel Boudart, *Evaporation Convection*
H. M. Tsuchiya, A. G. Fredrickson, and R. Aris, *Dynamics of Microbial Cell Populations*
Samuel Sideman, *Direct Contact Heat Transfer between Immiscible Liquids*
Howard Brenner, *Hydrodynamic Resistance of Particles at Small Reynolds Numbers*

Volume 7

Robert S. Brown, Ralph Anderson, and Larry J. Shannon, *Ignition and Combustion of Solid Rocket Propellants*
Knud Østergaard, *Gas–Liquid–Particle Operations in Chemical Reaction Engineering*
J. M. Prausnitz, *Thermodynamics of Fluid–Phase Equilibria at High Pressures*
Robert V. Macbeth, *The Burn-Out Phenomenon in Forced-Convection Boiling*
William Resnick and Benjamin Gal-Or, *Gas–Liquid Dispersions*

Volume 8

C. E. Lapple, *Electrostatic Phenomena with Particulates*
J. R. Kittrell, *Mathematical Modeling of Chemical Reactions*
W. P. Ledet and D. M. Himmelblau, *Decomposition Procedures for the Solving of Large Scale Systems*
R. Kumar and N. R. Kuloor, *The Formation of Bubbles and Drops*

Volume 9

Renato G. Bautista, *Hydrometallurgy*
Kishan B. Mathur and Norman Epstein, *Dynamics of Spouted Beds*
W. C. Reynolds, *Recent Advances in the Computation of Turbulent Flows*
R. E. Peck and D. T. Wasan, *Drying of Solid Particles and Sheets*

Volume 10

G. E. O'Connor and T. W. F. Russell, *Heat Transfer in Tubular Fluid–Fluid Systems*
P. C. Kapur, *Balling and Granulation*
Richard S. H. Mah and Mordechai Shacham, *Pipeline Network Design and Synthesis*
J. Robert Selman and Charles W. Tobias, *Mass-Transfer Measurements by the Limiting-Current Technique*

Volume 11

Jean-Claude Charpentier, *Mass-Transfer Rates in Gas–Liquid Absorbers and Reactors*
Dee H. Barker and C. R. Mitra, *The Indian Chemical Industry–Its Development and Needs*
Lawrence L. Tavlarides and Michael Stamatoudis, *The Analysis of Interphase Reactions and Mass Transfer in Liquid–Liquid Dispersions*
Terukatsu Miyauchi, Shintaro Furusaki, Shigeharu Morooka, and Yoneichi Ikeda, *Transport Phenomena and Reaction in Fluidized Catalyst Beds*

Volume 12

C. D. Prater, J. Wei, V. W. Weekman, Jr., and B. Gross, *A Reaction Engineering Case History: Coke Burning in Thermofor Catalytic Cracking Regenerators*
Costel D. Denson, *Stripping Operations in Polymer Processing*
Robert C. Reid, *Rapid Phase Transitions from Liquid to Vapor*
John H. Seinfeld, *Atmospheric Diffusion Theory*

Volume 13

Edward G. Jefferson, *Future Opportunities in Chemical Engineering*
Eli Ruckenstein, *Analysis of Transport Phenomena Using Scaling and Physical Models*
Rohit Khanna and John H. Seinfeld, *Mathematical Modeling of Packed Bed Reactors: Numerical Solutions and Control Model Development*
Michael P. Ramage, Kenneth R. Graziani, Paul H. Schipper, Frederick J. Krambeck, and Byung C. Choi, *KINPTR (Mobil's Kinetic Reforming Model): A Review of Mobil's Industrial Process Modeling Philosophy*

Volume 14

Richard D. Colberg and Manfred Morari, *Analysis and Synthesis of Resilient Heat Exchanger Networks*
Richard J. Quann, Robert A. Ware, Chi-Wen Hung, and James Wei, *Catalytic Hydrometallation of Petroleum*
Kent Davis, *The Safety Matrix: People Applying Technology to Yield Safe Chemical Plants and Products*

Volume 15

Pierre M. Adler, Ali Nadim, and Howard Brenner, *Rheological Models of Suspensions*
Stanley M. Englund, *Opportunities in the Design of Inherently Safer Chemical Plants*
H. J. Ploehn and W. B. Russel, *Interactions between Colloidal Particles and Soluble Polymers*

Volume 16

Perspectives in Chemical Engineering: Research and Education

Clark K. Colton, *Editor*

Historical Perspective and Overview

L. E. Scriven, *On the Emergence and Evolution of Chemical Engineering*
Ralph Landau, *Academic–Industrial Interaction in the Early Development of Chemical Engineering*
James Wei, *Future Directions of Chemical Engineering*

Fluid Mechanics and Transport

L. G. Leal, *Challenges and Opportunities in Fluid Mechanics and Transport Phenomena*
William B. Russel, *Fluid Mechanics and Transport Research in Chemical Engineering*
J. R. A. Pearson, *Fluid Mechanics and Transport Phenomena*

Thermodynamics

Keith E. Gubbins, *Thermodynamics*
J. M. Prausnitz, *Chemical Engineering Thermodynamics: Continuity and Expanding Frontiers*
H. Ted Davis, *Future Opportunities in Thermodynamics*

Kinetics, Catalysis, and Reactor Engineering

Alexis T. Bell, *Reflections on the Current Status and Future Directions of Chemical Reaction Engineering*
James R. Katzer and S. S. Wong, *Frontiers in Chemical Reaction Engineering*
L. Louis Hegedus, *Catalyst Design*

Environmental Protection and Energy

John H. Seinfeld, *Environmental Chemical Engineering*

T. W. F. Russell, *Energy and Environmental Concerns*
Janos M. Beer, Jack B. Howard, John P. Longwell, and Adel F. Sarofim, *The Role of Chemical Engineering in Fuel Manufacture and Use of Fuels*

Polymers

Matthew Tirrell, *Polymer Science in Chemical Engineering*
Richard A. Register and Stuart L. Cooper, *Chemical Engineers in Polymer Science: The Need for an Interdisciplinary Approach*

Microelectronic and Optical Materials

Larry F. Thompson, *Chemical Engineering Research Opportunities in Electronic and Optical Materials Research*
Klavs F. Jensen, *Chemical Engineering in the Processing of Electronic and Optical Materials: A Discussion*

Bioengineering

James E. Bailey, *Bioprocess Engineering*
Arthur E. Humphrey, *Some Unsolved Problems of Biotechnology*
Channing Robertson, *Chemical Engineering: Its Role in the Medical and Health Sciences*

Process Engineering

Arthur W. Westerberg, *Process Engineering*
Manfred Morari, *Process Control Theory: Reflections on the Past Decade and Goals for the Next*
James M. Douglas, *The Paradigm After Next*
George Stephanopoulous, *Symbolic Computing and Artificial Intelligence in Chemical Engineering: A New Challenge*

The Identity of Our Profession

Morton M. Denn, *The Identity of Our Profession*

Volume 17

Y. T. Shah, *Design Parameters for Mechanically Agitated Reactors*
Mooson Kwauk, *Particulate Fluidization: An Overview*

Volume 18

E. James Davis, *Microchemical Engineering: The Physics and Chemistry of the Microparticle*
Selim M. Senkan, *Detailed Chemical Kinetic Modeling: Chemical Reaction Engineering of the Future*
Lorenz T. Biegler, *Optimization Strategies for Complex Process Models*

Volume 19

Robert Langer, *Polymer Systems for Controlled Release of Macromolecules, Immobilized Enzyme Medical Bioreactors, and Tissue Engineering*
J. J. Linderman, P. A. Mahama, K. E. Forsten, and D. A. Lauffenburger, *Diffusion and Probability in Receptor Binding and Signaling*
Rakesh K. Jain, *Transport Phenomena in Tumors*
R. Krishna, *A Systems Approach to Multiphase Reactor Selection**
David T. Allen, *Pollution Prevention: Engineering Design at Macro-, Meso-, and Microscales*
John H. Seinfeld, Jean M. Andino, Frank M. Bowman, Hali J. L. Forstner, and Spyros Pandis, *Tropospheric Chemistry*

Volume 20

Arthur M. Squires, *Origins of the Fast Fluid Bed*
Yu Zhiqing, *Application Collocation*
Youchu Li, *Hydrodynamics*
Li Jinghai, *Modeling*
Yu Zhiqing and Jin Yong, *Heat and Mass Transfer*
Mooson Kwauk, *Powder Assessment*
Li Hongzhong, *Hardware Development*
Youchu Li and Xuyi Zhang, *Circulating Fluidized Bed Combustion*
Chen Junwu, Cao Hanchang, and Liu Taiji, *Catalyst Regeneration in Fluid Catalytic Cracking*

ISBN 0-12-008520-8